# SMART ANTENNAS

**THE ELECTRICAL ENGINEERING AND APPLIED SIGNAL PROCESSING SERIES**

*Edited by Alexander Poularikas*

*The Advanced Signal Processing Handbook: Theory and Implementation for Radar, Sonar, and Medical Imaging Real-Time Systems*
Stergios Stergiopoulos

*The Transform and Data Compression Handbook*
K.R. Rao and P.C. Yip

*Handbook of Multisensor Data Fusion*
David Hall and James Llinas

*Handbook of Neural Network Signal Processing*
Yu Hen Hu and Jenq-Neng Hwang

*Handbook of Antennas in Wireless Communications*
Lal Chand Godara

*Noise Reduction in Speech Applications*
Gillian M. Davis

*Signal Processing Noise*
Vyacheslav P. Tuzlukov

*Digital Signal Processing with Examples in MATLAB®*
Samuel Stearns

*Applications in Time-Frequency Signal Processing*
Antonia Papandreou-Suppappola

*The Digital Color Imaging Handbook*
Gaurav Sharma

*Pattern Recognition in Speech and Language Processing*
Wu Chou and Biing-Hwang Juang

*Propagation Handbook for Wireless Communication System Design*
Robert K. Crane

*Nonlinear Signal and Image Processing: Theory, Methods, and Applications*
Kenneth E. Barner and Gonzalo R. Arce

*Smart Antennas*
Lal Chand Godara

Forthcoming Titles

*Soft Computing with MATLAB®*
Ali Zilouchian

*Signal and Image Processing in Navigational Systems*
Vyacheslav P. Tuzlukov

*Wireless Internet: Technologies and Applications*
Apostolis K. Salkintzis and Alexander Poularikas

# SMART ANTENNAS

Lal Chand Godara

CRC PRESS

Boca Raton London New York Washington, D.C.

**Library of Congress Cataloging-in-Publication Data**

Godara, Lal Chand.
Smart antennas / Lal Chand Godara.
p. cm. — (Electrical engineering & applied signal processing)
Includes bibliographical references and index.
ISBN 0-8493-1206-X (alk. paper)
1. Adaptive antennas. I. Title. II. Electrical engineering and applied signal processing series; v. 15.

TK7871.67.A33G64 2004
621.382′4—dc22 2003065210



Direct all inquiries to CRC Press LLC, 2000 N.W. Corporate Blvd., Boca Raton, Florida 33431.


International Standard Book Number 0-8493-1206-X
Library of Congress Card Number 2003065210
Printed in the United States of America 1 2 3 4 5 6 7 8 9 0
Printed on acid-free paper

# *Dedication*

*With love to Saroj*

# Preface

Smart antennas involve processing of signals induced on an array of sensors such as antennas, microphones, and hydrophones. They have applications in the areas of radar, sonar, medical imaging, and communications.

Smart antennas have the property of spatial filtering, which makes it possible to receive energy from a particular direction while simultaneously blocking it from another direction. This property makes smart antennas a very effective tool in detecting and locating an underwater source of sound such as a submarine without using active sonar. The capacity of smart antennas to direct transmitting energy toward a desired direction makes them useful for medical diagnostic purposes. This characteristic also makes them very useful in canceling an unwanted jamming signal. In a communications system, an unwanted jamming signal is produced by a transmitter in a direction other than the direction of the desired signal. For a medical doctor trying to listen to the sound of a pregnant mother's heart, the jamming signal is the sound of the baby's heart.

Processing signals from different sensors involves amplifying each signal before combining them. The amount of gain of each amplifier dictates the properties of the antenna array. To obtain the best possible cancellation of unwanted interferences, the gains of these amplifiers must be adjusted. How to go about doing this depends on many conditions including signal type and overall objectives. For optimal processing, the typical objective is maximizing the output signal-to-noise ratio (SNR). For an array with a specified response in the direction of the desired signal, this is achieved by minimizing the mean output power of the processor subject to specified constraints. In the absence of errors, the beam pattern of the optimized array has the desired response in the signal direction and reduced response in the directions of unwanted interference.

The smart antenna field has been a very active area of research for over four decades. During this time, many types of processors for smart antennas have been proposed and their performance has been studied. Practical use of smart antennas was limited due to excessive amounts of processing power required. This limitation has now been overcome to some extent due to availability of powerful computers.

Currently, the use of smart antennas in mobile communications to increase the capacity of communication channels has reignited research and development in this very exciting field. Practicing engineers now want to learn about this subject in a big way. Thus, there is a need for a book that could provide a learning platform. There is also a need for a book on smart antennas that could serve as a textbook for senior undergraduate and graduate levels, and as a reference book for those who would like to learn quickly about a topic in this area but do not have time to perform a journal literature search for the purpose.

This book aims to provide a comprehensive and detailed treatment of various antenna array processing schemes, adaptive algorithms to adjust the required weighting on antennas, direction-of-arrival (DOA) estimation methods including performance comparisons, diversity-combining methods to combat fading in mobile communications, and effects of errors on array system performance and error-reduction schemes. The book brings almost all aspects of array signal processing together and presents them in a logical manner. It also contains extensive references to probe further.

After some introductory material in Chapter 1, the detailed work on smart antennas starts in Chapter 2 where various processor structures suitable for narrowband field are discussed. Behavior of both element space and beamspace processors is studied when their performance is optimized. Optimization using the knowledge of the desired signal direction as well as the reference signal is considered. The processors considered include conventional beamformer; null-steering beamformer; minimum-variance distortionless beamformer, also known as optimal beamformer; generalized side-lobe canceller; and postbeamformer interference canceler. Detailed analysis of these processors in the absence of errors is carried out by deriving expressions for various performance measures. The effect of errors on these processors has been analyzed to show how performance degrades because of various errors. Steering vector, weight vector, phase shifter, and quantization errors are discussed.

For various processors, solution of the optimization problem requires knowledge of the correlation between various elements of the antenna array. In practice, when this information is not available an estimate of the solution is obtained in real-time from received signals as these become available. There are many algorithms available in the literature to adaptively estimate the solution, with conflicting demands of implementation simplicity and speed with which the solution is estimated. Adaptive processing is presented in Chapter 3, with details on the sample matrix inversion algorithm, constrained and unconstrained least mean squares (LMS) algorithms, recursive LMS algorithm, recursive least squares algorithm, constant modulus algorithm, conjugate gradient method, and neural network approach. Detailed convergence analysis of many of these algorithms is presented under various conditions to show how the estimated solution converges to the optimal solution. Transient and steady-state behavior is analyzed by deriving expressions for various quantities of interest with a view to teach the underlying analysis tools. Many numerical examples are included to demonstrate how these algorithms perform.

Smart antennas suitable for broadband signals are discussed in Chapter 4. Processing of broadband signals may be carried out in the time domain as well as in the frequency domain. Both aspects are covered in detail in this chapter. A tapped-delay line structure behind each antenna to process the broadband signals in the time domain is described along with its frequency response. Various constraints to shape the beam of the broadband antennas are derived, optimization for this structure is considered, and a suitable adaptive algorithm to estimate the optimal solution is presented. Various realizations of time-domain broadband processors are discussed in detail along with the effect that the choice of origin has on performance. A detailed treatment of frequency-domain processing of broadband signals is presented and its relationship with time-domain processing is established. Use of the discrete Fourier transform method to estimate the weights of the time-domain structure and how its modular structure could help reduce real-time processing are described.

Correlation between a desired signal and unwanted interference exists in situations of multipath signals, deliberate jamming, and so on, and can degrade the performance of an antenna array processor. Chapter 5 presents models for correlated fields in narrowband and broadband signals. Analytical expressions for SNRs in both narrowband and broadband structures of smart antennas are derived, and the effects of several factors on SNR are explored, including the magnitude and phase of the correlation, number of elements in the array, direction and level of the interference source and the level of the uncorrelated noise. Many methods are described to decorrelate the correlated sources, and analytical expressions are derived to show the decorrelation effect of the proposed techniques.

In Chapter 6, various DOA estimation methods are described, followed by performance comparisons and sensitivity analyses. These estimation tools include spectral estimation methods, minimum variance distortionless response estimator, linear prediction method, maximum entropy method, maximum likelihood method, various eigenstructure methods including many versions of MUSIC algorithms, minimum norm methods, CLOSEST method, ESPRIT method, and weighted subspace fitting method. This chapter also contains discussion on various preprocessing and number-of-source estimation methods.

In the first six chapters, it is assumed that the directional signals arrive from point sources as plane wave fronts. In mobile communication channels, the received signal is a combination of many components arriving from various directions due to multipath propagation resulting in large fluctuation in the received signals. This phenomenon is called fading. In Chapter 7, a brief review of fading channels is presented, distribution of signal amplitude and received power on an antenna is developed, analysis of noise- and interference-limited single-antenna systems in Rayleigh and Nakagami fading channels is presented by deriving results for average bit error rate and outage probability. The results show how fading affects the performance of a single-antenna system.

Chapter 8 presents a comprehensive analysis of diversity combining, which is a process of combining several signals with independent fading statistics to reduce large attenuation of the desired signal in the presence of multipath signals. The diversity-combining schemes described and analyzed in this chapter include selection combiner, switched diversity combiner, equal gain combiner, maximum ratio combiner, optimal combiner, generalized selection combiner, cascade diversity combiner, and macroscopic diversity combiner. Both noise-limited and interference-limited systems are analyzed in various fading conditions by deriving results for average bit error rate and outage probability.

# *Acknowledgments*

I owe special thanks to Jill Paterson for her diligent and professional job of typing this book from my handwritten manuscript.

I am most grateful to my wife Saroj for her love, understanding, and patience, and my daughter Pankaj and son Vikas for their constant encouragement, which provided the necessary motivation to complete the project.

# *The Author*

**Lal Chand Godara, Ph.D.,** is Associate Professor at University College, the University of New South Wales, Australian Defense Force Academy, Canberra, Australia. He received the B.E. degree from Birla Institute of Technology and Science, Pilani, India in 1975; the M.Tech. degree from Indian Institute of Science, Banglore, India in 1977; the Ph.D. degree from the University of Newcastle, NSW, Australia in 1984; and the M.HEd. degree from The University of New South Wales, Australia.

Professor Godara has had visiting appointments at Stanford University, Yale University, and Syracuse University. His research interests include adaptive antenna array processing and their application to mobile communications. Included among his many publications are two significant papers in the *Proceedings of the IEEE*. Prof. Godara edited *Handbook of Antennas for Wireless Communications*, published by CRC Press in 2002.

Professor Godara is a Senior Member of the IEEE and a Fellow of the Acoustical Society of America. He was awarded the University College Teaching Excellence Award in 1998. Some of his activities/achievements in the IEEE included: Associate Editor, *IEEE Transactions on Signal Processing* (1998-2000); IEEE Third Millennium Medal (2000); and Member, SPS Sensor Array and Multichannel Technical Committee (2000-2002). He founded the IEEE Australian Capital Territory Section (1988) and served as its founding Chairman for three years (1988–1991). He served as Chairman of the IEEE Australian Council from 1995 to 1996.

# Contents

# 1

# *Introduction*



Widespread interest in smart antennas has continued for several decades due to their use in numerous applications. The first issue of *IEEE Transactions of Antennas and Propagation*, published in 1964 [IEE64], was followed by special issues of various journals [IEE76, IEE85, IEE86, IEE87a, IEE87b], books [Hud81, Mon80, Hay85, Wid85, Com88, God00], a selected bibliography [Mar86], and a vast number of specialized research papers. Some of the general papers in which various issues are discussed include [App76, d'A80, d'A84, Gab76, Hay92, Kri96, Mai82, Sch77, Sta74, Van88, Wid67].

The current demand for smart antennas to increase channel capacity in the fast-growing area of mobile communications has reignited the research and development efforts in this area around the world [God97]. This book aims to help researchers and developers by providing a comprehensive and detailed treatment of the subject matter. Throughout the book, references are provided in which smart antennas have been suggested for mobile communication systems. This chapter presents some introductory material and terminology associated with antenna arrays for those who are not familiar with antenna theory.

## 1.1 Antenna Gain

Omnidirectional antennas radiate equal amounts of power in all directions. Also known as isotropic antennas, they have equal gain in all directions. Directional antennas, on the other hand, have more gain in certain directions and less in others. A direction in which the gain is maximum is referred to as the antenna boresight. The gain of directional antennas in the boresight is more than that of omnidirectional antennas, and is measured with respect to the gain of omnidirectional antennas. For example, a gain of 10 dBi (some times indicated as dBic or simply dB) means the power radiated by this antenna is 10 dB more than that radiated by an isotropic antenna.

An antenna may be used to transmit or receive. The gain of an antenna remains the same in both the cases. The gain of a receiving antenna indicates the amount of power it delivers to the receiver compared to an omnidirectional antenna.

## 1.2 Phased Array Antenna

A phased array antenna uses an array of antennas. Each antenna forming the array is known as an element of the array. The signals induced on different elements of an array are combined to form a single output of the array.

This process of combining the signals from different elements is known as beamforming. The direction in which the array has maximum response is said to be the beam-pointing direction. Thus, this is the direction in which the array has the maximum gain. When signals are combined without any gain and phase change, the beam-pointing direction is broadside to the linear array, that is, perpendicular to the line joining all elements of the array.

By adjusting the phase difference among various antennas one is able to control the beam pointing direction. The signals induced on various elements after phase adjustment due to a source in the beam-pointing direction get added in phase. This results in array gain (or equivalently, gain of the combined antenna) equal to the sum of individual antenna gains.

## 1.3 Power Pattern

A plot of the array response as a function of angle is referred to as array pattern or antenna pattern. It is also called power pattern when the power response is plotted. It shows the power received by the array at its output from a particular direction due to a unit power source in that direction. A power pattern of an equispaced linear array of ten elements with half-wavelength spacing is shown in Figure 1.1. The angle is measured with respect to the line of the array. The beam-pointing direction makes a 90° angle with the line of the array. The power pattern has been normalized by dividing the number of elements in the array so that the maximum array gain in the beam-pointing direction is unity.

The power pattern drops to a low value on either side of the beam-pointing direction. The place of the low value is normally referred to as a null. Strictly speaking, a null is a position where the array response is zero. However, the term sometimes is misused to indicate the low value of the pattern. The pattern between the two nulls on either side of the beam-pointing direction is known as the main lobe (also called main beam or simply beam). The width of the main beam between the two half-power points is called the half-power beamwidth. A smaller beamwidth results from an array with a larger extent. The extent of the array is known as the aperture of the array. Thus, the array aperture is the distance between the two farthest elements in the array. For a linear array, the aperture is equal to the distance between the elements on either side of the array.

## 1.4 Beam Steering

For a given array the beam may be pointed in different directions by mechanically moving the array. This is known as mechanical steering. Beam steering can also be accomplished

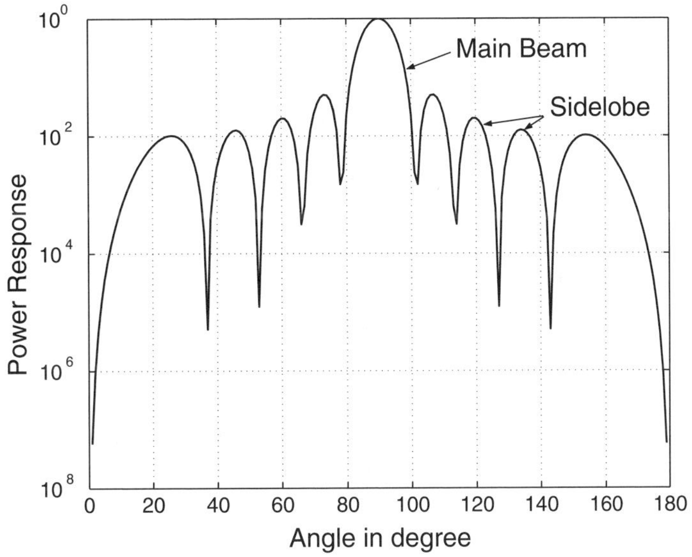


**FIGURE 1.1**
Power pattern of a ten-element linear array with half-wavelength spacing.

by appropriately delaying the signals before combining. The process is known as electronic steering, and no mechanical movement occurs. For narrowband signals, the phase shifters are used to change the phase of signals before combining.

The required delay may also be accomplished by inserting varying lengths of coaxial cables between the antenna elements and the combiner. Changing the combinations of various lengths of these cables leads to different pointing directions. Switching between different combinations of beam-steering networks to point beams in different directions is sometimes referred to as beam switching.

When processing is carried out digitally, the signals from various elements can be sampled, stored, and summed after appropriate delays to form beams. The required delay is provided by selecting samples from different elements such that the selected samples are taken at different times. Each sample is delayed by an integer multiple of the sampling interval; thus, a beam can only be pointed in selected directions when using this technique.

## 1.5 Degree of Freedom

The gain and phase applied to signals derived from each element may be thought of as a single complex quantity, hereafter referred to as the weighting applied to the signals. If there is only one element, no amount of weighting can change the pattern of that antenna. However, with two elements, when changing the weighting of one element relative to the other, the pattern may be adjusted to the desired value at one place, that is, you can place one minima or maxima anywhere in the pattern. Similarly, with three elements, two positions may be specified, and so on. Thus, with an L-element array, you can specify L – 1 positions. These may be one maxima in the direction of the desired signal and L – 2 minimas (nulls) in the directions of unwanted interferences. This flexibility of an L element array to be able to fix the pattern at L – 1 places is known as the degree of freedom of the array. For an equally spaced linear array, this is similar to an L – 1 degree polynomial of L – 1 adjustable coefficients with the first coefficient having the value of unity.

## 1.6 Optimal Antenna

An antenna is optimal when the weight of each antenna element is adjusted to achieve optimal performance of an array system in some sense. For example, assume that a communication system is operating in the presence of unwanted interferences. Furthermore, the desired signal and interferences are operating at the same carrier frequency such that these interferences cannot be eliminated by filtering. The optimal performance for a communication system in such a situation may be to maximize the signal-to-noise ratio (SNR) at the output of the system without causing any signal distortion. This would require adjusting the antenna pattern to cancel these interferences with the main beam pointed in the signal direction. Thus, the communication system is said to be employing an optimal antenna when the gain and the phase of the signal induced on each element are adjusted to achieve the maximum output SNR (sometimes also referred to as signal to interference and noise ratio, SINR).

## 1.7 Adaptive Antenna

The term adaptive antenna is used for a phased array when the weighting on each element is applied in a dynamic fashion. The amount of weighting on each channel is not fixed at the time of the array design, but rather decided by the system at the time of processing the signals to meet required objectives. In other words, the array pattern adapts to the situation and the adaptive process is under control of the system. For example, consider the situation of a communication system operating in the presence of a directional interference operating at the carrier frequency used by the desired signal, and the performance measure is to maximize the output SNR. As discussed previously, the output SNR is maximized by canceling the directional interference using optimal antennas. The antenna pattern in this case has a main beam pointed in the desired signal direction, and has a null in the direction of the interference. Assume that the interference is not stationary but moving slowly. If optimal performance is to be maintained, the antenna pattern needs to adjust so that the null position remains in the moving interference direction. A system using adaptive antennas adjusts the weighting on each channel with an aim to achieve such a pattern.

For adaptive antennas, the conventional antenna pattern concepts of beam width, side lobes, and main beams are not used, as the antenna weights are designed to achieve a set performance criterion such as maximization of the output SNR. On the other hand, in conventional phase-array design these characteristics are specified at the time of design.

## 1.8 Smart Antenna

The term smart antenna incorporates all situations in which a system is using an antenna array and the antenna pattern is dynamically adjusted by the system as required. Thus, a system employing smart antennas processes signals induced on a sensor array. A block diagram of such a system is shown in Figure 1.2.

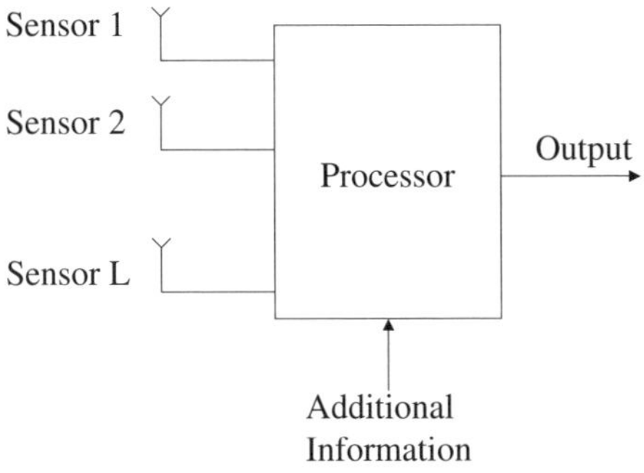


**FIGURE 1.2**
Block diagram of an antenna array system.

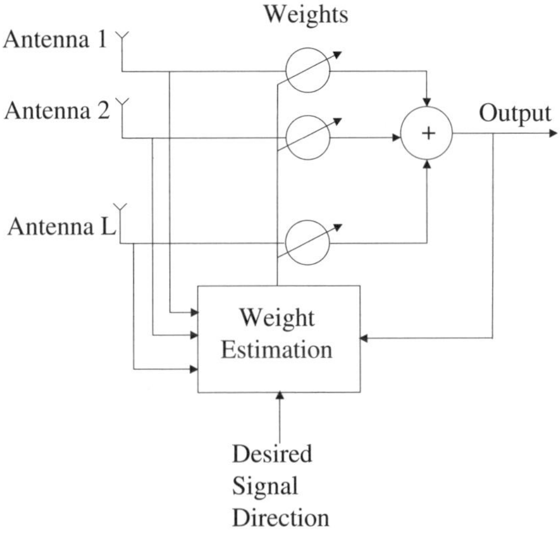


**FIGURE 1.3**
Block diagram of a communication system using an antenna array.

The type of sensors used and the additional information supplied to the processor depend on the application. For example, a communication system uses antennas as sensors and may use some signal characteristics as additional information. The processor uses this information to differentiate the desired signal from unwanted interference.

A block diagram of a narrowband communication system is shown in Figure 1.3 where signals induced on an antenna array are multiplied by adjustable complex weights and then combined to form the system output. The processor receives array signals, system output, and direction of the desired signal as additional information. The processor calculates the weights to be used for each channel.

## 1.9 Book Outline

Chapter 2 is dedicated to various narrowband processors and their performance. Adaptive processing of narrowband signals is discussed in Chapter 3. Descriptions and analyses of broadband-signal processors are presented in Chapter 4. In Chapter 5, situations are considered in which the desired signals and unwanted interference are not independent. Chapter 6 is focused on using the received signals on an array to identify the direction of a radiating source. Chapter 7 and Chapter 8 are focused on fading channels. Chapter 7 describes such channels and analyzes the performance of a single antenna system in a fading environment. Chapter 8 considers multiple antenna systems and presents various diversity-combining techniques.

## References


App76 Applebaum, S.P., Adaptive arrays, *IEEE Trans. Antennas Propagat.*, 24, 585–598, 1976.

Com88 Compton Jr., R.T., *Adaptive Antennas: Concepts and Performances*, Prentice Hall, New York, 1988.

d'A80 d'Assumpcao, H.A., Some new signal processors for array of sensors, *IEEE Trans. Inf. Theory*, 26, 441–453, 1980.

d'A84 d'Assumpcao, H.A. and Mountford, G.E., An overview of signal processing for arrays of receivers, *J. Inst. Eng. Aust. IREE Aust.*, 4, 6–19, 1984.

Gab76 Gabriel, W.F., Adaptive arrays: An introduction, *IEEE Proc.*, 64, 239–272, 1976.

God97 Godara, L.C., Application of antenna arrays to mobile communications. Part I: Performance improvement, feasibility and system considerations, *Proc. IEEE*, 85, 1031–1062, 1997.

God00 Godara, L.C., Ed., *Handbook of Antennas in Wireless Communications*, CRC Press, Boca Raton, FL, 2002.

Hay85 Haykin, S., Ed., *Array Signal Processing*, Prentice Hall, New York, 1985.

Hay92 Haykin, S. et al., Some aspects of array signal processing, *IEE Proc.*, 139, Part F, 1–19, 1992.

Hud81 Hudson, J.E., *Adaptive Array Principles*, Peter Peregrins, New York, 1981.

IEE64 IEEE, Special issue on active and adaptive antennas, *IEEE Trans. Antennas Propagat.*, 12, 1964.

IEE76 IEEE, Special issue on adaptive antennas, *IEEE Trans. Antennas Propagat.*, 24, 1976.

IEE85 IEEE, Special issue on beamforming, *IEEE J. Oceanic Eng.*, 10, 1985.

IEE86 IEEE, Special issue on adaptive processing antenna systems, *IEEE Trans. Antennas Propagat.*, 34, 1986.

IEE87a IEEE, Special issue on adaptive systems and applications, *IEEE Trans. Circuits Syst.*, 34, 1987.

IEE87b IEEE, Special issue on underwater acoustic signal processing, *IEEE J. Oceanic Eng.*, 12, 1987.

Kri96 Krim, H. and Viberg, M., Two decades of array signal processing: the parametric approach, *IEEE Signal Process. Mag.*, 13(4), 67–94, 1996.

Mai82 Maillous, R.J., Phased array theory and technology, *IEEE Proc.*, 70, 246–291, 1982.

Mar86 Marr, J.D., A selected bibliography on adaptive antenna arrays, *IEEE Trans. Aerosp. Electron. Syst.*, 22, 781–788, 1986.

Mon80 Monzingo, R. A. and Miller, T. W., *Introduction to Adaptive Arrays*, Wiley, New York, 1980.

Sch77 Schultheiss, P.M., Some lessons from array processing theory, in *Aspects of Signal Processing*, Part 1, Tacconi, G., Ed., D. Reidel, Dordrecht, 1977, p. 309–331.

Sta74 Stark, L., Microwave theory of phased-array antennas: a review, *IEEE Proc.*, 62, 1661–1701, 1974.

Van88 Van Veen, B.D. and Buckley, K.M., Beamforming: a versatile approach to spatial filtering, *IEEE ASSP Mag.*, 5, 4–24, 1988.

Wid67 Widrow, B. et al., Adaptive antenna systems, *IEEE Proc.*, 55, 2143–2158, 1967.

Wid85 Widrow, B. and Stearns, S.D., *Adaptive Signal Processing*, Prentice Hall, New York, 1985.

# 2

# *Narrowband Processing*

Consider the antenna array system consisting of L antenna elements shown in Figure 2.1, where signals from each element are multiplied by a complex weight and summed to form the array output. The figure does not show components such as preamplifiers, bandpass filters, and so on. It follows from the figure that an expression for the array output is given by

$$y(t) = \sum_{\ell=1}^{L} w_\ell^* \, x_\ell(t) \tag{2.1}$$

where * denotes the complex conjugate. The conjugate of complex weights is used to simplify the mathematical notation.

Denoting the weights of the array system using vector notation as

$$\mathbf{w} = \left[w_1, w_2, \ldots, w_L\right]^T \tag{2.2}$$

and signals induced on all elements as

$$\mathbf{x}(t) = \left[x_1(t), x_2(t), \ldots, x_L(t)\right]^T \tag{2.3}$$

the output of the array system becomes

$$y(t) = \mathbf{w}^H \mathbf{x}(t) \tag{2.4}$$

where superscript T and H, respectively, denote transposition and the complex conjugate transposition of a vector or matrix. Throughout the book **w** and **x**(t) are referred to as the weight vector and the signal vector, respectively. Note that to obtain the array output, you

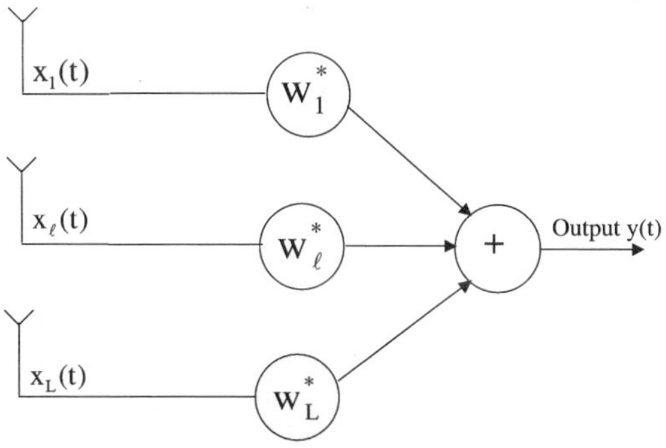


**FIGURE 2.1**
Antenna array system.

need to multiply the signals induced on all elements with the corresponding weights. In vector notation, this operation is carried out by taking the inner product of the weight vector with the signal vector as given by (2.4).

The output power of the array at any time t is given by the magnitude square of the array output, that is,

$$\begin{aligned} P(t) &= |y(t)|^2 \\ &= y(t)y^*(t) \end{aligned} \tag{2.5}$$

Substituting for y(t) from (2.4), the output power becomes

$$P(t) = \mathbf{w}^H \mathbf{x}(t) \mathbf{x}^H(t) \mathbf{w} \tag{2.6}$$

If the components of $\mathbf{x}(t)$ can be modeled as zero-mean stationary processes, then for a given $\mathbf{w}$ the mean output power of the array system is obtained by taking conditional expectation over $\mathbf{x}(t)$:

$$\begin{aligned} P(\mathbf{w}) &= E\left[\mathbf{w}^H \mathbf{x}(t) \mathbf{x}^H(t) \mathbf{w}\right] \\ &= \mathbf{w}^H E\left[\mathbf{x}(t) \mathbf{x}^H(t)\right] \mathbf{w} \\ &= \mathbf{w}^H R\, \mathbf{w} \end{aligned} \tag{2.7}$$

where $E[\cdot]$ denotes the expectation operator and R is the array correlation matrix defined by

$$R = E\left[\mathbf{x}(t)\, \mathbf{x}^H(t)\right] \tag{2.8}$$

Elements of this matrix denote the correlation between various elements. For example, $R_{ij}$ denotes the correlation between the ith and the jth element of the array.

Consider that there is a desired signal source in the presence of unwanted interference and random noise. The random noise includes both background and electronic noise. Let $\mathbf{x}_S(t)$, $\mathbf{x}_I(t)$, and $\mathbf{n}(t)$, respectively, denote the signal vector due to the desired signal source, unwanted interference, and random noise. The components of signal, interference, and random noise in the output $y_S(t)$, $y_I(t)$, and $y_n(t)$ are then obtained by taking the inner product of the weight vector with $\mathbf{x}_S(t)$, $\mathbf{x}_I(t)$, and $\mathbf{n}(t)$. These are given by

$$y_S(t) = \mathbf{w}^H \mathbf{x}_S(t) \tag{2.9}$$

$$y_I(t) = \mathbf{w}^H \mathbf{x}_I(t) \tag{2.10}$$

and

$$y_n(t) = \mathbf{w}^H \mathbf{n}(t) \tag{2.11}$$

Define the array correlation matrices due to the signal source, unwanted interference, and random noise, respectively, as

$$R_S = E\left[\mathbf{x}_S(t)\, \mathbf{x}_S^H(t)\right] \tag{2.12}$$

$$R_I = E\left[\mathbf{x}_I(t)\, \mathbf{x}_I^H(t)\right] \tag{2.13}$$

and

$$R_n = E\left[\mathbf{n}(t)\, \mathbf{n}^H(t)\right] \tag{2.14}$$

Note that R is the sum of these three matrices, that is,

$$R = R_S + R_I + R_n \tag{2.15}$$

Let $P_S$, $P_I$ and $P_n$ denote the mean output power due to the signal source, unwanted interference, and random noise, respectively. Following (2.7), these are given by

$$P_S = \mathbf{w}^H R_S \mathbf{w} \tag{2.16}$$

$$P_I = \mathbf{w}^H R_I \mathbf{w} \tag{2.17}$$

and

$$P_n = \mathbf{w}^H R_n \mathbf{w} \tag{2.18}$$

Let $P_N$ denote the mean power at the output of the array contributed by random noise and unwanted interference, that is,

$$P_N = P_I + P_n \tag{2.19}$$

We refer to $P_N$ as the mean noise power at the output of the array system. Note that the noise here includes random noise and contributions from all sources other than the desired signal. In some sources, this is also referred to as noise plus interference.

Substituting from (2.17) and (2.18) in (2.19),

$$\begin{aligned} P_N &= \mathbf{w}^H R_I \mathbf{w} + \mathbf{w}^H R_n \mathbf{w} \\ &= \mathbf{w}^H \left(R_I + R_n\right) \mathbf{w} \end{aligned} \tag{2.20}$$

Let $R_N$ denote the noise array correlation matrix, that is,

$$R_N = R_I + R_n \tag{2.21}$$

Then $P_N$, the mean noise power at the output of the system can be expressed in terms of weight vector and $R_N$ as

$$P_N = \mathbf{w}^H R_N \mathbf{w} \tag{2.22}$$

Let the output signal-to-noise ratio (SNR), sometimes also referred to as the signal to interference plus noise ratio (SINR), be defined as the ratio of the mean output signal power to the mean output noise power at the output of the array system, that is,

$$\text{SNR} = \frac{P_S}{P_N} \tag{2.23}$$

Substituting from (2.16) and (2.22) in (2.23), it follows that

$$\text{SNR} = \frac{\mathbf{w}^H R_S \mathbf{w}}{\mathbf{w}^H R_N \mathbf{w}} \tag{2.24}$$

The weights of the array system determine system performance. The selection process of these weights depends on the application and leads to various types of beamforming schemes.

In this chapter, various beamforming schemes are discussed, performance of a processor using these schemes is analyzed, and the effect of errors on processor performance is presented [God93, God97].

## 2.1 Signal Model

In this section, a signal model is described and expressions for the signal vector and the array correlation matrix required for the understanding of various beamforming schemes are written.

Assume that the array is located in the far field of directional sources. Thus, as far as the array is concerned, the directional signal incident on the array can be considered as a plane wave front. Also assume that the plane wave propagates in a homogeneous media and that the array consists of identical distortion-free omnidirectional elements. Thus, for the ideal case of nondispersive propagation and distortion free elements, the effect of propagation from a source to an element is a pure time delay.

Let the origin of the coordinate system be taken as the time reference as shown in Figure 2.2. Thus, the time taken by a plane wave arriving from the kth source in direction $(\phi_k, \theta_k)$ and measured from the $\ell$th element to the origin is given by

$$\tau_\ell(\phi_k, \theta_k) = \frac{\mathbf{r}_\ell \cdot \hat{\mathbf{v}}(\phi_k, \theta_k)}{c} \tag{2.1.1}$$

where $\mathbf{r}_\ell$ is the position vector of the $\ell$th element, $\hat{\mathbf{v}}(\phi_k, \theta_k)$ is the unit vector in direction $(\phi_k, \theta_k)$, c is the speed of propagation of the plane wave front, and the dot represents the

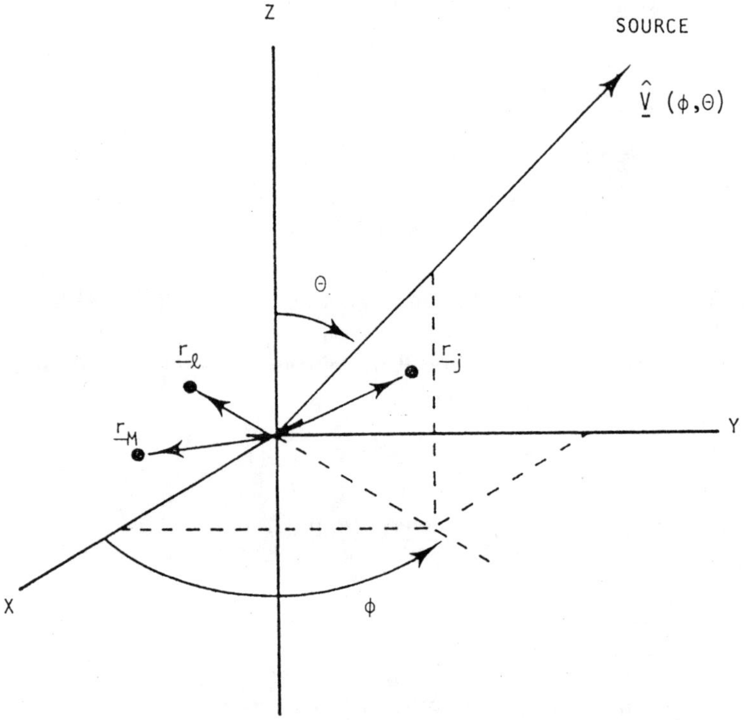


**FIGURE 2.2**
Coordinate system.

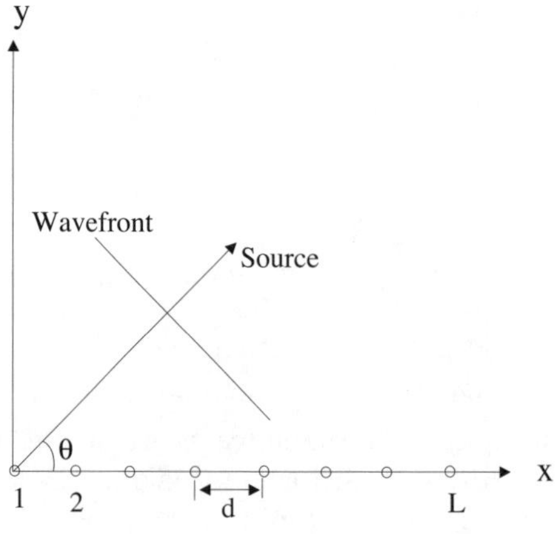


**FIGURE 2.3**
Linear array with element spacing d.

dot product. For a linear array of equispaced elements with element spacing d, aligned with the x-axis such that the first element is situated at the origin as shown in Figure 2.3, it becomes

$$\tau_\ell(\theta_k) = \frac{d}{c}(\ell - 1)\cos\theta_k \tag{2.1.2}$$

Note that when the kth source is broadside to the array, $\theta_k = 90°$. It follows from (2.1.2) that for this case, $\tau_\ell(\theta_k) = 0$ for all $\ell$. Thus, the wave front arrives at all the elements of the array at the same time and signals induced on all the elements due to this source are identical. For $\theta_k = 0°$, the wave front arrives at the $\ell$th element before it arrives at the origin, and the signal induced on the $\ell$th element leads to that induced on an element at the origin. The time delay given by (2.1.2) is

$$\tau_\ell(\theta_k) = \frac{d}{c}(\ell - 1) \tag{2.1.3}$$

On the other hand, for $\theta_k = 180°$, the time delay is given by

$$\tau_\ell(\theta_k) = -\frac{d}{c}(\ell - 1) \tag{2.1.4}$$

The negative sign is due to the definition of $\tau_\ell$. It is the time taken by the plane wave from the $\ell$th element to the origin. The negative sign indicates that the wave front arrives at the origin before it arrives at the $\ell$th element, and the signal induced on the $\ell$th element lags behind that induced on an element at the origin.

The signal induced on the reference element (an element at the origin) due to the kth source is normally expressed in complex notation as

$$m_k(t)\, e^{j2\pi f_0 t} \tag{2.1.5}$$

with $m_k(t)$ denoting the complex modulating function and $f_0$ denoting the carrier frequency. The structure of the modulating function reflects the particular modulation used in a communication system. For example, for frequency division multiple access (FDMA) systems it is a frequency-modulated signal given by $m_k(t) = A_k e^{j\xi_k(t)}$ with $A_k$ denoting the amplitude and $\xi_k(t)$ denoting the message. For time division multiple access (TDMA) systems, it is given by

$$m_k(t) = \sum_n d_k(n)p(t - n\Delta) \tag{2.1.6}$$

where p(t) is the sampling pulse, the amplitude $d_k(n)$ denotes the message symbol, and $\Delta$ is the sampling interval. For code division multiple access (CDMA) systems, $m_k(t)$ is given by

$$m_k(t) = d_k(t)g(t) \tag{2.1.7}$$

where $d_k(n)$ denotes the message sequence and g(t) is a pseudo random-noise binary sequence having the values +1 or −1.

In general, the modulating function is normally modeled as a complex low-pass process with zero mean and variance equal to the source power $p_k$ as measured at the reference element. Assuming that the wave front on the $\ell$th elements arrives $\tau_\ell(\phi_k,\theta_k)$ seconds before it arrives at the reference element, the signal induced on the $\ell$th element due to the kth source can be expressed as

$$m_k(t)\, e^{j2\pi f_0 (t + \tau_\ell(\phi_k, \theta_k))} \tag{2.1.8}$$

The expression is based upon the narrowband assumption for array signal processing, which assumes that the bandwidth of the signal is narrow enough and that the array dimensions are small enough for the modulating function to stay almost constant during $\tau_\ell(\phi_k,\theta_k)$ seconds, that is, the approximation $m_k(t) \cong m_k(t + \tau_\ell(\phi_k,\theta_k))$ holds.

Assume that there are M directional sources present. Let $x_\ell(t)$ denote the total signal induced due to all M directional sources and background noise on the $\ell$th element. Thus,

$$x_\ell(t) = \sum_{k=1}^{M} m_k(t)\, e^{j2\pi f_0 \left(t+\tau_\ell(\phi_k,\theta_k)\right)} + n_\ell(t) \tag{2.1.9}$$

where $n_\ell(t)$ is random noise component on the $\ell$th element, which includes background noise and electronic noise generated in the $\ell$th channel. It is assumed to be temporally white with zero mean and variance equal to $\sigma_n^2$. Furthermore, it is assumed to be uncorrelated with directional sources, that is,

$$E\left[m_k(t)n_\ell(t)\right] = 0 \tag{2.1.10}$$

The noise on different elements is also assumed to be uncorrelated, that is,

$$E\left[n_k(t)n_\ell(t)\right] = \begin{cases} 0 & \ell \neq k \\ \sigma_n^2 & \ell = k \end{cases} \tag{2.1.11}$$

It should be noted that if the elements were not omnidirectional, then the signal induced on each element due to a source is scaled by an amount equal to the response of the element under consideration in the direction of the source.

Substituting from (2.1.9) in (2.3), the signal vector becomes

$$\mathbf{x}(t) = \sum_{k=1}^{M} m_k(t) \begin{pmatrix} e^{j2\pi\tau_1(\phi_k,\theta_k)} \\ e^{j2\pi\tau_2(\phi_k,\theta_k)} \\ \cdot \\ e^{j2\pi\tau_L(\phi_k,\theta_k)} \end{pmatrix} + \mathbf{n}(t) \tag{2.1.12}$$

where the carrier term $e^{j2\pi f_0 t}$ has been dropped for the ease of notation as it plays no role in subsequent treatment and

$$\mathbf{n}(t) = \begin{pmatrix} n_1(t) \\ n_2(t) \\ \cdot \\ n_L(t) \end{pmatrix} \tag{2.1.13}$$

### 2.1.1 Steering Vector Representation

Steering vector is an L-dimensional complex vector containing responses of all L elements of the array to a narrowband source of unit power. Let $\mathbf{S}_k$ denote the steering vector associated with the kth source. For an array of identical elements, it is defined as

$$\mathbf{S}_k = \left[\exp\left(j2\pi f_0\tau_1\left(\phi_k, \theta_k\right)\right), \ldots, \exp\left(j2\pi f_0\tau_L\left(\phi_k, \theta_k\right)\right)\right]^T \tag{2.1.14}$$

Note that when the first element of the array is at the origin of the coordinate system $\tau_1(\phi_k,\theta_k) = 0$, the first element of the steering vector is identical to unity.

As the response of the array varies according to direction, a steering vector is associated with each directional source. Uniqueness of this association depends on array geometry [God81]. For a linear array of equally spaced elements with element spacing greater than half wavelength, the steering vector for every direction is unique.

For an array of identical elements, each component of this vector has unit magnitude. The phase of its ith component is equal to the phase difference between signals induced on the ith element and the reference element due to the source associated with the steering vector. As each component of this vector denotes the phase delay caused by the spatial position of the corresponding element of the array, this vector is also known as the space vector. It is also referred to as the array response vector as it measures the response of the array due to the source under consideration. In multipath situations such as in mobile communications, it also denotes the response of the array to all signals arising from the source [Nag94]. In this book, steering vector, space vector, and array response vector are used interchangeably.

Using (2.1.14) in (2.1.12), the signal vector can be compactly expressed as

$$\mathbf{x}(t) = \sum_{k=1}^{M} m_k(t)\mathbf{S}_k + \mathbf{n}(t) \tag{2.1.15}$$

Substituting for $\mathbf{x}(t)$ from (2.1.15) in (2.4), it follows that

$$\begin{aligned} y(t) &= \mathbf{w}^H\mathbf{x}(t) \\ &= \sum_{k=1}^{M} m_k(t)\mathbf{w}^H\mathbf{S}_k + \mathbf{w}^H\mathbf{n}(t) \end{aligned} \tag{2.1.16}$$

The first term on the right side of (2.1.16) is the contribution from all directional sources and the second term is the random noise contribution to the array output. Note that the contribution of all directional sources contained in the first term is the weighted sum of modulating functions of all sources. The weight applied to each source is the inner product of the processor weight vector and steering vector associated with that source, and denotes the complex response of the processor toward the source. Thus, the response of a processor with weight vector $\mathbf{w}$ toward a source in direction $(\phi,\theta)$ is given by

$$y(\phi,\theta) = \mathbf{w}^H\mathbf{S}(\phi,\theta) \tag{2.1.17}$$

An expression for the array correlation matrix is derived in terms of steering vectors. Substituting the signal vector $\mathbf{x}(t)$ from (2.1.15) in the definition of the array correlation matrix given by (2.8) leads to the following expression for the array correlation matrix:

$$R = E\left[\left(\sum_{k=1}^{M} m_k(t)\mathbf{S}_k + \mathbf{n}(t)\right)\left(\sum_{k=1}^{M} m_k(t)\mathbf{S}_k + \mathbf{n}(t)\right)^H\right]$$

$$= E\left[\left(\sum_{k=1}^{M} m_k(t)\mathbf{S}_k\right)\left(\sum_{k=1}^{M} m_k(t)\mathbf{S}_k\right)^H\right] + E\left[\mathbf{n}(t)\mathbf{n}^H(t)\right]$$

$$+ E\left[\left(\sum_{k=1}^{M} m_k(t)\mathbf{S}_k\right)\mathbf{n}^H(t)\right] + E\left[\mathbf{n}(t)\left(\sum_{k=1}^{M} m_k(t)\mathbf{S}_k\right)^H\right] \tag{2.1.18}$$

The first term on the right-hand side (RHS) of (2.1.18) simplifies to

$$E\left[\left(\sum_{k=1}^{M} m_k(t)\mathbf{S}_k\right)\left(\sum_{k=1}^{M} m_k(t)\mathbf{S}_k\right)^H\right] = \sum_{k,\ell=1}^{M} E\left[m_k(t)m_\ell^*(t)\right]\mathbf{S}_k\mathbf{S}_\ell^H \tag{2.1.19}$$

When sources are uncorrelated,

$$E\left[m_\ell(t)m_k^*(t)\right] = \begin{cases} 0 & \ell \neq k \\ p_k & \ell = k \end{cases} \tag{2.1.20}$$

where $p_k$ denotes the power of the kth source measured at one of the elements of the array. It should be noted that $p_k$ is the variance of the complex modulating function $m_k(t)$ when it is modeled as a zero-mean low-pass random process, as mentioned previously. Thus, for uncorrelated sources the first term becomes

$$E\left[\left(\sum_{k=1}^{M} m_k(t)\mathbf{S}_k\right)\left(\sum_{k=1}^{M} m_k(t)\mathbf{S}_k\right)^H\right] = \sum_{k=1}^{M} p_k\mathbf{S}_k\mathbf{S}_k^H \tag{2.1.21}$$

The fact that the directional sources and the white noise are uncorrelated results in the third and fourth terms on the RHS of (2.1.18) to be identical to zero. Using (2.1.11), the second term simplifies to $\sigma_n^2 I$ with I denoting an identity matrix. This along with (2.1.21) lead to the following expression for the array correlation matrix when directional sources are uncorrelated:

$$R = \sum_{k=1}^{M} p_k\mathbf{S}_k\mathbf{S}_k^H + \sigma_n^2 I \tag{2.1.22}$$

where I is the identity matrix and $\sigma_n^2 I$ denotes the component of the array correlation matrix due to random noise, that is

$$R_n = \sigma_n^2 I \tag{2.1.23}$$

Let $\mathbf{S}_0$ denote the steering vector associated with the signal source of power $p_S$. Then the array correlation matrix due to the signal source is given by

$$R_S = p_S\mathbf{S}_0\mathbf{S}_0^H \tag{2.1.24}$$

Similarly, the array correlation matrix due to an interference of power $p_I$ is given by

$$R_I = p_I \mathbf{S}_I \mathbf{S}_I^H \tag{2.1.25}$$

where $\mathbf{S}_I$ denotes the steering vector associated with the interference.

Using matrix notation, the correlation matrix R may be expressed in the following compact form:

$$R = ASA^H + \sigma_n^2 I \tag{2.1.26}$$

where columns of the $L \times M$ matrix A are made up of steering vectors, that is,

$$A = [\mathbf{S}_1, \mathbf{S}_2, \ldots, \mathbf{S}_M] \tag{2.1.27}$$

and $M \times M$ matrix S denote the source correlation. For uncorrelated sources, it is a diagonal matrix with

$$S_{ij} = \begin{cases} p_i & i = j \\ 0 & i \neq j \end{cases} \tag{2.1.28}$$

### 2.1.2 Eigenvalue Decomposition

Sometimes it is useful to express the array correlation matrix in terms of its eigenvalues and their associated eigenvectors. The eigenvalues of the array correlation matrix can be divided into two sets when the environment consists of uncorrelated directional sources and uncorrelated white noise.

The eigenvalues contained in one set are of equal value. Their value does not depend on directional sources and is equal to the variance of white noise. The eigenvalues contained in the second set are functions of directional source parameters and their number is equal to the number of these sources. Each eigenvalue of this set is associated with a directional source and its value changes with the change in the source power of this source. The eigenvalues of this set are bigger than those associated with the white noise. Sometimes these eigenvalues are referred to as the signal eigenvalues, and the others belonging to the first set are referred to as the noise eigenvalues. Thus, a correlation matrix of an array of L elements immersed in M uncorrelated directional sources and white noise has M signal eigenvalues and L – M noise eigenvalues.

Denoting the L eigenvalues of the array correlation matrix in descending order by $\lambda_\ell$, $\ell = 1, \ldots, L$ and their corresponding unit-norm eigenvectors by $\mathbf{U}_\ell$, $\ell = 1, \ldots, L$ the matrix takes the following form:

$$R = Q\Lambda Q^H \tag{2.1.29}$$

with a diagonal matrix

$$\Lambda = \begin{bmatrix} \lambda_1 & & 0 \\ & \ddots & \\ 0 & & \lambda_L \end{bmatrix} \tag{2.1.30}$$

and

$$Q = [\mathbf{U}_1 \ldots \mathbf{U}_L] \tag{2.1.31}$$

This representation is sometimes referred to as the spectral decomposition of the array correlation matrix. Using the fact that the eigenvectors form an orthonormal set,

$$QQ^H = I \tag{2.1.32}$$

and

$$Q^H Q = I \tag{2.1.33}$$

Thus,

$$Q^H = Q^{-1} \tag{2.1.34}$$

The orthonormal property of the eigenvectors leads to the following expression for the array correlation matrix:

$$R = \sum_{\ell=1}^{M} \lambda_\ell \mathbf{U}_\ell \mathbf{U}_\ell^H + \sigma_n^2 I \tag{2.1.35}$$

## 2.2 Conventional Beamformer

The conventional beamformer, sometimes also known as the delay-and-sum beamformer, has weights of equal magnitudes. The phases are selected to steer the array in a particular direction $(\phi_0, \theta_0)$, known as look direction. With $\mathbf{S}_0$ denoting the steering vector in the look direction, the array weights are given by

$$\mathbf{w}_c = \frac{1}{L}\mathbf{S}_0 \tag{2.2.1}$$

The response of a processor in a direction $(\phi,\theta)$ is obtained by using (2.1.17), that is, taking the dot product of the weight vector with the steering vector $\mathbf{S}(\phi,\theta)$. With the weights given by (2.2.1), the response $y(\phi,\theta)$ is given by

$$\begin{aligned} y(\phi, \theta) &= \mathbf{w}_c^H \mathbf{S}(\phi,\theta) \\ &= \frac{1}{L}\mathbf{S}_0^H \mathbf{S}(\phi,\theta) \end{aligned} \tag{2.2.2}$$

Next, the behavior of this processor is examined under various conditions. It is shown that the array with these weights has unity power response in the look direction, that is, the mean output power of the processor due to a source in the look direction is the same as the source power. An expression for the output SNR is also derived.

### 2.2.1 Source in Look Direction

Assume a source of power $p_S$ in the look direction, hereafter referred to as the signal source, with $m_S(t)$ denoting its modulating function. The signal induced on the $\ell$th element due to this source only is given by

$$x_\ell(t) = m_s(t)\, e^{j2\pi f_0 \left(t+\tau_\ell(\phi_0,\theta_0)\right)} \tag{2.2.3}$$

Thus, in vector notation, using steering vector to denote relevant phases, the array signal vector due to look direction signal becomes

$$\mathbf{x}(t) = m_s(t)\, e^{j2\pi f_0 t}\, \mathbf{S}_0 \tag{2.2.4}$$

The output of the processor is obtained by taking the inner product of weight vector $\mathbf{w}_c$ with the signal vector $\mathbf{x}(t)$ as in (2.4). Thus, the output of the processor is given by

$$y(t) = \mathbf{w}_c^H \mathbf{x}(t) \tag{2.2.5}$$

Substituting from (2.2.1) and (2.2.4), and noting that $\mathbf{S}_0^H\mathbf{S}_0 = L$, the output becomes

$$y(t) = m_s(t)\, e^{j2\pi f_0 t} \tag{2.2.6}$$

Thus, the output of the conventional processor is the same as the signal induced on an element positioned at the reference element. Next, look at its mean out power. As there is only the signal source present, the mean output power of the processor is the mean signal power given by (2.16), that is,

$$\begin{aligned} P(\mathbf{w}_c) &= P_S \\ &= \mathbf{w}_c^H R_S \mathbf{w}_c \end{aligned} \tag{2.2.7}$$

Since

$$R_S = p_S \mathbf{S}_0 \mathbf{S}_0^H \tag{2.2.8}$$

substituting from (2.2.1), (2.2.8) in (2.2.7), and noting that $\mathbf{S}_0^H\mathbf{S}_0 = L$,

$$P(\mathbf{w}_c) = p_S \tag{2.2.9}$$

Thus, the mean output power of the conventional processor steered in the look direction is equal to the power of the source in the look direction. The process is similar to mechanically steering the array in the look direction except that it is done electronically by adjusting the phases. This is also referred to as electronic steering, and phase shifters are used to adjust the required phases. It should be noted that the aperture of an electronically steered array is different from that of the mechanically steered array.

The concept of delay-and-sum beamformer can be further understood with the help of Figure 2.4, which shows an array with two elements separated by distance d. Assume that

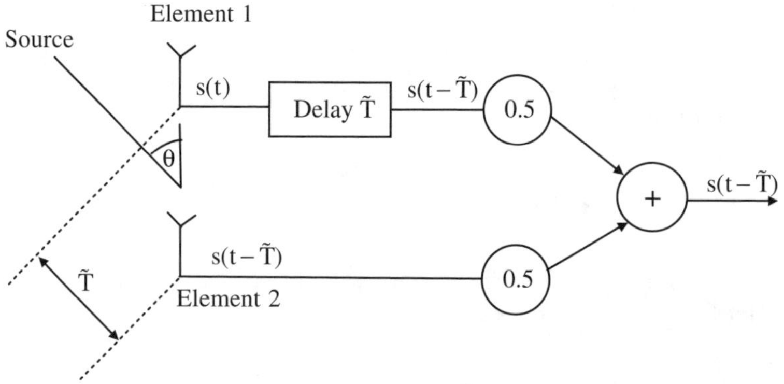


**FIGURE 2.4**
Delay-and-sum beamformer.

a plane wave arriving from direction $\theta$ induces voltage s(t) on the first element. As the wave arrives at the second element $\tilde{T}$ seconds later, with

$$\tilde{T} = \frac{d}{c}\cos\theta \tag{2.2.10}$$

the induced voltage on the second element equals $s(t-\tilde{T})$. If the signal induced at Element 1 is delayed by time $\tilde{T}$, the signal after the delay is $s(t-\tilde{T})$ and no delay is provided at Element 2, then both voltage wave forms are the same. The output of the processor is the sum of the two signals $s(t-\tilde{T})$. A scaling of each wave form by 0.5 provides the gain in direction $\theta$ equal to unity.

### 2.2.2 Directional Interference

Let only a directional interference of power $p_I$ be present in direction $(\phi_I,\theta_I)$. Let $m_I(t)$ and $\mathbf{S}_I$, respectively, denote the modulating function and the steering vector for the interference. The array signal vector for this case becomes

$$\mathbf{x}(t) = m_I(t)\, e^{j2\pi f_0 t}\, \mathbf{S}_I \tag{2.2.11}$$

The array output is obtained by taking the inner product of weight vector and the array signal vector. Thus,

$$\begin{aligned} y(t) &= \mathbf{w}_c^H \mathbf{x}(t) \\ &= m_I(t)\, e^{j2\pi f_0 t}\, \mathbf{w}_c^H \mathbf{S}_I \\ &= m_I(t)\, e^{j2\pi f_0 t}\, \frac{\mathbf{S}_0^H \mathbf{S}_I}{L} \end{aligned} \tag{2.2.12}$$

The quantity $1/L\ \mathbf{S}_0^H \mathbf{S}_0$ determines the amount of interference allowed to filter through the processor and thus is the response of the processor in the interference direction.

The amount of interference power at the output of a processor is given by (2.17). Thus, in the presence of interference only, an expression for the mean output power of the conventional processor becomes

$$P(\mathbf{w}_c) = P_I$$
$$= \mathbf{w}_c^H R_I \mathbf{w}_c \tag{2.2.13}$$

For a single source in the nonlook direction

$$R_I = p_I \mathbf{S}_I \mathbf{S}_I^H \tag{2.2.14}$$

Substituting for $R_I$ and $\mathbf{w}_c$ in (2.2.13),

$$P(\mathbf{w}_c) = p_I\,(1-\rho) \tag{2.2.15}$$

where

$$\rho = 1 - \frac{\mathbf{S}_0^H\ \mathbf{S}_I \mathbf{S}_I^H\ \mathbf{S}_0}{L^2} \tag{2.2.16}$$

and depends on the array geometry and the direction of the interference relative to the look direction.

The effect of the interference direction on parameter ρ is shown in Figure 2.5 and Figure 2.6 for two types of arrays, planar and linear. The planar array consists of two rings of five elements each, as shown in Figure 2.7, whereas the linear array consists of ten equispaced elements.

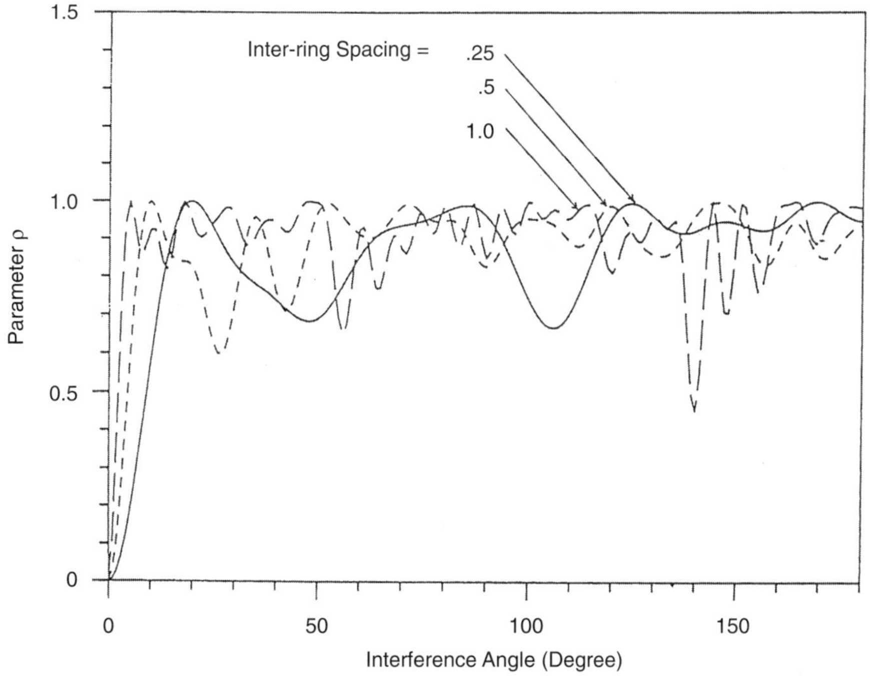


**FIGURE 2.5**
Parameter ρ vs. interference direction at three values of inter-ring spacing for the array geometry shown in Figure 2.7. From Godara, L.C., *J. Acoust. Soc. Am.*, 85, 202–213, 1989 [God89a]. With permission.)

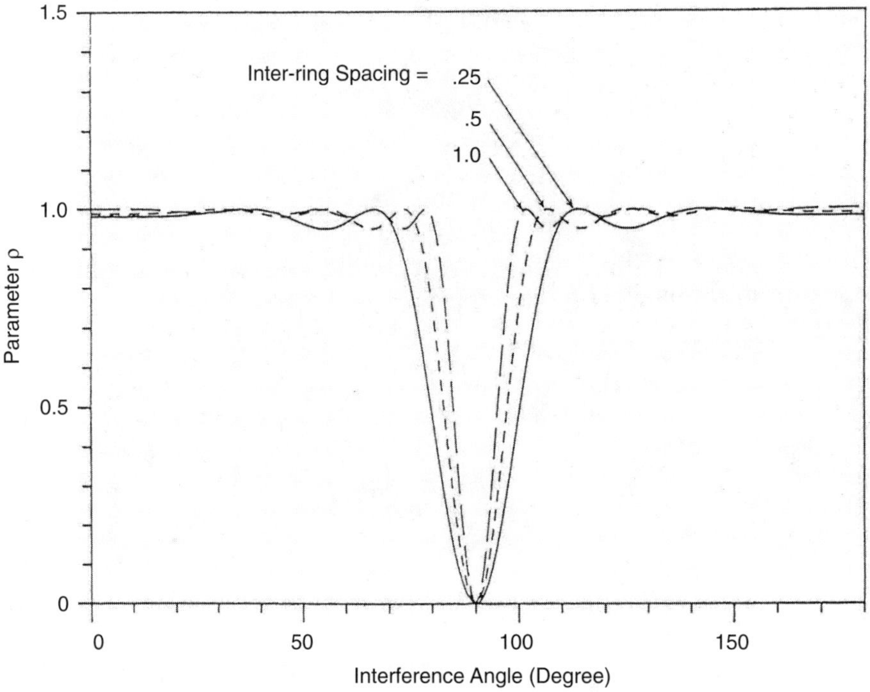


**FIGURE 2.6**
Parameter ρ vs. interference direction for a ten-element linear array. (From Godara, L.C., *J. Acoust. Soc. Am.*, 85, 202–213, 1989 [God89a]. With permission.)

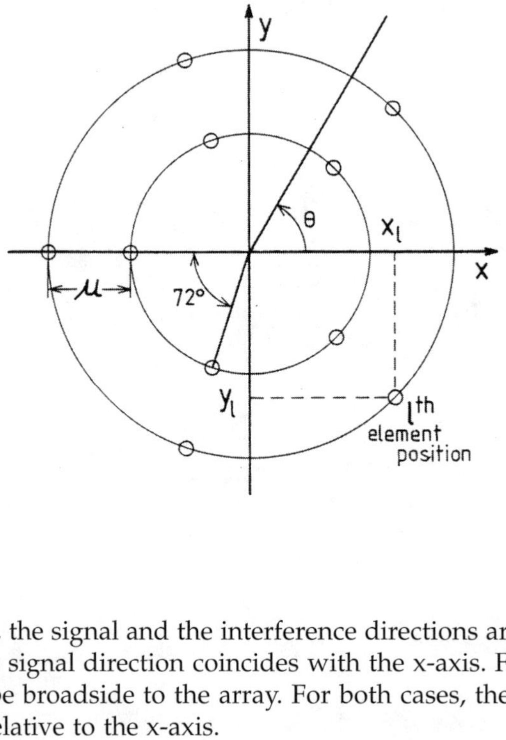


**FIGURE 2.7**
Structure of planar array.

For the planar array, the signal and the interference directions are assumed to be in the plane of the array; the signal direction coincides with the x-axis. For the linear array, the signal is assumed to be broadside to the array. For both cases, the direction of the interference is measured relative to the x-axis.

Figure 2.5 and Figure 2.6, respectively, show the values of $\rho$ for various interference directions at three values of inter-ring spacing $\mu$ and three values of inter-element spacing d. The parameters $\mu$ and d are expressed in terms of the wavelength of the narrowband sources. These figures show how $\rho$ depends on the array geometry for given interference and signal directions.

### 2.2.3 Random Noise Environment

Consider an environment consisting of uncorrelated noise of power $\sigma_n^2$. It is assumed that there is no directional source present. The array signal vector for this case becomes

$$\mathbf{x}(t) = \mathbf{n}(t) \tag{2.2.17}$$

The array output is obtained by taking the inner product of weight vector and the array signal vector. Thus,

$$y(t) = \mathbf{w}_c^H \mathbf{x}(t) \tag{2.2.18}$$

Substituting from (2.2.17) and (2.2.1), the output becomes

$$y(t) = \frac{\mathbf{S}_0^H \mathbf{n}(t)}{L} \tag{2.2.19}$$

The mean output noise power of a processor is given by (2.18). Thus, the mean output power of the conventional processor in the presence of uncorrelated noise only is given by

$$\begin{aligned} P(\mathbf{w}_c) &= P_n \\ &= \mathbf{w}_c^H R_n \mathbf{w}_c \end{aligned} \tag{2.2.20}$$

Since $R_n$ is given by

$$R_n = \sigma_n^2 I \tag{2.2.21}$$

substituting for $R_n$ and $\mathbf{w}_c$ in (2.2.20),

$$P(\mathbf{w}_c) = \frac{\sigma_n^2}{L} \tag{2.2.22}$$

Thus, the mean power at the output of the conventional processor is equal to the mean uncorrelated noise power at an element of the array divided by the number of elements in the array. In other words, the noise power at the array output is L times less than that present on each element.

### 2.2.4 Signal-to-Noise Ratio

Assume that the noise environment consists of the random noise of power $\sigma_n^2$ and a directional interference of power $p_I$ in the nonlook direction. Assume that there is a source of power $p_S$ in the look direction. Given that the interference and the signal are uncorrelated, the array signal vector for this case becomes

$$\mathbf{x}(t) = m_S(t)\, e^{j2\pi f_0 t}\, \mathbf{S}_0 + m_I(t)\, e^{j2\pi f_0 t}\, \mathbf{S}_I + \mathbf{n}(t) \tag{2.2.23}$$

Now we have two directional sources, a signal source, a directional interference, and the random noise. Thus, it follows from (2.1.22) that the array correlation matrix R is given by

$$R = p_S\, \mathbf{S}_0\, \mathbf{S}_0^H + p_I\, \mathbf{S}_I\, \mathbf{S}_I^H + \sigma_n^2\, I \tag{2.2.24}$$

The mean output power of the processor is given by

$$P(\mathbf{w}_c) = \mathbf{w}_c^H R \mathbf{w}_c \tag{2.2.25}$$

Substituting from (2.2.1), (2.2.24) and noting that $\mathbf{S}_0^H \mathbf{S}_0 = L$, the expression for the mean output power from (2.2.25) becomes

$$P(\mathbf{w}_c) = p_S + p_I\,(1-\rho) + \frac{\sigma_n^2}{L} \tag{2.2.26}$$

Note that the mean output power of the processor is the sum of the mean output powers due to signal source, directional interference, and uncorrelated noise.

The mean signal power at the output of the processor is equal to the mean power of the signal source, that is,

$$P_S = p_S \tag{2.2.27}$$

The mean noise power is the sum of the interference power and the uncorrelated noise power, that is,

$$P_N = p_I\,(1-\rho) + \frac{\sigma_n^2}{L} \tag{2.2.28}$$

The output signal to noise ratio is then given by

$$\begin{aligned} \mathrm{SNR} &= \frac{P_S}{P_N} \\ &= \frac{p_S}{p_I\,(1-\rho) + \frac{\sigma_n^2}{L}} \end{aligned} \tag{2.2.29}$$

Now consider a special case when no directional interference is present. For this case, the expression for the output SNR becomes

$$\mathrm{SNR} = \frac{p_S\, L}{\sigma_n^2} \tag{2.2.30}$$

As the input SNR is $p_S/\sigma_n^2$, this provides an array gain, which is defined as the ratio of the output SNR to the input SNR, equal to L, the number of elements in the array.

This processor provides maximum output SNR when no directional interference operating at the same frequency is present. It is not effective in the presence of directional interference, whether intentional or unintentional. The response of the processor toward a directional source is given by (2.2.2). The performance of the processor in the presence of one nonlook directional source indicated by SNR is given by (2.2.29). It is a function of the interference power and the parameter $\rho$ that in turn depends on the relative direction of two sources and array geometry.

## 2.3 Null Steering Beamformer

The null steering beamformer is used to cancel a plane wave arriving from a known direction and thus produces a null in the response pattern of the plane wave's direction of arrival. One of the earliest schemes, referred to as DICANNE [And69, And69a], achieves this by estimating the signal arriving from a known direction by steering a conventional beam in the direction of the source and then subtracting the output of this from each element. An estimate of the signal is made by delay-and-sum beamforming using shift registers to provide the required delay at each element, such that the signal arriving from the beam-steering direction appears in phase after the delay, and then sums these wave forms with equal weighting. This signal then is subtracted from each element after the delay. The process is very effective for canceling strong interference and could be repeated for multiple interference cancelation.

Although the process of subtracting the estimated interference using the delay-and-sum beamformer in the DICANNE scheme is easy to implement for single interference, it becomes cumbersome as the number of interferences grows. A beam with unity response in the desired direction and nulls in interference directions may be formed by estimating beamformer weights shown in Figure 2.1 using suitable constraints [d'As84, And69a]. Assume that $\mathbf{S}_0$ is the steering vector in the direction where unity response is required and that $\mathbf{S}_1, \ldots, \mathbf{S}_k$ are k steering vectors associated with k directions where nulls are required. The desired weight vector is the solution of the following simultaneous equations:

$$\mathbf{w}^H \mathbf{S}_0 = 1 \tag{2.3.1}$$

$$\mathbf{w}^H \mathbf{S}_i = 0, \; i = 1, \ldots, k \tag{2.3.2}$$

Using matrix notation, this becomes

$$\mathbf{w}^H A = \mathbf{e}_1^T \tag{2.3.3}$$

where A is a matrix with columns being the steering vectors associated with all directional sources including the look direction, that is,

$$A \triangleq [\mathbf{S}_0, \mathbf{S}_1, \ldots, \mathbf{S}_k] \tag{2.3.4}$$

and $\mathbf{e}_1$ is a vector of all zeros except the first element which is one, that is,

$$\mathbf{e}_1 = [1, 0, \ldots, 0]^T \tag{2.3.5}$$

For k = L – 1, A is a square matrix. Assuming that the inverse of A exists, which requires that all steering vectors are linearly independent [God81], the solution for the weight vector is given by

$$\mathbf{w}^H = \mathbf{e}_1^T A^{-1} \tag{2.3.6}$$

In case the steering vectors are not linearly independent, A is not invertible and its pseudo inverse can be used in its place.

It follows from (2.3.6) that due to the structure of the vector $\mathbf{e}_1$, the first row of the inverse of matrix A forms the weight vector. Thus, the weights selected as the first row of the inverse of matrix A have the desired properties of unity response in the look direction and nulls in the interference directions.

When the number of required nulls are less than L – 1, A is not a square matrix. A suitable estimate of weights may be produced using

$$\mathbf{w}^H = \mathbf{e}_1^T A^H \left(AA^H\right)^{-1} \tag{2.3.7}$$

Although the beam pattern produced by this beamformer has nulls in the interference directions, it is not designed to minimize the uncorrelated noise at the array output. It is possible to achieve this by selecting weights that minimize the mean output power subject to above constraints [Bre88].

An application of a null steering scheme for detecting amplitude-modulated signals by placing nulls in the known interference directions is described in [Cho93], which is able to cancel a strong jammer in a mobile communication system. The use of a null steering scheme for a transmitting array employed at a base station is discussed in [Chi94], which minimizes the interferences toward other co-channel mobiles. Performance analysis of a null steering algorithm is presented in [Fri89].

## 2.4 Optimal Beamformer

The null steering scheme described in the previous section requires knowledge of the directions of interference sources, and the beamformer using the weights estimated by this scheme does not maximize the output SNR. The optimal beamforming method described in this section overcomes these limitations and maximizes the output SNR in the absence of errors. It should be noted that the optimal beamformer, also known as the minimum variance distortionless response (MVDR) beamformer, described in this section does not require knowledge of directions and power levels of interferences as well as the level of the background noise power to maximize the output SNR. It only requires the direction of the desired signal.

In this section, first we discuss an optimal beamformer with its weights without any constraints, and then study its performance in the presence of one interference and uncorrelated noise [God86].

### 2.4.1 Unconstrained Beamformer

Let an L-dimensional complex vector $\hat{\mathbf{w}}$ represent the weights of the beamformer shown in Figure 2.1 that maximize the output SNR. For an array that is not constrained, an expression for $\hat{\mathbf{w}}$ is given by [App76, Ree74, Bre73]:

$$\hat{\mathbf{w}} = \mu_0 R_N^{-1} \mathbf{S}_0 \quad (2.4.1)$$

where $R_N$ is the array correlation matrix of the noise alone, that is, it does not contain any signal arriving from the look direction $(\phi_0, \theta_0)$ and $\mu_0$ is a constant.

Consider that the noise environment consists of the random noise of power $\sigma_n^2$ and a directional interference of power $p_I$ in nonlook direction. Assume that there is a source of power $p_S$ in the look direction, and that the interference and the signal are uncorrelated. For this case, the array correlation matrix R is given by

$$R = p_S \mathbf{S}_0 \mathbf{S}_0^H + p_I \mathbf{S}_I \mathbf{S}_I^H + \sigma_n^2 I \quad (2.4.2)$$

The mean output power of the processor is given by

$$\hat{P} = \hat{\mathbf{w}}^H R \hat{\mathbf{w}} \quad (2.4.3)$$

It follows from (2.4.2) and (2.4.3) that

$$\hat{P} = p_S \hat{\mathbf{w}}^H \mathbf{S}_0 \mathbf{S}_0^H \hat{\mathbf{w}} + p_I \hat{\mathbf{w}}^H \mathbf{S}_I \mathbf{S}_I^H \hat{\mathbf{w}} + \sigma_n^2 \hat{\mathbf{w}}^H \hat{\mathbf{w}} \quad (2.4.4)$$

Three terms on the RHS of (2.4.4) correspond to the output signal power, residual interference power, and output uncorrelated noise power of the unconstrained optimal beamformer. Let these be denoted by $\hat{P}_S$, $\hat{P}_I$ and $\hat{P}_n$, respectively. Thus, it follows that

$$\hat{P}_S = p_S \hat{\mathbf{w}}^H \mathbf{S}_0 \mathbf{S}_0^H \hat{\mathbf{w}} \quad (2.4.5)$$

$$\hat{P}_I = p_I \hat{\mathbf{w}}^H \mathbf{S}_I \mathbf{S}_I^H \hat{\mathbf{w}} \quad (2.4.6)$$

and

$$\hat{P}_n = \sigma_n^2 \hat{\mathbf{w}}^H \hat{\mathbf{w}} \quad (2.4.7)$$

Substituting for $\hat{\mathbf{w}}$ and noting that $\mathbf{S}_0^H \mathbf{S}_0 = L$, these equations become

$$\hat{P}_S = p_S \mu_0^2 \left(\mathbf{S}_0^H R_N^{-1} \mathbf{S}_0\right)^2 \quad (2.4.8)$$

$$\hat{P}_I = \mu_0^2 \mathbf{S}_0^H R_N^{-1} R_I R_N^{-1} \mathbf{S}_0 \quad (2.4.9)$$

and

$$\hat{P}_n = \sigma_n^2 \hat{\beta} \mu_0^2 \left(\mathbf{S}_0^H R_N^{-1} \mathbf{S}_0\right)^2 \quad (2.4.10)$$

where $R_I$ is the correlation matrix of interference and

$$\hat{\beta} = \frac{\mathbf{S}_0^H R_N^{-1} R_N^{-1} \mathbf{S}_0}{\left(\mathbf{S}_0^H R_N^{-1} \mathbf{S}_0\right)^2} \tag{2.4.11}$$

The total noise at the output is given by

$$\hat{P}_N = \hat{P}_I + \hat{P}_n \tag{2.4.12}$$

Substituting from (2.4.9) and (2.4.10), total noise becomes

$$\begin{aligned}\hat{P}_N &= \mu_0^2\left(\mathbf{S}_0^H R_N^{-1} R_I R_N^{-1} \mathbf{S}_0 + \sigma_n^2 \mathbf{S}_0^H R_N^{-1} R_N^{-1} \mathbf{S}_0\right)\\ &= \mu_0^2 \mathbf{S}_0^H R_N^{-1}\left(R_I + \sigma_n^2 I\right) R_N^{-1} \mathbf{S}_0\\ &= \mu_0^2 \mathbf{S}_0^H R_N^{-1} R_N R_N^{-1} \mathbf{S}_0\\ &= \mu_0^2 \mathbf{S}_0^H R_N^{-1} \mathbf{S}_0\end{aligned} \tag{2.4.13}$$

### 2.4.2 Constrained Beamformer

Let the array weights be constrained to have a unit response in the look direction, that is,

$$\hat{\mathbf{w}}^H \mathbf{S}_0 = 1 \tag{2.4.14}$$

Thus, it follows from (2.4.1) that constant $\mu_0$ is given by

$$\mu_0 = \frac{1}{\mathbf{S}_0^H R_N^{-1} \mathbf{S}_0} \tag{2.4.15}$$

Substituting this in (2.4.1) results in the following expression for the weight vector

$$\hat{\mathbf{w}} = \frac{R_N^{-1} \mathbf{S}_0}{\mathbf{S}_0^H R_N^{-1} \mathbf{S}_0} \tag{2.4.16}$$

Substituting for $\mu_0$ in (2.4.8), (2.4.9), (2.4.10) and (2.4.13) results in the following expressions for the output signal power, residual interference power, output uncorrelated noise power, and the total noise power of the constrained beamformer

$$\hat{P}_S = p_S \tag{2.4.17}$$

$$\hat{P}_I = \frac{\mathbf{S}_0^H R_N^{-1} R_I R_N^{-1} \mathbf{S}_0}{\left(\mathbf{S}_0^H R_N^{-1} \mathbf{S}_0\right)^2} \tag{2.4.18}$$

$$\hat{P}_n = \sigma_n^2 \hat{\beta} \tag{2.4.19}$$

and

$$\hat{P}_N = \frac{1}{\mathbf{S}_0^H R_N^{-1} \mathbf{S}_0} \tag{2.4.20}$$

Note from (2.4.19) that $\hat{\beta}$ is the ratio of the uncorrelated noise power at the output of the constrained beamformer to the uncorrelated noise power at its input.

As the weights for the optimal beamformer discussed above are computed using noise alone matrix inverse (NAME), the processor with these weights is referred to as the NAME processor [Cox73]. It is also known as the maximum likelihood (ML) filter [Cap69], as it finds the ML estimate of the power of the signal source, assuming all sources as interference. It should be noted $R_N$ may be not be invertible when the background noise is very small. In that case, it becomes a rank deficient matrix.

In practice when the estimate of the noise alone matrix is not available, the total array correlation matrix (signal plus noise) is used to estimate the weights and the processor is referred to as the SPNMI (signal-plus-noise matrix inverse) processor. An expression for the weights of the constrained processor for this case is given by

$$\hat{w} = \frac{R^{-1} \mathbf{S}_0}{\mathbf{S}_0^H R^{-1} S_0} \tag{2.4.21}$$

These weights are the solution of the following optimization problem:

$$\begin{aligned} &\underset{\mathbf{w}}{\text{minimize}} \quad \mathbf{w}^H R \mathbf{w} \\ &\text{subject to} \quad \mathbf{w}^H \mathbf{S}_0 = 1 \end{aligned} \tag{2.4.22}$$

Thus, the processor weights are selected by minimizing the mean output power of the processor while maintaining unity response in the look direction. The constraint ensures that the signal passes through the processor undistorted. Therefore, the output signal power is the same as the look direction source power. The minimization process then minimizes the total noise including interference and the uncorrelated noise. The minimization of the total output noise while keeping the output signal constant is the same as maximizing the output SNR.

It should be noted that the weights of the NAMI processor and the SPNAMI processor are identical; and in the absence of errors, the processor performs identically in both cases. This fact can be proved as follows. The Matrix Inversion Lemma for an invertible matrix A and a vector $\mathbf{x}$ states that

$$\left(A + \mathbf{x}\mathbf{x}^H\right)^{-1} = A^{-1} - \frac{A^{-1}\mathbf{x}\mathbf{x}^H A^{-1}}{1 + \mathbf{x}^H A^{-1}\mathbf{x}} \tag{2.4.23}$$

Since

$$R = p_S\, \mathbf{S}_0 \mathbf{S}_0^H + R_N \tag{2.4.24}$$

it follows from the Matrix Inversion Lemma that

$$R^{-1} = R_N^{-1} - \frac{p_S R_N^{-1}\mathbf{S}_0\mathbf{S}_0^H R_N^{-1}}{1 + S_0^H R_N^{-1}\mathbf{S}_0 p_S} \tag{2.4.25}$$

Hence

$$\begin{aligned} R^{-1}\mathbf{S}_0 &= R_N^{-1}\mathbf{S}_0 - \frac{p_S R_N^{-1}\mathbf{S}_0\mathbf{S}_0^H R_N^{-1}\mathbf{S}_0}{1 + S_0^H R_N^{-1}\mathbf{S}_0 p_S} \\ &= \frac{R_N^{-1}\mathbf{S}_0\left(1 + S_0^H R_N^{-1}\mathbf{S}_0 p_S\right) - p_S R_N^{-1}\mathbf{S}_0\mathbf{S}_0^H R_N^{-1}\mathbf{S}_0}{1 + S_0^H R_N^{-1}\mathbf{S}_0 p_S} \\ &= \frac{R_N^{-1}\mathbf{S}_0}{1 + S_0^H R_N^{-1}\mathbf{S}_0 p_S} \end{aligned} \tag{2.4.26}$$

and

$$\mathbf{S}_0^H R^{-1}\mathbf{S}_0 = \frac{\mathbf{S}_0^H R_N^{-1}\mathbf{S}_0}{1 + S_0^H R_N^{-1}\mathbf{S}_0 p_S} \tag{2.4.27}$$

Equations (2.4.21), (2.4.26), and (2.4.27) imply

$$\hat{w} = \frac{R_N^{-1}\,\mathbf{S}_0}{\mathbf{S}_0^H\, R_N^{-1}\, S_0} \tag{2.4.28}$$

Thus,

$$\hat{w} = \hat{\mathbf{w}} \tag{2.4.29}$$

and the optimal weights of the two processors are identical. The processor with these weights is referred to as the optimal processor. This is also known as MVDR beamformer.

### 2.4.3 Output Signal-to-Noise Ratio and Array Gain

The mean output power of the optimal processor is given by

$$\begin{aligned} \hat{P} &= \hat{\mathbf{w}}^H R \hat{\mathbf{w}} \\ &= \frac{1}{S_0^H R^{-1}\mathbf{S}_0} \end{aligned} \tag{2.4.30}$$

This power consists of the signal power, residual interference power, and uncorrelated noise power. Expressions for these quantities are given by (2.4.17), (2.4.18), and (2.4.19), respectively. The total noise at the output is the sum of residual interference and uncorrelated noise. The expression for total noise power is given by (2.4.20).

Let $\hat{\alpha}$ denote the SNR of the optimal beamformer, that is,

$$\hat{\alpha} = \frac{\hat{P}_S}{\hat{P}_N} \tag{2.4.31}$$

It follows from (2.4.17) and (2.4.20) that

$$\hat{\alpha} = p_S \mathbf{S}_0^H R_N^{-1} \mathbf{S}_0 \tag{2.4.32}$$

It should be noted that the same result also follows from (2.4.8) and (2.4.13), the expressions for the signal power and the total noise power at the output of unconstrained beamformer. Thus, the constrained as well as unconstrained beamformer results in the same output SNR.

The array gain of a beamformer is defined as the ratio of the output SNR to the input SNR. Let $\hat{G}$ denote the array gain of the optimal beamformer, that is,

$$\hat{G} = \frac{\hat{\alpha}}{\text{Input SNR}} \tag{2.4.33}$$

Let $p_N$ denote the total noise at the input. SNR at the input of the beamformer is then given by

$$\text{Input SNR} = \frac{p_S}{p_N} \tag{2.4.34}$$

It follows from (2.4.32), (2.4.33) and (2.4.34) that

$$\begin{aligned} \hat{G} &= p_N \mathbf{S}_0^H R_N^{-1} \mathbf{S}_0 \\ &= \frac{p_N}{\hat{P}_N} \end{aligned} \tag{2.4.35}$$

### 2.4.4 Special Case 1: Uncorrelated Noise Only

For a special case of the noise environment when no direction interference is present, the noise-only array correlation matrix is given by

$$R_N = \sigma_n^2 I \tag{2.4.36}$$

Substituting the matrix in (2.4.16), a simple calculation yields

$$\hat{\mathbf{w}} = \frac{\mathbf{S}_0}{L} \tag{2.4.37}$$

Thus, the weights of the optimal processor in the absence of errors are the same as those of the conventional processor, implying that the conventional processor is the optimal processor for this case. Thus, in the absence of directional interferences the conventional processor yields the maximum output SNR and the array gain. The output SNR $\hat{\alpha}$ and the array gain $\hat{G}$ of the optimal processor for this case are, respectively, given by

$$\hat{\alpha} = \frac{p_S L}{\sigma_n^2} \tag{2.4.38}$$

and

$$\hat{G} = L \tag{2.4.39}$$

These quantities are independent of array geometry and depend only on the number of elements in the array.

### 2.4.5 Special Case 2: One Directional Interference

Consider the case of a noise environment consisting of a directional interference of power $p_I$ and uncorrelated noise of power $\sigma_n^2$ on each element of the array. Let $\mathbf{S}_I$ denote the steering vector in the direction of interference. For this case, the noise-only array correlation matrix is given by

$$R_N = \sigma_n^2 I + p_I \mathbf{S}_I \mathbf{S}_I^H \tag{2.4.40}$$

Using the Matrix Inversion Lemma, this yields

$$R_N^{-1} = \frac{1}{\sigma_n^2}\left[ I - \frac{\mathbf{S}_I \mathbf{S}_I^H}{\dfrac{\sigma_n^2}{p_I} + 1} \right] \tag{2.4.41}$$

The substitution for $R_N^{-1}$, rearrangement, and algebraic manipulation leads to the following expression for the output SNR:

$$\hat{\alpha} = \frac{p_S L}{\sigma_n^2} \frac{\rho + \dfrac{\sigma_n^2}{p_I L}}{1 + \dfrac{\sigma_n^2}{p_I L}} \tag{2.4.42}$$

The array gain is given by

$$\hat{G} = \frac{p_I L}{\sigma_n^2} \frac{\left(1 + \dfrac{\sigma_n^2}{p_I}\right)\left(\rho + \dfrac{\sigma_n^2}{p_I L}\right)}{1 + \dfrac{\sigma_n^2}{p_I L}} \tag{2.4.43}$$

where

$$\rho = 1 - \frac{\mathbf{S}_0^H \mathbf{S}_I \mathbf{S}_I^H \mathbf{S}_0}{L^2} \tag{2.4.44}$$

is a scalar quantity and depends on the direction of the interference relative to the signal source and the array geometry, as discussed previously. It follows from (2.2.1) and (2.4.44) after rearrangement that

$$\rho = 1 - \mathbf{w}_c^H \mathbf{S}_I \mathbf{S}_I^H \mathbf{w}_c \tag{2.4.45}$$

Thus, this parameter is characterized by the weights of the conventional processor. As this parameter characterizes the performance of the optimal processor, it implies that the performance of the optimal processor in terms of its interference cancelation capability depends to a certain extent on the response of the conventional processor to interference. This fact has been further highlighted in [Gup82, Gup84].

An interesting special case is when the interference is much stronger compared to background noise, $p_I \gg \sigma_n^2$. For this case, these expressions may be approximated as

$$\hat{\alpha} \cong \frac{p_S L \rho}{\sigma_n^2} \tag{2.4.46}$$

and

$$\hat{G} \cong \frac{p_I L \rho}{\sigma_n^2} \tag{2.4.47}$$

When interference is away from the main lobe of the conventional processor $\rho \approx 1$, it follows that the output SNR of the optimal processor in the presence of a strong interference is the same as that of the conventional processor in the absence of interference. This implies that the processor has almost completely canceled the interference, yielding a very large array gain.

The performance of the processor in terms of its output SNR and the array gain is not affected by the look direction constraint, as it only scales the weights. Therefore, the treatment presented above is valid for the unconstrained processor.

For the optimal beamformer to operate as described above and to maximize the SNR by canceling interferences, the number of interferences must be less than or equal to $L-2$, as an array with L elements has $L-1$ degrees of freedom and one has been utilized by the constraint in the look direction. This may not be true in a mobile communications environment due to the existence of multipath arrivals, and the array beamformer may not be able to achieve the maximization of the output SNR by suppressing every interference. However, as argued in [Win84], the beamformer does not have to suppress interferences to a great extent and cause a vast increase in the output SNR to improve the performance of a mobile radio system. An increase of a few decibels in the output SNR can make possible a large increase in the system's channel capacity.

In the mobile communication literature, the optimal beamformer is often referred to as the optimal combiner. Discussion on the use of the optimal combiner to cancel interferences and to improve the performance of mobile communication systems can be found in [Win84, Win87, Sua93, Nag94a]. The optimal combiner is discussed in detail in a later chapter.

In the next section, a processor is described that requires a reference signal instead of the desired signal direction to estimate the optimal weights of the beamformer.

## 2.5 Optimization Using Reference Signal

A narrowband beamforming structure that employs a reference signal [App76, Wid67, Wid75, Zah73, App76a, Wid82] to estimate the weights of the beamformer is shown in

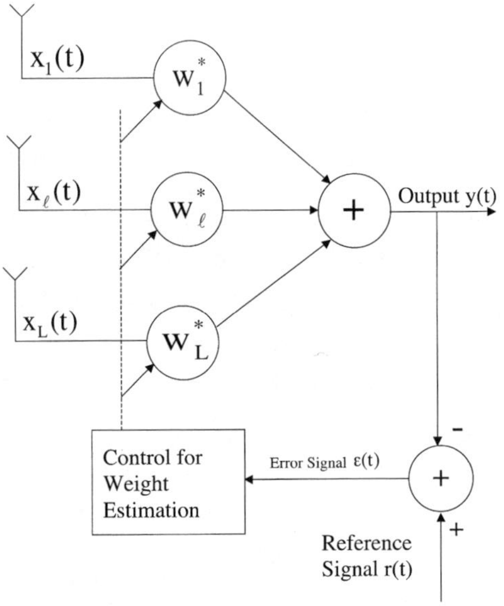


**FIGURE 2.8**
An array system using reference signal.

Figure 2.8. The array output is subtracted from an available reference signal r(t) to generate an error signal $\varepsilon(t) = r(t) - \mathbf{w}^H\mathbf{x}(t)$ that is used to control the weights. Weights are adjusted such that the mean squared error between the array output and the reference signal is minimized. The mean squared error $\xi(\mathbf{w})$ for a given $\mathbf{w}$ is given by

$$\begin{aligned} \xi(\mathbf{w}) &= E\left[|\varepsilon(t)|^2\right] \\ &= E\left[\varepsilon(t)\varepsilon(t)^*\right] \\ &= E\left[\{r(t) - \mathbf{w}^H\mathbf{x}(t)\}\{r(t) - \mathbf{w}^H\mathbf{x}(t)\}^*\right] \\ &= E\left[r(t)r(t)^* + \mathbf{w}^H\mathbf{x}(t)\mathbf{x}^H(t)\mathbf{w} - \mathbf{w}^H\mathbf{x}(t)r(t)^* - r(t)\mathbf{x}^H(t)\mathbf{w}\right] \\ &= E\left[|r(t)|^2\right] + \mathbf{w}^H R\mathbf{w} - \mathbf{w}^H\mathbf{z} - \mathbf{z}^H\mathbf{w} \end{aligned} \tag{2.5.1}$$

where

$$\mathbf{z} = E\left[\mathbf{x}(t)r(t)^*\right] \tag{2.5.2}$$

is the correlation between the reference signal and the array signals vector $\mathbf{x}(t)$.

The mean square error (MSE) surface is a quadratic function of $\mathbf{w}$ and is minimized by setting its gradient with respect to $\mathbf{w}$ equal to zero, with its solution yielding the optimal weight vector, that is,

$$\left.\frac{\partial \xi(\mathbf{w})}{\mathbf{w}}\right|_{\mathbf{w}=\hat{\mathbf{w}}_{\mathrm{MSE}}} = 0 \tag{2.5.3}$$

The gradient of MSE with respect to **w** is obtained by differentiating both sides of (2.5.1) with respect to **w**, yielding

$$\frac{\partial \xi(\mathbf{w})}{\mathbf{w}} = 2\mathrm{R}\mathbf{w} - 2\mathbf{z} \tag{2.5.4}$$

Substituting (2.5.4) in (2.5.3) and solving, you obtain the well-known Wiener–Hoff equation for optimal weights:

$$\hat{\mathbf{w}}_{\mathrm{MSE}} = \mathrm{R}^{-1}\,\mathbf{z} \tag{2.5.5}$$

The processor with these weights is also known as the Wiener filter. The minimum MSE $\hat{\xi}$ of the processor using these weights is obtained by substituting $\hat{\mathbf{w}}_{\mathrm{MSE}}$ for **w** in (2.5.1), resulting in

$$\hat{\xi} = \mathrm{E}\left[\left|\mathrm{r}(\mathrm{t})\right|^2\right] - \mathbf{z}^{\mathrm{H}}\mathrm{R}^{-1}\mathbf{z} \tag{2.5.6}$$

This scheme may be employed to acquire a weak signal in the presence of a strong jammer as discussed in [Zah73] by setting the reference signal to zero and initializing the weights to provide an omnidirectional pattern. The process starts to cancel strong interferences first and the weak signal later. Thus, intuitively, a time is expected when the output would consist of the signal, which has not been canceled too much, but strong interference has been reduced.

When an adaptive scheme (discussed in Chapter 3) is used to estimate $\hat{\mathbf{w}}_{\mathrm{MSE}}$, the strong jammer gets canceled first as the weights are adjusted to put a null in that direction to leave the signal-to-jammer ratio sufficient for acquisition.

Arrays using a reference signal equal to zero to adjust weights are referred to as power-inversion adaptive arrays [Com79]. The MSE minimization scheme (the Wiener filter) is a closed-loop method compared to the open-loop scheme of MVDR (the ML filter) described in the previous section. In general the Wiener filter provides higher-output SNR compared to the ML filter in the presence of a weak signal source. As the input signal power becomes large compared to the background noise, the two processors give almost the same results [Gri67]. This result is supported by a simulation study using mobile communications with two vehicles [Fli94]. The increased SNR by the Wiener filter is achieved at the cost of signal distortion caused by the filter. It should be noted that the optimal beamformer does not distort the signal.

The required reference signal for the Wiener filter may be generated in a number of ways depending on the application. In digital mobile communications, a synchronization signal may be used for initial weight estimation followed by the use of a detected signal as a reference signal. In systems using the TDMA scheme, a user-specific sequence may be part of every frame for this purpose [Win94]. The use of known symbols in every frame has also been suggested in [Cho92]. In other situations, use of an antenna for this purpose has been examined to show the suitability to provide a reference signal [Cho92].

Studies of mobile communication systems using reference signal to estimate array weights have also been reported in [And91, Geb95, Dio93].

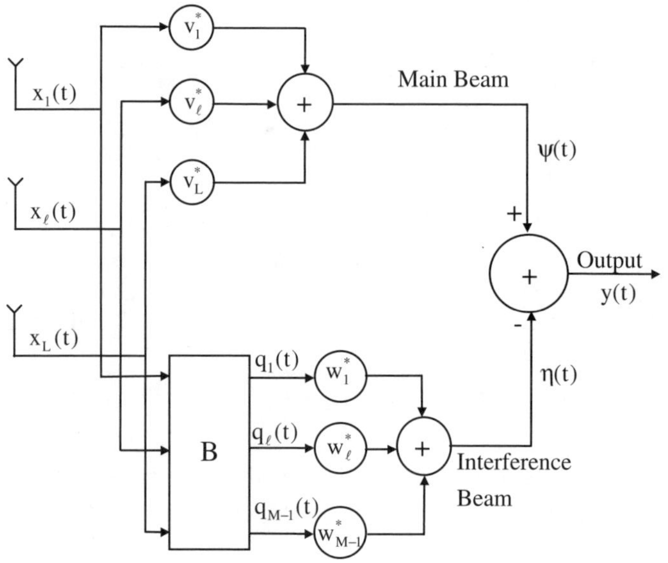


**FIGURE 2.9**
Beam-space processor structure.

## 2.6 Beam Space Processing

In contrast to the element space processing discussed in previous sections, where signals derived from each element are weighted and summed to produce the array output, the beam space processing is a two-stage scheme where the first stage takes the array signals as input and produces a set of multiple outputs, which are then weighted and combined to produce the array output. These multiple outputs may be thought of as the output of multiple beams. The processing done at the first stage is by fixed weighting of the array signals and amounts to produce multiple beams steered in different directions. The weighted sum of these beams is produced to obtain the array output and the weights applied to different beam outputs are then optimized to meet a specific optimization criterion.

In general, for an L-element array, a beam space processor consists of a main beam steered in the signal direction and a set of not more than L – 1 secondary beams. The weighted output of the secondary beams is subtracted from the main beam. The weights are adjusted to produce an estimate of the interference present in the main beam. The subtraction process then removes this interference. The secondary beams, also known as auxiliary beams, are designed such that they do not contain the desired signal from the look direction, to avoid signal cancelation in the subtraction process. A general structure of such a processor is shown in Figure 2.9. Beam space processors have been studied under many different names including the Howells–Applebaum array [App76, App76a, How76]; generalized side-lobe canceler (GSC) [Gri82, Gri77]; partitioned processor [Jim77, Can82]; partially adaptive arrays [Van87, Van89, Van90, Qia94, Qia95, Cha76, Mor78]; post-beamformer interference canceler [Can84, God86a, God89, God89a, God91]; adaptive-adaptive arrays [Bro86]; and multiple-beam antennas [May78, Kle75, Gob76].

The pattern of the main beam is normally referred to as the quiescent pattern and is chosen such that it has a desired shape. For a linear array of equispaced elements with equal weighting, the quiescent pattern has the shape of sin Lx/sin x with L being the number of elements in the array, whereas for Tschebysheff weighting (the weighting dependent on Tschebysheff polynomial coefficients), the pattern has equal side-lobe levels [Dol46]. The beam pattern of the main beam may be adjusted by applying various forms of constraints on the weights [App76a] and using various pattern synthesis techniques discussed in [Gri87, Tse92, Web90, Er93, Sim83, Ng02].

There are many schemes to generate the outputs of auxiliary beams such that no signal from the look direction is contained in them, that is, these beams have nulls in the look direction. In its simplest form, it can be achieved by subtracting the array signals from presteered adjacent pairs [Gab76, Dav67]. It relies on the fact that the component of the array signals induced from a source in the look direction is identical after the presteering, and this gets canceled in the subtraction process from the adjacent pairs. The process can be generalized to produce M – 1 beams from an L-element array signal $\mathbf{x}(t)$ using a matrix B such that

$$\mathbf{q}(t) = B^H \mathbf{x}(t) \tag{2.6.1}$$

where M – 1 dimensional vector $\mathbf{q}(t)$ denotes the outputs of M – 1 beams and the matrix B, referred to as the blocking matrix or the matrix prefilter, has the property that its M – 1 columns are linearly independent and the sum of elements of each column equals zero, implying that M – 1 beams are independent and have nulls in the look direction. For an array that is not presteered, the matrix needs to satisfy

$$B^H \mathbf{S}_0 = \mathbf{0} \tag{2.6.2}$$

where $\mathbf{S}_0$ is the steering vector associated with the look direction and $\mathbf{0}$ denotes a vector of zeros.

It is assumed in the above discussion that $M \leq L$, implying that the number of beams are less than or equal to the number of elements in the array. When the number of beams is equal to the number of elements in the array, the processing in the beam space has not reduced the degree of freedom of the array, that is, its null-forming capability has not been reduced. In this sense, these arrays are fully adaptive and have the same capabilities as that of the array using element space processing. In fact, in the absence of errors, both processing schemes produce identical results. On the other hand, when the number of beams is less than the number of elements, the arrays are referred to as partially adaptive. The null steering capabilities of these arrays have been reduced to equal the number of auxiliary beams. When adaptive schemes, discussed later, are used to estimate the weights, convergence is generally faster for these arrays. However, the MSE for these arrays is also high compared to fully adaptive arrays [Van91].

These arrays are useful in situations where the number of interferences are much less than the number of elements and offer computational advantage over element space processing, as you only need to adjust M – 1 weights compared to L weights for the element space case with $M < L$. Moreover, beam space processing requires less computation than the element space case to calculate the weights in general as it solves an unconstrained optimization compared to the constrained optimization problem solved in the latter case. It should be noted that for the element space processing case, constraints on the weights are imposed to prevent distortion of the signal arriving from the look direction and to make the array more robust against errors. For the beam space case, constraints are transferred to the main beam, leaving the adjustable weights free from constraints.

Auxiliary beamforming techniques other than the use of a blocking matrix described above include formation of M – 1 orthogonal beams and formation of beams in the direction of interference, if known. The beams are referred to as orthogonal beams to imply that the weight vectors used to form beams are orthogonal, that is, their dot product is equal to zero. The eigenvectors of the array correlation matrix taken as weights to generate auxiliary beams fall into this category. In situations where directions of arrival of interference are known, the formation of beams pointed in these directions may lead to more efficient interference cancelation [Bro86, Gab86].

Auxiliary beam outputs are weighted and summed, and the result is subtracted from the main beam output to cancel the unwanted interference present in the main beam. The weights are adjusted to cancel the maximum possible interference. This is normally done by minimizing the total mean output power after subtraction by solving the unconstrained optimization problem and leads to maximization of the output SNR in the absence of the desired signal in auxiliary channels. The presence of the signal in these channels causes signal cancelation from the main beam along with interference cancelation. A detailed discussion on the principles of signal cancelation in general and some possible cures is given in [Wid75, Wid82, Su86].

Use of multiple-beam array processing techniques for mobile communications has been reported in various studies [Jon95, Sak92], including development of an array system using digital hardware to study its feasibility [God02].

### 2.6.1 Optimal Beam Space Processor

It follows from the Figure 2.9 that the output of the main beam $\psi(t)$ is given by

$$\psi(t) = \mathbf{V}^H \mathbf{x}(t) \tag{2.6.3}$$

where the L-dimensional vector $\mathbf{V}$ is defined as

$$\mathbf{V} = \left[v_1, v_2, \ldots, v_L\right]^T \tag{2.6.4}$$

Let an M – 1 dimensional vector $\mathbf{q}(t)$ be defined as

$$\mathbf{q}(t) = \left[q_1, q_2, \ldots, q_{M-1}\right]^T \tag{2.6.5}$$

It denotes M – 1 auxiliary beams, output of matrix prefilter B, and is given by

$$\mathbf{q}(t) = B^H \mathbf{x}(t) \tag{2.6.6}$$

Let an M – 1 dimensional vector $\mathbf{w}$ denote the adjustable weights of the auxiliary beams. It follows from Figure 2.9 that the output $\eta(t)$ of the interference beam is given by

$$\eta(t) = \mathbf{w}^H \mathbf{q}(t) \tag{2.6.7}$$

The output y(t) of the overall beam space processor is obtained by subtracting the interference beam output from the main beam, and thus is given by

$$\begin{aligned} y(t) &= \psi(t) - \eta(t) \\ &= \psi(t) - \mathbf{w}^H \mathbf{q}(t) \end{aligned} \tag{2.6.8}$$

The mean output power P(**w**) of the processor for a given weight vector **w** is given by

$$\begin{aligned} P(\mathbf{w}) &= E\left[ y(t) y(t)^* \right] \\ &= E\left[ \{\psi(t) - \mathbf{w}^H \mathbf{q}(t)\} \{\psi(t) - \mathbf{w}^H \mathbf{q}(t)\}^* \right] \\ &= E\left[ \psi(t)\psi(t)^* + \mathbf{w}^H \mathbf{q}(t) \mathbf{q}^H(t) \mathbf{w} - \mathbf{w}^H \mathbf{q}(t) \psi(t)^* - \psi(t) \mathbf{q}^H(t) \mathbf{w} \right] \\ &= P_0 + \mathbf{w}^H R_{qq} \mathbf{w} - \mathbf{w}^H \mathbf{Z} - \mathbf{Z}^H \mathbf{w} \end{aligned} \tag{2.6.9}$$

where $P_0$ is the mean power of the main beam given by

$$P_0 = \mathbf{V}^H R\, \mathbf{V} \tag{2.6.10}$$

$R_{qq}$ is the correlation matrix of auxiliary beams defined as

$$R_{qq} = E\left[ \mathbf{q}(t) \mathbf{q}^H(t) \right] \tag{2.6.11}$$

and **Z** denotes the correlation between the output of auxiliary beams and the main beam. It is defined as

$$\mathbf{Z} = E\left[ \mathbf{q}(t) \psi(t)^* \right] \tag{2.6.12}$$

A substitution for **q**(t) and ψ(t) in (2.6.11) and (2.6.12) yields

$$\begin{aligned} R_{qq} &= E\left[ \mathbf{q}(t) \mathbf{q}^H(t) \right] \\ &= E\left[ B^H \mathbf{x}(t) \mathbf{x}^H(t) B \right] \\ &= B^H R B \end{aligned} \tag{2.6.13}$$

$$\begin{aligned} \mathbf{Z} &= E\left[ \mathbf{q}(t) \psi(t)^* \right] \\ &= E\left[ B^H \mathbf{x}(t) \mathbf{x}^H(t) \mathbf{V} \right] \\ &= B^H R \mathbf{V} \end{aligned} \tag{2.6.14}$$

Substituting for $P_0$, $R_{qq}$ and $\mathbf{Z}$ in (2.6.9), the expression for P(**w**) becomes

$$P(\mathbf{w}) = \mathbf{V}^H R\mathbf{V} + \mathbf{w}^H B^H RB\mathbf{w} - \mathbf{w}^H B^H R\mathbf{V} - \mathbf{V}^H RB\mathbf{w} \tag{2.6.15}$$

Note that P(**w**) is a quadratic function of **w** and has a unique minimum. Let $\hat{\mathbf{w}}$ denote weights that minimize P(**w**). Thus, it follows that

$$\left.\frac{\partial P(\mathbf{w})}{\partial \mathbf{w}}\right|_{\mathbf{w}=\hat{\mathbf{w}}} = 0 \tag{2.6.16}$$

Substituting (2.6.15) in (2.6.16) yields

$$B^H RB\hat{\mathbf{w}} - B^H R\mathbf{V} = 0 \tag{2.6.17}$$

As B has rank M – 1, $B^H RB$ is of full rank and its inverse exists. Thus, (2.6.17) yields

$$\hat{\mathbf{w}} = \left(B^H RB\right)^{-1} B^H R\mathbf{V} \tag{2.6.18}$$

Substituting for $\mathbf{w} = \hat{\mathbf{w}}$ from (2.6.18) in (2.6.15), you obtain the following expression for the mean output power of the optimal processor:

$$P(\hat{\mathbf{w}}) = \mathbf{V}^H R\mathbf{V} - \mathbf{V}^H RB\left(B^H RB\right)^{-1} B^H R\mathbf{V} \tag{2.6.19}$$

Expressions for the mean output signal power may be obtained by replacing the array correlation matrix R by the signal only array correlation matrix $R_S$ in (2.6.15), yielding

$$P_S(\mathbf{w}) = \mathbf{V}^H R_S\mathbf{V} + \mathbf{w}^H B^H R_S B\mathbf{w} - \mathbf{w}^H B^H R_S\mathbf{V} - \mathbf{V}^H R_S B\mathbf{w} \tag{2.6.20}$$

Since

$$R_S = p_S\mathbf{S}_0\mathbf{S}_0^H \tag{2.6.21}$$

and

$$B^H\mathbf{S}_0 = 0 \tag{2.6.22}$$

it follows from (2.6.20) that

$$P_S(\mathbf{w}) = \mathbf{V}^H R_S\mathbf{V} \tag{2.6.23}$$

Thus, when the blocking matrix B is selected such that $B^H\mathbf{S}_0 = 0$, there are no signal flows through the interference beam and the output signal power is present only in the main beam. When the main beam is taken as the conventional beam, that is,

$$\mathbf{V} = \frac{1}{L}\mathbf{S}_0 \tag{2.6.24}$$

the mean output signal power of the beam space processor becomes

$$P_S(\hat{\mathbf{w}}) = p_S \tag{2.6.25}$$

Note that the signal power is independent of **w**.

Similarly, an expression for the mean output noise power may be obtained by replacing the array correlation matrix R by the noise-only array correlation matrix $R_N$ in (2.6.15), yielding

$$P_N(\mathbf{w}) = \mathbf{V}^H R_N \mathbf{V} + \mathbf{w}^H B^H R_N B\mathbf{w} - \mathbf{w}^H B^H R_N \mathbf{V} - \mathbf{V}^H R_N B\mathbf{w} \tag{2.6.26}$$

Substituting for $\mathbf{w} = \hat{\mathbf{w}}$ from (2.6.18) in (2.6.26), you obtain the following expression for the mean output noise power of the optimal processor:

$$\begin{aligned} P_N(\hat{\mathbf{w}}) = \mathbf{V}^H R_N \mathbf{V} + \mathbf{V}^H RB(B^H RB)^{-1} B^H R_N B(B^H RB)^{-1} B^H R\mathbf{V} \\ - \mathbf{V}^H RB(B^H RB)^{-1} B^H R_N \mathbf{V} - \mathbf{V}^H R_N B(B^H RB)^{-1} B^H R\mathbf{V} \end{aligned} \tag{2.6.27}$$

The output SNR of the optimal beam space processor then becomes

$$\mathrm{SNR}(\hat{\mathbf{w}}) = \frac{p_S}{P_N(\hat{\mathbf{w}})} \tag{2.6.28}$$

These expressions cannot be simplified further without considering specific cases. In Section 2.6.3, a special case of beam space processor is considered where only one auxiliary beam is considered in the presence of one interference source to understand the behavior of beam space processors. The results are then compared with an element space processor. In the next section, a beam space processor referred to as the generalized side-lobe canceler (GSC) is considered. The main difference between the general beam space processor considered in this section and the GSC is that the GSC uses presteering delays.

### 2.6.2 Generalized Side-Lobe Canceler

A structure of the generalized side-lobe canceler is shown in Figure 2.10. The array is presteered by delaying received signals on all antennas such that the component of the received signal on all elements arriving from the look direction is in phase after presteering delays. Let $\alpha_\ell$, $\ell = 1, 2, \ldots, L$ denote the phase delays to steer the array in the look direction. These are given by

$$\alpha_\ell = 2\pi f_0 \tau_\ell(\phi_0, \theta_0) \quad \ell = 1, 2, \ldots, L \tag{2.6.29}$$

Let the received signals after presteering delays be denoted by $\mathbf{x}'(t)$. As these are delayed versions of $\mathbf{x}(t)$, it follows that their $\ell$th components are related by

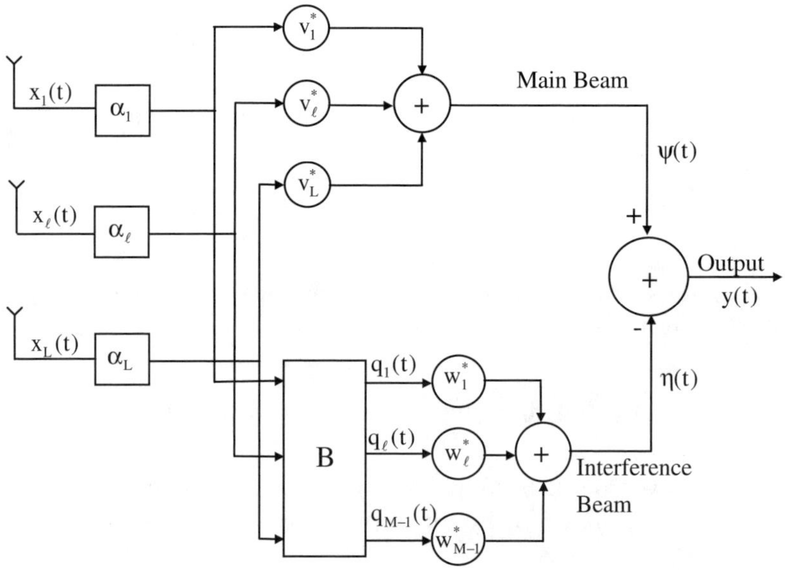


**FIGURE 2.10**
Generalized side-lobe canceler structure.

$$x'_\ell(t) = x_\ell\left(t - \tau_\ell(\phi_0, \theta_0)\right) \tag{2.6.30}$$

This along with (2.1.9) imply that $\mathbf{x}'(t)$ are related to $\mathbf{x}(t)$ by

$$\mathbf{x}'(t) = \Phi_0^H \mathbf{x}(t) \tag{2.6.31}$$

where $\Phi_0$ is a diagonal matrix defined as

$$\Phi_0 = \begin{bmatrix} e^{j\alpha_1} & & 0 \\ & e^{j\alpha_\ell} & \\ 0 & & e^{j\alpha_L} \end{bmatrix} \tag{2.6.32}$$

Note that $\Phi_0$ satisfies the relation, $\Phi_0^H \mathbf{S}_0 = 1$, where $\mathbf{1}$ is a vector of ones.

These signals are used to form the main beam as well as M – 1 interference beams. The main beam is formed using fixed weights on all channels. These weights are selected to be of equal to 1/L so that a unity response is maintained in the look direction. Let these be denoted by an L-dimensional vector $\mathbf{V}$ given by

$$\mathbf{V} = \frac{1}{L}\mathbf{1} \tag{2.6.33}$$

The M – 1 interference beams are formed using a blocking matrix B. Let these be denoted by an M – 1 dimensional vector $\mathbf{q}(t)$, given

$$\begin{aligned} \mathbf{q}(t) &= B^H \mathbf{x}'(t) \\ &= B^H \Phi_0^H \mathbf{x}(t) \end{aligned} \tag{2.6.34}$$

where matrix B has rank M – 1 and satisfies

$$B^H \mathbf{1} = \mathbf{0} \tag{2.6.35}$$

Expressions for the main beam, interference beams, and GSC output are then, respectively, given by

$$\psi(t) = \mathbf{V}^H \Phi_0^H \mathbf{x}(t) \tag{2.6.36}$$

$$\begin{aligned} \eta(t) &= \mathbf{w}^H \mathbf{q}(t) \\ &= \mathbf{w}^H B^H \Phi_0^H \mathbf{x}(t) \end{aligned} \tag{2.6.37}$$

and

$$\begin{aligned} y(t) &= \psi(t) - \eta(t) \\ &= \mathbf{V}^H \Phi_0^H \mathbf{x}(t) - \mathbf{w}^H B^H \Phi_0^H \mathbf{x}(t) \end{aligned} \tag{2.6.38}$$

It can easily be verified that an expression for the mean output power of the GSC for a given **w** is given by

$$\begin{aligned} P(\mathbf{w}) &= E\left[ y(t) y(t)^* \right] \\ &= E\left[ \left\{ \mathbf{V}^H \Phi_0^H \mathbf{x}(t) - \mathbf{w}^H B^H \Phi_0^H \mathbf{x}(t) \right\} \left\{ \mathbf{V}^H \Phi_0^H \mathbf{x}(t) - \mathbf{w}^H B^H \Phi_0^H \mathbf{x}(t) \right\}^* \right] \\ &= \mathbf{V}^H \Phi_0^H R \Phi_0 \mathbf{V} + \mathbf{w}^H B^H \Phi_0^H R \Phi_0 B \mathbf{w} - \mathbf{V}^H \Phi_0^H R \Phi_0 B \mathbf{w} - \mathbf{w}^H B^H \Phi_0^H R \Phi_0 \mathbf{V} \\ &= \mathbf{V}^H \tilde{R} \mathbf{V} + \mathbf{w}^H B^H \tilde{R} B \mathbf{w} - \mathbf{V}^H \tilde{R} B \mathbf{w} - \mathbf{w}^H B^H \tilde{R} \mathbf{V} \end{aligned} \tag{2.6.39}$$

where

$$\tilde{R} = \Phi_0^H R \Phi_0 \tag{2.6.40}$$

is the array correlation matrix after steering delays.

Comparing (2.6.15) and (2.6.39), one notes that the expression for the mean output power of the GSC for a given **w** is analogous to that given by (2.6.15), with **V** and $\tilde{R}$, respectively, given by (2.6.33) and (2.6.40) and B satisfying (2.6.35). Thus, the expression for GSC optimal weights is analogous to (2.6.18), with R replaced by $\tilde{R}$, that is,

$$\hat{\mathbf{w}} = \left( B^H \tilde{R} B \right)^{-1} B^H \tilde{R} \mathbf{V} \tag{2.6.41}$$

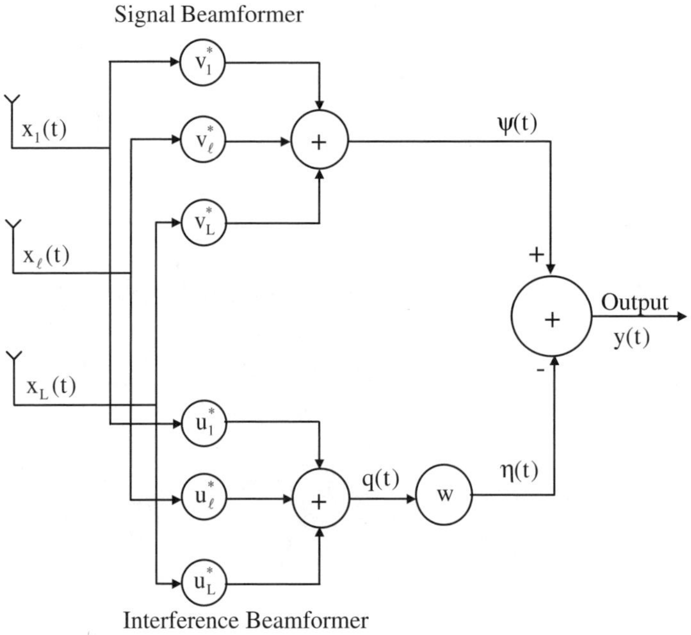


**FIGURE 2.11**
Post-beam former interference canceler structure.

The expression for the mean output noise power of the optimal GSC then becomes

$$\begin{aligned} P_N(\hat{\mathbf{w}}) = \mathbf{V}^H R_N \mathbf{V} + \mathbf{V}^H \tilde{R} B \left(B^H \tilde{R} B\right)^{-1} B^H R_N B \left(B^H \tilde{R} B\right)^{-1} B^H \tilde{R} \mathbf{V} \\ - \mathbf{V}^H \tilde{R} B \left(B^H \tilde{R} B\right)^{-1} B^H R_N \mathbf{V} - \mathbf{V}^H R_N B \left(B^H \tilde{R} B\right)^{-1} B^H \tilde{R} \mathbf{V} \end{aligned} \tag{2.6.42}$$

and the output SNR is given by (2.6.28).

### 2.6.3 Postbeamformer Interference Canceler

In this section, a processor with two beams referred to as the postbeamformer interference canceler (PIC) in previous studies [God86a, God89, God89a, God91] is examined in the presence of a look-direction signal of power $p_S$, an interference of power $p_I$, and uncorrelated noise of power $\sigma_n^2$.

As discussed previously for the general beam space processor, the two-beam processor processes the signals derived from an antenna array by forming two beams using fixed beamforming weights, as shown in Figure 2.11. One beam, referred to as the signal beam, is formed to have a fixed response in the look direction. The processed output of the second beam, referred to as the interference beam, is subtracted from the output of the signal beam to form the output of the PIC.

Let L-dimensional complex vectors **V** and **U** represent the fixed weights of the signal beamformer and the interference beamformer, respectively. It follows from Figure 2.11 that the output $\psi(t)$ of the signal beam and the output $q(t)$ of the interference beam are, respectively, given by

$$\psi(t) = \mathbf{V}^H \mathbf{x}(t) \tag{2.6.43}$$

and

$$q(t) = \mathbf{U}^H \mathbf{x}(t) \tag{2.6.44}$$

The output y(t) of the PIC processor is formed by subtracting the weighted output of the interference beam from the output of the signal beam, that is,

$$y(t) = \psi(t) - w\,q(t) \tag{2.6.45}$$

For a given weight w, the mean output power P(w) of the PIC processor is given by

$$P(w) = \mathbf{V}^H R \mathbf{V} + w^* w \mathbf{U}^H R \mathbf{U} - w^* \mathbf{V}^H R \mathbf{U} - w \mathbf{U}^H R \mathbf{V} \tag{2.6.46}$$

#### *2.6.3.1 Optimal PIC*

Let $\hat{w}$ represent the complex weight of the interference channel of the PIC that minimizes the mean output power of the PIC for given beamformer weights **V** and **U**. This weight $\hat{w}$ is referred to as the optimal weight, and the PIC with this weight is referred to as the optimal PIC.

From the definition of the optimal weight, it follows that

$$\left.\frac{\partial P(w)}{\partial w}\right|_{w=\hat{w}} = 0 \tag{2.6.47}$$

which along with (2.6.46), implies that

$$\hat{w} = \frac{\mathbf{V}^H R \mathbf{U}}{\mathbf{U}^H R \mathbf{U}} \tag{2.6.48}$$

The mean output power of the optimal PIC is given by

$$P(\hat{w}) = \mathbf{V}^H R \mathbf{V} - \mathbf{U}^H R \mathbf{V}\, \mathbf{V}^H R \mathbf{U} / \mathbf{U}^H R \mathbf{U} \tag{2.6.49}$$

In the following discussion, three different beamformer weights for the interference beam are considered. For these cases, the expressions for the signal power, residual interference power, and uncorrelated noise power at the output of the optimal PIC are derived in [God89a]. For the three cases considered, it is assumed that the signal beam is formed using the conventional beamforming weights, that is,

$$\mathbf{V} = \frac{\mathbf{S}_0}{L} \tag{2.6.50}$$

This choice of beamformer weights for the signal beam ensures that the response of the signal beam in the signal direction is unity.

### 2.6.3.2 *PIC with Conventional Interference Beamformer*

Let the interference beam be formed with the beamforming weights,

$$\mathbf{U} = \frac{\mathbf{S}_I}{L} \tag{2.6.51}$$

This choice of beamforming weights ensures that the response of the beam in the interference direction is unity. Note that these weights are not constrained to block the look direction signal passing through to the interference beam as was done using blocking matrix B in the previous discussion. This particular interference beam highlights the effect of the signal present in the auxiliary beams.

It follows from (2.6.50) and (2.6.51) that the response of the interference beam in the signal direction is the same as that of the signal beam in the interference direction. This implies that a large amount of the signal power leaks into the interference beam. This leads to a substantial amount of signal suppression and the presence of residual interference when the PIC is optimized. This aspect of the PIC is now considered and expressions for the mean output signal power and the mean output noise power of the optimal PIC are presented.

Substituting for **U** and **V** in (2.6.48), you obtain an expression for the weight $\hat{w}_c$ of the optimal PIC using the conventional interference beamformer (CIB):

$$\hat{w}_c = \frac{\mathbf{S}_0^H R \mathbf{S}_I}{\mathbf{S}_I^H R \mathbf{S}_I} \tag{2.6.52}$$

Substituting for R, this leads to

$$\hat{w}_c = \beta \frac{\left(1 + \frac{p_I}{p_S} + \frac{\sigma_n^2}{Lp_S}\right)}{\left(1 + \frac{p_I}{p_S} + \frac{\sigma_n^2}{Lp_S}\right) - \rho} \tag{2.6.53}$$

where $\beta$ is a normalized dot product of $\mathbf{S}_0$ and $\mathbf{S}_I$, defined as $\beta = \mathbf{S}_0^H \mathbf{S}_1 / L$.

Substituting for R equals $R_S$, $R_I$, $R_n$ and $R_N$, when $w = \hat{w}_c$ in (2.6.46), the following expressions are obtained for the output signal power, residual interference power, uncorrelated noise power, and output noise power, respectively:

$$P_S(\hat{w}_c) = p_S \rho^2 / (1 + \alpha_I)^2 \tag{2.6.54}$$

$$P_I(\hat{w}_c) = \frac{p_I \rho^2 / (1 - \rho)}{\left[1 + \frac{p_I/p_S + \sigma_n^2/Lp_S}{(1 - \rho)}\right]^2} \tag{2.6.55}$$

$$P_n(\hat{w}_c) = \frac{\sigma_n^2}{L} \rho \left(1 + \frac{\rho/(1 - \rho)}{\left[1 + \frac{p_I/p_S + \sigma_n^2/Lp_S}{(1 - \rho)}\right]^2}\right) \tag{2.6.56}$$

and

$$P_N(\hat{w}_c) = \frac{\sigma_n^2 \rho}{L} + \frac{\rho^2}{1-\rho} \frac{p_I + \sigma_n^2/L}{(1+1/\alpha_I)^2} \tag{2.6.57}$$

where

$$\alpha_I = \frac{(1-\rho)p_S}{(p_I + \sigma_n^2/L)} \tag{2.6.58}$$

is the SNR at the output of the interference beam. Since the SNR is a positive quantity and the parameter ρ is not more than unity, it follows from (2.6.54) that the signal power at the output of the optimal PIC using the CIB is less than the signal power at the output of the signal beam. Hence, the signal has been suppressed by the PIC. Furthermore, the signal suppression increases as (1)the parameter ρ, which depends on the array geometry and the relative directions of the two sources, decreases, and (2) the SNR at the output of the interference beam increases.

Since the SNR at the output of the interference beam is proportional to the input signal power, it follows that signal suppression increases as the input signal power increases. On the other hand, an increase in the interference power as well as the uncorrelated noise power at the input of the PIC decreases the SNR at the output of the interference beam and, hence, decreases the signal suppression of the optimal PIC using the CIB.

Physically, the signal suppression by the optimal PIC using the CIB arises from the leakage of the signal into the interference beam. The component of the signal in the interference beam is subtracted from the signal in the signal beam; in the process of minimization of total output power, this leads to signal suppression. Signal suppression increases as the parameter ρ decreases. The reason for this is that as ρ decreases, the response of the interference beam in the signal direction increases, which increases the signal leakage into the interference beam, causing more signal suppression.

To understand the dependency of the signal suppression on $\alpha_I$, the SNR at the output of the interference beam, rewrite (2.6.53) as

$$\hat{w}_c = \left(1 + \frac{\rho}{1-\rho} \frac{1}{1+\frac{1}{\alpha_I}}\right)\beta \tag{2.6.59}$$

It follows from (2.6.59) that as $\alpha_I$ increases, the magnitude of $\hat{w}_c$ increases, resulting in an increase of the signal suppression. In the limit, as $\alpha_I \to \infty$, $\hat{w}_c \to \beta/(1-\rho)$. It can easily be verified that for this value of $\hat{w}_c$, the output signal power reduces to zero, resulting in total signal suppression.

The behavior of the output noise power of the optimal PIC using the CIB is described by (2.6.57). The first term, which is proportional to the uncorrelated noise power at the input of the PIC, decreases as the number of elements in the array increases and the parameter ρ decreases. The second term, which is proportional to the total noise power at the output of the interference beam, also decreases as the parameter ρ decreases and depends on $\alpha_I$. As $\alpha_I$ increases, resulting in an increase of $\hat{w}_c$, the second term on the right side of (2.6.57) increases. This implies that the output noise power of the optimal PIC using the CIB increases as the input signal power increases.

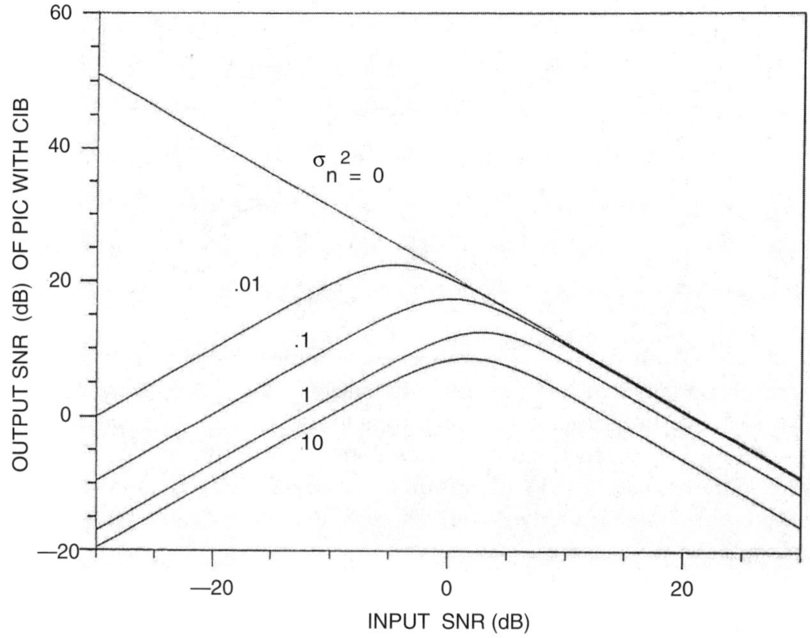


**FIGURE 2.12**
Output SNR of the PIC using CIB vs. input SNR for a ten-element linear array, $\theta_0 = 90°$, $p_I = 1$, $\theta_I = 30°$. (From Godara, L.C., *J. Acoust. Soc. Am.*, 85, 202–213, 1989 [God89a]. With permission.)

Let $\text{SNR}(\hat{w}_c)$ denote the output SNR of the optimal PIC using the CIB. Then, it follows from (2.6.54) and (2.6.57) that

$$\text{SNR}(\hat{w}_c) = \frac{\rho(1-\rho)p_S}{(1-\rho)(1+\alpha_I)^2(\sigma_n^2/L) + \rho\alpha_I^2(p_I + \sigma_n^2/L)} \tag{2.6.60}$$

For the special case when the noise environment consists of only directional sources, that is, when $\sigma_n^2 = 0$, (2.6.60) reduces to

$$\text{SNR}(\hat{w}_c) = 1/\alpha_I \tag{2.6.61}$$

which agrees with the results presented in [Wid75, Wid82] that in the absence of uncorrelated noise, the output SNR of an interference canceler is inversely proportional to the input SNR. In the presence of uncorrelated noise power, the behavior of $\text{SNR}(\hat{w}_c)$ is shown in Figure 2.12.

The results in Figure 2.12 are for an equally spaced linear array of ten elements, with inter-element spacing of one-half wavelength. The signal source is assumed to be broadside to the array, and an interference source of unity power is assumed 60° off broadside. For this array configuration and source scenario, the parameter $\rho$ is equal to 0.99. Figure 2.12 shows that the presence of uncorrelated noise changes the behavior of $\text{SNR}(\hat{w}_c)$ dramatically, particularly for low-input SNR. In the absence of uncorrelated noise, the PIC using the CIB is able to cancel most of the interference when the input SNR is small, resulting in high-output SNR. The presence of uncorrelated noise increases the total output noise significantly (see Equation 2.6.57), resulting in a substantial drop in the output SNR.

### 2.6.3.3 PIC with Orthogonal Interference Beamformer

Let the interference beam be formed using the beamforming weights

$$\mathbf{U} = \mathbf{U}_o \tag{2.6.62}$$

where $\mathbf{U}_o$ is a complex vector such that

$$\mathbf{U}_o^H \mathbf{S}_0 = 0 \tag{2.6.63}$$

The constraint specified by (2.6.63) ensures that the interference beam has a null in the signal direction. Thus, the interference beam does not contain any signal and the PIC using the orthogonal interference beamformer (OIB) does not suppress the signal. Note that the vector $\mathbf{U}_o$ may be a steering vector. This case corresponds to the parameter ρ taking on a value of unity.

Various expressions for optimal PIC using the OIB are now presented. It is assumed that the interference beam of the PIC using the OIB does not have a null in the interference direction. If the interference beam had a null in the interference direction, then there would be no interference present in this beam and no reduction in the interference from the signal beam would result by forming the PIC output by subtracting the weighted output of the interference beam from the signal beam.

From (2.6.48), (2.6.50) and (2.6.62), it follows that the optimal weight $\hat{w}_o$ of the PIC using the OIB is given by

$$\hat{w}_o = \frac{\mathbf{S}_o^H R \mathbf{U}_0}{L \mathbf{U}_o^H R \mathbf{U}_o} \tag{2.6.64}$$

Substituting for R in (2.6.64), one obtains, after manipulation,

$$\hat{w}_o = \frac{\mathbf{S}_o^H \mathbf{S}_I \mathbf{S}_I^H \mathbf{U}_0}{L^2 \beta_o \left(\gamma_o + \sigma_n^2 / L p_I\right)} \tag{2.6.65}$$

where

$$\beta_o = \mathbf{U}_o^H \mathbf{U}_o \tag{2.6.66}$$

and

$$\gamma_o = \frac{\mathbf{U}_o^H \mathbf{S}_I \mathbf{S}_I^H \mathbf{U}_o}{L \mathbf{U}_o^H \mathbf{U}_o} \tag{2.6.67}$$

Note that $\gamma_o$, as defined by (2.6.67), is a positive real scalar, with

$$0 < \gamma_o \le 1 \tag{2.6.68}$$

and represents the normalized power response of the interference beam in the direction of the interference.

The expressions for the signal power, the residual interference power, the uncorrelated noise power, and total noise power at the output of the optimal PIC using the OIB are, respectively, obtained by substituting for R equals $R_S$, $R_I$, $R_n$ and $R_N$, and $w = \hat{w}_0$ in (2.6.46). These are given by

$$P_S(\hat{w}_o) = p_S \tag{2.6.69}$$

$$P_I(\hat{w}_o) = \frac{p_I(1-\rho)}{\left[1+\gamma_o\left(Lp_I/\sigma_n^2\right)\right]^2} \tag{2.6.70}$$

$$P_n(\hat{w}_o) = \frac{\sigma_n^2}{L} + \frac{\sigma_n^2}{L}\left[\frac{(1-\rho)\gamma_o}{\left(\gamma_o + \sigma_n^2/Lp_I\right)^2}\right] \tag{2.6.71}$$

and

$$P_N(\hat{w}_o) = \frac{\sigma_n^2}{L}\left[1 + \frac{(1-\rho)}{\gamma_o + \sigma_n^2/Lp_I}\right] \tag{2.6.72}$$

From expressions (2.6.69) to (2.6.72), the following observations can be made:

1. The optimal PIC using the OIB does not suppress the signal. This is because there is no leakage of the signal into the interference beam.
2. The residual interference power of the optimal PIC using the OIB depends on $p_I/\sigma_n^2$. For a given array geometry and noise environment, the normalized residual interference power $P_I(\hat{w}_0)/p_I$ decreases as $p_I/\sigma_n^2$ increases. In a noise environment with a very high $p_I/\sigma_n^2$, the residual interference power of the optimal PIC using the OIB becomes very small. In the limit, as

$$\frac{p_I}{\sigma_n^2} \to \infty \qquad \hat{w}_o \to \frac{\mathbf{S}_o^H \mathbf{S}_I \mathbf{S}_I^H \mathbf{U}_o}{L^2 \beta_o \gamma_o} \tag{2.6.73}$$

which lead to full cancelation of the interference (see Equation 2.6.70). On the other hand, as

$$\frac{p_I}{\sigma_n^2} \to 0 \qquad \hat{w}_o \to 0 \tag{2.6.74}$$

and no cancelation of the interference takes place.
3. The uncorrelated noise power at the output of the PIC is more than the uncorrelated noise power at the output of the signal beam. This follows from (2.6.71). The RHS of (2.6.71) consists of two terms. The first term is the same as the uncorrelated noise power at the output of the signal beam and the second term is proportional to the uncorrelated noise power at the output of the signal beam; the proportionality constant in the square brackets depends on the $p_I/\sigma_n^2$. As $p_I/\sigma_n^2$ increases, the quantity in the square brackets increases. This is due to the fact that $\hat{w}_o$ increases as $p_I/\sigma_n^2$ increases. In the limit, the maximum increase in the uncorrelated noise power caused by the optimal PIC using the OIB is $\sigma_n^2/L\ (1-\rho)/\gamma_o$.
4. The total noise power $P_N(\hat{w}_o)$ at the output of the optimal PIC using the OIB does not depend on the signal power. It is proportional to the uncorrelated noise power at the output of the signal beam and decreases as $p_1/\sigma_n^2$ decreases. The uncorrelated noise dominates the total noise at the output of the optimal PIC.

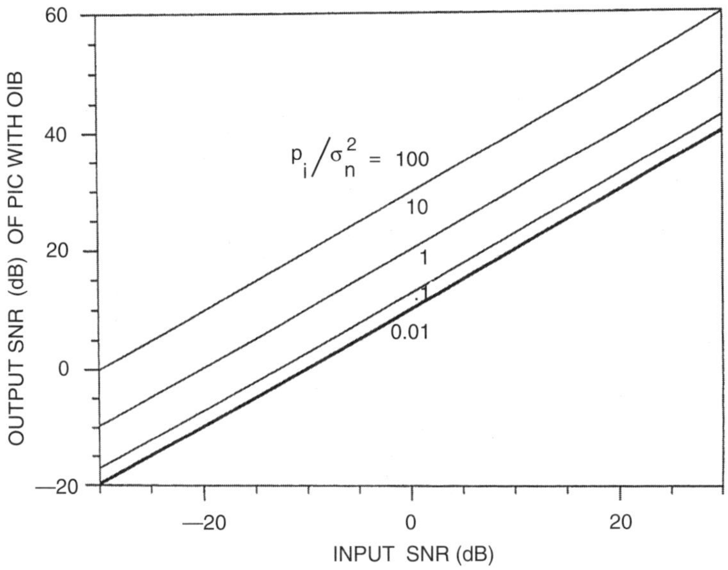


**FIGURE 2.13**
Output SNR of the PIC using OIB vs. input SNR for a ten-element linear array, $\theta_0 = 90°$, $p_I = 1$, $\theta_I = 30°$. (From Godara, L.C., *J. Acoust. Soc. Am.*, 85, 202–213, 1989 [God89a]. With permission.)

Now the output SNR of the optimal PIC using the OIB is examined. Let this ratio be denoted by SNR($\hat{w}_o$). It follows from (2.6.69) and (2.6.72) that

$$\text{SNR}(\hat{w}_o) = \frac{\dfrac{Lp_S}{\sigma_n^2}\left(\gamma_o + \dfrac{\sigma_n^2}{Lp_I}\right)}{\gamma_o + \dfrac{\sigma_n^2}{Lp_I} + 1 - \rho} \tag{2.6.75}$$

Thus, the output SNR of the optimal PIC using the OIB is proportional to the number of elements and $p_S/\sigma_n^2$; and depends on $p_I/\sigma_n^2$. As

$$\frac{p_I}{\sigma_n^2} \to \infty \qquad \text{SNR}(\hat{w}_o) \to \frac{Lp_S}{\sigma_n^2}\frac{\gamma_o}{(1+\gamma_o-\rho)} \tag{2.6.76}$$

Figure 2.13 shows SNR($\hat{w}_o$) vs. input SNR for various $p_I/\sigma_n^2$. The array geometry and noise environment used for this example is the same as that used for Figure 2.12. The interference beam is formed using the steering vector in the endfire direction. The parameter $\gamma_o$ for this case is 0.17. From Figure 2.13, for a given input SNR the output SNR increases as $p_I/\sigma_n^2$ increases.

### 2.6.3.4 PIC with Improved Interference Beamformer

As discussed in previous sections, the output of the optimal PIC contains residual interference power and uncorrelated noise power. This section presents and analyzes the optimal PIC using an interference beamformer that eliminates all interference in the output while simultaneously reducing the contribution of uncorrelated noise in the output. For this case, let the interference beam be formed with the beamforming weights

$$\mathbf{U} = \frac{R^{-1}\mathbf{S}_I}{\mathbf{S}_I^H R^{-1}\mathbf{S}_I} \tag{2.6.77}$$

Note that the above expression is similar to the expression for the weights of constrained optimal beamformer except that in this case the beam is constrained in the direction of the interference rather than the look direction. Thus, it can easily be verified that the interference beam formed with these weights has unity response in the interference direction and has a reduced response in the signal direction. The response of the interference beam in the signal direction depends on the signal source power and uncorrelated noise power. It can be shown that this choice of beamforming weights minimizes the sum of signal power and uncorrelated noise power in the interference channel output.

A substitution for **V** and **U** in (2.6.48) from (2.6.50) and (2.6.77), respectively, leads to the following expression for $\hat{w}_1$, the weight of the optimal PIC using the improved interference beamformer (IIB):

$$\hat{w}_I = \mathbf{S}_0^H \mathbf{S}_I / L \tag{2.6.78}$$

It follows from (2.6.78) that the weight, which minimizes the output power of PIC using the IIB is independent of the signal, the interference, and the uncorrelated noise powers. This weight depends only on the array geometry and relative directions of the two sources.

The expressions for the signal power and the noise power at the output of the optimal PIC using the IIB are, respectively, given by

$$P_S(\hat{w}_I) = p_S \rho^2 \left( \frac{\left(1 + \sigma_n^2 / Lp_S\right)}{\left(\rho + \sigma_n^2 / Lp_S\right)} \right)^2 \tag{2.6.79}$$

and

$$P_N(\hat{w}_I) = \frac{\sigma_n^2}{L} \left\{ \rho \left( \frac{\left(1 + \sigma_n^2 / Lp_S\right)}{\left(\rho + \sigma_n^2 / Lp_S\right)} \right)^2 \right\} \tag{2.6.80}$$

One observes from expressions (2.6.79) and (2.6.80) that the output signal power and the output noise power of the optimal PIC using the IIB are independent of the interference power. Thus, the optimal PIC using the IIB has completely suppressed the interference. Furthermore, the output signal power and output noise power depend on $\sigma_n^2/Lp_S$ (ratio of uncorrelated noise power to signal power at signal beam output). The output signal power increases as $\sigma_n^2/Lp_S$ decreases, and approaches the input signal power in the limit. Thus, in the presence of a strong signal source, the signal suppression by the optimal PIC using the IIB is negligible. The signal suppression becomes further reduced as the number of elements in the array is increased.

The total noise power at the output of the optimal PIC using the IIB is equal to the uncorrelated noise power at the output of the signal beam when $\rho = 1$. To investigate the effect of $\sigma_n^2/Lp_S$ on the output noise power when $\rho < 1$, you can rewrite the quantity in the braces on the right side of (2.6.80) in the following form:

$$\rho \left( \frac{\left(1 + \sigma_n^2 / Lp_S\right)}{\left(\rho + \sigma_n^2 / Lp_S\right)} \right)^2 = 1 + (1 - \rho) \frac{\rho - \left(\sigma_n^2 / Lp_S\right)^2}{\left(\rho + \sigma_n^2 / Lp_S\right)^2} \tag{2.6.81}$$

**TABLE 2.1**

Comparison of Normalized Signal Power, Interference Power, Uncorrelated Noise Power and SNR at the Output of the Optimal PIC Forming Interference Beam with CIB, OIB and IIB, $\gamma_0 = \frac{\mathbf{U}_0^H \mathbf{S}_I \mathbf{S}_I^H \mathbf{U}_0}{L\mathbf{U}_0^H \mathbf{U}_0}$ and $\rho = 1 - \frac{\mathbf{S}_0^H \mathbf{S}_I \mathbf{S}_I^H \mathbf{S}_0}{L^2}$

| Optimal PIC with | CIB | OIB | IIB |
|---|---|---|---|
| Normalized output signal power | $\frac{\rho^2}{\left[1+\frac{1-\rho}{p_I/p_S+\sigma_n^2/Lp_S}\right]^2}$ | 1 | $\left[1-(1-\rho)\frac{\sigma_n^2/Lp_S}{\rho+\sigma_n^2/Lp_S}\right]^2$ |
| Normalized residual interference power | $\frac{\rho^2}{(1-\rho)^2}\frac{1}{\left[1+\frac{p_I/p_S+\sigma_n^2/Lp_S}{(1-\rho)}\right]^2}$ | $\frac{1}{\left[1+\gamma_0\frac{Lp_I}{\sigma_n^2}\right]^2}$ | 0 |
| Normalized uncorrelated noise power | $\rho+\frac{\rho^2}{(1-\rho)}\frac{1}{\left[1+\frac{p_I/p_S+\sigma_n^2/Lp_S}{(1-\rho)}\right]^2}$ | $1+\frac{(1-\rho)\gamma_0}{\left[\gamma_0+\sigma_n^2/Lp_I\right]^2}$ | $1+(1-\rho)\frac{\rho-(\sigma_n^2/Lp_S)^2}{(\rho+\sigma_n^2/Lp_S)^2}$ |
| Output SNR | $\frac{\rho}{\rho\frac{(1-\rho)}{p_I/p_S+\sigma_n^2/Lp_S}+\sigma_n^2/Lp_S\left[1+\frac{(1-\rho)^2}{p_I/p_S+\sigma_n^2/Lp_S}\right]}$ | $\frac{Lp_S/\sigma_n^2}{1+\frac{(1-\rho)}{\gamma_0+\sigma_n^2/Lp_I}}$ | $\frac{Lp_S\rho}{\sigma_n^2}$ |

Since $\rho < 1$, it follows from (2.6.81) that the second term on the RHS is negative if $\rho < (\sigma_n^2/Lp_S)^2$. Thus, under this condition the quantity in the braces on the right side of (2.6.80) is less than unity and, hence, the uncorrelated noise power at the output of the PIC is less than the uncorrelated noise power at the output of the signal beam. Thus, the optimal PIC using the IIB reduces the uncorrelated noise when $\rho < (\sigma_n^2/Lp_S)^2$. On the other hand, when $\rho > (\sigma_n^2/Lp_S)^2$, the quantity in the braces on the right side of (2.6.80) is more than unity and the optimal PIC using the IIB increases the uncorrelated noise power. Note that at the output of the optimal PIC using the IIB, total noise consists of uncorrelated noise only: it increases as $\sigma_n^2/Lp_S$ decreases and in the limit approaches $\sigma_n^2/Lp_S$.

Now the output SNR of the optimal PIC using the IIB is examined. Let this ratio be denoted by $\text{SNR}(\hat{w}_I)$. It follows then from (2.6.79) and (2.6.80) that

$$\text{SNR}(\hat{w}_I) = \frac{p_S \rho L}{\sigma_n^2} \tag{2.6.82}$$

Thus, the output SNR of the optimal PIC using the IIB is proportional to the input signal to uncorrelated noise ratio, the number of elements in the array, and the parameter $\rho$.

### *2.6.3.5 Discussion and Comments*

A comparison of the various results is presented in Table 2.1. The output signal power, residual interference power, and output uncorrelated noise power of the optimal PIC are, respectively, normalized by $p_S$, $p_I(1-\rho)$, and $\sigma_n^2/L$. These quantities correspond to the signal power, the interference power, and the uncorrelated noise power at the output of the signal beam. This particular form of normalization is chosen to facilitate the comparison between the performance of the PIC using the OIB, IIB, and CIB, and that of an element space processor using conventional weights (the signal beam is formed using conventional weights).

It follows from Table 2.1 that the SNR of the optimal PIC for the three cases is the same when $\rho$ is equal to unity or, equivalently, when the steering vectors in the signal and interference directions are orthogonal to each other. The case of $\rho < 1$ is now considered. For this situation, the results of the optimal PIC with the three interference beamformers are discussed and some examples are presented. All examples presented here are for a linear array of ten equally spaced elements with one-half wavelength spacing. The signal direction is broadside to the array, and the uncorrelated noise power on each element is equal to 0.01. The interference beam for the OIB case is formed using the steering vector in the endfire direction. Thus, knowledge of the interference direction is not used in selecting $\mathbf{U}_o$.

### 2.6.3.5.1 *Signal Suppression*

From Table 2.1, the following observations about the normalized output signal power of the optimal PIC for the three cases can be made:

1. The optimal PIC using the OIB does not suppress the signal; in the other two cases the signal is suppressed. The signal suppression by the optimal PIC using the CIB is more than that by the PIC using the IIB. This follows from the following expression for the difference of the normalized output signal powers:

$$\frac{P_S(\hat{w}_c)}{p_S} - \frac{P_S(\hat{w}_I)}{p_S} = -\rho^2 \frac{\left[\left(\rho + \sigma_n^2/Lp_S\right) + \left(1 + \sigma_n^2/Lp_S\right)\left(1 + \alpha_I\right)\right]\left[\left(1 - \rho\right) + \alpha_I\left(1 + \sigma_n^2/Lp_S\right)\right]}{\left(1 + \alpha_I\right)^2\left(\rho + \sigma_n^2/Lp_S\right)^2} \tag{2.6.83}$$

   Physically, the interference beam rejects more of the signal in the IIB than in the CIB and rejects all of the signal in the OIB. This leads to no suppression of signal by the PIC using the OIB and less suppression in the case of the IIB than that of the CIB.

2. The normalized output signal power of the optimal PIC using the IIB is independent of the interference power. In the case of the optimal PIC using the CIB, it increases as the interference power increases. Thus, it follows that the difference between the normalized output signal power for the two cases decreases as the interference power increases. In the limit the difference approaches

$$-\rho^2(1-\rho)\frac{1+\rho+2\left(\sigma_n^2/Lp_S\right)}{\left(\rho+\sigma_n^2/Lp_S\right)^2}$$

3. The normalized output signal power depends on the input signal power for both the CIB and IIB cases. In the case of the optimal PIC using the CIB, it decreases as the input signal power increases. Thus, the signal suppression increases as the input signal power increases. However, in the case of the optimal PIC using the IIB, the normalized output signal power increases as the input signal power increases, approaching unity in the limit. Thus, the signal suppression is negligibly small when the input signal to uncorrelated noise ratio is large.

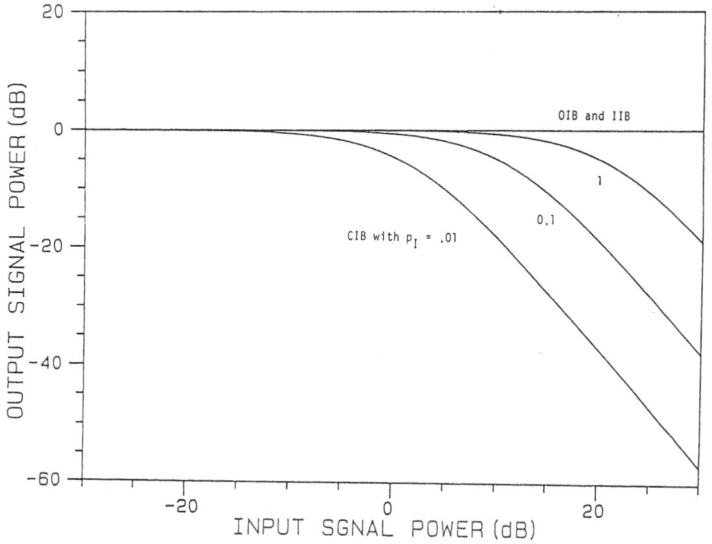


**FIGURE 2.14**
Normalized output signal power of the PIC using the OIB with $p_I = 1$; the IIB with $p_I = 1$; and the CIB with $p_I$ = 1, 0.1 and 0.01 vs. input signal power for a ten- element linear array, $\theta_0 = 90°$, $\sigma_n^2 = 0.01$, $\theta_I = 30°$. (From Godara, L.C., *J. Acoust. Soc. Am.*, 85, 202–213, 1989 [God89a]. With permission.)

Figure 2.14 and Figure 2.15 show plots of the normalized output signal power of the optimal PIC using OIB and IIB when the interference power is 1.0 and using the CIB when the interference powers are 0.01, 0.1, and 1.0. For Figure 2.14, the interference is at an angle of 60° from the signal while for Figure 2.15, the angle is at 5°. The parameter ρ for these cases is 0.99 and 0.48, respectively. Note that for both the cases the normalized output signal power of the PIC using the CIB increases as the interference power increases. Signal suppression by the PIC using the CIB increases as the input signal power increases in both cases, but the signal suppression is greater in Figure 2.15 (ρ = 0.48). This is because more signal leaks into the interference beam for the scenario of Figure 2.15 than for Figure 2.14.

### 2.6.3.5.2 *Residual Interference*

The following observations about the residual interference can be made:

1. The output of the optimal PIC using the IIB does not contain any residual interference; in the OIB and CIB cases, residual interference is present.
2. For the optimal PIC using the OIB, the normalized output residual interference depends on $p_I/\sigma_n^2$ and the number of elements in the array. As $p_I/\sigma_n^2$ increases, the normalized residual interference decreases and approaches zero in the limit. As this ratio decreases, the normalized residual interference increases but never exceeds unity. Thus, the optimal PIC using the OIB always cancels some of the interference present at the output of the signal beam. The interference cancelation increases as $p_I/\sigma_n^2$ and the number of elements in the array increase.
3. As presented in Table 2.1, the expression for the normalized residual interference at the output of the optimal PIC using the CIB is a product of two terms. The first term depends on the parameter ρ, which in turn is controlled by the array geometry and the relative directions of the two sources: for ρ greater than one-half, the

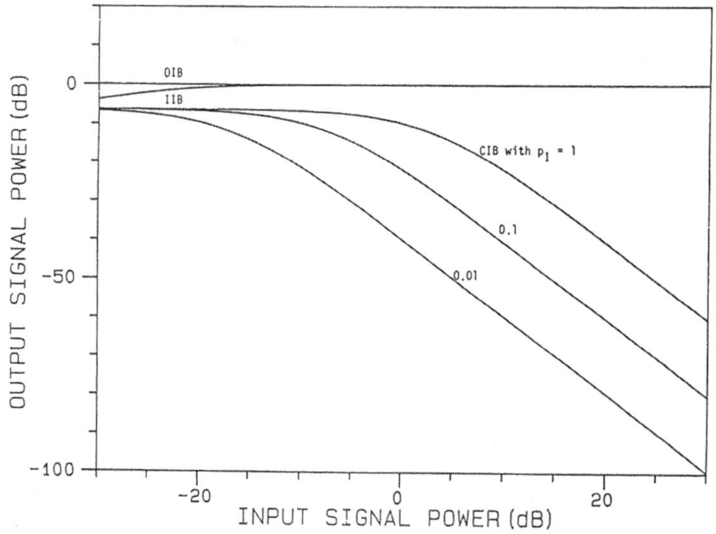


**FIGURE 2.15**
Normalized output signal power of the PIC using the OIB with $p_I = 1$; the IIB with $p_I = 1$; and the CIB with $p_I$ = 1, 0.1 and 0.01 vs. input signal power for a ten-element linear array, $\theta_0 = 90°$, $\sigma_n^2 = 0.01$, $\theta_I = 85°$. (From Godara, L.C., *J. Acoust. Soc. Am.*, 85, 202–213, 1989 [God89a]. With permission.)

term exceeds unity. The second term depends on $\sigma_n^2/Lp_S$ and $p_I/p_S$, and increases as these parameters decrease (stronger signal), in the limit approaching unity. It follows that the normalized residual interference at the output of the optimal PIC using the CIB increases as the signal power increases, and approaches a limit that is more than unity when $\rho < 0.5$. Thus, in certain cases, the interference power at the output of the optimal PIC using the CIB may be more than the interference power at the output of the signal beam.

Comparisons of the normalized residual interference at the output of the optimal PIC using the CIB and OIB are shown in Figure 2.16 and Figure 2.17. The interference directions are 5° and 60° off broadside, respectively. The signal power is assumed to be unity. These figures show plots of the interference power at the output of the optimal PIC normalized by the interference power at the output of signal beam. Thus, the interference level above the 0 dB line indicates an increase in the interference power from that present in the signal beam.

Figure 2.16 (the interference and signal are 5° apart, $\rho = 0.48$) shows that the optimal PIC in both cases cancels some interference present in the signal beam. However, the cancelation is very small for the lower range of the input interference and increases as the input interference increases. For the lower range of the input interference power, the optimal PIC using the CIB cancels slightly more interference than that using the OIB. The reverse is true at the other end of the input interference range. The optimal PIC using the OIB cancels about 10 dB more interference than that using the CIB when the input interference power is unity.

Figure 2.17 shows the normalized output interference of the optimal PIC using the OIB and CIB when the interference and the signal are 60° apart ($\rho = 0.99$). The figure shows that for the lower range of the input interference, the residual interference at the output of the optimal PIC using the CIB is about 40 dB more than the interference contents in the signal beam. Thus, the optimal PIC using the CIB does not suppress weak interference, but increases its level. In the case of the optimal PIC using the OIB, when the input

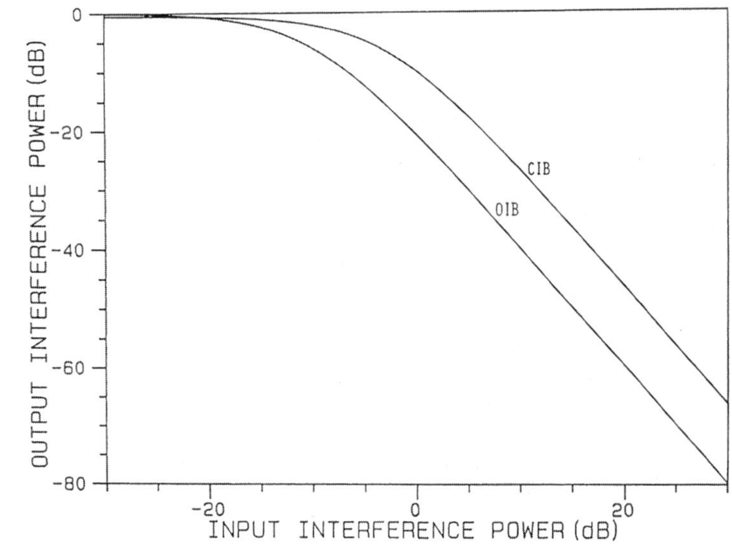


**FIGURE 2.16**
Normalized residual interference power of the PIC using the OIB and the CIB vs. input interference power for a ten-element linear array, $\theta_0 = 90°$, $p_S = 1.0$, $\sigma_n^2 = 0.01$, $\theta_I = 85°$. (From Godara, L.C., *J. Acoust. Soc. Am.*, 85, 202–213, 1989 [God89a]. With permission.)

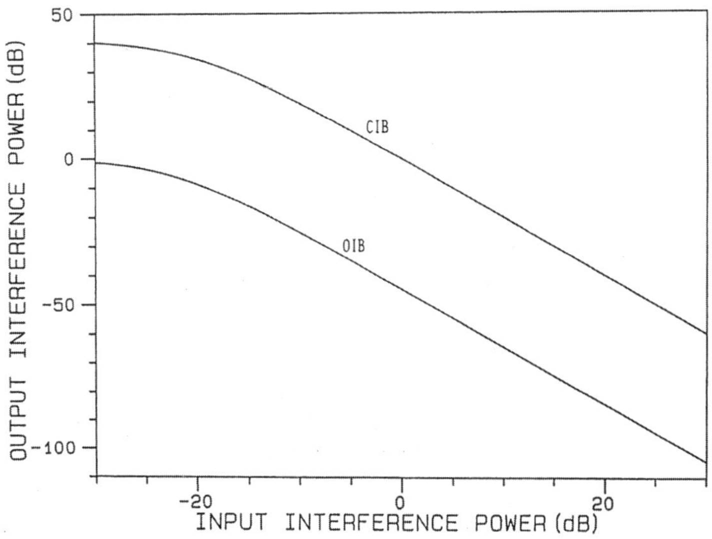


**FIGURE 2.17**
Normalized residual interference power of the PIC using the OIB and the CIB vs. input interference power for a ten-element linear array, $\theta_0 = 90°$, $p_S = 1.0$, $\sigma_n^2 = 0.01$, $\theta_I = 30°$. (From Godara, L.C., *J. Acoust. Soc. Am.*, 85, 202–213, 1989 [God89a]. With permission.)

interference power is very small, some interference reduction takes place. The reduction is about 2 dB.

For both cases, the normalized output interference decreases as the input interference power increases. For the entire range of input interference level, the residual interference at the output of the optimal PIC using the CIB is about 42 dB more than that using the OIB.

*2.6.3.5.3 Uncorrelated Noise Power*

A comparison of the normalized uncorrelated noise power at the output of the optimal PIC for the CIB, OIB, and IIB is shown in Table 2.1. The table shows that the normalized uncorrelated noise power at the output of the optimal PIC using the OIB is greater than unity. In other words, the optimal PIC has increased the uncorrelated noise.

For the case of the optimal PIC using the IIB, the decrease or increase in the uncorrelated noise power depends on the difference between the parameter $\rho$ and the square of the uncorrelated noise to signal ratio at the output of the signal beam $(\sigma_n^2/Lp_S)^2$. The normalized uncorrelated noise power at the output of the PIC is more than unity when $\rho > (\sigma_n^2/Lp_S)^2$. Thus, in the presence of a relatively stronger signal source, the optimal PIC using the IIB increases the uncorrelated noise power.

*2.6.3.5.4 Signal-to-Noise Ratio*

First a comparison between the SNRs of the PIC using the IIB and OIB is considered. It follows from (2.6.75) and (2.6.82) that

$$\frac{\mathrm{SNR}(\hat{w}_o)}{\mathrm{SNR}(\hat{w}_I)} = \frac{1}{\rho}\frac{\gamma_o + \dfrac{\sigma_n^2}{Lp_I}}{\gamma_o + \dfrac{\sigma_n^2}{Lp_I} + 1 - \rho} \tag{2.6.84}$$

which implies that

$$\mathrm{SNR}(\hat{w}_I) > \mathrm{SNR}(\hat{w}_o) \tag{2.6.85}$$

Furthermore, for $\rho \approx 1$

$$\mathrm{SNR}(\hat{w}_o) \approx \mathrm{SNR}(\hat{w}_I) \tag{2.6.86}$$

Now consider the PIC using the IIB and CIB. It follows from (2.6.60) and (2.6.82) that

$$\frac{\mathrm{SNR}(\hat{w}_c)}{\mathrm{SNR}(\hat{w}_I)} = \frac{1}{(1+\alpha_1)^2 + \dfrac{\rho}{1-\rho}\alpha_1^2\left(1 + Lp_1/\sigma_n^2\right)} \tag{2.6.87}$$

Thus

$$\mathrm{SNR}(\hat{w}_I) > \mathrm{SNR}(\hat{w}_c) \tag{2.6.88}$$

Furthermore, for low values of $\alpha_1$

$$\mathrm{SNR}(\hat{w}_I) \cong \mathrm{SNR}(\hat{w}_c) \tag{2.6.89}$$

Note that

$$\alpha_1 = \frac{(1-\rho)p_S}{\left(p_1 + \sigma_n^2/L\right)} \tag{2.6.90}$$

is the SNR at the output of the interference beam.

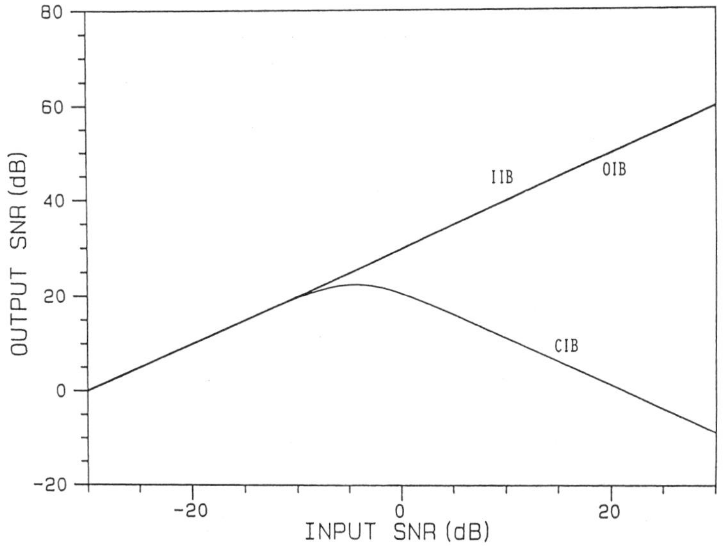


**FIGURE 2.18**
Output SNR of the PIC using the OIB, the IIB and the CIB vs. input SNR for a ten-element linear array, $\theta_0 = 90°$, $p_I = 1.0$, $\sigma_n^2 = 0.01$, $\theta_I = 30°$. (From Godara, L.C., *J. Acoust. Soc. Am.*, 85, 202–213, 1989 [God89a]. With permission.)

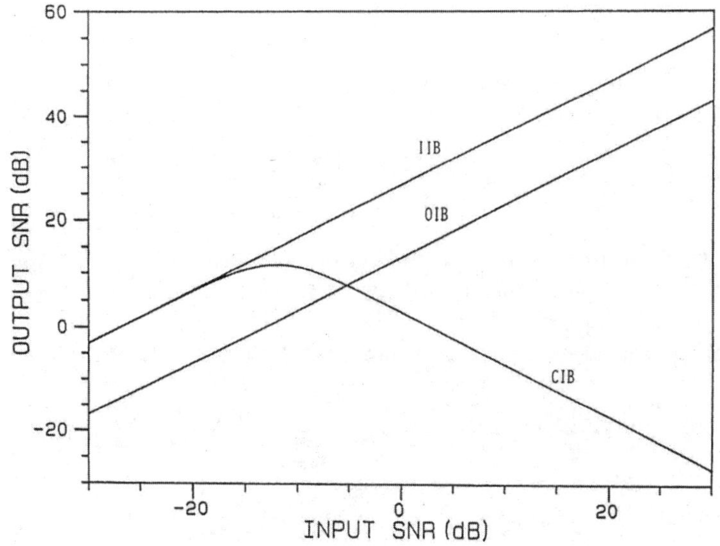


**FIGURE 2.19**
Output SNR of the PIC using the OIB, the IIB and the CIB vs. input SNR for a ten-element linear array, $\theta_0 = 90°$, $p_S = 1.0$, $\sigma_n^2 = 0.01$, $\theta_I = 85°$. (From Godara, L.C., *J. Acoust. Soc. Am.*, 85, 202–213, 1989 [God89a]. With permission.)

The above discussion agrees with the comparison of the output SNRs for the IIB, OIB, and CIB cases shown in Figure 2.18 and Figure 2.19. For these cases, a unit power interference is assumed to be present. The direction of the interference is 60° from broadside in Figure 2.18 and 5° from broadside in Figure 2.19. The parameter $\rho$ is 0.99 and 0.48, respectively, and the parameter $\gamma_o$ is 0.17 and 0.01, respectively. One observes from these figures that in the case of the CIB, the output SNR decreases as the input SNR increases beyond –8 dB in Figure 2.18 and beyond –16 dB in Figure 2.19. However, in the other two cases the output SNR increases as the input SNR increases, resulting in array gains of the order of 20 to 30 dB.

In the next two sections, a comparison of the optimal element space processor (ESP) and the optimal PIC with an OIB is presented. It should be noted that the ESP is optimized to minimize the mean output power subject to a unity constraint in the look direction and the PIC is optimized to minimize the mean output power with the interference beam having a null in the look direction.

### 2.6.4 Comparison of Postbeamformer Interference Canceler with Element Space Processor

Performance of the optimal ESP is a function of $\rho$, and the performance of the optimal PIC with an OIB is dependent on $\rho$ and $\gamma_o$. Thus, performance comparison of the two processors depends on the relative values of these two constants.

First, consider a case where the precise interference direction is known. Let the interference beam be formed using an OIB given by

$$\hat{\mathbf{U}}_o = P\mathbf{S}_I \tag{2.6.91}$$

where

$$P = I - \left(\mathbf{S}_0\mathbf{S}_0^H\right)/L \tag{2.6.92}$$

A simple calculation indicates that for the interference beamformer weights given by (2.6.91) and (2.6.92), $\gamma_o$ attains its maximum value and

$$\gamma_o = \rho \tag{2.6.93}$$

A comparison of the results derived in Sections 2.4 and 2.6.3 reveals that for this case the output powers and the SNRs of the two processors are identical (see (2.4.42) and (2.6.75)). Thus, if the interference beam of the PIC is formed by an OIB for which (2.6.93) holds, then the performance of the optimal PIC is identical to the performance of the optimal ESP.

However, if the interference beam of the PIC is formed by an OIB for which

$$\gamma_o < \rho \tag{2.6.94}$$

then a comparison of the results for the two processors (an expression for $\hat{P}_N$ results using (2.4.20) and (2.4.41) reveals that

$$P_N(\hat{w}_o) > \hat{P}_N \tag{2.6.95}$$

and

$$\mathrm{SNR}(\hat{w}_o) < \hat{\alpha} \tag{2.6.96}$$

Thus, the total noise power at the output of the optimal PIC in this case is more than the total noise power at the output of the optimal ESP, and the SNR achievable by the

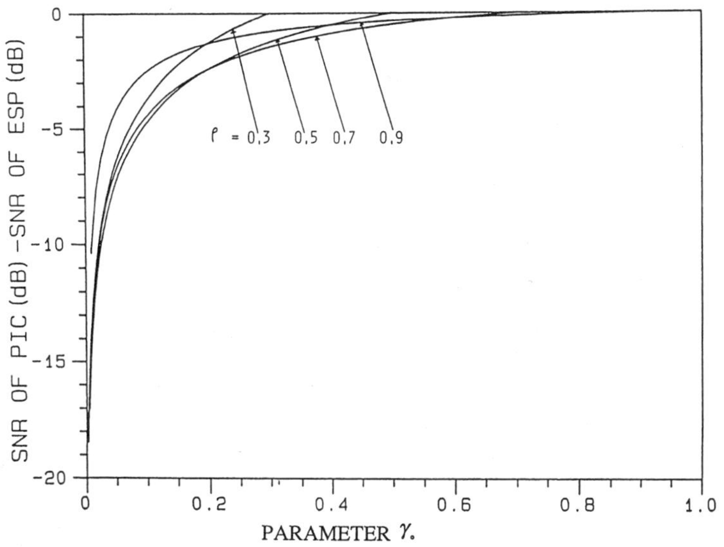


**FIGURE 2.20**
Difference in the SNRs of the two processors calculated using (2.6.98) as a function of ρ and $\gamma_0$. (From Godara, L.C., *IEEE Trans. Circuits Syst.*, 34, 721–730, 1987. ©IEEE. With permission.)

optimal PIC is less than that achievable by the optimal ESP. It follows from (2.4.42) and (2.6.75) that the ratio of the two SNRs is given by

$$\frac{\mathrm{SNR}(\hat{w}_o)}{\mathrm{SNR}(\hat{\mathbf{w}})} = \frac{\sigma_n^2 + Lp_I}{\sigma_n^2 + (1+\gamma_o-\rho)Lp_I}\,\frac{\sigma_n^2 + \gamma_o Lp_I}{\sigma_n^2 + \rho Lp_I} \tag{2.6.97}$$

For $\sigma_n^2/Lp_1 \ll \gamma_o$, this ratio reduces to

$$\frac{\mathrm{SNR}(\hat{w}_o)}{\mathrm{SNR}(\hat{\mathbf{w}})} \approx \frac{\gamma_o}{(1+\gamma_o-\rho)\rho} \tag{2.6.98}$$

and depends on the relative values of ρ and $\gamma_o$. Furthermore, if $\rho \cong 1$, then it follows from (2.6.98) that the output SNRs of the two processors are approximately the same. Plots of (2.6.98) for four values of ρ as a function of $\gamma_o$ are shown in Figure 2.20. The figure shows that the difference in the output SNRs of the two processors is smaller for the larger values of these constants and increases as these constants decrease.

### 2.6.5 Comparison in Presence of Look Direction Errors

Knowledge of the look direction is used to constrain the array response in the direction of the signal such that the signal arriving from the look direction is passed through the array processor undistorted. The array weights of the element space optimal beamformer are estimated by minimizing the mean out power subject to the look direction constraint. The processor maximizes the output signal to noise ratio by canceling all interference. A direction source is treated as interference if it is not in the look direction. This shows the importance of the accuracy of the look direction. An error occurs when the look direction is not the same as the desired signal direction. For this case, the processor treats the desired signal source as interference and attenuates it. The amount of attenuation depends on the power of the signal and the amount of error [Ows73, Cox73, God87, Zah72]. A stronger signal is canceled more and a larger error causes more cancelation of the signal.

The solution to the look direction error, also known as the beam-pointing error, is to make the beam broader so that when the signal is not precisely in the direction where it should be (the look direction), its cancelation does not take place. The various methods of broadening the beam include multiple linear constraints [Ows73, Ste83] and norm constraints. Norm constraints prohibit the main beam blowing out as is the case in the presence of pointing error. In the process of canceling a source close to the point constraint in the look direction, the array response gets increased in the direction opposite to the pointing error. A scheme to reduce the effect of pointing error, which does not require broadening of the main beam, has been reported in [Pon96]. It makes use of direction finding techniques combined with a reduced dimensional maximum likelihood formulation to accurately estimate the direction of the desired signal. The effectiveness of this scheme in mobile communications has been demonstrated using computer simulations. Other schemes to remedy pointing error problems may be found in [Lo90, Muc81, Roc87].

In this section, the performance of the optimal element space processor and the beam space processor in the presence of beam-pointing error is compared [God87]. The comparison presented here indicates that beam space processors in general are more robust to pointing errors than elements space processors.

It is assumed for this analysis that the actual signal direction is different from the known signal direction. Let the steering vector in the actual signal direction be denoted by $\mathbf{S}_0$. The array correction matrix R in this case is given by

$$R = p_S \mathbf{S}_0 \mathbf{S}_0^H + p_I \mathbf{S}_I \mathbf{S}_I^H + \sigma_n^2 I \tag{2.6.99}$$

and the weights $\hat{\mathbf{w}}$ of the optimal ESP and $\hat{w}_o$ of the optimal PIC with an OIB estimated from the known signal direction are given by

$$\hat{\mathbf{w}} = \left(R^{-1}\mathbf{S}_0\right) / \mathbf{S}_0^H R^{-1} \mathbf{S}_0 \tag{2.6.100}$$

and

$$\hat{w}_o = \left(\mathbf{V}^H \mathbf{R} \mathbf{U}_o\right) / \mathbf{U}_o^H \mathbf{R} \mathbf{U}_o \tag{2.6.101}$$

where $\mathbf{V}$ is given by (2.6.50) and $\mathbf{U}_o$ satisfies (2.6.63).

The output power $\hat{P}$ of the ESP is given by

$$\hat{P} = \hat{\mathbf{w}}^H R \hat{\mathbf{w}} \tag{2.6.102}$$

and the output power $P(\hat{w}_o)$ of the PIC processor is given by

$$P(\hat{w}_o) = \mathbf{V}^H \mathbf{R} \mathbf{V} + \hat{w}_o^* \hat{w}_o \mathbf{U}_o^H \mathbf{R} \mathbf{U}_o - \hat{w}_o \mathbf{U}_o^H \mathbf{R} \mathbf{V} - \hat{w}_o^* \mathbf{V}^H \mathbf{R} \mathbf{U}_o \tag{2.6.103}$$

A detailed comparative study of the performance of the two processors in the presence of the signal direction error (SDE) is presented in Figure 2.21 to Figure 2.28. These figures show how the direction of the interference source, the number of elements in the array, and the uncorrelated noise power level affect the performance of the two processors as a

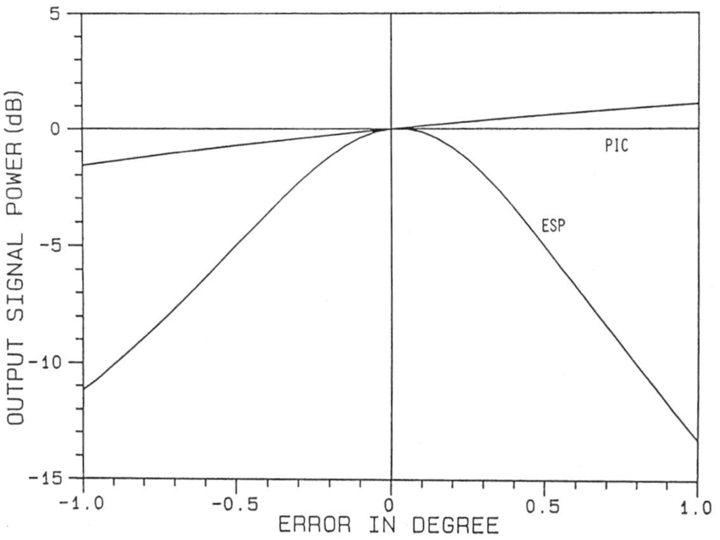


**FIGURE 2.21**
Output signal power vs. the SDE for a ten-element linear array, $p_I = 100$, $\sigma_n^2 = 0.01$, $\theta_I = 85°$. (From Godara, L.C., *IEEE Trans. Circuits Syst.*, 34, 721–730, 1987. ©IEEE. With permission.)

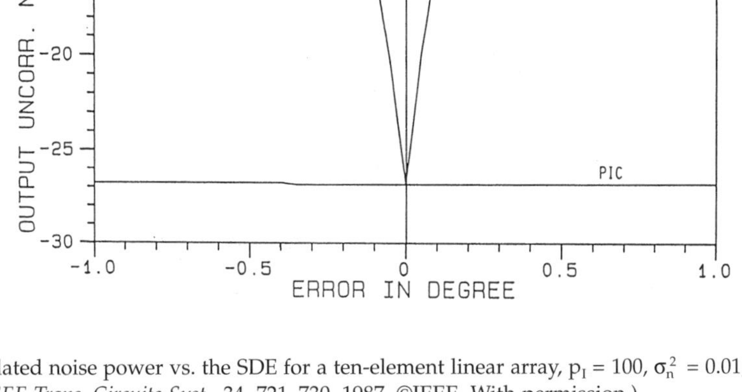

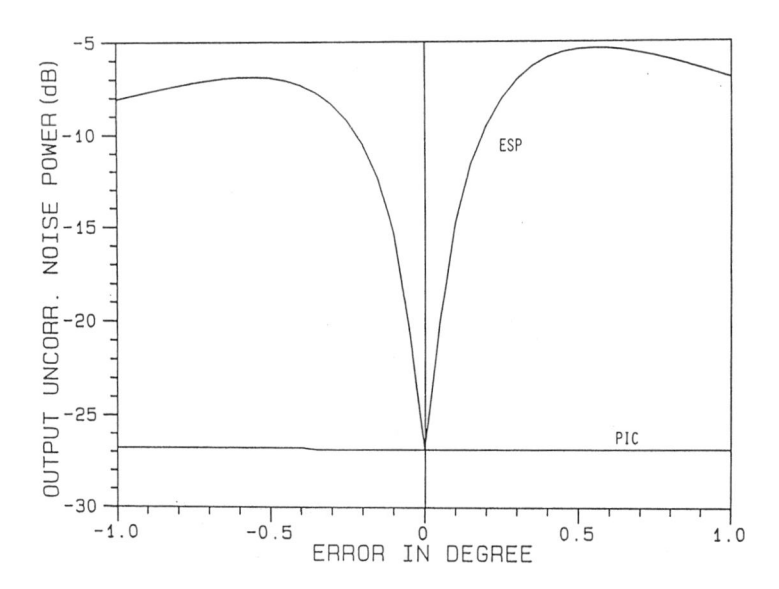


**FIGURE 2.22**
Output uncorrelated noise power vs. the SDE for a ten-element linear array, $p_I = 100$, $\sigma_n^2 = 0.01$, $\theta_I = 85°$. (From Godara, L.C., *IEEE Trans. Circuits Syst.*, 34, 721–730, 1987. ©IEEE. With permission.)

function of the error in the signal direction. For all these figures, interference power is taken to be 20 dB more than signal power.

Figure 2.21 to Figure 2.24 show, respectively, the comparison of the output signal powers, the output uncorrelated noise powers, the power patterns, and the output SNRs of two processors when the assumed look direction is broadside to a ten-element linear array with half-wavelength spacing. The direction of the interference is 85° relative to the line of the array, and the uncorrelated noise power level is 20 dB below the signal level. The

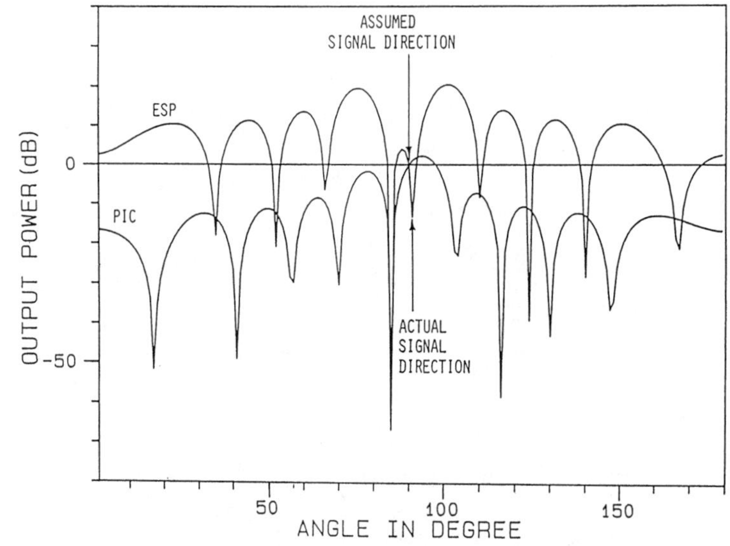


**FIGURE 2.23**
Power pattern of a ten-element linear array when SDE = 1°, $p_I = 100$, $\sigma_n^2 = 0.01$, $\theta_I = 85°$. (From Godara, L.C., *IEEE Trans. Circuits Syst.*, 34, 721–730, 1987. ©IEEE. With permission.)

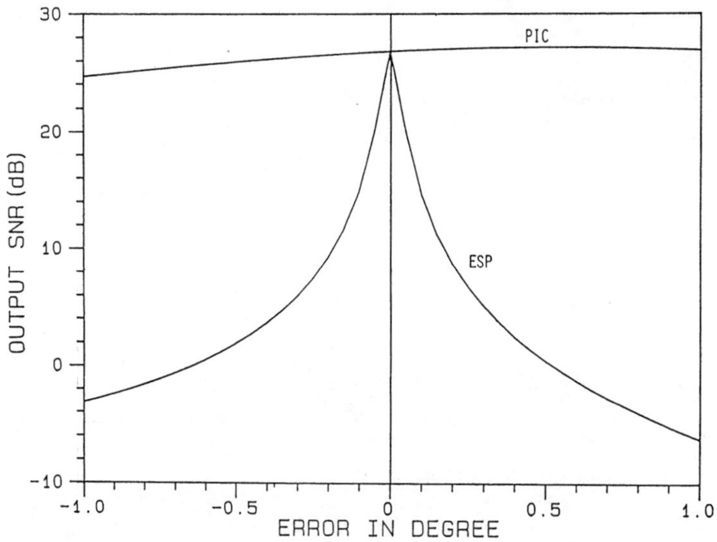


**FIGURE 2.24**
Output SNR vs. the SDE for a ten-element linear array, $p_I = 100$, $\sigma_n^2 = 0.01$, $\theta_I = 85°$. (From Godara, L.C., *IEEE Trans. Circuits Syst.*, 34, 721–730, 1987. ©IEEE. With permission.)

interference beamforming weights of the PIC processor are calculated using (2.6.91) and (2.6.92).

Figure 2.21 shows that the output signal powers of the two processors are the same in the absence of the SDE. As the SDE increases, the signal suppression by the ESP increases, and it suppresses more than 11 dB signal power in the presence of a 1° error in the signal direction. Note that the error in the signal direction is measured relative to the look direction and is assumed to be positive in the counterclockwise direction. Thus, –1° and

1° error, respectively, means that the signal direction is 89° and 91° relative to the line of the array. Furthermore, the line of the array, interference direction, and signal direction are in the same plane.

The signal suppression of the PIC processor is substantially less than that of the ESP. It reduces the output signal power less than 2 dB in comparison to 11 dB of the ESP when the error is –1° and increases the output signal power by about 1 dB when the error is 1°, in which case the ESP suppresses more than 13 dB of signal.

A comparison of the uncorrelated noise powers of the two processors is shown in Figure 2.22. This figure shows that there is no noticeable effect on the output uncorrelated noise power of the PIC processor due to the presence of the SDE. However, there is a significant increase in the uncorrelated noise output power of the ESP. A small SDE, of the order of 0.4°, causes an increase of the order of 20 dB in the uncorrelated noise output power.

Figure 2.23 shows the power patterns of the two processors when the error is 1°. The reduced response in the signal direction and an increased response to the uncorrelated noise are clearly visible from the pattern of the ESP.

Figure 2.24 compares the output SNRs of the two processors. The performance of the two processors is the same in the absence of errors. The effect of the SDE on the output SNR of the PIC is a slight reduction for a –1° error and a slight increase for a 1° error. However, the error causes a significant reduction in the output SNR of the ESP.

Figure 2.25 compares the output SNRs of the two processors when the interference direction is 25° relative to the line of the array. This figure demonstrates that the output SNR of the ESP in Figure 2.25 is reduced by more than 20 dB by 0.1° error in the signal direction. On the other hand, the effect of the SDE on the output SNR of the PIC is negligibly small. It should be noted that the constant ρ attains values of 0.99 and 0.48, respectively, for the scenarios of Figure 2.24 and Figure 2.25. A comparison of these figures shows how the direction of the interference affects the output SNR of the ESP for a given SDE. One observes that the performance of the ESP in a noise configuration with a higher value of ρ is poorer than that with a lower value of ρ.

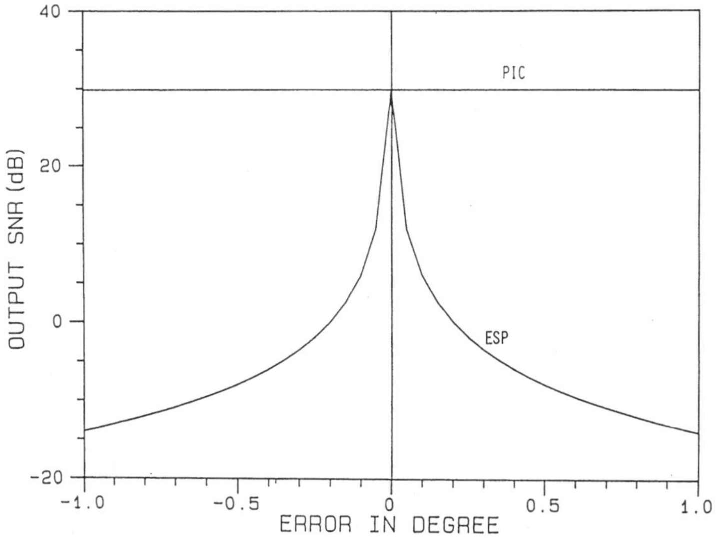


**FIGURE 2.25**
Output SNR vs. the SDE for a ten-element linear array, $p_I = 100$, $\sigma_n^2 = 0.01$, $\theta_I = 25°$. (From Godara, L.C., *IEEE Trans. Circuits Syst.*, 34, 721–730, 1987. ©IEEE. With permission.)

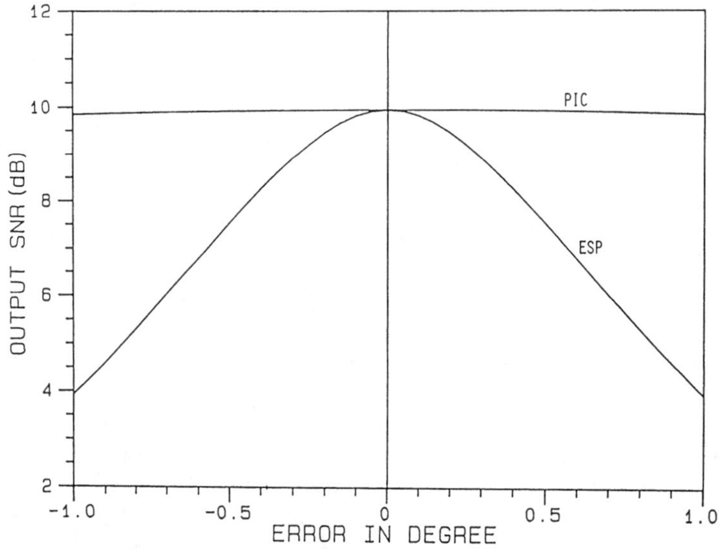


**FIGURE 2.26**
Output SNR vs. the SDE for a ten-element linear array, $p_I = 100$, $\sigma_n^2 = 1$, $\theta_I = 25°$. (From Godara, L.C., *IEEE Trans. Circuits Syst.*, 34, 721–730, 1987. ©IEEE. With permission.)

Figure 2.26 shows the output SNR plots of the two processors when the uncorrelated noise level is raised to that of the signal level. Other noise and array parameters are the same as in Figure 2.25. The effect of the raised uncorrelated noise level on the ESP in the presence of the SDE is that the processor becomes less sensitive to the error. The output SNR of the ESP in the presence of a 1° error is about 4 dB, in comparison to about 10 dB of the PIC processor. The output SNR of 10dB is the level achievable by the two processors in the absence of the error.

For a given uncorrelated noise level, the output SNRs of the two processors in the absence of errors can be increased by increasing the number of elements, as shown in Figure 2.27, where the number of elements of the linear array is increased from 10 to 20. Comparing Figure 2.27 with Figure 2.26, an increase of about 3 dB in the output SNRs of the two processors in the absence of SDE is noticeable. One also observes from the two figures that the ESP is more sensitive to the SDE in the presence of an array with a greater number of elements. With an array of 20 elements, the output SNR of the ESP in the presence of 1° SDE is about –4 dB, in comparison to 4 dB when the number of elements in the array is ten.

All the above results are for a linear array. Similar results were reported in [God87] when a planar array was used.

The above results show that in the absence of errors both processors produce identical results, whereas in the presence of look direction errors the beam space processor produces superior performance. The situation arises when the known direction of the signal is different from the actual direction. Now let us look at the reason for this difference in the performance of the two processors.

The weights of the processor are constrained with the known look direction. When the actual signal direction is different from the one used to constrain weights, the ESP cancels this signal as if it were interference close to the look direction. The beam space processor, on the other hand, is designed to have the main beam steered in the known look direction and the auxiliary beams are formed to have nulls in this direction. The response of the main beam does not alter much as one moves slightly away from the look direction, and

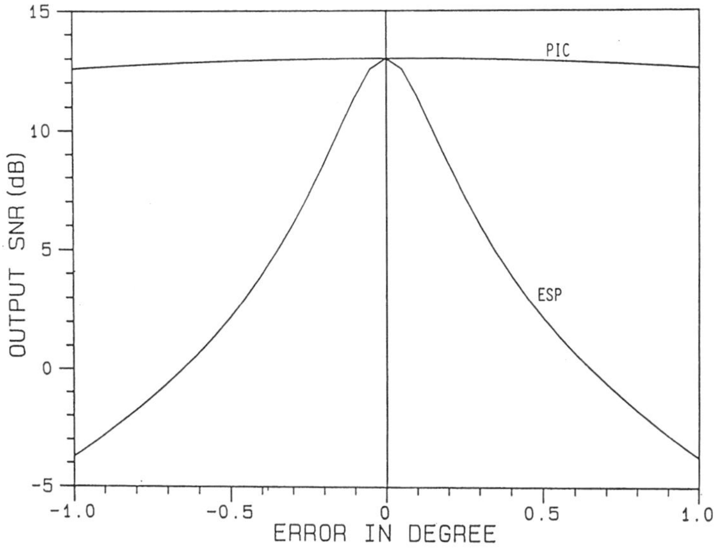


**FIGURE 2.27**
Output SNR vs. the SDE for a 20-element linear array, $p_I = 100$, $\sigma_n^2 = 1$, $\theta_I = 25°$. (From Godara, L.C., *IEEE Trans. Circuits Syst.*, 34, 721–730, 1987. ©IEEE. With permission.)

thus the signal level in the main beam is not affected. Similarly, when a null of the auxiliary beams is placed in the known look direction, a very small amount of the signal leaks in the auxiliary beam due to a source very close to the null and thus the subtraction process does not affect the signal level in the main beam, yielding a very small signal cancelation in the beam space processing compared to the ESP. For details of the effect of other errors on the beam space processors, particularly GSC, see, for example [Jab86].

A comparison of the performance of the PIC with the tamed element space processor is presented in Figure 2.28 for the scenario of Figure 2.27. For the tamed array, as discussed in [Tak86], the weights of the optimal ESP are calculated using the array correlation matrix $R_T$, given by

$$R_T = R + \alpha_0^2 I \tag{2.6.104}$$

where $\alpha_0^2$ is a control variable. The performance of the tamed array is optimized for

$$\alpha_0^2 = Lp_S/\sqrt{2} \tag{2.6.105}$$

Figure 2.27 and Figure 2.28 show that the performance of ESP in the presence of SDE has improved substantially using this procedure. However, the PIC performs better than the tamed ESP.

## 2.7 Effect of Errors

The optimal weights of an antenna array, computed using the steering vector in the direction of arrival of the desired signal and the noise-only array correlation matrix or the

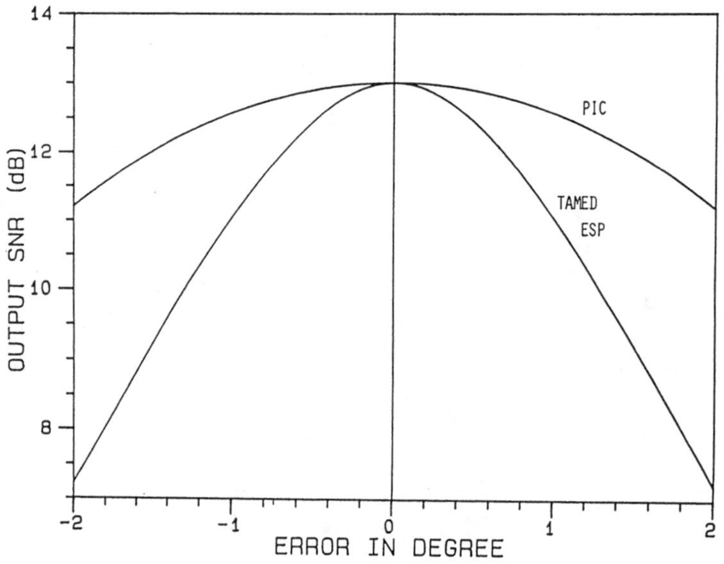


**FIGURE 2.28**
Output SNR vs. the SDE for a 20-element linear array, $p_I = 100$, $\sigma_n^2 = 1$, $\theta_I = 25°$. (From Godara, L.C., *IEEE Trans. Circuits Syst.*, 34, 721–730, 1987. ©IEEE. With permission.)

total array correlation matrix, maximizes the output SNR in the absence of errors. In practice, the estimated optimal weights are corrupted by random errors that arise due to imperfect knowledge of the array correlation matrix, errors in steering vector estimation caused by imperfect knowledge of the array element positions, and error due the finite word-length arithmetic used, and so on. Thus, it is important to know how these errors degrade array performance. The effect of some of these errors on the performance of the optimal processor is discussed in the following sections.

### 2.7.1 Weight Vector Errors

Array weights are calculated using ideal conditions and then stored in memory, and are implemented using amplifiers and phase shifters. Theoretical study of system performance assumes the ideal error-free weights, whereas the actual performance of the system is dependent on the implemented weights. The amplitude as well as the phase of these weights are different from the ideal ones, and these differences arise from many types of errors caused at various points in the system, starting from the deviation in the assumption that a plane wave arrives at the array, uncertainty in the positions and the characteristics of array elements, error in the knowledge of the array correlation matrix caused by its estimation from finite number of samples, error in the steering vector or the reference signal used to calculate weights, computational error caused by finite precision arithmetic, quantization error in converting the analog weights into digital form for storage, and implementation error caused by component variation. Studies of weight errors have been conducted in which these errors are modeled as random fluctuations in weights [God86, Lan85, Ber77, Hud77, Nit76, Ard88], or by modeling them as errors in amplitude and phase [Ram80, Far83, Qua82, Kle80, Cox88, DiC78]. Performance indices to measure the effect of errors include the array gain [God86, Far83], reduction in null depth [Lan85], reduction in interference rejection capability [Nit76], change in side-lobe level [Ram80, Qua82, Kle80], and bias in the angle of arrival estimation [Cox 88], and so on.

The array gain is the ratio of the output SNR to the input SNR. The effect of random weight fluctuation is to cause reduction in the array gain. The effect is sensitive to the number of elements in the array and the array gain of the error-free system [God86]. For an array with a large number of elements and with a large error-free gain, a large weight fluctuation could reduce its array gain to unity, which implies that output SNR becomes equal to the input SNR and no array gain is obtainable.

In this section, the effects of random errors in the weights of the processors on the output signal power, output noise power, output SNR, and array gain are analyzed [God86]. It is assumed that the estimated weights are different from the optimal weights by additive random noise components. Let these errors be represented by an L-dimensional vector $\boldsymbol{\Gamma}$ with the following statistics:

$$
\begin{aligned}
&E(\Gamma_i)=0 \qquad i=1,2,\ldots,L\\
&E\left[\Gamma_i\Gamma_j^*\right]=\begin{cases}\sigma_w^2 & i=j\\ 0 & i\neq j\end{cases} \qquad i,j=1,2,\ldots,L
\end{aligned}
\tag{2.7.1}
$$

Let an L-dimensional complex vector $\overline{\mathbf{w}}$ represent the estimated weights of the processor. Thus,

$$
\overline{\mathbf{w}}=\frac{R_N^{-1}\mathbf{S}_0}{\mathbf{S}_0^H R_N^{-1}\mathbf{S}_0}+\boldsymbol{\Gamma} \tag{2.7.2}
$$

#### 2.7.1.1 Output Signal Power

The output signal power of the processor with estimated weights $\overline{\mathbf{w}}$ is given by

$$
P_S(\overline{\mathbf{w}})=p_S\overline{\mathbf{w}}^H\mathbf{S}_0\mathbf{S}_0^H\overline{\mathbf{w}} \tag{2.7.3}
$$

Substituting for $\overline{\mathbf{w}}$ and taking the mean value on both sides, this becomes, after manipulation,

$$
\overline{P}_S=p_S\left(1+\sigma_w^2L\right) \tag{2.7.4}
$$

Thus, the output signal power increases due to the random errors in the weight vector. This increase is proportional to the input signal power, variance of errors, and number of elements in the array.

#### 2.7.1.2 Output Noise Power

The output noise power of the processor with estimated weights is given by

$$
P_N(\overline{\mathbf{w}})=\overline{\mathbf{w}}^H R_N\overline{\mathbf{w}} \tag{2.7.5}
$$

Substituting for $\overline{\mathbf{w}}$, taking the expected value on both sides, and recognizing the fact that

$$\begin{aligned} E\left[\boldsymbol{\Gamma}^H R_N \boldsymbol{\Gamma}\right] &= E\left[\mathrm{Tr}\left(\boldsymbol{\Gamma}\boldsymbol{\Gamma}^H R_N\right)\right] \\ &= \mathrm{Tr}\left(E\left[\boldsymbol{\Gamma}\boldsymbol{\Gamma}^H\right] R_N\right) \\ &= \sigma_w^2 \mathrm{Tr}\left(R_N\right) \\ &= \sigma_w^2 p_N L \end{aligned} \tag{2.7.6}$$

after manipulation, the result is

$$\begin{aligned} \overline{P}_N &= \hat{P}_N + \sigma_w^2 L p_N \\ &= \hat{P}_N \left[1 + \sigma_w^2 L \frac{p_N}{\hat{P}_N}\right] \\ &= \hat{P}_N \left[1 + \sigma_w^2 L \hat{G}\right] \end{aligned} \tag{2.7.7}$$

where Tr[.] denotes the trace of [.] and $p_N$ is the total input noise power that includes directional interferences as well as uncorrelated noise.

Thus, the output noise power increases due to the presence of random errors in the weights of the processor. The increase is proportional to the error variance, number of elements in the array, and total input noise power of the processor.

#### 2.7.1.3 Output SNR and Array Gain

Let $\alpha_w$ and $G_w$ denote the output SNR and the array gain of the processor with the random errors in the weights. It follows from (2.7.4) and (2.7.7) that

$$\alpha_w = \hat{\alpha} \frac{\left(1 + \sigma_w^2 L\right)}{1 + \sigma_w^2 L \hat{G}} \tag{2.7.8}$$

where $\hat{\alpha}$ is the output SNR of the error-free beamformer. Equation (2.7.8) describes the behavior of the output SNR as a function of the variance of the random errors, number of elements in the array, output SNR, and array gain of the optimal processor.

Dividing both sides of this expression by the input SNR leads to an expression for $G_w$, that is,

$$G_w = \hat{G} \frac{1 + \sigma_w^2 L}{1 + \sigma_w^2 L \hat{G}} \tag{2.7.9}$$

From this expression the following observations can be made:

1. The array gain $G_w$ of the processor with the random additive errors in the weights is a monotonically decreasing function of the variance of the random errors.
2. In the absence of errors in the weights, $G_w$ is equal to $\hat{G}$, the array gain of the optimal processor.
3. As $\sigma_w^2$ increases very high $G_w$ approaches unity. Thus, for finite variance in the random errors the output SNR is more than the input SNR.

An analysis similar to that presented here shows that in the presence of weight vector error (WVE), the expressions for the output signal power and output noise power of the SPNMI processor are the same as those of the NAMI processor. Hence, the presence of the signal component in the array correlation matrix, which is used to estimate the optimal weights, has not affected the performance of the processor in the presence of WVE. However, as shown in the next section, this is not the case for steering vector error (SVE).

### 2.7.2 Steering Vector Errors

The known look direction appears in the optimal weight calculation through the steering vector. The optimal weight calculation for the constrained beamforming requires knowledge of the array correlation matrix and the steering vector in the look direction. Thus, the pointing error causes an error to occur in the steering vector, which is used for weight calculation.

The steering vector may also be erroneous due to other factors such as imperfect knowledge of array element positions, errors caused by finite word-length arithmetic, and so on. The effect of steering vectors has been reported in [God86, Muc81, Com82]. An analytical study by modeling the error as an additive random error indicates [God86] that the effect of error is severe in the SPNMI processor, that is, when the array correlation matrix, which is used to estimate the weights, contains the signal.

As the signal power increases, the performance of the processor deteriorates further due to errors. By estimating the weights using a combination of a reference signal and a steering vector, sensitivity of a processor to the SVE may be reduced [Hon87].

In this section, the effect of SVE on optimal beamformer performance is considered [God86]. It is assumed that each component of the estimated steering vector $\bar{\mathbf{S}}$ is different from $\mathbf{S}_0$ by an additive error component, that is,

$$\bar{\mathbf{S}} = \mathbf{S}_0 + \boldsymbol{\Gamma}_S \tag{2.7.10}$$

where

$$E[\Gamma_{Si}] = 0 \qquad i = 1, 2, \ldots, L \tag{2.7.11}$$

and

$$E[\Gamma_{Si}\Gamma_{Sj}^*] = \begin{cases} \sigma_S^2 & i = j \\ 0 & i \neq j \end{cases} \qquad i, j = 1, 2, \ldots, L \tag{2.7.12}$$

The analysis presented here is for processors without constraints. The NAMI processor is first considered.

#### *2.7.2.1 Noise-Alone Matrix Inverse Processor*

Let an L-dimensional vector $\tilde{\mathbf{w}}$ represent the estimated weights of the processor when $\bar{\mathbf{S}}$ rather than $\mathbf{S}_0$ is used in estimating the optimal weights. The expression for the estimated weights of the processor in this case becomes

$$\begin{aligned} \tilde{\mathbf{w}} &= \tilde{\mu}\ R_N^{-1}\bar{\mathbf{S}} \\ &= \tilde{\mu}\ R_N^{-1}(\mathbf{S}_0 + \boldsymbol{\Gamma}_S) \end{aligned} \tag{2.7.13}$$

where $\tilde{\mu}$ is a constant.

The expected value of the mean output signal power and the mean output noise power are given below. The expectation is taken over the randomness in the steering vectors.

*2.7.2.1.1 Output Signal Power*

The output signal power of a processor with weights $\tilde{\mathbf{w}}$ is given by

$$P_S(\tilde{\mathbf{w}}) = p_S \tilde{\mathbf{w}}^H \mathbf{S}_0 \mathbf{S}_0^H \tilde{\mathbf{w}} \tag{2.7.14}$$

Substituting for $\tilde{\mathbf{w}}$ from (2.7.13), the signal power becomes

$$\begin{aligned} P_S(\tilde{\mathbf{w}}) &= p_S\left[\tilde{\mu} R_N^{-1}(\mathbf{S}_0 + \boldsymbol{\Gamma}_S)\right]^H \mathbf{S}_0 \mathbf{S}_0^H \left[\tilde{\mu} R_N^{-1}(\mathbf{S}_0 + \boldsymbol{\Gamma}_S)\right] \\ &= p_S \tilde{\mu}^2 \mathbf{S}_0^H R_N^{-1} \mathbf{S}_0 \mathbf{S}_0^H R_N^{-1} \mathbf{S}_0 \\ &\quad + p_S \tilde{\mu}^2 \left[\mathbf{S}_0^H R_N^{-1} \mathbf{S}_0 \mathbf{S}_0^H R_N^{-1} \boldsymbol{\Gamma}_S + \boldsymbol{\Gamma}_S^H R_N^{-1} \mathbf{S}_0 \mathbf{S}_0^H R_N^{-1} \mathbf{S}_0\right] \\ &\quad + p_S \tilde{\mu}^2 \mathbf{S}_0^H R_N^{-1} \boldsymbol{\Gamma}_S \boldsymbol{\Gamma}_S^H R_N^{-1} \mathbf{S}_0 \end{aligned} \tag{2.7.15}$$

Taking the expected value on both sides of (2.7.15) and using (2.7.11) and (2.7.12), after rearrangements,

$$\begin{aligned} \tilde{P}_S &= p_S \tilde{\mu}^2 \left[\left(\mathbf{S}_0^H R_N^{-1} \mathbf{S}_0\right)^2 + \sigma_S^2 \mathbf{S}_0^H R_N^{-1} R_N^{-1} \mathbf{S}_0\right] \\ &= p_S \left[1 + \sigma_S^2 \frac{\mathbf{S}_0^H R_N^{-1} R_N^{-1} \mathbf{S}_0}{\left(\mathbf{S}_0^H R_N^{-1} \mathbf{S}_0\right)^2}\right] \tilde{\mu}^2 \left(\mathbf{S}_0^H R_N^{-1} \mathbf{S}_0\right)^2 \\ &= p_S\left[1 + \hat{\beta}\sigma_s^2\right] \tilde{\mu}^2 / \hat{P}_N^2 \end{aligned} \tag{2.7.16}$$

where $\hat{\beta}$ is the ratio of uncorrelated noise power at the output to the uncorrelated noise power at the input of the optimal processor and $\hat{P}_N$ is the mean output noise power of the optimal processor.

*2.7.2.1.2 Total Output Noise Power*

The output noise power of the processor with weight vector $\tilde{\mathbf{w}}$ is given by

$$\tilde{P}_N = \tilde{\mathbf{w}}^H R_N \tilde{\mathbf{w}} \tag{2.7.17}$$

Substituting for $\tilde{\mathbf{w}}$ from (2.7.13), it becomes

$$\begin{aligned} \tilde{P}_N &= \tilde{\mu}^2 \mathbf{S}_0^H R_N^{-1} \mathbf{S}_0 \\ &\quad + \tilde{\mu}^2 \left[\mathbf{S}_0^H R_N^{-1} \boldsymbol{\Gamma}_S + \boldsymbol{\Gamma}_S^H R_N^{-1} \mathbf{S}_0\right] \\ &\quad + \tilde{\mu}^2 \boldsymbol{\Gamma}_S^H R_N^{-1} \boldsymbol{\Gamma}_S \end{aligned} \tag{2.7.18}$$

Taking the expected value on both sides of (2.7.18) and using (2.7.11) and (2.7.12), after rearrangements,

$$\begin{aligned}\tilde{P}_N &= \tilde{\mu}^2\left[\mathbf{S}_0^H R_N^{-1}\mathbf{S}_0 + \sigma_S^2 \mathrm{Tr}\left(R_N^{-1}\right)\right] \\ &= \tilde{\mu}^2\left[1+\sigma_S^2\frac{\mathrm{Tr}\left(R_N^{-1}\right)}{\mathbf{S}_0^H R_N^{-1}\mathbf{S}_0}\right]\mathbf{S}_0^H R_N^{-1}\mathbf{S}_0 \\ &= \left[1+\sigma_s^2\kappa\right]\frac{\tilde{\mu}^2}{\hat{P}_N}\end{aligned} \tag{2.7.19}$$

where

$$\kappa \triangleq \frac{\mathrm{Tr}\left(R_N^{-1}\right)}{\left(\mathbf{S}_0^H R_N^{-1}\mathbf{S}_0\right)} \tag{2.7.20}$$

Since $\kappa > 0$, it follows from (2.7.19) that the output noise power increases in proportion to the variance of the random errors in the steering vector.

#### 2.7.2.1.3 *Output SNR and Array Gain*

Let $\alpha_s$ and $G_S$ denote the output SNR and the array gain of the NAMI processor with SVE. It follows then from (2.7.16) and (2.7.19) that

$$\alpha_s = \hat{\alpha}\frac{1+\sigma_s^2\hat{\beta}}{1+\sigma_s^2\kappa} \tag{2.7.21}$$

and

$$G_s = \hat{G}\frac{1+\sigma_s^2\hat{\beta}}{1+\sigma_s^2\kappa} \tag{2.7.22}$$

It follows from these two equations that the behavior of the output SNR and the array gain of the NAMI processor with SVE depend on the relative magnitudes of $\hat{\beta}$ and $\kappa$. It can be shown that $\kappa \geq \hat{\beta}$, and thus the array gain of the NAMI processor with the random errors in the steering vector is a monotonically decreasing function of the error variance.

### 2.7.2.2 Signal-Plus-Noise Matrix Inverse Processor

Let an L-dimensional vector $\hat{\mathbf{w}}$ represent the estimated weights of the SPNMI processor when $\bar{\mathbf{S}}$ rather than $\mathbf{S}_0$ is used in estimating the optimal weights. The expression for $\hat{\mathbf{w}}$ in this case becomes

$$\hat{\mathbf{w}} = \hat{\mu}\, R^{-1}\left[\mathbf{S}_0 + \boldsymbol{\Gamma}_S\right] \tag{2.7.23}$$

where

$$R = R_N + p_S\mathbf{S}_0\mathbf{S}_0^H \tag{2.7.24}$$

Using the Matrix Inversion Lemma,

$$R^{-1} = R_N^{-1} - a_0 R_N^{-1}\mathbf{S}_0\mathbf{S}_0^H R_N^{-1} \tag{2.7.25}$$

where

$$a_0 = \frac{p_S}{1 + p_S \mathbf{S}_0^H R_N^{-1}\mathbf{S}_0} \tag{2.7.26}$$

From (2.7.23) and (2.7.25), it follows that

$$\hat{\mathbf{w}} = \hat{\mu}\, R_N^{-1}\left[\mathbf{S}_0 + \mathbf{\Gamma}_S\right] - \hat{\mu}\, a_0 R_N^{-1}\mathbf{S}_0\mathbf{S}_0^H R_N^{-1}\left(\mathbf{S}_0 + \mathbf{\Gamma}_S\right) \tag{2.7.27}$$

Comparing (2.7.13) with (2.7.27) one notes that the second term in (2.7.27) is due to the presence of the signal component in the array correlation matrix that is used in estimating the optimal weights. As the signal component goes to zero, the second term goes to zero because $a_0$ goes to zero, and thus $\hat{\mathbf{w}}$ becomes $\bar{\mathbf{w}}$.

The effect of SVE on the output signal power, the output noise power, the output SNR, and the array gain is now examined.

#### 2.7.2.2.1 *Output Signal Power*

Following a procedure similar to that used for the NAMI processor, an expression for the mean output signal power of the SPNMI processor in the presence of the SVE becomes

$$\hat{P}_S = \frac{\hat{\mu}^2}{\left(\hat{P}_N + p_S\right)^2} p_S\left[1 + \sigma_s^2\hat{\beta}\right] \tag{2.7.28}$$

Comparing (2.7.16) with (2.7.28), in the presence of SVE the output signal power of both processors increases and the increase is proportional to the output signal power of the respective error-free processor and the parameter $\hat{\beta}$, which is the ratio of the uncorrelated noise powers at the output of the optimal processor to its input. Hence, the effect of the random SVE on both processors is the same. Thus, the presence of the signal component in the array correlation matrix has not altered the effects of SVE on output signal power. In the next section, it is shown that this is not the case for the output noise power.

#### 2.7.2.2.2 *Total Output Noise Power*

Following a procedure similar to that used for the NAMI processor, an expression for the mean output noise power of the processor with weight vector $\hat{\mathbf{w}}$ becomes

$$\hat{P}_N = \frac{\hat{\mu}^2}{\left(\hat{P}_N + p_S\right)^2}\hat{P}_N\left\{1 + \sigma_s^2\left[\kappa + \left(\hat{\alpha}^2 + 2\hat{\alpha}\right)\left(\kappa - \hat{\beta}\right)\right]\right\} \tag{2.7.29}$$

Since $\kappa \geq \hat{\beta}$, it follows from (2.7.29) that the output noise power of the SPNMI processor increases with the increase in the variance of random errors $\sigma_s^2$, and the increase is enhanced by the input signal power due to the presence of product terms $\sigma_s^2\hat{\alpha}^2$ and $\sigma_s^2\hat{\alpha}$. Note that the third term in (2.7.29), which contains these terms, is missing from the expression of the output noise power given by (2.7.19) when the array correlation matrix of noise only is used in the calculation of the optimal weights.

*2.7.2.2.3 Output SNR*

Let $\breve{\alpha}_s$ denote the output SNR of the SPNMI processor in the presence of SVE. Then it follows that

$$\breve{\alpha}_s = \frac{\hat{\alpha}\left(1+\sigma_s^2\hat{\beta}\right)}{1+\sigma_s^2\left[\kappa+\left(\hat{\alpha}^2+2\hat{\alpha}\right)\left(\kappa-\hat{\beta}\right)\right]} \tag{2.7.30}$$

which describes the behavior of the output SNR of the SPNMI processor in the presence of random SVE. Comparing this with (2.7.21), the expression for the output SNR of the NAMI processor, one observes the presence of $\hat{\alpha}^2$ and $\hat{\alpha}$ in the denominator of (2.7.30). As the output SNR of the optimal processor $\hat{\alpha}$ is directly proportional to the input SNR of the processor, it follows that:

1. The effect of SVE on output SNR of the SPNMI processor is very sensitive to the input signal power.
2. The output SNR of the SPNMI processor drops faster than the output SNR of the NAMI processor as the error variance increases.
3. For a given level of SVE, the output SNR of the SPNMI processor is less than the output SNR of the NAMI processor, and the difference increases as the power of the signal source increases.

It should be noted here that the above observations are true for any array geometry and noise environment. However, the array geometry and the noise environment would affect the results as $\hat{\alpha}$, $\kappa$, and $\hat{\beta}$ depend on them.

Now the array gain of the SPNMI processor $\breve{G}_S$ is compared with the array gain of the NAMI processor $G_S$ in the presence of SVE. For this case, array gain is given by

$$\breve{G}_S = \frac{\left(1+\sigma_s^2\hat{\beta}\right)\hat{G}}{1+\sigma_s^2\left[\kappa+\left(\hat{\alpha}^2+2\hat{\alpha}\right)\left(\kappa-\hat{\beta}\right)\right]} \tag{2.7.31}$$

Since $\kappa \geq \hat{\beta}$, it follows from (2.7.22) and (2.7.31) that for a given $\sigma_s^2$, the array gain $\breve{G}_S$ of the SPNMI processor is less than the array gain $G_S$ of the NAMI processor, and $\breve{G}_S$ falls more rapidly than $G_S$ as the variance of the random SVE increases. The fall in $\breve{G}_S$ is greater at a higher input SNR than at a lower input SNR.

### 2.7.2.3 Discussion and Comments

Table 2.2 compares the various results on SVE and WVE. All quantities are normalized with their respective error-free values to facilitate observation of the effect of errors. The following observations can be made from the table:

**TABLE 2.2**

Comparison of the SVE and WVE*

| | Normalized Mean Output Signal Power | Normalized Mean Output Noise Power | Normalized Array Gain |
|---|---|---|---|
| Effect of SVE on NAMI processor | $1+\hat{\beta}\sigma_s^2$ | $1+\sigma_s^2\kappa$ | $\frac{1+\sigma_s^2\hat{\beta}}{1+\sigma_s^2\kappa}$ |
| Effect of SVE on SPAMI processor | $1+\hat{\beta}\sigma_s^2$ | $1+\sigma_s^2\left[\kappa+\left(\hat{\alpha}^2+2\hat{\alpha}\right)\left(\kappa-\hat{\beta}\right)\right]$ | $\frac{\left(1+\sigma_s^2\hat{\beta}\right)}{1+\sigma_s^2\left[\kappa+\left(\hat{\alpha}^2+2\hat{\alpha}\right)\left(\kappa-\hat{\beta}\right)\right]}$ |
| Effect of WVE on both processors | $1+\sigma_w^2 L$ | $1+\sigma_w^2 L\hat{G}$ | $\frac{\left(1+\sigma_w^2 L\right)}{1+\sigma_w^2 L\hat{G}}$ |

* $\hat{\beta}$: Ratio of the uncorrelated noise at the output to the input of the optimal beamformer; $\hat{G}$: array gain of the optimal beamformer; $\hat{\alpha}$: output SNR of the optimal beamformer; $\sigma_s^2$: variance of the additive random steering vector errors; $\sigma_w^2$: variance of the additive random weight vector errors; $\kappa = \frac{\mathrm{Tr}\left(R_N^{-1}\right)}{\left(\mathbf{S}_0^H R_N^{-1}\mathbf{S}_0\right)}$.

1. The output signal power in all cases increases with the increase in error variance. For WVE case, the increase depends only on the number of elements, whereas for SVE it depends on the array geometry and the noise environment.
2. The output noise power in all cases increases with the increase in error variance. For the WVE case, the increase depends on $\hat{G}$ and is independent of signal power. For the SVE case, the increase in the output noise power is dependent on the input signal power for the SPNMI processor, and is independent of the signal power for the NAMI processor.
3. The array gain in all cases decreases with the increase in the error variance. In the case of WVE, the decrease in the array gain depends on $\hat{G}$. The greater $\hat{G}$ is, the faster the array gain drops as the error variance increases. In the SVE case, the array gain of the SPNMI processor is dependent on the output SNR of the optimal processor ($\hat{\alpha}$), and it drops as $\hat{\alpha}$ is increased. Note that $\hat{\alpha}$ is directly proportional to the input signal power. The effect of SVE on the NAMI processor is not affected by the input signal power.

Two special cases of the noise environment are considered below to study the effect of array elements, uncorrelated noise power, direction, and power of the interference source.

#### 2.7.2.3.1 *Special Case 1: Uncorrelated Noise Only*

Consider the case of a noise environment where only uncorrelated noise is present. Let $\Delta$ denote the ratio of the input signal power to the uncorrelated noise power on each element. For this case,

$$\hat{G} = L \tag{2.7.32}$$

$$\hat{\beta} = \frac{1}{L} \tag{2.7.33}$$

$$\hat{\alpha} = L\Delta \tag{2.7.34}$$

**TABLE 2.3**

Comparison of Array Gain in the Presence of SVE and WVE with No Interference Present*

| | |
|---|---|
| Array gain NAMI processor in SVE | $\dfrac{L+\sigma_s^2}{1+\sigma_s^2}$ |
| Array gain of SPAMI processor in SVE | $\dfrac{L+\sigma_s^2}{1+\sigma_s^2\left[1+L\Delta^2(L-1)+2\Delta(L-1)\right]}$ |
| Array gain of both processors in WVE | $\dfrac{L+\sigma_w^2L^2}{1+\sigma_w^2L^2}$ |

* $\sigma_s^2$: Variance of steering vector error; $\sigma_w^2$: variance of weight vector error, $\Delta = p_S/\sigma_n^2$.

and

$$\kappa = 1 \tag{2.7.35}$$

The expressions for the array gains of the two processors in the presence of SVE and WVE are shown in Table 2.3. From the table, the following observations can be made.

1. For a given error level, say $\sigma_S = \sigma_{S0}$, the array gain of the NAMI processor increases as the number of elements in the array increases. Thus, for a given error level and input SNR, the output SNR of the NAMI processor increases as L increases.
2. The array gain of the NAMI processor decreases as the error level is increased, and it does not depend on the ratio of the input signal to the uncorrelated noise power, $\Delta$. However, the behavior of the array gain $\hat{G}_S$ of the SPNMI processor in the presence of SVE depends on $\Delta$. For a given L, $\hat{G}_S$ drops faster at a higher $\Delta$ than at a lower $\Delta$ as the SVE level is increased.
3. For $\Delta \ll 1$, the expression for $\hat{G}_S$ becomes

$$\hat{G}_S \cong \frac{L+\sigma_s^2}{1+\sigma_s^2} \tag{2.7.36}$$

   and for a given level of errors the array gain increases with the increase in the number of elements, as in the case of the NAMI processor.
4. For $\Delta \gg 1$, the expression for $\hat{G}_S$ becomes

$$\hat{G}_S \cong \frac{L+\sigma_s^2}{1+\sigma_s^2\Delta^2L(L-1)} \tag{2.7.37}$$

   Thus, for a given $\sigma_S$, the array gain decreases with the increase in the number of elements for a very high input signal to uncorrelated noise ratio.
5. The plots of $\hat{G}_S$ vs. the input SNR for various values of L are shown in Figure 2.29 for error variance equal to 0.01. The results displayed in the figure are in agreement with the above observations.
6. A comparison of the expressions for the array gain in the presence of the SVE and the WVE reveal that $G_w$, the array gain of both processors in the presence of WVE,

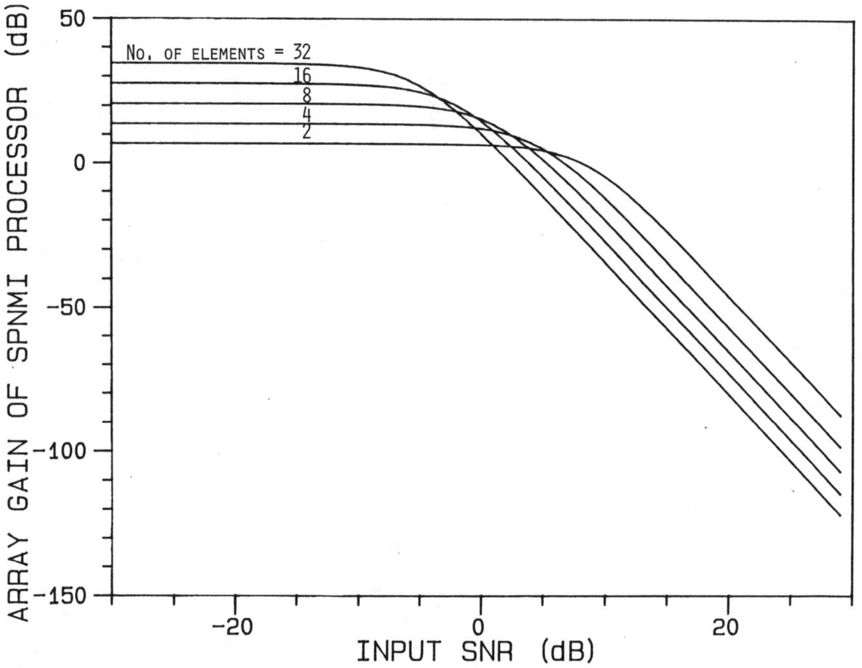


**FIGURE 2.29**
Array gain of SPNMI processor vs. SNR, no interference, and $\sigma_s^2 = 0.01$. (From Godara, L.C., *IEEE Trans. Aerosp. Electron. Syst.*, 22, 395–409, 1986. ©IEEE. With permission.)

behaves similarly to $G_S$, the array gain of NAMI processor in the presence of SVE. For a given error level, both $G_w$ and $G_S$ increase with the increase in L. However, for the same error level, say $\sigma_S = \sigma_w = \sigma_0$,

$$G_S - G_w = \frac{\sigma_0^2(L-1)\left(L^2-1\right)}{\left(1+\sigma_0^2\right)\left(1+\sigma_0^2 L^2\right)} > 0 \tag{2.7.38}$$

### 2.7.2.3.2 *Special Case 2: One Directional Interference*

Consider the case of a noise environment consisting of a directional interference of power $p_I$ and uncorrelated noise of power $\sigma_n^2$ on each element of the array. For this case, $\hat{G}$ and $\hat{\alpha}$ are, respectively, given by (2.4.43) and (2.4.42),

$$\hat{\beta} = \frac{1}{L} + \frac{\rho L(1-\rho)}{\left(\rho L + \varepsilon_0\right)^2} \tag{2.7.39}$$

and

$$\kappa = 1 + \frac{L(1-\rho)-1}{\rho L + \varepsilon_0} \tag{2.7.40}$$

where

$$\varepsilon_0 = \frac{\sigma_n^2}{p_I} \tag{2.7.41}$$

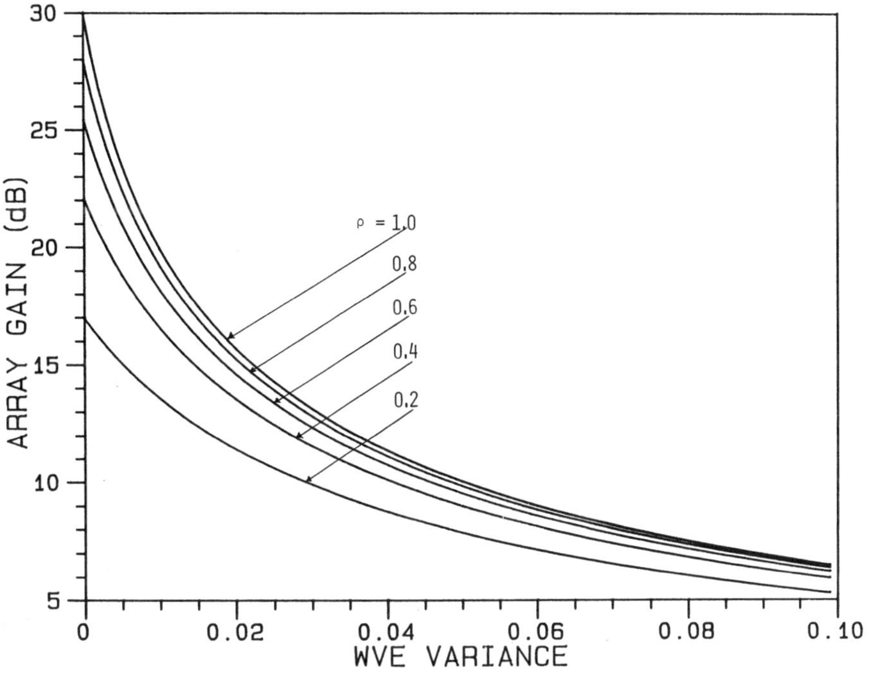


**FIGURE 2.30**
$G_w$ vs. $\sigma_w^2$, $\sigma_n^2/p_I$ = 0 dB, and ten-element array. (From Godara, L.C., *IEEE Trans. Aerosp. Electron. Syst.*, 22, 395–409, 1986. ©IEEE. With permission.)

The effect of a variation in ρ on the array gain of the two processors in the presence of WVE and SVE is shown in Figure 2.30 to Figure 2.33. The number of elements in the array for these figures is taken to be ten.

Figure 2.30 shows $G_w$ vs. $\sigma_w^2$ for five values of ρ. One observes from the figure that $G_w$, which denotes the array gain of both the processors in the WVE, decreases faster at higher values of ρ than at lower values of ρ, as the variance of the errors is increased. The result is expected, since $\hat{G}$ increases as ρ increases.

Figure 2.31 and Figure 2.32 show the effect of ρ on the array gain of the SPNMI processor in the presence of SVE for $\sigma_n^2/p_I$ = 0 dB and $\sigma_n^2/p_I$ = –40 dB, respectively. These figures show that as the error variance is increased, the array gain falls more rapidly at higher values of ρ than at lower values of ρ. The result is expected, since $\hat{\alpha}$ increases as ρ increases.

A comparison of Figure 2.31 and Figure 2.32 reveals that the effect of SVE on the array gain is not altered significantly by increasing the interference power. The result is predictable from the expression for $\overset{\frown}{G}_s$, since for $L\rho \gg \sigma_n^2/p_I$ the constants $\hat{\beta}$, κ, and $\hat{\alpha}$ are independent of interference power.

The effect of ρ on the array gain of the NAMI processor in the presence of the SVE is shown in Figure 2.33. The figure demonstrates that the effect of the SVE on the array gain of the NAMI processor is almost the same for all values of ρ. This observation implies that the array geometry and direction of interference do not significantly influence the effect of SVE on the NAMI processor unless the interference direction is very close to the look direction.

Figure 2.34 and Figure 2.35 compare the three array gains $G_w$, $G_S$, and $\overset{\frown}{G}_s$ for the case of weak interference $\sigma_n^2/p_I$ = 0 dB, and strong interference, dB. For these figures, input signal power is equal to uncorrelated noise power. These figures show that the array gains of both processors in the presence of the SVE are not affected by the interference power, whereas $G_w$, the array gain of two processors in the presence of WVE, is highly dependent

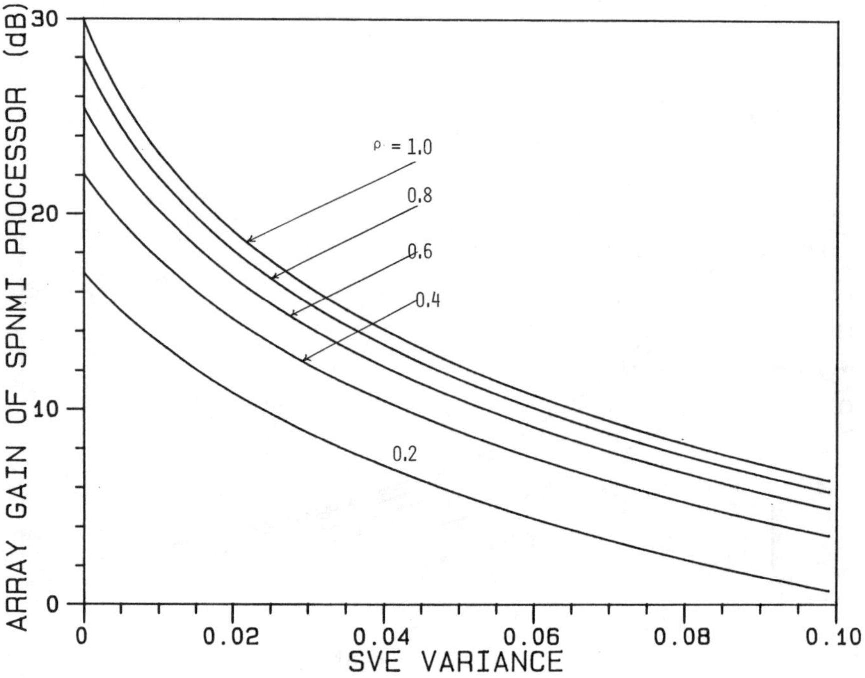


**FIGURE 2.31**
$\hat{G}_s$ vs. $\sigma_s^2$, $\sigma_n^2/p_I = 0$ dB, and ten-element array. (From Godara, L.C., *IEEE Trans. Aerosp. Electron. Syst.*, 22, 395–409, 1986. ©IEEE. With permission.)

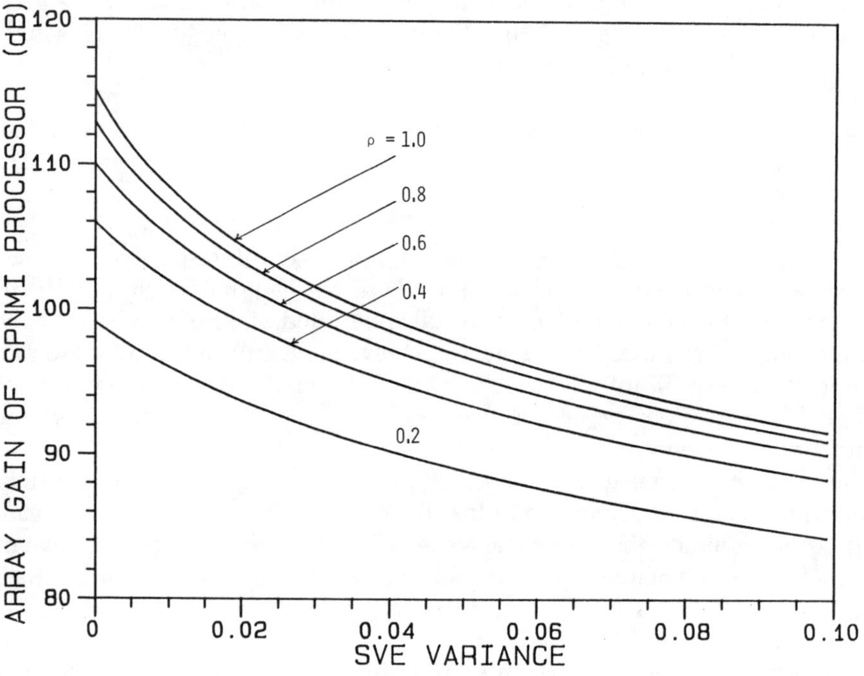


**FIGURE 2.32**
$\hat{G}_s$ vs. $\sigma_s^2$, $\sigma_n^2/p_I = -40$ dB, and ten-element array. (From Godara, L.C., *IEEE Trans. Aerosp. Electron. Syst.*, 22, 395–409, 1986. ©IEEE. With permission.)

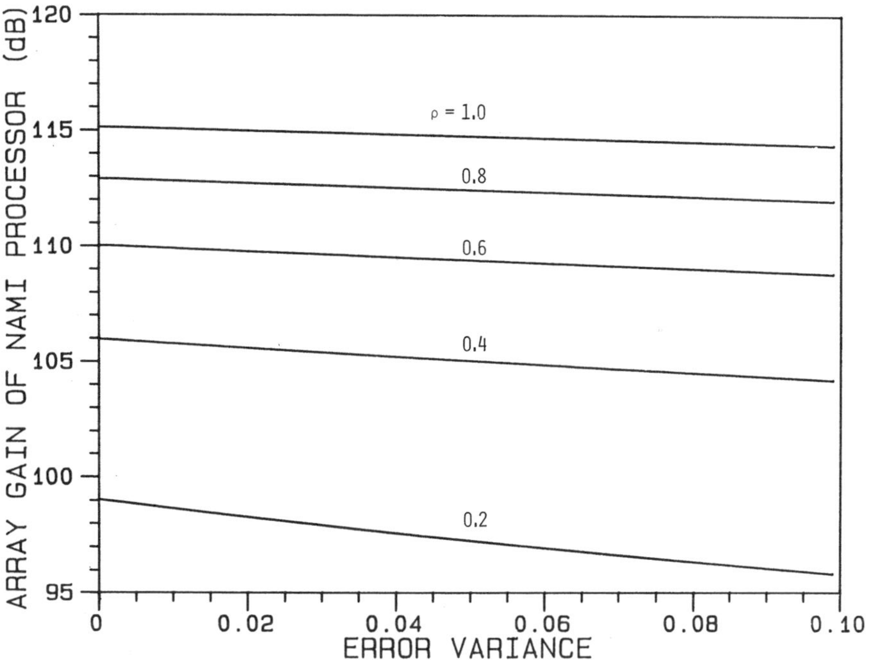


**FIGURE 2.33**
$\hat{G}_s$ vs. $\sigma_s^2$, $\sigma_n^2/p_I = -40$ dB, and ten-element array. (From Godara, L.C., *IEEE Trans. Aerosp. Electron. Syst.*, 22, 395–409, 1986. ©IEEE. With permission.)

on the interference power. It drops faster as $\sigma_w^2$ is increased in the presence of the interference and the rate of drop increases with the increase in the interference power. Note the difference in the vertical scales of the two figures.

### 2.7.3 Phase Shifter Errors

The phase of the array weight is an important parameter and an error in the phase may cause an estimate of the source to appear in a wrong direction when an array is used for finding directions of sources, such as in [Cox88]. The phase control of signals is used to steer the main beam of the array in desired positions, as in electronic steering. A device normally used for this purpose is a phase shifter. Commonly available types are ferrite phase shifters and diode phase shifters [Mai82, Sta70]. One of the specifications that concerns an array designer is the root mean square (RMS) phase error.

Analysis of the RMS phase error shows that it causes the output SNR of the constrained optimal processor to suppress the desired signal, and the suppression is proportional to the product of the signal power and the random error variance [God85]. Furthermore, suppression is maximum in the absence of directional interferences. Quantization error occurs in digital phase shifters. In a p-bit digital phase shifter, the minimum value of the phase that can be changed equals $2\pi/2^p$. Assuming that the error is distributed uniformly between $\pi/2^p$ to $\pi/2^p$, the variance of this error equals $\pi^2/3 \times 2^{2p}$.

In this section, the effect of random phase errors on the performance of the optimal processor is analyzed [God85]. To facilitate this analysis, the phase shifters are separated from the weights as shown in Figure 2.36 and are selected to steer the array in the look direction.

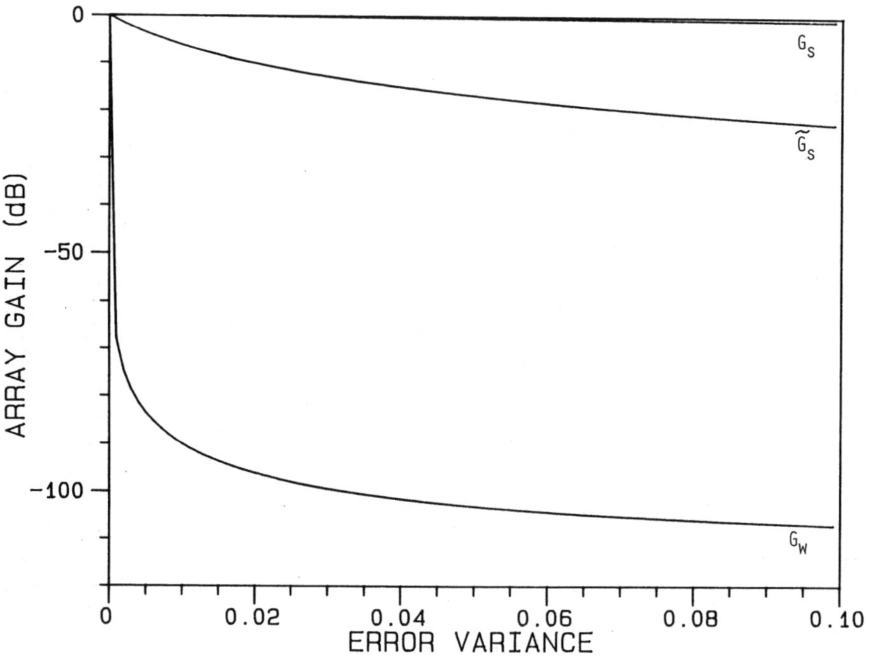


**FIGURE 2.34**
$G_w$, $\hat{G}_s$, and $G_s$ vs. $\sigma_s^2$, $\sigma_n^2/p_I$ = –40 dB, ρ = 0.9, and ten-element array. (From Godara, L.C., *IEEE Trans. Aerosp. Electron. Syst.*, 22, 395–409, 1986. ©IEEE. With permission.)

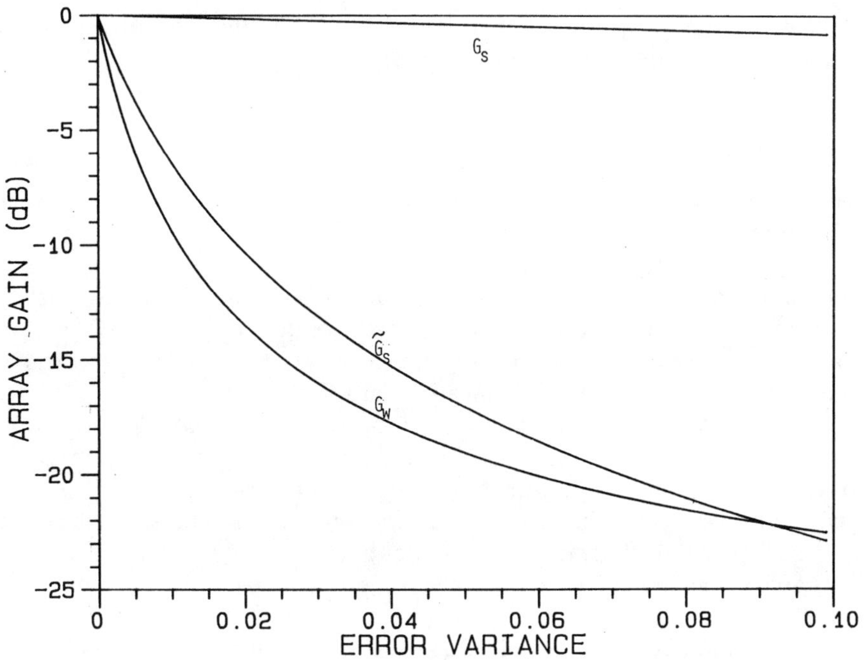


**FIGURE 2.35**
$G_w$, $\hat{G}_s$, and $G_s$ vs. $\sigma_s^2$, $\sigma_n^2/p_I$ = 0 dB, ρ = 0.9, and ten-element array. (From Godara, L.C., *IEEE Trans. Aerosp. Electron. Syst.*, 22, 395–409, 1986. ©IEEE. With permission.)

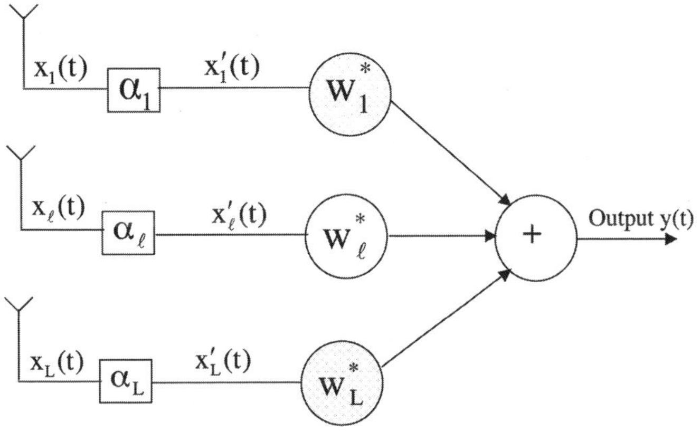


**FIGURE 2.36**
Beamformer structure showing phase shifters.

Let the optimal weights of the beamformer of Figure 2.36, referred to as the beamformer using phase shifters, be denoted by $\breve{\mathbf{w}}$. It follows from the figure that the output of the optimal beamformer using phase shifters is given by

$$y(t) = \breve{\mathbf{w}}^H \mathbf{x}'(t) \tag{2.7.42}$$

where $\mathbf{x}'(t)$ is the array signal received after the phase shifters and are given by (2.6.31). Thus, using (2.6.31), (2.7.42) becomes

$$y(t) = \breve{\mathbf{w}}^H \Phi_0^H \mathbf{x}(t) \tag{2.7.43}$$

Now a relationship between $\breve{\mathbf{w}}$ and $\hat{\mathbf{w}}$, the weights of the optimal beamformer without using phase shifters discussed in Section 2.4, is established. The output of the optimal beamformer without using phase shifters is given by

$$y(t) = \hat{\mathbf{w}}^H \mathbf{x}(t) \tag{2.7.44}$$

Since the outputs of both structures are identical, it follows from (2.7.43) and (2.7.44) that $\breve{\mathbf{w}}$ and $\hat{\mathbf{w}}$ are related as follows:

$$\hat{\mathbf{w}} = \Phi_0 \breve{\mathbf{w}} \tag{2.7.45}$$

An expression for $\breve{w}$ may be obtained from (2.4.16) and (2.7.45) and is given by

$$\breve{\mathbf{w}} = \frac{\Phi_0^H R_N^{-1} \mathbf{S}_0}{\mathbf{S}_0^H R_N^{-1} \mathbf{S}_0} \tag{2.7.46}$$

#### *2.7.3.1 Random Phase Errors*

In this section, the effect of random phase errors on optimal processor performance is examined. Phase shifters with random phase errors are termed "actual phase shifters," and the processor in this case is termed "optimal processor with phase errors" (OPPE). It is assumed that random phase errors that exist in the phase shifters can be modeled as stationary processes of zero mean and equal variance and are not correlated with each other.

Let $\delta_\ell$, $\ell = 1, 2, \ldots, L$ represent the phase error in the $\ell$th phase shifter. By assumption,

$$E[\delta_\ell] = 0, \qquad \ell = 1, 2, \ldots, L \tag{2.7.47}$$

and

$$E[\delta_\ell \delta_k] = \begin{cases} \sigma^2 \text{ if } \ell = k, \\ 0 \quad \text{otherwise,} \end{cases} \quad \ell, k = 1, 2, \ldots, L \tag{2.7.48}$$

Let $\tilde{\alpha}_\ell$, $\ell = 1, 2, \ldots, L$ represent the phase delays of the actual phase shifters. Then

$$\tilde{\alpha}_\ell = \alpha_\ell + \delta \qquad \ell = 1, 2, \ldots, L \tag{2.7.49}$$

where $\alpha_\ell$, $\ell = 1, 2, \ldots, L$ are the phase delays of error-free phase shifters, and are given by (2.6.29).

Let a diagonal matrix $\Phi$ be defined as

$$\Phi_{\ell\ell} = \exp(j\tilde{\alpha}_\ell), \quad \ell = 1, 2, \ldots, L \tag{2.7.50}$$

It follows from (2.7.43) that an expression for the mean output power of the optimal beamformer using phase shifters is given by

$$\begin{aligned} \hat{P} &= E[y(t)y^*(t)] \\ &= \breve{\mathbf{w}}^H \Phi_0^H E[\mathbf{x}(t)\mathbf{x}^H(t)] \Phi_0 \breve{\mathbf{w}} \\ &= \breve{\mathbf{w}}^H \Phi_0^H R \Phi_0 \breve{\mathbf{w}} \end{aligned} \tag{2.7.51}$$

Similarly, the mean signal power, interference power, and uncorrelated noise power, respectively, are given by

$$\hat{P}_S = \breve{\mathbf{w}}^H \Phi_0^H R_S \Phi_0 \breve{\mathbf{w}} \tag{2.7.52}$$

$$\hat{P}_I = \breve{\mathbf{w}}^H \Phi_0^H R_I \Phi_0 \breve{\mathbf{w}} \tag{2.7.53}$$

and

$$\begin{aligned} \hat{P}_n &= \sigma_n^2 \breve{\mathbf{w}}^H \breve{\mathbf{w}} \\ &= \sigma_n^2 \hat{\beta} \end{aligned} \tag{2.7.54}$$

where $\hat{\beta}$ is defined by (2.4.11). The last step in (2.7.54) follows from using (2.7.46).

Note that the mean output uncorrelated noise power given by (2.7.54) is not a function of phase angles and is not affected by the random errors in the phase shifters.

The effect of random phase errors on the output signal power and output interference power is now examined. Substituting $\Phi$ for $\Phi_0$ in (2.7.52) and (2.7.53) and taking expectation over random phase errors, expressions for the mean output signal power $\breve{P}_S$ and interference power $\breve{P}_I$ of the OPPE follow:

$$\begin{aligned}\breve{P}_S &= \breve{\mathbf{w}}^H E\left[\Phi^H R_S \Phi\right]\breve{\mathbf{w}} \\ &= p_S \breve{\mathbf{w}}^H E\left[\Phi^H \mathbf{S}_0 \mathbf{S}_0^H \Phi\right]\breve{\mathbf{w}}\end{aligned} \tag{2.7.55}$$

and

$$\breve{P}_I = \breve{\mathbf{w}}^H E\left[\Phi^H R_I \Phi\right]\breve{\mathbf{w}} \tag{2.7.56}$$

#### 2.7.3.2 *Signal Suppression*

Rewrite (2.7.55) in the following form:

$$\breve{P}_S = p_S \sum_{\ell,k}^{L} \breve{\mathbf{w}}_\ell^* E\left[\Phi_{\ell\ell}^* \mathbf{S}_{0\ell} \mathbf{S}_{0k}^* \Phi_{kk}\right]\breve{\mathbf{w}}_k \tag{2.7.57}$$

Substituting for $\Phi$ and $\mathbf{S}_0$ in (2.7.57), after rearrangement,

$$\begin{aligned}\breve{P}_S &= p_S \sum_{\ell,k}^{L} \breve{\mathbf{w}}_\ell^* \breve{\mathbf{w}}_k E\left[\exp\left(-j\left(\delta_\ell - \delta_k\right)\right)\right] \\ &= p_S \sum_{\substack{\ell,k \\ \ell\neq k}}^{L} \breve{\mathbf{w}}_\ell^* \breve{\mathbf{w}}_k E\left[\exp -j\left(\delta_\ell - \delta_k\right)\right] + p_S \sum_{\substack{\ell,k \\ \ell=k}}^{L} \breve{\mathbf{w}}_\ell^* \breve{\mathbf{w}}_k\end{aligned} \tag{2.7.58}$$

Using the expansion

$$\exp(z) = 1 + z + \frac{z^2}{2!} + \frac{z^3}{3!} + \cdots \tag{2.7.59}$$

the first term on the RHS of (2.7.58) becomes

$$\begin{aligned}&p_S \sum_{\substack{\ell,k \\ \ell\neq k}}^{L} \breve{\mathbf{w}}_\ell^* \breve{\mathbf{w}}_k E\left[\exp -j\left(\delta_\ell - \delta_k\right)\right] \\ &= p_S \sum_{\ell\neq k}^{L} \breve{\mathbf{w}}_\ell^* \breve{\mathbf{w}}_k E\left[1 + j\left(\delta_k - \delta_\ell\right) - \frac{\left(\delta_k - \delta_\ell\right)^2}{2!} - \frac{j\left(\delta_k - \delta_\ell\right)^3}{3!} + \cdots\right]\end{aligned} \tag{2.7.60}$$

Assuming that the contribution of the higher-order terms is negligibly small, using (2.7.47) and (2.7.48), (2.7.60) results in

$$p_s\sum_{\ell\neq k}\breve{\mathbf{w}}_\ell^*\breve{\mathbf{w}}_k E[1+j(\delta_k-\delta_\ell)] = p_S\sum_{\ell\neq k}\breve{\mathbf{w}}_\ell^*\breve{\mathbf{w}}_k(1-\sigma^2)$$

$$= p_S(1-\sigma^2)\sum_{\ell,k}\breve{\mathbf{w}}_\ell^*\breve{\mathbf{w}}_k - p_S(1-\sigma^2)\sum_{\ell=k}\breve{\mathbf{w}}_\ell^*\breve{\mathbf{w}}_k \tag{2.7.61}$$

$$= p_S(1-\sigma^2)\breve{\mathbf{w}}^H\mathbf{1}\mathbf{1}^T\breve{\mathbf{w}} - p_S(1-\sigma^2)\hat{\beta}$$

Noting that the second term on the RHS of (2.7.58) is $p_S\hat{\beta}$, and the fact that $\breve{\mathbf{w}}^H\mathbf{1} = 1$, (2.7.58) and (2.7.61) yield

$$\breve{P}_S = p_S - p_S\sigma^2(1-\hat{\beta}) \tag{2.7.62}$$

Note that in the absence of directional interferences $\hat{\beta} = 1/L$, (2.7.62) becomes

$$\breve{P}_S = p_S - p_S\frac{L-1}{L}\sigma^2 \tag{2.7.63}$$

Thus, the output signal power of OPPE is suppressed. The suppression of the output signal power is proportional to the input signal power and random error variance. In the presence of directional interference $\hat{\beta}$ increases and thus the reduction in the signal power is less than otherwise. In other words, signal suppression is maximum in the absence of directional interference, and is given by the second term on the RHS of (2.7.63).

#### 2.7.3.3 *Residual Interference Power*

Rewrite (2.7.56) in the following form:

$$\breve{P}_I = \sum_{\ell,k}^{L}\breve{w}_\ell^* E[\Phi_{\ell\ell}^* R_{I\ell k}\Phi_{kk}]\breve{w}_k \tag{2.7.64}$$

Using (2.6.32), (2.7.49), and (2.7.50) in (2.7.64),

$$\begin{aligned}\breve{P}_I &= \sum_{\ell,k}^{L}\breve{w}_\ell^*\Phi_{0\ell\ell}^* R_{I\ell k}\Phi_{0kk}\breve{w}_k E[\exp(j(\delta_k-\delta_\ell))]\\ &= \sum_{\ell=k}^{L}\breve{w}_\ell^*\Phi_{0\ell\ell}^* R_{I\ell k}\Phi_{0kk}\breve{w}_k E[\exp(j(\delta_k-\delta_\ell))]\\ &\quad+\sum_{\ell\neq k}\breve{w}_\ell^*\Phi_{0\ell\ell}^* R_{I\ell k}\Phi_{0kk}\breve{w}_k E[\exp(j(\delta_k-\delta_\ell))]\end{aligned} \tag{2.7.65}$$

Noting that the diagonal entries of $R_I$ are the sum of all directional interference power $p_I$, the first term in the RHS of (2.7.65) reduces to $p_I\hat{\beta}$. Following steps (2.7.59) to (2.7.61), the second term in the RHS of (2.7.65) becomes $(1-\sigma^2)[\breve{\mathbf{w}}^H\Phi_0^H R_I\Phi_0\breve{\mathbf{w}} - p_I\hat{\beta}]$. Thus,

$$\breve{P}_I = \left(1-\sigma^2\right)\left[\breve{\mathbf{w}}^H \Phi_0^H R_I \Phi_0 \breve{\mathbf{w}} + \sigma^2 p_I \hat{\beta}\right] \tag{2.7.66}$$

Substituting for $\breve{\mathbf{w}}$ from (2.7.46) in (2.7.66), it follows that

$$\breve{P}_I = \hat{P}_I + \sigma^2\left(\hat{\beta} p_I - \hat{P}_I\right) \tag{2.7.67}$$

where $p_I$ is the total power of all directional interferences at the input of the processor and $\hat{P}_I$ is the residual interference power of the optimal processor given by (2.4.18).

### 2.7.3.4 Array Gain

In this section, the effect of random phase errors on the array gain of OPPE is examined. Let $\breve{SNR}_0$ be the output SNR of OPPE. Thus,

$$\breve{SNR}_0 = \frac{\breve{P}_S}{\breve{P}_N} \tag{2.7.68}$$

where

$$\breve{P}_N = \breve{P}_I + \breve{P}_n \tag{2.7.69}$$

is the total mean output noise power of OPPE.

Since the uncorrelated mean output noise power is not affected by the random phase errors, it follows from (2.7.54) that

$$\begin{aligned} \breve{P}_n &= \hat{P}_n \\ &= \sigma_n^2 \hat{\beta} \end{aligned} \tag{2.7.70}$$

Substituting from (2.7.70) and (2.7.67) in (2.7.69), using (2.4.12) and (2.4.35), after manipulation,

$$\breve{P}_N = \hat{P}_N\left[1+\sigma^2\left(\hat{\beta}\hat{G}-1\right)\right] \tag{2.7.71}$$

where $\hat{G}$ is the array gain of the optimal processor.

From (2.7.62), (2.7.68), and (2.7.71) it follows that

$$\breve{SNR}_0 = \frac{p_S}{\hat{P}_N}\left[\frac{1+\sigma^2\left(\hat{\beta}-1\right)}{1+\sigma^2\left(\hat{G}\hat{\beta}-1\right)}\right] \tag{2.7.72}$$

If $\breve{G}$ denotes the array gain of OPPE, then it follows from (2.7.72), (2.4.34), and (2.4.35) that

$$\breve{G} = \hat{G}\frac{1+\sigma^2\left(\hat{\beta}-1\right)}{1+\sigma^2\left(\hat{G}\hat{\beta}-1\right)} \tag{2.7.73}$$

**TABLE 2.4**

**Comparison of Steering Vector Errors and Phase Shifter Errors ***

| Type of Error | Phase-Shifter Error | Steering-Vector Error |
|---|---|---|
| Normalized output signal power | $1-\sigma^2\left(1-\hat{\beta}\right)$ | $1+\sigma_s^2\hat{\beta}$ |
| Normalized total output noise power | $1-\sigma^2\left(1-\hat{\beta}\hat{G}\right)$ | $1+\sigma_s^2\kappa$ |
| Normalized array gain | $\dfrac{1-\sigma^2\left(1-\hat{\beta}\right)}{1-\sigma^2\left(1-\hat{\beta}\hat{G}\right)}$ | $\dfrac{1+\sigma_s^2\hat{\beta}}{1+\sigma_s^2\kappa}$ |

* $\hat{\beta}$: Ratio of the uncorrelated noise at the output to the input of the optimal processor; $\hat{G}$: array gain of the optimal processor; $\sigma^2$: variance of the additive random phase shifter errors; $\sigma_s^2$: variance of the additive random steering vector errors, $\kappa = \dfrac{\mathrm{Tr}\left(R_N^{-1}\right)}{\left(\mathbf{S}_0^H R_N^{-1}\mathbf{S}_0\right)}$.

Let

$$\breve{G}_1 = \breve{G}\Big|_{\sigma=\sigma_1} \tag{2.7.74}$$

and

$$\breve{G}_2 = \breve{G}\Big|_{\sigma=\sigma_2} \tag{2.7.75}$$

A simple algebraic manipulation using (2.7.73) to (2.7.75) shows that for $\sigma_2 > \sigma_1$,

$$\breve{G}_2 < \breve{G}_1 \tag{2.7.76}$$

Thus, the array gain of the optimal processor with random phase errors is a monotonically decreasing function of the error variance.

#### *2.7.3.5 Comparison with SVE*

Now, a comparison between the effect of the random phase shifter errors and the effect of random SVE on optimal processor performance is made. SVE is discussed in Section 2.7.2.

Table 2.4 compares results. For purposes of the comparison, the results in both cases are normalized with corresponding error-free values and thus are referred to as normalized. The mean output signal power decreases with the increase in variance of phase shifter error if $\hat{\beta} < 1$, whereas in the case of SVE it is a monotonically increasing function of the variance of the errors. Note that for white noise only, $\hat{\beta} < 1/L$. The total mean output noise power is a monotonically increasing function of SVE variance, whereas it decreases with the increase in variance of the phase-shifter error if $\hat{\beta}\hat{G} < 1$. The array gains in both the cases are monotonically decreasing functions of the variance of random errors.

### 2.7.4 Phase Quantization Errors

In this section, a special case of random error, namely the phase quantisation error, which arises in digital phase shifters, is considered. In a p-bit phase shifter, the minimum value of phase that can be changed is $2\pi/2^p$. Thus, it is assumed that the error which exists in a p-bit digital phase shifter is uniformly distributed between $-\pi/2^p$ and $\pi/2^p$.

For a uniformly distributed random variable x in the interval (–C, C), it can easily be verified that

$$E\left[x^2\right]=\frac{C^2}{3} \tag{2.7.77}$$

Substituting for $C = \pi/2^p$ in (2.7.77), the variance $\sigma_p^2$ of the error in a p-bit phase shifter, is given by

$$\sigma_p^2=\frac{\pi^2}{3.2^{2p}} \tag{2.7.78}$$

Substituting $\sigma_p$ for $\sigma$ in expressions for the mean output signal power, mean output noise power, output SNR, and the array gain, the following expressions for these quantities are obtained as a function of the variance of the phase quantization error:

$$\breve{P}_S = p_S\left[1-\sigma_p^2\left(1-\hat{\beta}\right)\right] \tag{2.7.79}$$

$$\breve{P}_N = \hat{P}_N\left[1-\sigma_p^2\left(1-\hat{\beta}\hat{G}\right)\right] \tag{2.7.80}$$

$$\breve{SNR}_0 = \frac{p_S\left[1-\sigma_p^2\left(1-\hat{\beta}\right)\right]}{\hat{P}_N\left[1-\sigma_p^2\left(1-\hat{\beta}\hat{G}\right)\right]} \tag{2.7.81}$$

and

$$\breve{G} = \hat{G}\frac{\left[1-\sigma_p^2\left(1-\hat{\beta}\right)\right]}{\left[1-\sigma_p^2\left(1-\hat{\beta}\hat{G}\right)\right]} \tag{2.7.82}$$

### 2.7.5 Other Errors

Uncertainty about the position of an array element causes degradation in the array performance in general [Gof87, Kea80, She87, Ram80, Gil55], and particularly when the array beam pattern is determined by constrained beamforming. As discussed previously, element position uncertainty causes SVE, which in turn leads to a lower array gain. The effect of position uncertainty on the beam pattern is to create a background beam pattern similar to that of a single element, in addition to the normal pattern of the array [Gil55]. A general discussion on the effect of various errors on the array pattern is provided in [Ste76].

A calibration process is normally used to determine the position of an antenna element in an array. It requires auxiliary sources in known locations [Dor80]. A procedure that does not require the location of these sources is described in [Roc87, Roc87a].

The element failure tends to cause an increase in side levels and the weights estimated for the full array no longer remain optimal [She87]. This requires recalculation of the optimal weight with the known failed elements taken into account [She87, Ram80].

The effect of perturbation in the medium, which causes the wave front to deviate from the plane wave propagation assumption, and related topics are found in [Vur79, Hin80, Ste82]. The effect of a finite number of samples used in weight estimation is considered in [Bor80, Ber86, Rag92] and how bandwidth affects narrowband beamformer performance are discussed in [God86a, May79]. Effects of amplitude and phase errors on a mobile satellite communication system using a spherical array employing digital beamforming has also been studied [Chu92].

### 2.7.6 Robust Beamforming

The perturbation of many array parameters from ideal conditions under which the theoretical system performance is predicted, causes degradation in system performance by reducing the array gain and altering the beam pattern. Various schemes have been proposed to overcome these problems and to enhance array system performance operating under nonideal conditions [God87, Cox87, Eva82, Kim92, You93, Er85, Er93, Er93a, Er94, Tak86]. Many of these schemes impose various kinds of constraints on the beam pattern to alleviate the problem caused by parameter perturbation. A survey of robust signal-processing techniques in general is conducted in [Kas85]. It contains an excellent reference list and discusses various issues concerning robustness.

## Acknowledgments

Edited versions of Sections 2.6.3, 2.6.4, 2.6.5, 2.7.1, 2.7.2, and Table 2.1 are reprinted from Godara, L.C., Constrained beamforming and adaptive algorithms, in *Handbook of Statistics,* Vol. 10, Bose, N.K and Rao, C.R., Eds., Copyright 1993, with permission from Elsevier. Edited versions of Sections 2.1 to 2.5 are reprinted from Godara, L.C., Application of antenna arrays to mobile communications, I. Beamforming and DOA considerations, *Proc. IEEE,* 85(8), 1195-1247, 1997.

## Notation and Abbreviations

| | |
|---|---|
| E[.] | Expectation operator |
| **1** | vector of ones |
| $(.)^H$ | Hermitian transposition of vector or matrix (.) |
| $(.)^T$ | Transposition of vector or matrix (.) |
| A | Matrix of steering vectors |
| B | Matrix prefilter |
| c | Speed of propagation |
| CIB | conventional interference beamformer |
| d | element spacing |
| ESP | element space processor |
| $f_0$ | carrier frequency |
| $\hat{G}$ | array gain of optimal beamformer |

| | |
|---|---|
| $G_w$ | array gain of optimal beamformer with WVE |
| $G_s$ | array gain of NAMI beamformer with SVE |
| $\frown{G}_S$ | array gain of SPNMI processor with SVE |
| $\breve{G}$ | array gain of OPPE |
| IIB | improved interference beamformer |
| L | number of elements in array |
| MSE | mean square error |
| MMSE | minimum mean square error |
| M | number of directional sources |
| $m_k(t)$ | complex modulating function of kth source |
| $m_s(t)$ | complex modulating function of signal source |
| $m_I(t)$ | complex modulating function of interference source |
| NAMI | noise alone matrix inverse |
| $n_\ell(t)$ | random noise on $\ell$th antenna |
| $\mathbf{n}(t)$ | signal vector due to random noise |
| OIB | orthogonal interference beamformer |
| OPPE | optimal processor with phase errors |
| PIC | postbeamformer interference canceler |
| $p_k$ | power of kth source |
| $p_I$ | power of interference source |
| $p_S$ | power of signal source |
| $p_N$ | total noise at input |
| $P(\mathbf{w})$ | mean output power for given $\mathbf{w}$ |
| $\hat{P}$ | mean output power of optimal beamformer |
| $\hat{\underline{P}}$ | mean output power of optimal beamformer when known look direction is in error |
| $P(\hat{\mathbf{w}})$ | mean output power of optimal beam-space processor |
| $P(\hat{w})$ | mean output power of optimal PIC processor |
| $P_S$ | mean output signal power |
| $\hat{P}_S$ | mean output signal power of optimal beamformer |
| $\overline{P}_S$ | mean output signal power of optimal beamformer in presence of WVE |
| $\tilde{P}_S$ | mean output signal power of NAMI processor in presence of SVE |
| $\frown{P}_S$ | mean output signal power of SPNMI processor in presence of SVE |
| $\breve{P}_S$ | mean output signal power of OPPE |
| $P_S(\mathbf{w})$ | mean output signal power of beam-space processor for given $\mathbf{w}$ |
| $P_S(\hat{w})$ | mean output signal power of optimal PIC processor with weight $\hat{w}$ |
| $P_I$ | mean output interference power |
| $\hat{P}_I$ | mean output interference power of optimal beamformer |
| $\check{P}_I$ | mean output interference power of OPPE |
| $P_I(\hat{w})$ | mean output interference power of optimal PIC processor with weight $\hat{w}$ |
| $P_n$ | mean output uncorrelated noise power |

| | |
|---|---|
| $\hat{P}_n$ | mean output uncorrelated noise power of optimal beamformer |
| $\breve{P}_n$ | mean output uncorrelated noise power of OPPE |
| $P_n(\hat{w})$ | mean output uncorrelated noise power of optimal PIC processor with weight $\hat{w}$ |
| $P_N$ | mean output noise power |
| $\hat{P}_N$ | mean output noise power of optimal beamformer |
| $\bar{P}_N$ | mean output noise power of optimal processor in presence of WVE |
| $\tilde{P}_N$ | mean output noise power of NAMI processor in presence of SVE |
| $\overset{\frown}{P}_N$ | mean output noise power of SPNMI processor in presence of SVE |
| $\breve{P}_N$ | mean output noise power of OPPE |
| $P_N(\mathbf{w})$ | mean output noise power of beam-space processor for given $\mathbf{w}$ |
| $P_N(\hat{w})$ | mean output noise power of optimal PIC processor with weight $\hat{w}$ |
| $P_0$ | mean power of main beam |
| Q | matrix of eigenvectors |
| $\mathbf{q}(t)$ | outputs of $M - 1$ auxiliary beams |
| $q(t)$ | output of interference beam |
| $r(t)$ | reference signal |
| $\mathbf{r}_\ell$ | position vector of $\ell$th antenna |
| R | array correlation matrix |
| $R_N$ | noise-only array correlation matrix |
| $R_n$ | random noise-only array correlation matrix |
| $R_I$ | interference-only array correlation matrix |
| RMS | root mean square |
| $R_S$ | signal-only array correlation matrix |
| $R_T$ | array correlation matrix used in tamed array |
| $R_{\mathbf{qq}}$ | correlation matrix of auxiliary beams |
| $\tilde{R}$ | array correlation matrix after steering delays |
| $\mathrm{R}\!\!\!-$ | actual array correlation matrix when known look direction is in error |
| SPNMI | signal-plus-noise matrix inverse |
| SNR | signal-to-noise ratio |
| $SNR(\hat{\mathbf{w}})$ | signal-to-noise ratio of optimal beam-space processor |
| $SNR(\hat{w})$ | signal-to-noise ratio of optimal PIC processor with weight $\hat{w}$ |
| $\breve{SNR}_0$ | SNR of OPPE |
| SVE | steering vector error |
| $s(t)$ | signal induces on reference element |
| $s(t - \tilde{T})$ | signal delayed by $\tilde{T}$ |
| S | source correlation matrix |
| $\mathbf{S}_k$ | steering vector associated with kth source |
| $\mathbf{S}_0$ | steering vector associated with known look direction |
| $\mathbf{S}\!\!\!-_0$ | steering vector associated with actual look direction when known look direction is in error |

| | |
|---|---|
| $\bar{\mathbf{S}}$ | steering vector associated with known look direction in presence of SVE |
| $\mathbf{S}_I$ | steering vector associated with interference |
| $\mathbf{S}(\phi,\theta)$ | steering vector associated with direction $(\phi,\theta)$ |
| $\tilde{T}$ | delay time |
| $Tr[\cdot]$ | Trace of $[\cdot]$ |
| $\mathbf{U}$ | weight vector of interference beam of PIC |
| $\mathbf{U}_o$ | weight vector of interference beam of PIC with OIB |
| $\mathbf{U}_\ell$ | eigenvector associated with $\ell$th eigenvalue |
| $\mathbf{V}$ | main beam weight vector |
| $\hat{\mathbf{v}}(\phi_k, \theta_k)$ | unit vector in direction $(\phi_k,\theta_k)$ |
| WVE | weight vector error |
| $\hat{w}$ | weight of optimal PIC |
| $\hat{w}_c$ | weight of optimal PIC using CIB |
| $\hat{w}_I$ | weight of optimal PIC using IIB |
| $\hat{w}_o$ | weight of optimal PIC using OIB |
| $\hat{\hat{w}}_o$ | weight of optimal PIC using OIB when known look direction is in error |
| $\mathbf{w}_c$ | weights of conventional beamformer |
| $w_\ell$ | weight on $\ell$th antenna |
| $\mathbf{w}$ | weight vector |
| $\hat{\mathbf{w}}$ | weights of optimal beamformer |
| $\hat{\hat{\mathbf{w}}}$ | weights of optimal beamformer when known look direction is in error |
| $\bar{\mathbf{w}}$ | weights of optimal beamformer in presence of weight errors |
| $\tilde{\mathbf{w}}$ | weights of NAMI processor in presence of SVE |
| $\widehat{\mathbf{w}}$ | weights of SPNMI processor in presence of SVE |
| $\breve{\mathbf{w}}$ | weights of optimal beamformer using phase shifters to steer array |
| $\hat{\mathbf{w}}_{MSE}$ | optimal weights of beamformer using reference signal |
| $x_\ell(t)$ | signal induced on $\ell$th antenna |
| $\mathbf{x}(t)$ | element signal vector |
| $\mathbf{x}'(t)$ | element signal vector after presteering delay |
| $\mathbf{x}_S(t)$ | element signal vector due to desired signal source |
| $\mathbf{x}_I(t)$ | element signal vector due to interference source |
| $y(t)$ | array output |
| $y_S(t)$ | signal component in array output |
| $y_I(t)$ | interference component in array output |
| $y_n(t)$ | random noise component in array output |
| $y(\phi,\theta)$ | response of a beamformer in $(\phi,\theta)$ |
| $\mathbf{z}$ | correlation between reference signal and $\mathbf{x}(t)$ |
| $\mathbf{Z}$ | correlation between outputs of auxiliary beams and main beam |
| $\alpha_\ell$ | phase delays on $\ell$th channel to steer array in look direction, phase delays of error-free phase shifter on $\ell$th channel |
| $\tilde{\alpha}_\ell$ | phase delays of actual phase shifter (including error) on $\ell$th channel |

| | |
|---|---|
| $\hat{\alpha}$ | SNR of optimal beamformer |
| $\alpha_w$ | output SNR of optimal beamformer with WVE |
| $\alpha_s$ | output SNR of NAMI processor with SVE |
| $\hat{\alpha}_s$ | output SNR of SPNMI processor with SVE |
| $\alpha_0^2$ | control variable used in tamed arrays |
| $\alpha_I$ | SNR at output of interference beam of PIC processor |
| $\hat{\beta}$ | ratio of uncorrelated noise power at out of optimal beamformer to input uncorrelated noise power |
| $\beta$ | normalized dot product of $\mathbf{S}_0$ and $\mathbf{S}_I$ |
| $\beta_p$ | phase of parameter $\beta$ |
| $\beta_o$ | Euclidian norm of $\mathbf{U}_o$ |
| $\gamma_o$ | normalized power response of interference beam in interference direction |
| $\delta_\ell$ | phase error in $\ell$th phase shifter |
| $\varepsilon(t)$ | error signal |
| $\varepsilon_0$ | ratio of uncorrelated noise to interference power at input of beamformer |
| $\xi(\mathbf{w})$ | MSE for given $\mathbf{w}$ |
| $\hat{\xi}$ | minimum MSE |
| $\kappa$ | scalar parameter defined by (2.7.20) |
| $\psi(t)$ | output of main beam |
| $\eta(t)$ | output of interference beam |
| $\mu_0$ | scalar constant |
| $\rho$ | scalar parameter function of array geometry, $\theta_S$ and $\theta_I$ |
| $\mathbf{\Gamma}$ | vector of random errors in weights |
| $\mathbf{\Gamma}_S$ | vector of random errors in steering vectors |
| $\Lambda$ | diagonal matrix of eigenvalues |
| $\lambda_\ell$ | $\ell$th eigenvalue of array correlation matrix |
| $\sigma_n^2$ | power of random noise induced on element |
| $\sigma_w^2$ | variance of weight vector errors |
| $\sigma_s^2$ | variance of steering vector errors |
| $\sigma^2$ | variance of phase shifter errors |
| $\sigma_p^2$ | variance of phase error in p-bit phase shifter |
| $(\phi_k,\theta_k)$ | direction of kth source using three-dimensional notation |
| $(\phi_0,\theta_0)$ | look direction using three-dimensional notation |
| $(\theta_k)$ | direction of kth source using two-dimensional notation |
| $\theta_I$ | direction of interference source using two-dimensional notation |
| $\theta_S$ | direction of signal source using two-dimensional notation |
| $\tau_\ell(\phi_k,\theta_k)$ | propagation delay on $\ell$th antenna from source in $(\phi_k,\theta_k)$ |
| $\tau_\ell(\theta_k)$ | propagation delay on $\ell$th antenna from source in $(\theta_k)$ |
| $\Phi_0$ | diagonal matrix of error-free phase delays |
| $\Phi$ | diagonal matrix of phase delays (including errors) |

## References


And69 Anderson, V.C. and Rudnick, P., Rejection of a coherent arrival at an array, *J. Acoust. Soc. Am.*, 45, 406–410, 1969.

And69a Anderson, V.C., DICANNE, a realizable adaptive process, *J. Acoust. Soc. Am.*, 398–405, 1969.

And91 Anderson, S. et al., An adaptive array for mobile communication systems, *IEEE Trans. Vehicular Technol.*, 40, 230–236, 1991.

Ard88 Ardalan, S., On the sensitivity of transversal RLS algorithms to random perturbations in the filter coefficients, *IEEE Trans. Acoust. Speech Signal Process.*, 36, 1781–1783, 1988.

App76 Applebaum, S.P., Adaptive arrays, *IEEE Trans. Antennas Propagat.*, 24, 585–598, 1976.

App76a Applebaum, S.P. and Chapman, D.J., Adaptive arrays with main beam constraints, *IEEE Trans. Antennas Propagat.*, AP-24, 650–662, 1976.

Ber77 Berni, A.J., Weight jitter phenomena in adaptive array control loop, *IEEE Trans. Aerosp. Electron. Syst.*, 13, 355–361, 1977.

Ber86 Bershad, N.J., On the optimum data nonlinearity in LMS adaptation, *IEEE Trans. Acoust. Speech Signal Process.*, 34, 69–76, 1986.

Bor80 Boroson, D.M., Sample size considerations for adaptive arrays, *IEEE Trans. Aerosp. Electron. Syst.*, 16, 446–451, 1980.

Bre88 Bresler, Y., Reddy, V.U. and Kailath, T., Optimum beamforming for coherent signal and interferences, *IEEE Trans. Acoust. Speech Signal Process.*, 36, 833–843, 1988.

Bre73 Brennan, L.E. and Reed, I.S., Theory of adaptive radar, *IEEE Trans. Aerosp. Electron. Syst.*, 9, 237–252, 1973.

Bro86 Brookner, E. and Howell, J.M., Adaptive-adaptive array processing, *IEEE Proc.*, 74, 602–604, 1986.

Can82 Cantoni, A. and Godara, L.C., Fast algorithms for time domain broadband adaptive array processing, *IEEE Trans. Aerosp. Electron. Syst.*, 18, 682–699, 1982.

Can84 Cantoni, A. and Godara, L.C., Performance of a postbeamformer interference canceller in the presence of broadband directional signals, *J. Acoust. Soc. Am.*, 76, 128–138, 1984.

Cap69 Capon, J., High-resolution frequency-wave number spectrum analysis, *IEEE Proc.*, 57, 1408–1418, 1969.

Cha76 Chapman, D.J., Partial adaptivity for the large array, *IEEE Trans. Antennas Propagat.*, 24, 685–696, 1976.

Chi94 Chiba, I., Takahashi, T. and Karasawa, Y., Transmitting null beam forming with beam space adaptive array antennas, in *Proceedings of IEEE 44th Vehicular Technology Conference*, 1498–1502, 1994.

Cho92 Choi, S. and Sarkar, T.K., Adaptive antenna array utilizing the conjugate gradient method for multipath mobile communication, *Signal Process.*, 29, 319–333, 1992.

Cho93 Choi, S., Sarkar, T.K. and Lee, S.S., Design of two-dimensional Tseng window and its application to antenna array for the detection of AM signal in the presence of strong jammers in mobile communications, *Signal Process.*, 34, 297–310, 1993.

Chu92 Chujo, W. and Kashiki, K., Spherical array antenna using digital beamforming techniques for mobile satellite communications, *Electronics & Communications in Japan, Part I: Communications*, 75, 76–86, 1992 (English translation of *Denshi Tsushin Gakkai Ronbunshi*).

Com79 Compton Jr., R.T., The power-inversion adaptive array: concepts and performance, *IEEE Trans. Aerosp. Electron. Syst.*, 15, 803–814, 1979.

Com82 Compton Jr., R.T., The effect of random steering vector errors in the Applebaum adaptive array, *IEEE Trans. Aerosp. Electron. Syst.*, 18, 392–400, 1982.

Cox73 Cox, H., Resolving power and sensitivity to mismatch of optimum array processors, *J. Acoust. Soc. Am.*, 54, 771–785, 1973.

Cox87 Cox, H., Zeskind, R.M. and Owen, M.M., Robust adaptive beamforming, *IEEE Trans. Acoust. Speech Signal Process.*, 35, 1365–1376, 1987.

Cox 88 Cox, H., Zeskind, R.M. and Owen, M.M., Effects of amplitude and phase errors on linear predictive array processors, *IEEE Trans. Acoust. Speech Signal Process.*, 36, 10–19, 1988.

dAs84 d'Assumpcao, H.A. and Mountford, G.E., An overview of signal processing for arrays of receivers, *J. Inst. Eng. Aust. IREE Aust.*, 4, 6–19, 1984.

Dav67 Davies, D.E.N., Independent angular steering of each zero of the directional pattern for a linear array, *IEEE Trans. Antennas Propagat. (Commn.)*, 15, 296–298, 1967.

Dio93 Diouris, J.F., Feuvrie, B. and Saillard, J., Adaptive multisensor receiver for mobile communications, *Ann. Telecommn.*, 48, 35–46, 1993.

DiC78 DiCarlo, D.M. and Compton Jr., R.T., Reference loop phase shift in adaptive arrays, *IEEE Trans. Aerosp. Electron. Syst.*, 14, 599–607, 1978.

Dol46 Dolf, C. L., A current distribution for broadside arrays which optimizes the relationship between beamwidth and sidelobe levels, *IRE Proc.*, 34, 335–348, 1946.

Dor80 Dorny, C.N. and Meaghr, B.S., Cohering of an experimental nonrigid array by self-survey, *IEEE Trans. Antennas Propagat.*, 28, 902–904, 1980.

Er85 Er, M.H. and Cantoni, A., An alternative formulation for an optimum beamformer with robustness capability, *IEE Proc.*, 132, Pt. F, 447–460, 1985.

Er93 Er, M.H., Sim, S.L. and Koh, S.N., Application of constrained optimization techniques to array pattern synthesis, *Signal Process.*, 34, 327–334, 1993.

Er93a Er, M.H. and Ng, B.C., A robust method for broadband beamforming in the presence of pointing error, *Signal Process.*, 30, 115–121, 1993.

Er94 Er, M.H. and Ng, B.C., A new approach to robust beamforming in the presence of steering vector errors, *IEEE Trans. Signal Process.*, 42, 1826–1829, 1994.

Eva82 Evans, R.J. and Ahmed, K.M., Robust adaptive array antennas, *J. Acoust. Soc. Am.*, 71, 384–394, 1982.

Far83 Farrier, D.R., Gain of an array of sensors subjected to processor perturbation, *IEE Proc.*, 130, Pt. H, 251–254, 1983.

Fli94 Flieller, A., Larzabal, P. and Clergeot, H., Applications of high resolution array processing techniques for mobile communication system, in *Proceedings of IEEE Intelligent Vehicles Symposium*, 606–611, 1994.

Fri89 Friedlander, B. and Porat, B., Performance analysis of a null-steering algorithm based on direction-of-arrival estimation, *IEEE Trans. Acoust. Speech, Signal Process.*, 37, 461–466, 1989.

Gab76 Gabriel, W.F., Adaptive arrays: an introduction, *IEEE Proc.*, 64, 239–272, 1976.

Gab86 Gabriel, W.F., Using spectral estimation techniques in adaptive processing antenna systems, *IEEE Trans. Antennas Propagat.*, 34, 291–300, 1986.

Geb95 Gebauer, T. and Gockler, H.G., Channel-individual adaptive beamforming for mobile satellite communications, *IEEE J. Selected Areas Commn.*, 13, 439–48, 1995.

Gil55 Gilbert, E.N. and Morgan, S.P., Optimum design of directive antenna arrays subject to random variations, *Bell Syst. Technol. J.*, 431–663, 1955.

Gob76 Gobert, J., Adaptive beam weighting, *IEEE Trans. Antennas Propagat.*, 24, 744–749, 1976.

God81 Godara, L.C. and Cantoni, A. Uniqueness and linear independence of steering vectors in array space, *J. Acoust. Soc. Am.*, 70, 467–475, 1981.

God85 Godara, L.C., The effect of phase-shifter errors on the performance of an antenna-array beamformer, *IEEE J. Oceanic Eng.*, 10, 278–284, 1985.

God86 Godara, L.C., Error analysis of the optimal antenna array processors, *IEEE Trans. Aerosp. Electron. Syst.*, 22, 395–409, 1986.

God86a Godara, L.C. and Cantoni, A., The effect of bandwidth on the performance of post beamformer interference canceller, *J. Acoust. Soc. Am.*, 80, 794–803, 1986.

God87 Godara, L.C., A robust adaptive array processor, *IEEE Trans. Circuits Syst.*, 34, 721–730, 1987.

God89 Godara, L.C., Analysis of transient and steady state weight covariance in adaptive postbeamformer interference canceller, *J. Acoust. Soc. Am.*, 85, 194–201, 1989.

God89a Godara, L.C., Postbeamformer interference canceller with improved performance, *J. Acoust. Soc. Am.*, 85, 202–213, 1989.

God91 Godara, L.C., Adaptive postbeamformer interference canceller with improved performance in the presence of broadband directional sources, *J. Acoust. Soc. Am.*, 89, 266–273, 1991.

God93 Godara, L.C., Constrained beamforming and adaptive algorithms, in *Handbook of Statistics*, Vol. 10, N.K. Bose and C.R. Rao, Eds., Elsevier, Amsterdam, New York, 1993, pp. 269–354.

God97 Godara, L.C., Application to antenna arrays to mobile communications. Part II: Beamforming and direction of arrival considerations, *Proc. IEEE*, 85, 1195–1247, 1997.

God02 Godara, L.C., Ed., *Handbook of Antennas in Wireless Communications*, CRC Press, Boca Raton, FL, 2002.

Gof87 Goffer, A. and Langholz, G., The performance of adaptive array in a dynamic environment, *IEEE Trans. Aerosp. Electron. Syst.*, 23, 485–492, 1987.

Gri67 Griffiths, L.J., A comparison of multidimensional Weiner and maximum-likelihood filters for antenna arrays, *IEEE Proc. Lett.*, 55, 2045–2047, 1967.

Gri77 Griffiths, L.J. An adaptive beamformer which implements constraints using an auxiliary array processor, in *Aspects of Signal Processing*, Part 2, G. Tacconi, Ed., Reidel, Dordrecht, 517–522, 1977.

Gri82 Griffiths, L.J. and Jim, C.W., An alternative approach to linearly constrained adaptive beamforming, *IEEE Trans. Antennas Propagat.*, 30, 27–34, 1982.

Gri87 Griffiths, L.J. and Buckley, K.M., Quiescent pattern control in linearly constrained adaptive arrays, *IEEE Trans. Acoust. Speech Signal Process.*, 35, 917–926, 1987.

Gup82 Gupta, I.J. and Ksienski, A.A., Dependence of adaptive array performance on conventional array design, *IEEE Trans. Antennas Propagat.*, 30, 549–553, 1982.

Gup84 Gupta, I.J., Effect of jammer power on the performance of adaptive arrays, *IEEE Trans. Antennas Propagat.*, 32, 933–934, 1984.

Hin80 Hinich, M.J., Beamforming when the sound velocity is not precisely known, *J. Acoust. Soc. Am.*, 68, 512–515, 1980.

Hon87 Hong, Y.J., Ucci, D.R. and Yeh, C.C., The performance of an adaptive array combining reference signal and steering vector, *IEEE Trans. Antennas Propagat.*, 35, 763–770, 1987.

How76 Howells, P.W., Explorations in fixed and adaptive resolution at GE and SURC, *IEEE Trans. Antennas Propagat.*, 24, 575–584, 1976.

Hud77 Hudson, J.E., The effect of signal and weight coefficient quantization in adaptive array processors, in *Aspects of Signal Process*ing, Part 2, G. Tacconi, Ed., Reidel, Dordrecht, 423–428, 1977.

Jab86 Jablon, N.K., Adaptive beamforming with the generalized sidelobe canceller in the presence of array imperfections, *IEEE Trans. Antennas Propagat.*, 34, 996–1012, 1986.

Jim77 Jim, C.W., A comparison of two LMS constrained optimal array structures, *IEEE Proc.*, 65, 1730–1731, 1977.

Jon95 Jones, M.A. and Wickert, M. A., Direct sequence spread spectrum using directionally constrained adaptive beamforming to null interference, *IEEE J. Selected Areas Comm.*, 13, 71–79, 1995.

Kas85 Kassam, S.A. and Poor, H.V., Robust techniques for signal processing: a survey, *IEEE Proc.*, 73, 433–481, 1985.

Kea80 Keating, P.N., The effect of errors on frequency domain adaptive interference rejection, *J. Acoust. Soc. Am.*, 68, 1690–1695, 1980.

Kim92 Kim, J.W. and Un, C.K., An adaptive array robust to beam pointing error, *IEEE Trans. Signal Process.*, 40, 1582–1584, 1992.

Kle75 Klemm, R., Suppression of jammers by multiple beam signal processing, *IEEE Int. Radar Conf.*, 176–180, 1975.

Kle80 Kleinberg, L.I., Array gain for signals and noise having amplitude and phase fluctuations, *J. Acoust. Soc. Am.*, 67, 572–577, 1980.

Lan85 Lang, R.H., Din, M.A.B. and Pickhotz, R.L., Stochastic effects in adaptive null-steering antenna array performance, *IEEE J. Selected Areas Comm.*, 3, 767–778, 1985.

Lo90 Lo, K.W., Reducing the effect of pointing errors on the performance of an adaptive array, *IEE Electron. Lett.*, 26, 1646–1647, 1990.

Mai82 Maillous, R.J., Phased array theory and technology, *IEEE Proc.*, 70, 246–291, 1982.

May78 Mayhan, J.T., Adaptive nulling with multiple beam antennas, *IEEE Trans. Antennas Propagat.*, 26, 267–273, 1978.

May79 Mayhan, J.T., Some techniques for evaluating the bandwidth characteristics of adaptive nulling systems, *IEEE Trans. Antennas Propagat.*, 27, 363–378, 1979.

Mor78 Morgan, D.R., Partially adaptive array techniques, *IEEE Trans. Antennas Propagat.*, 26, 823–833, 1978.

Muc81 Mucci, R.A. and Pridham, R.G., Impact of beam steering errors on shifted sideband and phase shift beamforming techniques, *J. Acoust. Soc. Am.*, 69, 1360–1368, 1981.

Nag94 Naguib, A.F., Paulraj, A. and Kailath, T., Capacity improvement with base-station antenna arrays in cellular CDMA, *IEEE Trans. Vehicular Technol.*, 43, 691–698, 1994.

Nag94a Naguib, A.F. and Paulraj, A. Performance of CDMA cellular networks with base-station antenna arrays, in *Proceedings of IEEE International Zurich Seminar on Communications*, 87–100, 1994.

Nit76 Nitzberg, R., Effect of errors in adaptive weights, *IEEE Trans. Aerosp. Electron. Syst.*, 12, 369–373, 1976.

Ng02 Ng, B.P. and Er, M.H., Basic array theory and pattern synthesis techniques, in Godara, L.C., Ed., *Handbook of Antennas in Wireless Communications*, CRC Press, Boca Raton, FL, 2002, pp. 16.1–16.42.

Ows73 Owsley, N.L., A recent trend in adaptive spatial processing for sensor arrays: Constrained adaptation, in *Signal Processing*, J.W.R. Griffiths et al., Eds., Academic Press, New York, 1973.

Pon96 Ponnekanti, S. and Sali, S., Effective adaptive antenna scheme for mobile communications, *IEE Electron. Lett.*, 32, 417–418, 1996.

Qua82 Quazi, A.H., Array beam response in the presence of amplitude and phase fluctuations, *J. Acoust. Soc. Am.*, 72, 171–180, 1982.

Qia94 Qian, F. and Van Veen, B.D., Partially adaptive beamformer design subject to worst case performance constraints, *IEEE Trans. Signal Process.*, 42, 1218–1221, 1994.

Qia95 Qian, F. and Van Veen, B.D., Partially adaptive beamforming for correlated interference rejection, *IEEE Trans. Signal Process.*, 43, 506–515, 1995.

Rag92 Raghunath, K.J. and Reddy, U.V., Finite data performance analysis of MVDR beamformer with and without spatial smoothing, *IEEE Trans. Acoust. Speech Signal Process.*, 40, 2726–2736, 1992.

Ram80 Ramsdale, D.J. and Howerton, R.A., Effect of element failure and random errors in amplitude and phase on the sidelobe level attainable with a linear array, *J. Acoust. Soc. Am.*, 68, 901–906, 1980.

Ree74 Reed, I.S., Mallett, J.D. and Brennan, L.E., Rapid convergence rate in adaptive arrays, *IEEE Trans. Aerosp. Electron. Syst.*, 10, 853–863, 1974.

Roc87 Rockah, Y. and Schultheiss, P.M., Array shape calibration using sources in unknown locations. Part 1: Far field sources, *IEEE Trans. Acoust. Speech Signal Process.*, 35, 286–299, 1987.

Roc87a Rockah, Y. and Schultheiss, P.M., Array shape calibration using sources in unknown locations. Part 2: Near field sources and estimator implementation, *IEEE Trans. Acoust. Speech Signal Process.*, 35, 724–735, 1987.

Sak92 Sakagami, S., et al., Vehicle position estimates by multibeam antennas in multipath environments, *IEEE Trans. Vehicular Technol.*, 41, 63–68, 1992.

She87 Shernill, M.S. and Streit, L.R., In situ optimal reshading of arrays with failed elements, *IEEE J. Oceanic Eng.*, 12, 155–162, 1987.

Sim83 Simaan, M., Optimum array filters for array data signal processing, *IEEE Trans. Acoust. Speech Signal Process.*, 31, 1006–10015, 1983.

Sta70 Stark, L., Burns, R.W. and Clark, W.P., Phase shifters for arrays, in *Radar Hand Book*, M.I. Skolnik, Ed., McGraw-Hill, New York, 1970.

Ste76 Steinberg, B.D., *Principles of Aperture and Array System Design*, Wiley, New York, 1976.

Ste82 Steinberg, B.D. and Luthra, A.J., Simple theory of the effects of medium turbulence upon scanning with an adaptive phased array, *J. Acoust. Soc. Am.*, 71, 630–634, 1982.

Ste83 Steel, A.K., Comparison of directional and derivative constraints for beamformers subject to multiple linear constraints, *IEEE Proc. H*, 130, 41–45, 1983.

Su86 Su, Y.L., Shan, T.J. and Widrow, B., Parallel spatial processing: a cure for signal cancellation in adaptive arrays, *IEEE Trans. Antennas Propagat.*, 34, 347–355, 1986.

Sua93 Suard, B., et al., Performance of CDMA mobile communication systems using antenna arrays, *IEEE Int. Conf. Acoust., Speech, Signal Process.*, 153–156, 1993.

Tak86 Takao, K. and Kikuma, N., Tamed adaptive antenna arrays, *IEEE Trans. Antennas Propagat.*, 34, 388–394, 1986.

Tse92 Tseng, C.Y. and Griffiths, L.J., A simple algorithm to achieve desired patterns for arbitrary arrays, *IEEE Trans. Signal Process.*, 40, 2737–2746, 1992.

Van87 Van Veen, B.D. and Roberts, R.A., Partially adaptive beamformer design via output power minimization, *IEEE Trans. Acoust. Speech Signal Process.*, 35, 1524–1532, 1987.

Van89 Van Veen, B.D., An analysis of several partially adaptive beamformer designs, *IEEE Trans. Acoust. Speech Signal Process.*, 37, 192–203, 1989.

Van90 Van Veen, B.D., Optimization of quiescent response in partially adaptive beamformers, *IEEE Trans. Acoust. Speech Signal Process.*, 38, 471–477, 1990.

Van91 Van Veen, B.D., Adaptive convergence of linearly constrained beamformers based on the sample covariance matrix, *IEEE Trans. Signal Process.*, 39, 1470–1473, 1991.

Vur79 Vural, A.M., Effects of perturbations on the performance of optimum/adaptive arrays, *IEEE Trans. Aerosp. Electron. Syst.*, 15, 76–87, 1979.

Web90 Webster, R.J. and Lang, T.N., Prescribed sidelobes for the constant beam width array, *IEEE Trans. Acoust. Speech Signal Process.*, 38, 727–730, 1990.

Wid67 Widrow, B., et al., Adaptive antenna systems, *IEEE Proc.*, 55, 2143–2158, 1967.

Wid75 Widrow, B., et al., Adaptive noise canceling: principles and applications, *IEEE Proc.*, 63, 1692–1716, 1975.

Wid82 Widrow, B. et al., Signal cancellation phenomena in adaptive antennas: causes and cures, *IEEE Trans. Antennas Propagat.*, 30, 469–478, 1982.

Win84 Winters, J.H., Optimum combining in digital mobile radio with cochannel interference, *IEEE J. Selected Areas Commn.*, 2, 528–539, 1984.

Win87 Winters, J.H., Optimum combining for indoor radio systems with multiple users, *IEEE Trans. Commn.*, 35, 1222–1230, 1987.

Win94 Winters, J.H., Salz, J. and Gitlin, R.D., The impact of antenna diversity on the capacity of wireless communication systems, *IEEE Trans. Commn.*, 42, 1740–51, 1994.

You93 Youn, W.S. and Un, C.K., A linearly constrained beamforming robust to array imperfections, *IEEE Trans. Signal Process.*, 41, 1425–1428, 1993.

Zah72 Zahm, C.L., Effects of errors in the direction of incidence on the performance of an antenna array, *IEEE Proc.*, 1008–1009, 1972.

Zah73 Zahm, C.L., Application of adaptive arrays to suppress strong jammers in the presence of weak signals, *IEEE Trans. Aerosp. Electron. Syst.*, 9, 260–271, 1973.

# 3

# *Adaptive Processing*

The weights of an element-space antenna array processor that has a unity response in the look direction and maximizes the output SNR in the absence of errors are given by

$$\hat{\mathbf{w}} = \frac{R_N^{-1}\mathbf{S}_0}{\mathbf{S}_0^H R_N^{-1}\mathbf{S}_0} \tag{3.1}$$

where $R_N$ is the array correlation matrix with no signal present, and is referred to as the noise-only array correlation matrix, and $\mathbf{S}_0$ is the steering vector associated with the look direction. When the noise-only array correlation matrix is not available, the array correlation matrix R is used to calculate the optimal weights. For this case the expression becomes

$$\hat{\mathbf{w}} = \frac{R^{-1}\mathbf{S}_0}{\mathbf{S}_0^H R^{-1}\mathbf{S}_0} \tag{3.2}$$

The weights of the processor that minimizes the mean square error (MSE) between the array output and a reference signal are given by

$$\hat{\mathbf{w}}_{MSE} = R^{-1}\mathbf{z} \tag{3.3}$$

where $\mathbf{z}$ denotes the correlation between the reference signal and the array signals vector $\mathbf{x}(t)$.

In practice, neither the array correlation matrix nor the noise-alone matrix is available to calculate optimal weights of the array. Thus, the weights are adjusted by some other means using the available information derived from the array output, array signals, and so on to make an estimate of the optimal weights. There are many such schemes and these are normally referred to adaptive algorithms. Some are described in this chapter, and their characteristics such as the speed of adaption, mean and variance of estimated weights, and parameters affecting these characteristics are discussed. Both element space and beam space processors are considered.

## 3.1 Sample Matrix Inversion Algorithm

This algorithm estimates array weights by replacing correlation matrix R by its estimate [God97]. An unbiased estimate of R using N samples of the array signals may be obtained using a simple averaging scheme as follows:

$$\hat{R}(N) = \frac{1}{N}\sum_{n=0}^{N-1} \mathbf{x}(n)\,\mathbf{x}^H(n) \tag{3.1.1}$$

where $\hat{R}(N)$ denotes the estimated array correlation matrix using N samples, and $\mathbf{x}(n)$ denotes the array signal sample also known as the array snapshot at the nth instant of time with t replaced by nT with T denoting the sampling time. The sampling time T has been omitted for ease of notation.

Let $\hat{R}(n)$ denote the estimate of array correlation matrix and $\mathbf{w}(n)$ denote the array weights at the nth instant of time. The estimate of R may be updated when the new samples arrive using

$$\hat{R}(n+1) = \frac{n\hat{R}(n) + \mathbf{x}(n+1)\,\mathbf{x}^H(n+1)}{n+1} \tag{3.1.2}$$

and a new estimate of the weights $\mathbf{w}(n + 1)$ at time instant n + 1 may be made.

Let P(n) denote the output power at the nth instant of time given by

$$P(n) = \mathbf{w}^H(n)\mathbf{x}(n)\,\mathbf{x}^H(n)\mathbf{w}(n) \tag{3.1.3}$$

When N samples are used to estimate the array correlation matrix and the processor has K degree of freedom the mean output power is given by [Van91]

$$E[P(n)] = \frac{N-K}{N}\hat{P} \tag{3.1.4}$$

where $\hat{P}$ denotes the mean output power of the processor with the optimal weights, that is,

$$\hat{P} = \hat{\mathbf{w}}^H R \hat{\mathbf{w}} \tag{3.1.5}$$

The factor (N – K)/N represents the loss due to estimate of R and determines the convergence behavior of the mean output power.

It should be noted that as the number of samples grows, the matrix update approaches its true value and thus the estimated weights approaches optimal weights, that is, as $n \to \infty$, $R(n) \to R$ and $\hat{\mathbf{w}}(n) \to \hat{\mathbf{w}}$ or $\hat{\mathbf{w}}_{MSE}$, as the case may be.

The expression of optimal weights requires the inverse of the array correlation matrix, and this process of estimating R and then its inverse may be combined to update the inverse of the array correlation matrix from array signal samples using the Matrix Inversion Lemma as follows:

$$\hat{R}^{-1}(n) = \hat{R}^{-1}(n-1) - \frac{\hat{R}^{-1}(n-1)\,\mathbf{x}(n)\,\mathbf{x}^H(n)\,\hat{R}^{-1}(n-1)}{1+\mathbf{x}^H(n)\,\hat{R}^{-1}(n-1)\,\mathbf{x}(n)} \tag{3.1.6}$$

with

$$\hat{R}^{-1}(0) = \frac{1}{\varepsilon_0} I, \quad \varepsilon_0 > 0 \tag{3.1.7}$$

This scheme of estimating weights using the inverse update is referred to as the recursive least squares (RLS) algorithm, which is further discussed in Section 3.9. More discussion on the simple matrix inversion (SMI) algorithm is found in [Ree74, Van91, Hor79].

Application of SMI to estimate the weights of an array to operate in mobile communication systems has been considered in many studies [Win94, Geb95, Lin95, Vau88, Has93, Pas96]. One of these studies [Lin95] considers beamforming for GSM signals using a variable reference signal as available during the symbol interval of the time-division multiple access (TDMA) system. Applications discussed include vehicular mobile communications [Vau88], reducing delay spread in indoor radio channels [Pas96], and mobile satellite communication systems [Geb95].

## 3.2 Unconstrained Least Mean Squares Algorithm

Application of least mean squares (LMS) algorithm to estimate optimal weights of an array is widespread and its study has been of considerable interest for some time. The algorithm is referred to as the constrained LMS algorithm when the weights are subjected to constraints at each iteration, whereas it is referred to as the unconstrained LMS algorithm when weights are not constrained at each iteration. The latter is applicable mainly when weights are updated using reference signals and no knowledge of the direction of the signal is utilized, as is the case for the constrained case.

The algorithm updates the weights at each iteration by estimating the gradient of the quadratic MSE surface, and then moving the weights in the negative direction of the gradient by a small amount. The constant that determines this amount is referred to as the step size. When this step size is small enough, the process leads these estimated weights to the optimal weights. The convergence and transient behavior of these weights along with their covariance characterize the LMS algorithm, and the way the step size and the process of gradient estimation affect these parameters are of great practical importance. These and other issues are discussed in detail in the following.

A real-time unconstrained LMS algorithm for determining optimal weight $\hat{\mathbf{w}}_{MSE}$ of the system using the reference signal has been studied by many authors [Wid67, Gri69, Wid76, Wid76a, Hor81, Ilt85, Cla87, Feu85, Gar86, Bol87, Fol88, Sol89, Jag90, Sol92, God97] and is given by

$$\mathbf{w}(n+1) = \mathbf{w}(n) - \mu \mathbf{g}(\mathbf{w}(n)) \tag{3.2.1}$$

where $\mathbf{w}(n+1)$ denotes the new weights computed at the $(n+1)$th iteration, $\mu$ is a positive scalar (gradient step size) that controls the convergence characteristic of the algorithm (i.e., how fast and how close the estimated weights approach the optimal weights), and $\mathbf{g}(\mathbf{w}(n))$ is an unbiased estimate of the MSE gradient. For a given $\mathbf{w}(n)$, the MSE is given by (2.5.1), that is,

$$\xi(\mathbf{w}(n)) = E\left[|r(n+1)|^2\right] + \mathbf{w}^H(n)R\mathbf{w}(n) - \mathbf{w}^H(n)\mathbf{z} - \mathbf{z}^H\mathbf{w}(n) \tag{3.2.2}$$

The MSE gradient at the nth iteration is obtained by differentiating (3.2.2) with respect to $\mathbf{w}$, yielding

$$\nabla_{\mathbf{w}}\xi(\mathbf{w})\Big|_{\mathbf{w}=\mathbf{w}(n)} = 2R\,\mathbf{w}(n) - 2\mathbf{z} \tag{3.2.3}$$

Note that at the (n + 1)th iteration, the array is operating with weights $\mathbf{w}(n)$ computed at the previous iteration; however, the array signal vector is $\mathbf{x}(n+1)$, the reference signal sample is $r(n+1)$, and the array output

$$y(\mathbf{w}(n)) = \mathbf{w}^H(n)\mathbf{x}(n+1) \tag{3.2.4}$$

### 3.2.1 Gradient Estimate

In its standard form, the LMS algorithm uses an estimate of the gradient by replacing R and $\mathbf{z}$ by their noisy estimates available at the (n + 1)th iteration, leading to

$$\mathbf{g}(\mathbf{w}(n)) = 2\mathbf{x}(n+1)\,\mathbf{x}^H(n+1)\,\mathbf{w}(n) - 2\mathbf{x}(n+1)\,r^*(n+1) \tag{3.2.5}$$

Since the error $\varepsilon(\mathbf{w}(n))$ between the array output and the reference signal is given by

$$\varepsilon(\mathbf{w}(n)) = r(n+1) - \mathbf{w}^H(n)\mathbf{x}(n+1) \tag{3.2.6}$$

it follows from (3.2.5) that

$$\mathbf{g}(\mathbf{w}(n)) = -2\mathbf{x}(n+1)\,\varepsilon^*(\mathbf{w}(n)) \tag{3.2.7}$$

Thus, the estimated gradient is a product of the error between the array output and the reference signal and the array signals after the nth iteration. Taking the conditional expectation on both sides of (3.2.5), it can easily be established that the mean of the gradient estimate for a given $\mathbf{w}(n)$ becomes

$$\bar{\mathbf{g}}(\mathbf{w}(n)) = 2R\mathbf{w}(n) - 2\mathbf{z} \tag{3.2.8}$$

where $\bar{\mathbf{g}}(\mathbf{w}(n))$ denotes the mean of the gradient estimate for a given $\mathbf{w}(n)$. From (3.2.3) and (3.2.8) it follows that the gradient estimate is unbiased.

### 3.2.2 Covariance of Gradient

A particular characteristic of the gradient estimate, which is important in determining the performance of the algorithm, is the covariance of the gradient estimate used. To obtain results on the covariance of the gradient estimate given by (3.2.5), making an additional Gaussian assumption about the sequence {**x**(k)} is necessary. Thus, it is assumed that {**x**(k)} is an independent indentically distributed (i.i.d.) complex Gaussian sequence.

The following result is useful for the analysis to obtain a fourth-order moment of complex variables. The result, based on the Gaussian moment–factoring theorem, states that [Ree62] when $x_1$, $x_2$, $x_3$, and $x_4$ are zero mean, complex jointly Gaussian random variables, the following relationship holds:

$$E\left[x_1 x_2^* x_3 x_4^*\right] = E\left[x_1 x_2^*\right] E\left[x_3 x_4^*\right] + E\left[x_1 x_4^*\right] E\left[x_2^* x_3\right] \tag{3.2.9}$$

Now consider the covariance of the gradient estimate given by (3.2.5). By definition, the covariance of the gradient for a given **w**(n) is given by

$$\begin{aligned} V_g(\mathbf{w}(n)) &= E\left[\{\mathbf{g}(\mathbf{w}(n)) - \bar{\mathbf{g}}(\mathbf{w}(n))\}\{\mathbf{g}(\mathbf{w}(n)) - \bar{\mathbf{g}}(\mathbf{w}(n))\}^H\right] \\ &= E\left[\mathbf{g}(\mathbf{w}(n))\mathbf{g}^H(\mathbf{w}(n))\right] - E\left[\mathbf{g}(\mathbf{w}(n))\bar{\mathbf{g}}^H(\mathbf{w}(n))\right] \\ &\quad - E\left[\bar{\mathbf{g}}(\mathbf{w}(n))\mathbf{g}^H(\mathbf{w}(n))\right] - E\left[\bar{\mathbf{g}}(\mathbf{w}(n))\bar{\mathbf{g}}^H(\mathbf{w}(n))\right] \\ &= E\left[\mathbf{g}(\mathbf{w}(n))\mathbf{g}^H(\mathbf{w}(n))\right] - \bar{\mathbf{g}}(\mathbf{w}(n))\bar{\mathbf{g}}^H(\mathbf{w}(n)) \end{aligned} \tag{3.2.10}$$

The second term on the RHS of (3.2.10) is obtained by taking the outer product of (3.2.8), yielding

$$\bar{\mathbf{g}}(\mathbf{w}(n))\bar{\mathbf{g}}^H(\mathbf{w}(n)) = 4R\mathbf{w}(n)\mathbf{w}^H(n)R - 4R\mathbf{w}(n)\mathbf{z}^H - 4\mathbf{z}\mathbf{w}^H(n)R + 4\mathbf{z}\mathbf{z}^H \tag{3.2.11}$$

To evaluate the first term on the RHS of (3.2.10), take the outer product of (3.2.5):

$$\begin{aligned} \mathbf{g}(\mathbf{w}(n))\mathbf{g}^H(\mathbf{w}(n)) = 4\{&\mathbf{x}(n+1)\,\mathbf{x}^H(n+1)\,\mathbf{w}(n)\mathbf{w}^H(n)\mathbf{x}(n+1)\mathbf{x}^H(n+1) \\ &- \mathbf{x}(n+1)\,\mathbf{x}^H(n+1)\,\mathbf{w}(n)r(n+1)\mathbf{x}^H(n+1) \\ &- \mathbf{x}(n+1)\,r^*(n+1)\mathbf{w}(n)\mathbf{x}(n+1)\,\mathbf{x}^H(n+1) \\ &+ \mathbf{x}(n+1)\,r^*(n+1)r(n+1)\mathbf{x}^H(n+1)\} \end{aligned} \tag{3.2.12}$$

Taking the conditional expectation on both sides one obtains, for a given **w**(n),

$$\begin{aligned} E\left[\mathbf{g}(\mathbf{w}(n))\mathbf{g}^H(\mathbf{w}(n))\right] = &\,4E\left[\mathbf{x}(n+1)\,\mathbf{x}^H(n+1)\,\mathbf{w}(n)\mathbf{w}^H(n)\mathbf{x}(n+1)\mathbf{x}^H(n+1)\right] \\ &- 4E\left[\mathbf{x}(n+1)\,\mathbf{x}^H(n+1)\,\mathbf{w}(n)r(n+1)\mathbf{x}^H(n+1)\right] \\ &- 4E\left[\mathbf{x}(n+1)\,r^*(n+1)\mathbf{w}^H(n)\mathbf{x}(n+1)\,\mathbf{x}^H(n+1)\right] \\ &+ 4E\left[\mathbf{x}(n+1)\,r^*(n+1)r(n+1)\mathbf{x}^H(n+1)\right] \end{aligned} \tag{3.2.13}$$

Consider the fourth term on the RHS of (3.2.13), and define a matrix:

$$A = E\left[\mathbf{x}(n+1)\, r^*(n+1) r(n+1) \mathbf{x}^H(n+1)\right] \tag{3.2.14}$$

It follows from (3.2.14) and (3.2.9) that

$$\begin{aligned} A_{ij} &= E\left[x_i(n+1)\, r^*(n+1) r(n+1) x_j^*(n+1)\right] \\ &= E\left[x_i(n+1)\, r^*(n+1)\right] E\left[r(n+1) x_j^*(n+1)\right] \\ &\quad + E\left[x_i(n+1) x_j^*(n+1)\right] E\left[r^*(n+1) r(n+1)\right] \end{aligned} \tag{3.2.15}$$

This along with (2.5.2) implies that

$$A = \mathbf{z}\mathbf{z}^H + R p_r \tag{3.2.16}$$

where

$$p_r = E\left[r^*(n) r(n)\right] \tag{3.2.17}$$

is the mean power of the reference signal.

Similarly evaluating the other terms on the RHS of (3.2.13),

$$\begin{aligned} E\left[\mathbf{g}(\mathbf{w}(n)) \mathbf{g}^H(\mathbf{w}(n))\right] &= 4\mathbf{w}^H(n) R \mathbf{w}(n) R + 4R\mathbf{w}(n)\mathbf{w}^H(n) R \\ &\quad - 4R\mathbf{w}(n)\mathbf{z}^H - 4R\mathbf{z}^H\mathbf{w}(n) \\ &\quad - 4\mathbf{z}\mathbf{w}^H(n) R - 4R\mathbf{w}^H(n)\mathbf{z} \\ &\quad + 4\mathbf{z}\mathbf{z}^H + 4R p_r \end{aligned} \tag{3.2.18}$$

Subtracting (3.2.11) from (3.2.18) and using (3.2.2),

$$\begin{aligned} V_{\mathbf{g}}(\mathbf{w}(n)) &= 4R\left\{\mathbf{w}^H(n) R \mathbf{w}(n) - \mathbf{z}^H\mathbf{w}(n) - \mathbf{w}^H(n)\mathbf{z} + p_r\right\} \\ &= 4R\,\xi(\mathbf{w}(n)) \end{aligned} \tag{3.2.19}$$

where $\xi(\mathbf{w}(n))$ is the MSE given by (3.2.2).

### 3.2.3 Convergence of Weight Vector

In this section, it is shown that the mean value of the weights estimated by (3.2.1) using the gradient estimate given by (3.2.5) approaches the optimal weights in the limit as the number of iterations grows large. For this discussion, it is assumed that the successive array signal samples are uncorrelated. This is usually achieved by having a sufficiently long iteration cycle of the algorithm. Substituting from (3.2.5) in (3.2.1), it follows that

$$\mathbf{w}(n+1) = \mathbf{w}(n) - 2\mu\mathbf{x}(n+1)\, \mathbf{x}^H(n+1)\, \mathbf{w}(n) + 2\mu\mathbf{x}(n+1)\, r^*(n+1) \tag{3.2.20}$$

Equation (3.2.20) shows that $\mathbf{w}(n)$ is only a function of $\mathbf{x}(0)$, $\mathbf{x}(1)$, …, $\mathbf{x}(n)$. This along with the assumption that the successive array samples are uncorrelated implies that $\mathbf{w}(n)$ and $\mathbf{x}(n+1)$ are uncorrelated. Hence, taking the expected value on both sides of (3.2.20),

$$\begin{aligned}\overline{\mathbf{w}}(n+1) &= \overline{\mathbf{w}}(n) - 2\mu E\left[\mathbf{x}(n+1)\,\mathbf{x}^H(n+1)\right]\overline{\mathbf{w}}(n) + 2\mu E\left[\mathbf{x}(n+1)\,r^*(n+1)\right] \\ &= \overline{\mathbf{w}}(n) - 2\mu R\,\overline{\mathbf{w}}(n) + 2\mu\mathbf{z} \\ &= [I - 2\mu R]\overline{\mathbf{w}}(n) + 2\mu\mathbf{z}\end{aligned} \tag{3.2.21}$$

where

$$\overline{\mathbf{w}}(n) = E[\mathbf{w}(n)] \tag{3.2.22}$$

Define a mean error vector $\overline{\mathbf{v}}(n)$ as

$$\overline{\mathbf{v}}(n) = \overline{\mathbf{w}}(n) - \hat{\mathbf{w}}_{MSE} \tag{3.2.23}$$

where $\hat{\mathbf{w}}_{MSE}$ is the optimal weight given by (3.3), that is,

$$\hat{\mathbf{w}}_{MSE} = R^{-1}\mathbf{z} \tag{3.2.24}$$

It follows from (3.2.23) that $\overline{\mathbf{w}}(n)$ is given by

$$\overline{\mathbf{w}}(n) = \overline{\mathbf{v}}(n) + \hat{\mathbf{w}}_{MSE} \tag{3.2.25}$$

Substituting for $\overline{\mathbf{w}}(n)$ in (3.2.21),

$$\overline{\mathbf{v}}(n+1) = [I - 2\mu R]\overline{\mathbf{v}}(n) - 2\mu R\hat{\mathbf{w}}_{MSE} + 2\mu\mathbf{z} \tag{3.2.26}$$

Noting from (3.2.24) that

$$\mathbf{z} = R\,\hat{\mathbf{w}}_{MSE} \tag{3.2.27}$$

it follows from (3.2.26) that

$$\begin{aligned}\overline{\mathbf{v}}(n+1) &= [I - 2\mu R]\overline{\mathbf{v}}(n) \\ &= [I - 2\mu R]^{n+1}\overline{\mathbf{v}}(0)\end{aligned} \tag{3.2.28}$$

The behavior of the RHS of (3.2.28) can be explained better by converting it in diagonal form, which can be done by using the eigenvalue decomposition of R given by (2.1.29). In the following, (2.1.29) is rewritten:

$$R = Q\Lambda Q^H \tag{3.2.29}$$

where $\Lambda$ is a diagonal matrix of the eigenvalues of R and Q is given by (2.1.31). It is a matrix, with columns being the eigenvectors of R.

Substituting for R in (3.2.28),

$$\bar{\mathbf{v}}(n+1) = \left[I - 2\mu Q \Lambda Q^H\right]^{n+1} \bar{\mathbf{v}}(0) \tag{3.2.30}$$

Equation (3.2.30) may be rewritten in the following form, using indexing:

$$\bar{\mathbf{v}}(n+1) = Q[I - 2\mu\Lambda]^{n+1} Q^H \bar{\mathbf{v}}(0) \tag{3.2.31}$$

For n = 0, it follows from (3.2.30) using $QQ^H = I$ that

$$\begin{aligned} \bar{\mathbf{v}}(1) &= \left[I - 2\mu Q\Lambda\, Q^H\right] \bar{\mathbf{v}}(0) \\ &= Q[I - 2\mu\Lambda] Q^H \bar{\mathbf{v}}(0) \end{aligned} \tag{3.2.32}$$

Thus, (3.2.31) holds for n = 0. For n = 1, it follows from (3.2.30) using $Q^H Q = I$ that

$$\begin{aligned} \bar{\mathbf{v}}(2) &= \left[I - 2\mu Q\Lambda\, Q^H\right]^2 \bar{\mathbf{v}}(0) \\ &= \left[I - 2\mu\, Q\Lambda\, Q^H - 2\mu\, Q\Lambda\, Q^H + 4\mu^2 Q\Lambda\, Q^H\, Q\Lambda\, Q^H\right] \bar{\mathbf{v}}(0) \\ &= Q\left[I - 2\mu\Lambda - 2\mu\Lambda + 4\mu^2\Lambda^2\right] Q^H\, \bar{\mathbf{v}}(0) \\ &= Q[I - 2\mu\Lambda]^2\, Q^H\, \bar{\mathbf{v}}(0) \end{aligned} \tag{3.2.33}$$

Thus, (3.2.31) holds for n = 1. If (3.2.31) holds for any n, that is,

$$\bar{\mathbf{v}}(n) = Q[I - 2\mu\Lambda]^n\, Q^H\, \bar{\mathbf{v}}(0) \tag{3.2.34}$$

then

$$\begin{aligned} \bar{\mathbf{v}}(n+1) &= \left[I - 2\mu Q\Lambda\, Q^H\right]^{n+1} \bar{\mathbf{v}}(0) \\ &= \left[I - 2\mu Q\Lambda\, Q^H\right]^n \left[I - 2\mu Q\Lambda\, Q^H\right] \bar{\mathbf{v}}(0) \\ &= Q[I - 2\mu\Lambda]^n\, Q^H \left[I - 2\mu Q\Lambda\, Q^H\right] \bar{\mathbf{v}}(0) \\ &= \left\{Q[I - 2\mu\Lambda]^n\, Q^H - 2\mu Q[I - 2\mu\Lambda]^n\, Q^H\, Q\Lambda Q^H\right\} \bar{\mathbf{v}}(0) \\ &= \left\{Q[I - 2\mu\Lambda]^n\, Q^H - 2\mu Q[I - 2\mu\Lambda]^n\, \Lambda Q^H\right\} \bar{\mathbf{v}}(0) \\ &= \left\{Q[I - 2\mu\Lambda]^n\, [I - 2\mu\Lambda] Q^H\right\} \bar{\mathbf{v}}(0) \\ &= Q[I - 2\mu\Lambda]^{n+1} Q^H\, \bar{\mathbf{v}}(0) \end{aligned} \tag{3.2.35}$$

and it holds for n + 1. Thus, by indexing it is proved that (3.2.30) may be rewritten in the form of (3.2.31). The quantity in the square bracket on the RHS of (3.2.31) is a diagonal matrix with each entry $(1 - 2\mu\lambda_i)$, with $\lambda_i$, i = 1, …, L being the L eigenvalues of R.

For $\mu < 1/\lambda_{max}$, with $\lambda_{max}$ denoting the maximum eigenvalue of R, the magnitude of each diagonal element is less than 1, that is,

$$\left|1 - 2\mu\lambda_i\right| < 1 \forall i \tag{3.2.36}$$

Hence, as the iteration number increases, each diagonal element of the matrix in the square bracket diminishes, yielding

$$\lim_{n\to\infty} \bar{\mathbf{v}}(n) = 0 \tag{3.2.37}$$

This along with (3.2.23) implies that

$$\lim_{n\to\infty} \bar{\mathbf{w}}(n) = \hat{\mathbf{w}}_{MSE} \tag{3.2.38}$$

Thus, for $\mu < 1/\lambda_{max}$, the algorithm is stable and the mean value of the estimated weights converges to the optimal weights. As the sum of all eigenvalues of R equals its trace, the sum of its diagonal elements, the gradient step size μ can be selected in terms of measurable quantities using $\mu < 1/\mathrm{Tr}(R)$, with Tr(R) denoting the trace of R. It should be noted that each diagonal element of R is equal to the average power measured on the corresponding element of the array. Thus, for an array of identical elements, the trace of R equals the power measured on any one element times the number of elements in the array.

### 3.2.4 Convergence Speed

The convergence speed of the algorithm refers to the speed by which the mean of the estimated weights (ensemble average of many trials) approaches the optimal weights, and is normally characterized by L trajectories along L eigenvectors of R. To obtain the convergence time constant along an eigenvector of R, consider the initial mean error vector $\bar{\mathbf{v}}(0)$ and express it as a linear combination of L eigenvectors of R, that is,

$$\bar{\mathbf{v}}(0) = \sum_{i=1}^{L} \alpha_i \mathbf{U}_i \tag{3.2.39}$$

where $\alpha_i$, i = 1, 2, …, L are scalars and $\mathbf{U}_i$, i = 1, 2, …, L are eigenvectors corresponding to L eigenvalues of R.

Substituting from (3.2.39) in (3.2.31) yields

$$\bar{\mathbf{v}}(n+1) = Q[I - 2\mu\Lambda]^{n+1} Q^H \sum_{i=1}^{L} \alpha_i \mathbf{U}_i \tag{3.2.40}$$

Since eigenvectors of R are orthogonal, (3.2.40) can be expressed as

$$\begin{aligned}\bar{\mathbf{v}}(n+1) &= \sum_{i=1}^{L} \alpha_i Q[I - 2\mu\Lambda]^{n+1} Q^H \mathbf{U}_i \\ &= \sum_{i=1}^{L} \alpha_i [1 - 2\mu\lambda_i]^{n+1} \mathbf{U}_i\end{aligned} \tag{3.2.41}$$

The convergence of the mean weight vector to the optimal weight vector along the ith eigenvector is therefore geometric, with geometric ratio $1 - 2\mu\lambda_i$. If an exponential envelope of the time constant $\tau_i$ is fitted to the geometric sequence of (3.2.41), then

$$\tau_i = \frac{-1}{\ell n(1 - 2\mu\lambda_i)} \tag{3.2.42}$$

where $\ell n$ denotes the natural logarithm and the unit of time is assumed to be one iteration. The negative sign in (3.2.42) appears due to the fact that the quantity in parentheses is less than unity and the logarithm of that is a negative quantity.

Note that if

$$2\mu\lambda_i \ll 1 \tag{3.2.43}$$

the time constant of the ith trajectory may be approximated to

$$\tau_i = \frac{1}{2\mu\lambda_i} \tag{3.2.44}$$

Thus, these time constants are functions of the eigenvalues of the array correlation matrix, the smallest one dependent on $\lambda_{max}$, which normally corresponds to the strongest source and the largest one controlled by the smallest eigenvalue that corresponds to the weakest source or the background noise. Therefore, the larger the eigenvalue spread, the longer it takes for the algorithm to converge. In terms of interference rejection capability, this means canceling the strongest source first and the weakest last.

The convergence speed of an algorithm is an important property and its importance for mobile communications is highlighted in [Nag94] by discussing how the LMS algorithm does not perform as well as some other algorithms due to its slow convergence speed in situations of fast-changing signal characteristics. Time availability for an algorithm to converge in mobile communication systems not only depends on the system design, which dictates duration of the user signal present such as the user slot duration in a TDMA system, it is also affected by the speed of mobiles, which changes the rate at which a signal fades. For example, a mobile on foot would cause the signal to fade at a rate of about 5 Hz, whereas it would be of the order of about 50 Hz for a vehicle mobile, implying that an algorithm needs to converge faster in a system being used by vehicle mobiles compared to the one used in a handheld portable [Win87]. Some of these issues for the IS-54 system are discussed in [Win94] where the convergence of the LMS and the SMI algorithms in mobile communication situations is compared.

Even when the mean of the estimated weights converges to optimal weights, they have finite covariance, that is, their covariance matrix is not identical to a matrix with all elements equal to zero. This causes the average of the MSE not to converge to the minimum

MSE (MMSE) and leads to excess MSE. Convergence of the weight covariance matrix and the excess MSE is discussed in following sections.

### 3.2.5 Weight Covariance Matrix

The covariance matrix of the weights at the nth iteration is given by

$$\begin{aligned} k_{ww}(n) &= E\left[\left(\mathbf{w}(n)-\overline{\mathbf{w}}(n)\right)\left(\mathbf{w}(n)-\overline{\mathbf{w}}(n)\right)^H\right] \\ &= E\left[\mathbf{w}(n)\mathbf{w}^H(n)\right]-\left[\mathbf{w}(n)\overline{\mathbf{w}}^H(n)\right] \\ &\quad + E\left[\overline{\mathbf{w}}(n)\mathbf{w}^H(n)\right]-E\left[\overline{\mathbf{w}}(n)\overline{\mathbf{w}}^H(n)\right] \\ &= R_{ww}(n)-\overline{\mathbf{w}}(n)\overline{\mathbf{w}}^H(n) \end{aligned} \tag{3.2.45}$$

where expectation is unconditional and taken over **w**,

$$\overline{\mathbf{w}}(n) = E\left[\mathbf{w}(n)\right] \tag{3.2.46}$$

and

$$R_{ww}(n) = E\left[\mathbf{w}(n)\mathbf{w}^H(n)\right] \tag{3.2.47}$$

In this section, a recursive relationship for the weight covariance matrix is derived. The relationship is useful for understanding the transient behavior of the matrix.

It follows from (3.2.45) that

$$k_{ww}(n+1) = R_{ww}(n+1)-\overline{\mathbf{w}}(n+1)\overline{\mathbf{w}}^H(n+1) \tag{3.2.48}$$

and from (3.2.47) that

$$R_{ww}(n+1) = E\left[\mathbf{w}(n+1)\mathbf{w}^H(n+1)\right] \tag{3.2.49}$$

Substituting from (3.2.1) in (3.2.49),

$$\begin{aligned} R_{ww}(n+1) &= R_{ww}(n)+\mu^2 E\left[\mathbf{g}(\mathbf{w}(n))\mathbf{g}^H(\mathbf{w}(n))\right] \\ &\quad -\mu E\left[\mathbf{g}(\mathbf{w}(n))\mathbf{w}^H(n)\right]-\mu E\left[\mathbf{w}(n)\mathbf{g}^H(\mathbf{w}(n))\right] \end{aligned} \tag{3.2.50}$$

Taking unconditional expectation on both sides of (3.2.10), and rearranging, it follows that

$$E\left[\mathbf{g}(\mathbf{w}(n))\mathbf{g}^H(\mathbf{w}(n))\right] = E\left[V_g(\mathbf{w}(n))\right]+E\left[\overline{\mathbf{g}}(\mathbf{w}(n))\overline{\mathbf{g}}^H(\mathbf{w}(n))\right] \tag{3.2.51}$$

where $\overline{\mathbf{g}}(\mathbf{w}(n))$ is the mean value of the gradient estimate for a given $\mathbf{w}(n)$. An expression for $\overline{\mathbf{g}}(\mathbf{w}(n))$ is given by (3.2.8). From (3.2.8), taking the outer product of $\overline{\mathbf{g}}(\mathbf{w}(n))$,

$$E\left[\bar{\mathbf{g}}(\mathbf{w}(n))\bar{\mathbf{g}}^H(\mathbf{w}(n))\right] = E\left[\{2R\mathbf{w}(n) - 2\mathbf{z}\}\{2R\mathbf{w}(n) - 2\mathbf{z}\}^H\right]$$
$$= 4\{RR_{ww}(n)R + \mathbf{z}\mathbf{z}^H - R\bar{\mathbf{w}}(n)\mathbf{z}^H - \mathbf{z}\bar{\mathbf{w}}^H R\} \quad (3.2.52)$$

From (3.2.5),

$$E\left[\mathbf{w}(n)\mathbf{g}^H(\mathbf{w}(n))\right] = 2R_{ww}(n)R - 2\bar{\mathbf{w}}(n)\mathbf{z}^H \quad (3.2.53)$$

and

$$E\left[\mathbf{g}(\mathbf{w}(n))\mathbf{w}^H(n)\right] = 2RR_{ww}(n) - 2\mathbf{z}\bar{\mathbf{w}}^H(n) \quad (3.2.54)$$

From (3.2.50) to (3.2.54) it follows that

$$R_{ww}(n+1) = R_{ww}(n) + \mu^2 E\left[V_g(\mathbf{w}(n))\right]$$
$$+ 4\mu^2\{RR_{ww}(n)R + \mathbf{z}\mathbf{z}^H - R\bar{\mathbf{w}}(n)\mathbf{z}^H - \mathbf{z}\bar{\mathbf{w}}^H R\} \quad (3.2.55)$$
$$- 2\mu\{RR_{ww}(n) - \mathbf{z}\bar{\mathbf{w}}^H(n) + R_{ww}(n)R - \bar{\mathbf{w}}(n)\mathbf{z}^H\}$$

Evaluation of (3.2.48) requires the outer product of $\bar{\mathbf{w}}(n + 1)$. From (3.2.1) and (3.4.8),

$$\bar{\mathbf{w}}(n+1) = \bar{\mathbf{w}}(n) - 2\mu R\,\bar{\mathbf{w}}(n) + 2\mu\mathbf{z} \quad (3.2.56)$$

and thus

$$\bar{\mathbf{w}}(n+1)\,\bar{\mathbf{w}}^H(n+1) = \bar{\mathbf{w}}(n)\bar{\mathbf{w}}^H(n)$$
$$+ 4\mu^2\{R\bar{\mathbf{w}}(n)\bar{\mathbf{w}}^H(n)R + \mathbf{z}\mathbf{z}^H - R\bar{\mathbf{w}}(n)\mathbf{z}^H - \mathbf{z}\bar{\mathbf{w}}^H(n)R\} \quad (3.2.57)$$
$$- 2\mu\{\bar{\mathbf{w}}(n)\bar{\mathbf{w}}^H(n)R + R\bar{\mathbf{w}}(n)\bar{\mathbf{w}}^H(n) - \bar{\mathbf{w}}(n)\mathbf{z}^H - \mathbf{z}\bar{\mathbf{w}}^H(n)\}$$

Subtracting (3.2.57) from (3.2.55) and using (3.2.48),

$$k_{ww}(n+1) = k_{ww}(n) + 4\mu^2 Rk_{ww}(n)R - 2\mu\{Rk_{ww}(n) + k_{ww}(n)R\}$$
$$+ \mu^2 E\left[V_g(\mathbf{w}(n))\right] \quad (3.2.58)$$

Thus, at each iteration the weight covariance matrix depends on the mean value of the gradient covariance used at the previous iteration. Equation (3.2.58) may be further simplified by substituting for $V_g(\mathbf{w}(n))$ from (3.2.19). Taking the expectation over $\mathbf{w}$ on both sides of (3.2.19),

$$E\left[V_g(\mathbf{w}(n))\right] = 4RE\left[\mathbf{w}^H(n)R\mathbf{w}(n) - \mathbf{z}^H\mathbf{w}(n) - \mathbf{w}^H(n)\mathbf{z} + p_r\right]$$
$$= 4R\{E\left[\mathbf{w}^H(n)R\mathbf{w}(n)\right] - \mathbf{z}^H\bar{\mathbf{w}}(n) - \bar{\mathbf{w}}^H(n)\mathbf{z} + p_r\} \quad (3.2.59)$$

Using

$$\begin{aligned} E\left[\mathbf{w}^H(n)R\mathbf{w}(n)\right] &= E\left[Tr\left[R\mathbf{w}(n)\mathbf{w}^H(n)\right]\right] \\ &= Tr\left[E\left[R\mathbf{w}(n)\mathbf{w}^H(n)\right]\right] \\ &= Tr\left[RR_{ww}(n)\right] \\ &= Tr\left[Rk_{ww}(n) + R\overline{\mathbf{w}}(n)\overline{\mathbf{w}}^H(n)\right] \\ &= Tr\left[Rk_{ww}(n)\right] + \overline{\mathbf{w}}^H(n)R\overline{\mathbf{w}}(n) \end{aligned} \tag{3.2.60}$$

(3.2.59) becomes

$$E\left[V_g(\mathbf{w}(n))\right] = 4RTr\left[Rk_{ww}(n)\right] + 4R\xi(\overline{\mathbf{w}}(n)) \tag{3.2.61}$$

where

$$\xi(\overline{\mathbf{w}}(n)) = \overline{\mathbf{w}}^H(n)R\overline{\mathbf{w}}(n) - \mathbf{z}^H\overline{\mathbf{w}}(n) - \overline{\mathbf{w}}^H(n)\mathbf{z} + p_r \tag{3.2.62}$$

and Tr[.] denotes the trace of [.].

Substituting (3.2.61) in (3.2.58),

$$\begin{aligned} k_{ww}(n+1) &= k_{ww}(n) + 4\mu^2 Rk_{ww}(n)R + 4\mu^2 RTr\left[Rk_{ww}(n)\right] \\ &\quad - 2\mu\left\{Rk_{ww}(n) + k_{ww}(n)R\right\} + \mu^2 4R\xi(\overline{\mathbf{w}}(n)) \end{aligned} \tag{3.2.63}$$

Thus, at the (n + 1)st iteration the weight covariance matrix is a function of $\xi(\overline{\mathbf{w}}(n))$.

### 3.2.6 Transient Behavior of Weight Covariance Matrix

In this section, the transient behavior of the weight covariance matrix is studied by deriving an expression for $k_{ww}(n)$ and its limit as $n \to \infty$. Define

$$\Sigma(n) = Q^H k_{ww}(n)Q \tag{3.2.64}$$

By pre- and postmultiplying by $Q^H$ and Q on both sides of (3.2.63), and using

$$QQ^H = I \tag{3.2.65}$$

it follows that

$$\begin{aligned} Q^H k_{ww}(n+1)Q &= Q^H k_{ww}(n)Q + 4\mu^2 Q^H RQQ^H k_{ww}(n)QQ^H RQ \\ &\quad + 4\mu^2 Q^H RQTr\left[Q^H RQQ^H k_{ww}(n)Q\right] \\ &\quad - 2\mu\left\{Q^H RQQ^H k_{ww}(n)Q + Q^H k_{ww}(n)QQ^H RQ\right\} \\ &\quad + \mu^2 4Q^H RQ\xi(\overline{\mathbf{w}}(n)) \end{aligned} \tag{3.2.66}$$

Using $Q^H RQ = \Lambda$ and (3.2.64) in (3.2.66), the following matrix difference equation is derived:

$$\Sigma(n+1) = \Sigma(n) + 4\mu^2\Lambda^2\Sigma(n) + 4\mu^2\Lambda \mathrm{Tr}[\Lambda\Sigma(n)] \\ - 4\mu\Lambda\Sigma(n) + \mu^2 4\Lambda\xi(\overline{\mathbf{w}}(n)) \tag{3.2.67}$$

Now it is shown by induction that $\Sigma(n)$, $n \ge 0$ is a diagonal matrix. Consider $n = 0$. Since the initial weight vector is $\mathbf{w}(0)$, it follows from (3.2.45) to (3.2.47) and (3.2.64) that

$$\Sigma(0) = 0 \tag{3.2.68}$$

From (3.2.67) and (3.2.68),

$$\Sigma(1) = \mu^2 4\Lambda\xi(\overline{\mathbf{w}}(0)) \tag{3.2.69}$$

As $\Lambda$ is a diagonal matrix, it follows from (3.2.69) that $\Sigma(1)$ is a diagonal matrix. Thus, $\Sigma(n)$ is diagonal for $n = 0$ and 1. If $\Sigma(n)$ is diagonal for any n, then it follows from (3.2.67) that it is diagonal for $n + 1$. Thus, $\Sigma(n)$, $n \ge 0$ is a diagonal matrix.

As Q is a unitary transformation, it follows that the diagonal elements of $\Sigma(n)$ are the eigenvalues of $k_{\mathbf{ww}}(n)$. Let these be denoted by $\eta_\ell(n)$, $\ell = 1, \ldots, L$. Defining

$$\boldsymbol{\lambda} = [\lambda_1, \ldots, \lambda_L]^T \tag{3.2.70}$$

and

$$\boldsymbol{\eta}(n) = [\eta_1(n), \ldots, \eta_L(n)]^T \tag{3.2.71}$$

to denote the eigenvalues of R and $\Sigma(n)$, respectively,

$$\mathrm{Tr}[\Lambda\Sigma(n)] = \boldsymbol{\lambda}^T\boldsymbol{\eta}(n) \tag{3.2.72}$$

Substituting (3.2.72) in (3.2.67), the vector difference equation for the eigenvalues of $\Sigma(n)$ is

$$\boldsymbol{\eta}(n+1) = \{I - 4\mu\Lambda + 4\mu^2\Lambda^2 + 4\mu^2\boldsymbol{\lambda}\boldsymbol{\lambda}^T\}\boldsymbol{\eta}(n) + 4\mu^2\xi(\overline{\mathbf{w}}(n))\boldsymbol{\lambda} \tag{3.2.73}$$

With

$$H = 4\mu\Lambda - 4\mu^2\Lambda^2 - 4\mu^2\boldsymbol{\lambda}\boldsymbol{\lambda}^T \tag{3.2.74}$$

equation (3.2.73) has the solution

$$\boldsymbol{\eta}(n) = \{I - H\}^n\boldsymbol{\eta}(0) + 4\mu^2\boldsymbol{\lambda}\sum_{i=1}^{n}\{I - H\}^{i-1}\xi(\overline{\mathbf{w}}(n-i)) \tag{3.2.75}$$

Since Q diagonalizes $k_{ww}(n)$, it follows that

$$k_{ww}(n) = \sum_{\ell=1}^{L} \eta_\ell(n)\mathbf{U}_\ell \mathbf{U}_\ell^H \tag{3.2.76}$$

where $\mathbf{U}_\ell$, $\ell = 1, \ldots, L$ are the eigenvectors of R. Equation (3.2.76) describes the transient behavior of the weight covariance matrix.

The next section shows that $\lim_{n\to\infty} \boldsymbol{\lambda}^T \boldsymbol{\eta}(n)$ exists under the conditions noted there. This, along with the fact that $0 < \lambda_i < \infty \forall_i$, implies that $\lim_{n\to\infty} \boldsymbol{\eta}(n)$ exists. It then follows from (3.2.73) and (3.2.74) that

$$\lim_{n\to\infty} \boldsymbol{\eta}(n) = 4\mu^2\hat{\xi}H^{-1}\boldsymbol{\lambda}$$

$$= \frac{\mu\hat{\xi}}{1-\mu\sum_{i=1}^{L}\frac{\lambda_i}{1-\mu\lambda_i}} \begin{bmatrix} \frac{1}{1-\mu\lambda_1} \\ \cdot \\ \cdot \\ \frac{1}{1-\mu\lambda L} \end{bmatrix} \tag{3.2.77}$$

where $\hat{\xi}$ is the minimum MSE given by (2.5.6). This along with (3.2.76) implies that an expression for the steady-state weight covariance matrix is given by

$$\lim_{n\to\infty} k_{ww}(n) = \frac{\mu\hat{\xi}}{1-\mu\sum_{i=1}^{L}\frac{\lambda_i}{1-\mu\lambda_i}} \sum_{\ell=1}^{L}\frac{1}{1-\mu\lambda_\ell}\mathbf{U}_\ell \mathbf{U}_\ell^H \tag{3.2.78}$$

### 3.2.7 Excess Mean Square Error

From the expressions of MSE given by (2.5.1), it follows that for a given $\mathbf{w}(n)$,

$$\xi(\mathbf{w}(n)) = \hat{\xi} + \mathbf{v}^H(n)R\mathbf{v}(n) \tag{3.2.79}$$

where $\hat{\xi}$ is the minimum MSE, $\mathbf{v}(n)$ is the error vector at the nth iteration denoting the difference between estimated weights $\mathbf{w}(n)$ and the optimal weights $\hat{\mathbf{w}}$, and $\mathbf{v}^H(n)R\mathbf{v}(n)$ is the excess MSE.

Taking the expected value over $\mathbf{w}$ on both sides of (3.2.79), the average value of the MSE at the nth iteration is derived, that is,

$$\bar{\xi}(n) = \hat{\xi} + E\left[\mathbf{v}^H(n)R\mathbf{v}(n)\right] \tag{3.2.80}$$

where

$$\bar{\xi}(n) = E\left[\xi(\mathbf{w}(n))\right] \tag{3.2.81}$$

and $E[\mathbf{v}^H(n)R\mathbf{v}(n)]$ denotes average excess MSE at the nth iteration.

Taking the limit as $n \to \infty$ yields the steady-state MSE, that is,

$$\begin{aligned}\bar{\xi}(\infty) &= \lim_{n\to\infty} \bar{\xi}(n) \\ &= \hat{\xi} + \lim_{n\to\infty} E\left[\mathbf{v}^H(n)R\mathbf{v}(n)\right]\end{aligned} \tag{3.2.82}$$

Note that as $n \to \infty$ $E[\mathbf{v}(n)] \to \mathbf{0}$ but the average value of the excess MSE does not approach zero, that is, $\lim_{n\to\infty} E[\mathbf{v}^H(n)R\mathbf{v}(n)] \neq 0$. Now let us discuss the meaning of this quantity.

Substituting for $\mathbf{v}(n)$ in (3.2.79),

$$\begin{aligned}E\left[\mathbf{v}^H(n)R\mathbf{v}(n)\right] &= E\left[\mathbf{w}^H(n)R\mathbf{w}(n)\right] + \hat{\mathbf{w}}^H(n)R\hat{\mathbf{w}}(n) \\ &\quad - \hat{\mathbf{w}}^H(n)R\overline{\mathbf{w}}(n) - \overline{\mathbf{w}}^H(n)R\hat{\mathbf{w}}(n)\end{aligned} \tag{3.2.83}$$

Consider the mean output power of the processor for a given $\mathbf{w}$, that is,

$$P(\mathbf{w}(n)) = \mathbf{w}^H(n)R\mathbf{w}(n)$$

Taking the expectation over $\mathbf{w}$, it gives the mean output power at the nth iteration $\overline{P}(n)$, that is,

$$\begin{aligned}\overline{P}(n) &= E\left[P(\mathbf{w}(n))\right] \\ &= E\left[\mathbf{w}^H(n)R\mathbf{w}(n)\right]\end{aligned} \tag{3.2.84}$$

This along with (3.2.83) yields

$$\begin{aligned}E\left[\mathbf{v}^H(n)R\mathbf{v}(n)\right] &= \overline{P}(n) + \hat{\mathbf{w}}^H(n)R\hat{\mathbf{w}}(n) \\ &\quad - \hat{\mathbf{w}}^H(n)R\overline{\mathbf{w}}(n) - \overline{\mathbf{w}}^H(n)R\hat{\mathbf{w}}(n)\end{aligned} \tag{3.2.85}$$

that in the limit becomes

$$\lim_{n\to\infty} E\left[\mathbf{v}^H(n)R\mathbf{v}(n)\right] = \overline{P}(\infty) - \hat{\mathbf{w}}^H(n)R\hat{\mathbf{w}}(n) \tag{3.2.86}$$

Thus, the steady-state average excess MSE is the difference between the mean output power of the processor in the limit $\overline{P}(\infty)$ and the mean output power of the optimal processor, $\hat{\mathbf{w}}^H(n)R\hat{\mathbf{w}}(n)$. It is the excess power contributed by the weight variance in the steady state.

Next, an independent expression for the steady-state average excess MSE is derived. Using (3.2.60) and the notation of the previous section, it follows that

$$\begin{aligned}E\left[\mathbf{w}^H(n)R\mathbf{w}(n)\right] &= \mathrm{Tr}\left[Rk_{\mathbf{ww}}(n)\right] + \overline{\mathbf{w}}^H(n)R\overline{\mathbf{w}}(n) \\ &= \mathrm{Tr}\left[Q^H RQQ^H k_{\mathbf{ww}}(n)Q\right] + \overline{\mathbf{w}}^H(n)R\overline{\mathbf{w}}(n)\end{aligned} \tag{3.2.87}$$

$$\begin{aligned} &= \mathrm{Tr}\left[\Lambda\Sigma(n)\right] + \overline{\mathbf{w}}^H(n)R\overline{\mathbf{w}}(n) \\ &= \boldsymbol{\lambda}^T\boldsymbol{\eta}(n) + \overline{\mathbf{w}}^H(n)R\overline{\mathbf{w}}(n) \end{aligned} \tag{3.2.87}$$

Substituting from (3.2.87) in (3.2.83),

$$E\left[\mathbf{v}^H(n)R\mathbf{v}(n)\right] = \boldsymbol{\lambda}^T\boldsymbol{\eta}(n) + \overline{\mathbf{v}}^H(n)R\overline{\mathbf{v}}(n) \tag{3.2.88}$$

Taking the limits on both sides, this becomes

$$\lim_{n\to\infty} E\left[\mathbf{v}^H(n)R\mathbf{v}(n)\right] = \lim_{n\to\infty} \boldsymbol{\lambda}^T\boldsymbol{\eta}(n) \tag{3.2.89}$$

It should be noted that (3.2.89) only holds in the limit. At the nth iteration, the average excess MSE $\mathbf{v}^H(n)R\mathbf{v}(n)$ is not equal to $\boldsymbol{\lambda}^T\boldsymbol{\eta}(n)$. A relationship between the two quantities is given by (3.2.88). Appendix 3.1 shows that

$$\lim_{n\to\infty} \boldsymbol{\lambda}^T\boldsymbol{\eta}(n) = \frac{\hat{\xi}\mu \sum_{i=1}^{L} \frac{\lambda_i}{1-\mu\lambda_i}}{1-\mu\sum_{i=1}^{L}\frac{\lambda_i}{1-\mu\lambda_i}} \tag{3.2.90}$$

Thus, we have the following result for the steady-state average excess MSE, $\lim_{n\to\infty} E[\mathbf{v}^H(n)R\mathbf{v}(n)]$. If $\mu$ satisfies

$$0 < \mu < \frac{1}{2\lambda_{max}} \tag{3.2.91}$$

and

$$\sum_{i=1}^{L} \frac{\mu\lambda_i}{(1-\mu\lambda_i)} < 1 \tag{3.2.92}$$

then

$$\lim_{n\to\infty} E\left[\mathbf{v}^H(n)R\mathbf{v}(n)\right] = \frac{\hat{\xi}\mu \sum_{i=1}^{L} \frac{\lambda_i}{1-\mu\lambda_i}}{1-\mu\sum_{i=1}^{L}\frac{\lambda_i}{1-\mu\lambda_i}} \tag{3.2.93}$$

### 3.2.8 Misadjustment

The difference between the weights estimated by the adaptive algorithm and optimal weights is further characterized by the ratio of the average excess steady-state MSE and the MMSE. It is referred to as the misadjustment [Wid66]. It is a dimensionless parameter and measures the performance of the algorithm. The misadjustment is a kind of noise, and is caused by the use of the noisy estimate of the gradient. This noise is referred to as the misadjustment noise.

Let M denote the misadjustment. Thus, by definition

$$M = \frac{\bar{\xi}(\infty) - \hat{\xi}}{\hat{\xi}} \tag{3.2.94}$$

It follows from (3.2.94), (3.2.82), (3.2.93) that when the gradient is estimated by multiplying the array signals with the error between the array output and the reference signal, and the gradient step size selected such that (3.2.91) and (3.2.92) hold, then the misadjustment $M_U$ for the unconstrained LMS algorithm is given by

$$M_U = \frac{\mu \sum_{i=1}^{L} \frac{\lambda_i}{1-\mu\lambda_i}}{1 - \mu \sum_{i=1}^{L} \frac{\lambda_i}{1-\mu\lambda_i}} \tag{3.2.95}$$

For a sufficiently small $\mu$, this results in

$$M_U = \mu \mathrm{Tr}[R] \tag{3.2.96}$$

It follows from this expression that increasing $\mu$ increases the misadjustment noise. On the other hand, an increase in $\mu$ causes the algorithm to converge faster as discussed earlier. Thus, the selection of the gradient step size requires satisfying conflicting demands of reaching the vicinity of the solution point quicker but wandering around over a larger region causing a bigger misadjustment and arriving near the solution point slowly with the smaller movement in the weights at the end. The latter causes an additional problem, particularly in nonstationary environments, say when interference is slowly moving, where the optimal solution moves, causing slowly adapting estimated weights to lag behind the optimal weights. This phenomenon is referred to as the weight vector lag.

Many schemes including variable step size have been suggested to overcome this problem [Soo91, Pri91, Yas87, Eva93, Kwo92, Kwo92a, Har86, Che90]. Some of these are briefly discussed.

The adaptive algorithm estimates the weights by minimizing the MSE. Thus, in schemes where a variable step size is used, it reflects the value of the MSE at that iteration, going up and down as the MSE goes up and down such that it stays between the maximum permissible value for convergence and the minimum value based on the allowed misadjustment. It may be truly variable or may be allowed to switch between a few preselected values for the ease of implementation as well as by shifting by one bit left or right where digital implementation is used. The step size may also be adjusted to reflect the change in the direction of the error surface gradient at each iteration [Har86].

The optimal value of the step size at each step is suggested in [Yas87] such that it minimizes the MSE at each iteration. This is a function of the value of the true gradient at each iteration and the array correlation matrix. In practice, these may be replaced by their instantaneous values, leading to a suboptimal value.

Instead of having a single step size for a whole weight vector, a variable step size can be selected for each weight separately, leading to increased convergence of the algorithm [Eva93]. The convergence speed of the algorithm may also be increased by adjusting weights such that interferences are canceled one at a time [Ko93, Ko93a], and by using a scheme known as block processing [Ben92]. For broadband signals, an implementation in the frequency domain may help increase the speed of convergence.

The application of frequency domain beamforming to estimate the weights using the LMS algorithm for the case when a reference signal is available shows [Den78, Nar81, Flo88, Ber86] how the frequency domain approach yields improved convergence and reduced computational complexities compared to the time domain approach. Improved convergence normally arises from the use of different gradient step sizes in different bins. For the constrained LMS case, this is likely to cause deterioration in the steady-state performance of the algorithm. This, however, does not affect the performance of the unconstrained algorithm [Feu93].

The "sign algorithm," in which the error between the array output and the reference signal is replaced by its sign, is computationally less complex than the LMS algorithm, as discussed in [Che90, Mat87].

The algorithm is usually analyzed assuming that successive samples are uncorrelated. This assumption helps in simplifying the mathematics by allowing expectations of data products to be replaced by the products of their expectations. Discussion of correlated samples in nonstationary environment may be found in [Ber84, Ber85, Ewe90]. Applications of the unconstrained LMS algorithm to mobile communication systems using an array include base mobile communication systems [Win84], indoor radio systems [Win87], and satellite-to-satellite communication systems [Jon95].

## 3.3 Normalized Least Mean Squares Algorithm

This algorithm is a variation of the constant-step-size LMS algorithm and uses a data-dependent step size at each iteration [God97]. At the nth iteration, the step size is given by

$$\mu(n) = \frac{\mu_0}{\mathbf{x}^H(n)\,\mathbf{x}(n)} \tag{3.3.1}$$

where $\mu_0$ is a constant. The algorithm and its convergence using various types of data have been studied widely [Nit85, Nit86, Ber86a, Slo93, Rup93]. It avoids the need for estimating the eigenvalues of the correlation matrix or its trace for selection of the maximum permissible step size. The algorithm normally has better convergence performance and less signal sensitivity compared to the normal LMS algorithm. See [Bar94] for discussion of its application to mobile communications.

## 3.4 Constrained Least Mean Squares Algorithm

A real-time constrained algorithm [Hud81, Fro72, Can80, God83, God86, God89, God90, God93, God97, Mos70] for determining the optimal weight vector $\hat{\mathbf{w}}$ is

$$\mathbf{w}(n+1) = P\{\mathbf{w}(n) - \mu\mathbf{g}(\mathbf{w}(n))\} + \frac{\mathbf{S}_0}{L} \tag{3.4.1}$$

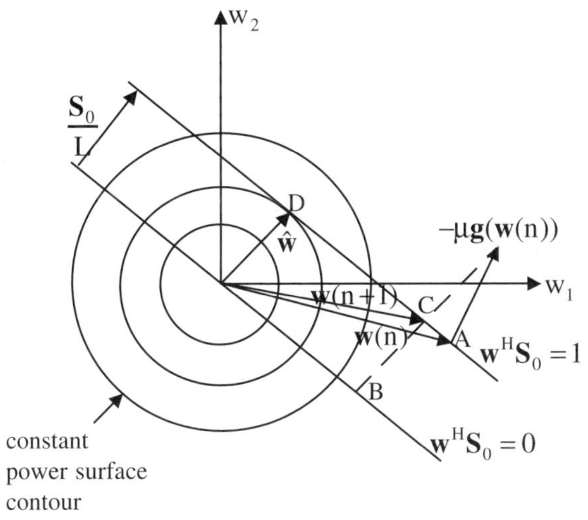


**FIGURE 3.1**
Constrained LMS algorithm: a pictorial view of projection process.

where

$$P = I - \frac{\mathbf{S}_0 \mathbf{S}_0^H}{L} \tag{3.4.2}$$

is a projection operator, $\mathbf{g}(\mathbf{w}(n))$ is an unbiased estimate of the gradient of the power surface $\mathbf{w}^H(n)R\mathbf{w}(n)$ with respect to $\mathbf{w}(n)$ after the nth iteration; $\mu$ is the gradient step size, a positive scalar constant that controls the characteristics of the adaptive algorithm; and $\mathbf{S}_0$ is the steering vector in the look direction.

The algorithm is called constrained because the weight vector satisfies the constraint at every iteration, that is, $\mathbf{w}^H(n)\mathbf{S}_0 = 1$, $\forall n$. The process of imposing constraints may be understood from Figure 3.1. It shows how weights are updated and how the projection system works using a vector diagram for a two-weight system [Fro72]. The figure shows constant power contours; the constraint surface (a line $\mathbf{w}^H\mathbf{S}_0 = 1$ for a two-dimensional system); a surface parallel to the constraint surface passing through the origin ($\mathbf{w}^H\mathbf{S}_0 = 0$); weight vectors $\mathbf{w}(n)$, $\mathbf{w}(n + 1)$, and $\hat{\mathbf{w}}$; and the gradient at the nth iteration.

Point A on the diagram indicates the position of the weight after completion of the nth iteration. It is the cross-section of the constraint equation $\mathbf{w}^H\mathbf{S}_0 = 1$ and the power surface $\mathbf{w}^H(n)\mathbf{R}w\mathbf{(n)}$ (not shown in the figure). The weights are perturbed by adding a small amount $-\mu\mathbf{g}(\mathbf{w}(n))$ and then are projected on $\mathbf{w}^H\mathbf{S}_0 = 0$ using projection operator P. This is indicated by point B on the diagram. Note that $P\mathbf{S}_0 = \mathbf{0}$; thus the projection operator projects the weights orthogonal to $\mathbf{S}_0$. The vector $\mathbf{S}_0/L$ is added to restore the constraint. This action moves the updated weights $\mathbf{w}(n + 1)$ to point C. The process continues by moving the estimated weights toward point D, the optimal solution.

The effect of the gradient step size $\mu$ on the convergence speed and misadjustment noise may also be understood using Figure 3.1. A larger step size means that the weight vector moves faster toward point D, the solution point, but wanders around it over a larger region, not closely approaching and causing more misadjustment.

The gradient of $\mathbf{w}^H(n)R\mathbf{w}(n)$ with respect to $\mathbf{w}(n)$ is given by

$$\nabla_w \left(\mathbf{w}^H R \mathbf{w}\right)\Big|_{\mathbf{w}=\mathbf{w}(n)} = 2R\mathbf{w}(n) \tag{3.4.3}$$

and its computation using this expression requires knowledge of R, which normally is not available in practice. A typical scheme to estimate the required gradient is to replace R by its noisy sample $\mathbf{x}(n+1)\,\mathbf{x}^H(n+1)$ available at time instant $(n + 1)$.

There are a number of schemes used for estimating the required gradient [Fro72, Can80, God83, God86, God90, God89]. Even though the estimated gradient in each case is unbiased, the covariance of the estimated gradient obtained with each method is different, and thus the transient and steady-state behavior of the constrained algorithm is different in each case. In the following sections, some of these methods are described and the behavior of the algorithm in each case is examined.

First, the normal gradient estimation scheme where R is replaced by its noisy sample is discussed, and the algorithm in this case is referred to as the standard LMS algorithm to differentiate it from the algorithm when a gradient estimated by different methods is used.

In the next section, the gradient estimation scheme used by the standard LMS algorithm is described, and then some properties of the gradient are discussed along with the convergence of the weights estimated by the algorithm to the optimal weights and the study of the misadjustment [God86, God93].

### 3.4.1 Gradient Estimate

When all receiver outputs are accessible, the usual estimate of the gradient is made by multiplying the array output by the receiver output, that is,

$$\mathbf{g}(\mathbf{w}(n)) = 2\mathbf{x}(n+1)y^*(\mathbf{w}(n)) \tag{3.4.4}$$

In obtaining this estimate, the array correlation matrix has been replaced by $\mathbf{x}(n+1)\mathbf{x}^H(n+1)$, which is a noisy sample of the array correlation matrix at the time instant $(n + 1)$.

If $\{\mathbf{x}(n)\}$ is a zero-mean, stationary complex vector process, then for a given $\mathbf{w}(n)$ the estimate of the gradient defined by (3.4.4) is unbiased, that is,

$$\begin{aligned} E\left[\mathbf{g}(\mathbf{w}(n))\middle|\mathbf{w}(n)\right] &= E\left[2\mathbf{x}(n+1)y^*\left[\mathbf{w}(n)\right]\right] \\ &= 2E\left[\mathbf{x}(n+1)\mathbf{x}^H(n+1)\mathbf{w}(n)\right] \\ &= 2R\mathbf{w}(n) \end{aligned} \tag{3.4.5}$$

### 3.4.2 Covariance of Gradient

The covariance of the gradient estimate used in the weight update equation is important in determining the performance of the algorithm, as was discussed previously. To obtain results on the covariance of the gradient estimate defined by (3.4.4), it is necessary to make an additional Gaussian assumption about the sequence $\{\mathbf{x}(k)\}$. Thus, if $\{\mathbf{x}(k)\}$ is an i.i.d. complex Gaussian sequence, then $V_{gs}(\mathbf{w}(n))$, the covariance of the gradient estimated by this method for a given $\mathbf{w}(n)$, is given by

$$V_{gs}(\mathbf{w}(n)) = 4\mathbf{w}^H(n)R\mathbf{w}(n)R \tag{3.4.6}$$

A derivation of (3.4.6) is presented in Appendix 3.2.

It follows from the expression that the covariance at the nth iteration is proportional to the mean output power of the processor for a given **w**(n), the quantity that the gradient algorithm is trying to minimize. Thus, the gradient estimate improves as the weight vector approaches the optimal value.

### 3.4.3 Convergence of Weight Vector

In this section, results on convergence of the estimated weights to the optimal weights are presented. The derivation of these results appears in Appendix 3.3.

Let $\hat{\lambda}_{max}$ denote the maximum eigenvalue of PRP and $\hat{\lambda}_i$ denote the ith eigenvalue of PRP. If $\{\mathbf{x}(k)\}$ is an i.i.d. Gaussian sequence, and $\mathbf{w}^H(0)\mathbf{w}(0) < \infty$ and

$$0 < \mu < \frac{1}{\hat{\lambda}_{max}} \tag{3.4.7}$$

then

$$\lim_{n\to\infty} E[\mathbf{w}(n)] = \hat{\mathbf{w}} \tag{3.4.8}$$

and the convergence of E[**w**(n)] to $\hat{\mathbf{w}}$ along the ith eigenvector of PRP has the following time constant:

$$\tau_i = \frac{-1}{\ell n\left[1 - 2\mu\hat{\lambda}_i\right]} \tag{3.4.9}$$

where $\ell n[\cdot]$ denotes the natural logarithm.

Thus, the mean value of the estimated weights converges to the optimal weights in the limit provided that one starts with a bounded initial weight vector and the gradient step size is small enough to satisfy the condition (3.4.7). It should be noted that upper limit on the gradient step size, as well as convergence speed, depend on PRP. It follows from

$$R = p_S \mathbf{S}_0 \mathbf{S}_0^H + R_N \tag{3.4.10}$$

and

$$P\mathbf{S}_0 = \mathbf{0} \tag{3.4.11}$$

that PRP = $PR_NP$, and hence the convergence speed of the mean value of weights characterized by the time constants and the upper limit on the gradient step size only depend on the eigenvalues of $PR_NP$, indicating that the signal arriving from the look direction does not affect these quantities. The eigenvalues of $PR_NP$ are functions of the directions and powers of directional sources as well as the array geometry with the maximum eigenvalue being controlled by the strongest source governing the initial convergence speed. The latter part of the convergence is controlled by the smaller eigenvalues associated with weak sources or background noise, and thus the overall speed of the algorithm depends on the eigenvalue spread of $PR_NP$.

The discussion thus far has concentrated on the convergence of the mean value of weights to optimal weights. The variance of these weights is an important parameter and the transient and steady-state behavior of the weight covariance matrix $k_{ww}(n)$ are indicators of algorithm performance as discussed previously for the unconstrained LMS algorithm.

### 3.4.4 Weight Covariance Matrix

The weight covariance matrix is defined as

$$k_{ww}(n) = E\left[\left(\mathbf{w}(n) - \overline{\mathbf{w}}(n)\right)\left(\mathbf{w}(n) - \overline{\mathbf{w}}(n)\right)^H\right] \tag{3.4.12}$$

where

$$\overline{\mathbf{w}}(n) = E\left[\mathbf{w}(n)\right] \tag{3.4.13}$$

Appendix 3.4 shows that the matrix satisfies the following recursive relations. If $V_g\mathbf{w}(n))$ denotes the covariance of the gradient used in the constrained LMS algorithm for a given $\mathbf{w}(n)$, and $k_{ww}(n)$ denotes the covariance of $\mathbf{w}(n)$, then

$$\begin{aligned} k_{ww}(n+1) &= Pk_{ww}(n)P - 2\mu P\left[Rk_{ww}(n) + k_{ww}(n)R\right]P \\ &\quad + 4\mu^2 PRk_{ww}(n)RP + \mu^2 PE\left[V_g\left(\mathbf{w}(n)\right)\right]P \end{aligned} \tag{3.4.14}$$

where the expectation is taken over $\mathbf{w}$.

The weight covariance matrix at each iteration depends on the mean value of the covariance of the gradient used at the previous iteration. Equation (3.4.14) may be further simplified by substituting for $V_g(\mathbf{w}(n))$. Taking expectation over $\mathbf{w}(n)$, pre- and post-multiplying by P on both sides of the expression for the covariance of the gradient given by (3.4.6) and using (3.2.60),

$$\begin{aligned} PE\left[V_{gs}\left(\mathbf{w}(n)\right)\right]P &= 4PRPE\left[\mathbf{w}^H(n)R\mathbf{w}(n)\right] \\ &= 4PRP\left\{Tr\left[Rk_{ww}(n)\right] + k_0(n)\right\} \end{aligned} \tag{3.4.15}$$

where

$$k_0(n) = \overline{\mathbf{w}}^H(n)R\overline{\mathbf{w}}(n) \tag{3.4.16}$$

Equations (3.4.14) and (3.4.15) imply that

$$\begin{aligned} k_{ww}(n+1) &= PK_{ww}(n)P - 2\mu\left\{PRK_{ww}(n) + k_{ww}(n)RP\right\} \\ &\quad + 4\mu^2 PRk_{ww}(n)RP + 4\mu^2 PRP\left\{Tr\left[Rk_{ww}(n)\right] + k_0(n)\right\} \end{aligned} \tag{3.4.17}$$

### 3.4.5 Transient Behavior of Weight Covariance Matrix

The study of the convergence and transient behavior of the weight covariance matrix presented here requires that the matrix be diagonalized. Conditions required for diagonalization of the weight covariance matrix by the transformation, which also diagonalizes PRP, are described below.

The necessary and the sufficient condition for the diagonalization of $k_{ww}(n + 1)$, $n \geq 0$, and PRP by the same unitary transformation is that the unitary transformation also diagonalizes $PE[V_g(\mathbf{w}(n))]P$ for all n, where $V_g(\mathbf{w}(n))$ is the covariance of $\mathbf{g}(\mathbf{w}(n))$ for a given $\mathbf{w}(n)$ and the expectation is taken over $\mathbf{w}$. A proof of the diagonalization conditions is presented in Appendix 3.5.

Thus, to verify that the weight covariance matrix for the standard algorithm is diagonalizable by the same unitary transformation that diagonalizes PRP, we need to test if this transformation diagonalizes $PE[V_{g_S}(\mathbf{w}(n))]P$. Since PRP is a Hermitian matrix, a unitary matrix $\hat{Q}$ exists, such that

$$\hat{Q}^H PRP\hat{Q} = \hat{\Lambda} \tag{3.4.18}$$

where $\hat{\Lambda}$ is a diagonal matrix with its diagonal elements being the eigenvalues of PRP.

It follows from (3.4.15) and (3.4.18) that

$$\hat{Q}^H PE\left[V_{g_S}(\mathbf{w}(n))\right]P\hat{Q} = 4\hat{\Lambda}\mathrm{Tr}\left[Rk_{ww}(n)\right] + 4\hat{\Lambda}k_0(n) \tag{3.4.19}$$

This implies that $V_{g_S}(w(n))$ satisfy the conditions required for the diagonalization of $k_{ww}(n)$. Thus, $\hat{Q}^H k_{ww}(n)\hat{Q}$ is a diagonal matrix when the covariance of the gradient used for updating $\mathbf{w}(n)$ is given by (3.4.6). Let this be denoted by diagonal matrix $\Sigma(n)$, that is,

$$\Sigma(n) = \hat{Q}^H k_{ww}(n)\hat{Q} \tag{3.4.20}$$

Now the transient behavior of $\Sigma(n)$ is analyzed. To study the transient behavior of $\Sigma(n)$, a matrix difference equation for $\Sigma(n)$ is developed, a vector difference equation for its diagonal terms is derived, and its solution is presented.

Pre- and postmultiplying (3.4.17) by $\hat{Q}^H$ and $\hat{Q}$, noting that

$$P^2 = P \tag{3.4.21}$$

$$k_{ww}(n) = Pk_{ww}(n)P \tag{3.4.22}$$

and using (3.4.20), the following matrix difference equation is derived:

$$\begin{aligned}\Sigma(n+1) = \Sigma(n) - 4\mu\hat{\Lambda}\Sigma(n) + 4\mu^2\hat{\Lambda}^2\Sigma(n) \\ + 4\mu^2\Lambda\left\{\mathrm{Tr}\left(\hat{\Lambda}\Sigma(n)\right) + k_0(n)\right\}\end{aligned} \tag{3.4.23}$$

Let the two L-dimensional vectors $\hat{\boldsymbol{\lambda}}$ and $\boldsymbol{\eta}(n)$ represent the L eigenvalues of PRP and $k_{ww}(n)$, respectively, that is,

$$\hat{\boldsymbol{\lambda}} = \left[\hat{\lambda}_1, \hat{\lambda}_2, \ldots, \hat{\lambda}_L\right]^T \tag{3.4.24}$$

and

$$\boldsymbol{\eta}(n) = \left[\eta_1, \eta_2, \ldots, \eta_L\right]^T \tag{3.4.25}$$

where $\hat{\lambda}_i$ and $\eta_i(n)$, i = 1, 2, …, L, are the eigenvalues of PRP and $k_{\mathbf{ww}}(n)$, respectively.

From (3.4.23) to (3.4.25) and the fact that

$$\mathrm{Tr}\left[\hat{\Lambda}\Sigma(n)\right] = \hat{\boldsymbol{\lambda}}^T\boldsymbol{\eta}(n) \tag{3.4.26}$$

the following vector difference equation for the eigenvalues of $k_{\mathbf{ww}}(n)$ is derived:

$$\boldsymbol{\eta}(n+1) = \left[1 - 4\mu\hat{\Lambda} + 4\mu^2\hat{\Lambda}^2 + 4\mu^2\hat{\boldsymbol{\lambda}}\hat{\boldsymbol{\lambda}}^T\right]\boldsymbol{\eta}(n) + 4\mu^2 k_0(n)\hat{\boldsymbol{\lambda}} \tag{3.4.27}$$

Since

$$\lim_{n\to\infty} \overline{\mathbf{w}}(n) = \hat{\mathbf{w}} \tag{3.4.28}$$

it follows from (3.4.16) that

$$\lim_{n\to\infty} k_0(n) = \hat{\mathbf{w}}^H R\hat{\mathbf{w}} \tag{3.4.29}$$

With

$$H = 4\mu\hat{\Lambda} - 4\mu^2\hat{\Lambda}^2 - 4\mu^2\hat{\boldsymbol{\lambda}}\hat{\boldsymbol{\lambda}}^T \tag{3.4.30}$$

(3.4.27) has the solution

$$\boldsymbol{\eta}(n) = (I - H)^n\boldsymbol{\eta}(0) + 4\mu^2\sum_{i=1}^{n} k_0(n-i)(I-H)^{i-1}\hat{\boldsymbol{\lambda}} \tag{3.4.31}$$

where $\boldsymbol{\eta}(0)$ denotes the eigenvalues of $k_{\mathbf{ww}}(0)$. Since $\hat{Q}$ diagonalizes $k_{\mathbf{ww}}(n)$, it follows that

$$k_{\mathbf{ww}}(n) = \sum_{\ell=1}^{L} \eta_\ell(n)\hat{\mathbf{Q}}_\ell\hat{\mathbf{Q}}_\ell^H \tag{3.4.32}$$

where $\hat{\mathbf{Q}}_\ell$, $\ell$ = 1, 2, …, L are the eigenvectors of PRP.

Equations (3.4.31) and (3.4.32) completely describe the transient behavior of the weight covariance.

### 3.4.6 Convergence of Weight Covariance Matrix

In this section, the convergence of the weight covariance matrix is examined. Consider the following equation:

$$\boldsymbol{\eta}(n+1) = (I - H)\boldsymbol{\eta}(n) + \mu^2 4k_0(n)\hat{\boldsymbol{\lambda}} \tag{3.4.33}$$

This represents a set of L difference equations. Before studying the convergence, these equations are reduced to a set of $L-1$ difference equations by showing that one of the components in each of the vectors is identical to zero.

Let $\lambda_{min}(.)$ denote the minimum eigenvalue of a matrix (.). Based on (3.4.22) and $\lambda_{min}(P) = 0$, $\lambda_{min}(k_{ww}(n)) = 0$. Also, $\hat{\lambda}_{min} = 0$. Let

$$\hat{\lambda}_\ell \triangleq \hat{\lambda}_{min} = 0 \tag{3.4.34}$$

and $\hat{\mathbf{Q}}_\ell$ be the eigenvector corresponding to $\hat{\lambda}_\ell$. Since $\hat{Q}$ diagonalizes $k_{ww}(n)$ and P, $\hat{\mathbf{Q}}_\ell$ must also be the eigenvector corresponding to the zero eigenvalue of $k_{ww}(n)$ and P. Thus,

$$\eta_\ell(n) = 0 \tag{3.4.35}$$

It follows from (3.4.34) and (3.4.35) that the $\ell$th difference equation in (3.4.33) is identical to zero. Thus, these reduce to a set of $L-1$ difference equations. Define $L-1$ dimensional vectors $\hat{\boldsymbol{\lambda}}'$ and $\boldsymbol{\eta}'(n)$ such that the ith component is given by

$$(\cdot)_i' \triangleq \begin{cases} (\cdot)_i, & i = 1, 2, \ldots, \ell - 1 \\ (\cdot)_{i+1}, & i = \ell, \ell+1, \ldots, L-1 \end{cases} \tag{3.4.36}$$

where $(\cdot)'$ denotes the $L-1$ dimensional vectors $\hat{\boldsymbol{\lambda}}'$ and $\boldsymbol{\eta}'(n)$, and (.) denotes the corresponding L-dimensional vectors $\hat{\boldsymbol{\lambda}}$ and $\boldsymbol{\eta}(n)$. Similarly, define an $L-1 \times L-1$ dimensional matrices $H'$ by dropping the column of zeros and the row of zeros from H.

With $\hat{\Lambda}'$ denoting the diagonal matrix of $L-1$ nonzero eigenvalues of PRP, it follows from (3.4.30) and the definition of the above $L-1$ dimensional vectors that

$$H' = 4\mu\hat{\Lambda}' - 4\mu^2\hat{\Lambda}'^2 - 4\mu^2\hat{\boldsymbol{\lambda}}'\hat{\boldsymbol{\lambda}}'^T \tag{3.4.37}$$

It follows from (3.4.33) to (3.4.37) that

$$\boldsymbol{\eta}'(n+1) = (I - H')\boldsymbol{\eta}'(n) + 4\mu^2 k_0(n)\boldsymbol{\lambda}' \tag{3.4.38}$$

It can be shown that $\lim_{n\to\infty} \boldsymbol{\eta}'(n)$ exists under certain conditions (see Appendix 3.6) and is given by

$$\lim_{n\to\infty} \boldsymbol{\eta}'(n) = 4\mu^2\hat{\mathbf{w}}^H R\hat{\mathbf{w}} H'^{-1}\hat{\boldsymbol{\lambda}}' \tag{3.4.39}$$

Substituting for the inverse of $H'$,

$$\lim_{n\to\infty} \boldsymbol{\eta}'(n) = \frac{\mu \hat{\mathbf{w}}^H R \hat{\mathbf{w}}}{1-\mu \sum_{i=1}^{L-1} \frac{\hat{\lambda}_i'}{1-\mu\hat{\lambda}'}} \left[ \frac{1}{1-\mu\hat{\lambda}_i'}, \ldots, \frac{1}{1-\mu\hat{\lambda}_{L-1}'} \right]^T \tag{3.4.40}$$

Substituting for the eigenvalues of $k_{ww}(n)$ from (3.4.40) in (3.4.32) yields the steady-state expressions for the covariance matrices.

$$\lim_{n\to\infty} k_{ww}(n) = \frac{\mu \hat{\mathbf{w}}^H R \hat{\mathbf{w}}}{1-\mu \sum_{i=1}^{L-1} \frac{\hat{\lambda}_i'}{1-\mu\hat{\lambda}_i'}} \sum_{i=1}^{L-1} \frac{1}{1-\mu\hat{\lambda}_i'} \hat{Q}_i \hat{Q}_i^H \tag{3.4.41}$$

### 3.4.7 Misadjustment

Misadjustment is a dimensionless measure of algorithm performance near the convergence point as discussed previously. It is a normalized difference between the adaptive and optimal performance of a processor. It is defined as the ratio of the excess mean output power to the optimal power, that is,

$$M = \lim_{n\to\infty} \frac{E\left[\mathbf{w}^H(n)\mathbf{x}(n+1)\mathbf{x}^H(n+1)\mathbf{w}(n)\right] - \hat{\mathbf{w}}^H R \hat{\mathbf{w}}}{\hat{\mathbf{w}}^H R \hat{\mathbf{w}}} \tag{3.4.42}$$

Noting that $\mathbf{w}(n)$ and $\mathbf{x}(n + 1)$ are independent, the expectation over $\mathbf{w}(n)$ and $\mathbf{x}(n + 1)$ in (3.4.42) can be taken independently. Taking the conditional expectation for a given $\mathbf{w}(n)$, it follows that

$$\begin{aligned} E\left[\mathbf{w}^H(n)\mathbf{x}(n+1)\mathbf{x}^H(n+1)\mathbf{w}(n) \middle| \mathbf{w}(n)\right] &= \mathbf{w}^H(n) R \mathbf{w}(n) \\ &= \mathrm{Tr}\left[\mathbf{w}(n)\mathbf{w}^H(n) R\right] \end{aligned} \tag{3.4.43}$$

Since

$$\begin{aligned} E\left[\mathbf{w}(n)\mathbf{w}^H(n)\right] &= R_{ww}(n) \\ &= k_{ww}(n) + \overline{\mathbf{w}}(n)\overline{\mathbf{w}}^H(n) \end{aligned} \tag{3.4.44}$$

it follows from (3.4.43), after taking unconditional expectation on both sides, that

$$\begin{aligned} E\left[\mathbf{w}^H(n)\mathbf{x}(n+1)\mathbf{x}^H(n+1)\mathbf{w}(n)\right] &= \mathrm{Tr}\left[R_{ww}(n) R\right] \\ &= \mathrm{Tr}\left[k_{ww}(n) R + \overline{\mathbf{w}}(n)\overline{\mathbf{w}}^H(n) R\right] \\ &= \mathrm{Tr}\left[k_{ww}(n) R\right] + \overline{\mathbf{w}}^H(n) R \overline{\mathbf{w}}(n) \end{aligned} \tag{3.4.45}$$

and (3.4.42) becomes

$$M = \lim_{n\to\infty} \frac{\mathrm{Tr}\left[k_{ww}(n)R\right] + \overline{\mathbf{w}}^H(n)R\overline{\mathbf{w}}(n) - \hat{\mathbf{w}}^H R\hat{\mathbf{w}}}{\hat{\mathbf{w}}^H R\hat{\mathbf{w}}} \tag{3.4.46}$$

The contribution of the second and third terms in (3.4.46) is zero in the limit because of (3.4.8). Since $k_{ww}(n) = Pk_{ww}(n)P$, it follows that

$$\begin{aligned}\mathrm{Tr}\left[k_{ww}(n)R\right] &= \mathrm{Tr}\left[Pk_{ww}(n)PR\right]\\ &= \mathrm{Tr}\left[k_{ww}(n)PRP\right]\\ &= \mathrm{Tr}\left[k_{ww}(n)\hat{Q}\hat{\Lambda}\hat{Q}^H\right]\\ &= \mathrm{Tr}\left[\hat{Q}^H k_{ww}(n)\hat{Q}\hat{\Lambda}\right]\\ &= \hat{\boldsymbol{\lambda}}^T \mathbf{d}(n)\end{aligned} \tag{3.4.47}$$

where

$$\mathbf{d}(n) = \mathrm{Diag}\left[\hat{Q}^H k_{ww}(n)\hat{Q}\right] \tag{3.4.48}$$

Thus, (3.4.46) becomes

$$M = \lim_{n\to\infty} \frac{\hat{\boldsymbol{\lambda}}^T \mathbf{d}(n)}{\hat{\mathbf{w}}^H R\hat{\mathbf{w}}} \tag{3.4.49}$$

Appendix 3.6 proves that for the standard LMS algorithm, if

$$0 < \mu < \frac{1}{2\hat{\lambda}_{max}} \tag{3.4.50}$$

and

$$\mu \sum_{i=1}^{L-1} \frac{\hat{\lambda}_i}{1-\mu\hat{\lambda}_i} < 1 \tag{3.4.51}$$

then the misadjustment is given by

$$M_S = \frac{\mu \displaystyle\sum_{i=1}^{L-1} \frac{\hat{\lambda}_i}{1-\mu\hat{\lambda}_i}}{1-\mu \displaystyle\sum_{i=1}^{L-1} \frac{\hat{\lambda}_i}{1-\mu\hat{\lambda}_i}} \tag{3.4.52}$$

For sufficiently small $\mu$, this results in

$$M_S \approx \mu \sum_{i=1}^{L-1} \hat{\lambda}_i = \mu \mathrm{Tr}(PRP) \tag{3.4.53}$$

## 3.5 Perturbation Algorithms

The LMS algorithm discussed in previous sections requires that the signals on all elements are accessible. In some situations this may not be possible. For example, in a large radio frequency array it may not be economical to provide a coherent channel on all elements in the array and thereby make the required signal inaccessible. In situations like this, one needs to estimate the required gradient by other means if the LMS algorithm is to be used for weight updating.

In this section, a method to estimate the required gradient for the LMS algorithm when the signals on all elements are not accessible is described using three different processor structures. One structure uses a single receiver to measure the power of the processor and is referred to as a single-receiver system. The other two structures use two receivers to measure the output power, one using dual perturbation and the other using a reference receiver. The gradient estimate obtained using three different structures is unbiased [Can80].

LMS algorithm performance using the gradient estimate by this method can be analyzed using an approach similar to that used in previous sections. However, the results on the mean and covariance of the gradient, and the covariance of the weights and misadjustments are stated in this section. The method described in this section is for updating weights of the constrained optimal beamformer. The methods applicable to other processors can easily be developed using a similar approach.

The method uses orthogonal sequences to perturb the weights of the processor, and then measures the output power of the processor to estimate the required gradient. The LMS algorithm using the gradient estimated by this method is referred to as perturbation algorithm [Can80]. The perturbation algorithm requires more array samples and thus more time than the LMS algorithm discussed in previous sections. A weight iteration cycle in this case includes a complete weight perturbation cycle occupying, say, M time instants to estimate the required gradient. Thus, the weight iteration index and the time index are not the same in the perturbation algorithm, as may be the case for standard LMS algorithm. Details on the algorithm and its analyses may be found in [Can80, God83, God86, God93].

Consider some useful definitions required to understand the material discussed in this section. Let S denote a complex vector sequence defined as

$$S = \{\boldsymbol{\delta}(1), \boldsymbol{\delta}(2), \ldots, \boldsymbol{\delta}(M)\} \tag{3.5.1}$$

where $\boldsymbol{\delta}(\ell)$, $\ell = 1, 2, \ldots, M$ are L-dimensional complex vectors.

The sequence S is said to be an orthogonal complex vector sequence if

$$\frac{1}{M}\sum_{i=1}^{M} \mathrm{Re}[\boldsymbol{\delta}(i)]\mathrm{Re}[\boldsymbol{\delta}^H(i)] = I \tag{3.5.2}$$

$$\frac{1}{M}\sum_{i=1}^{M}\mathrm{Im}[\boldsymbol{\delta}(i)]\mathrm{Im}[\boldsymbol{\delta}^{H}(i)] = I \tag{3.5.3}$$

$$\frac{1}{M}\sum_{i=1}^{M}\mathrm{Re}[\boldsymbol{\delta}(i)]\mathrm{Im}[\boldsymbol{\delta}^{H}(i)] = 0 \tag{3.5.4}$$

and

$$\frac{1}{M}\sum_{i=1}^{M}\mathrm{Im}[\boldsymbol{\delta}(i)]\mathrm{Re}[\boldsymbol{\delta}^{H}(i)] = 0 \tag{3.5.5}$$

The sequence S is said to be of zero mean if

$$\frac{1}{M}\sum_{i=1}^{M}\boldsymbol{\delta}(i) = 0 \tag{3.5.6}$$

and is said to have odd symmetry if for every i, $1 \le i \le M$, there exists a j, $1 \le j \le M$, such that $\boldsymbol{\delta}(i) = -\boldsymbol{\delta}(j)$.

The next section discusses a scheme to generate the required perturbation sequences.

### 3.5.1 Time Multiplex Sequence

Perturbation sequences with the required properties for obtaining an unbiased gradient estimate can be constructed in a number of ways. However, for a time multiplex sequence it is possible to evaluate certain expressions in closed form. A procedure to construct a time multiplex sequence is given below. Let

$$h_i(j) = \begin{cases} \sqrt{2L}, & j = 2i - 1 \\ -\sqrt{2L}, & j = 2i \\ 0 & \text{elsewhere in the range } 1 \le j \le 4L \end{cases} \qquad i = 1, 2, \ldots, 2L \tag{3.5.7}$$

A multiplex sequence can be defined in terms of $h_i(j)$ as follows:

$$\left.\begin{aligned} \mathrm{Re}(\delta_i(j)) &= h_i(j) \\ \mathrm{Im}(\delta_i(j)) &= h_{i+L}(j) \end{aligned}\right\} i = 1, 2, \ldots, L, \quad j = 1, 2, \ldots, 4L \tag{3.5.8}$$

where $\delta_i(j)$ denotes the ith element of the column vector $\boldsymbol{\delta}(j)$.

It can be verified that S has zero-mean odd symmetry and satisfies the required orthogonality properties. The time multiplex sequence defined above has length $M = 4L$ and can be used to obtain an unbiased gradient estimate for all three structures. However, in the case of a dual receiver with dual perturbation, a time multiplex sequence of length $M = 2L$ can be constructed, which provides an unbiased estimate of the gradient [Can80].

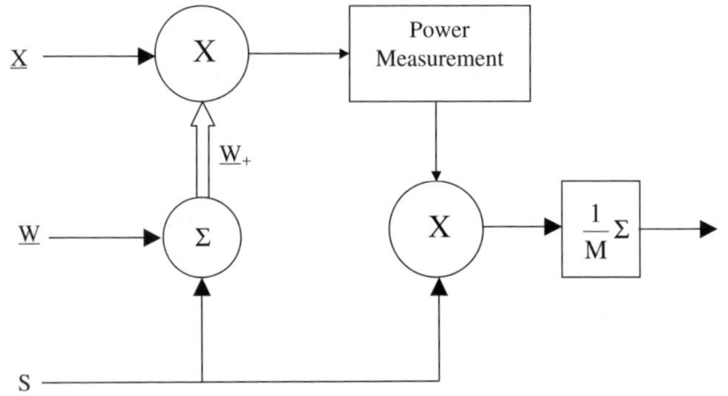


**FIGURE 3.2**
Schematic diagram of a single receiver system.

### 3.5.2 Single-Receiver System

In this section, a gradient estimate scheme using a single receiver system is described. Figure 3.2 shows a schematic diagram of a single receiver system. The sequence S is used to perturb the array weights about their nominal value **w**(n). The instantaneous output power is then correlated with the sequence S and an estimate of the required gradient is made.

At the ith instant within the perturbation cycle, $1 \le i \le M$, the weight vector is given by

$$\mathbf{w}_+(\mathbf{w}(n), i) = \mathbf{w}(n) + \gamma\boldsymbol{\delta}(i), \quad 1 \le i \le M \tag{3.5.9}$$

where $\gamma$ is a real positive scalar and denotes the perturbation step size. An estimate of the gradient is given by

$$\mathbf{g}_1(\mathbf{w}(n)) = \frac{1}{\gamma M}\sum_{i=1}^{M} f_1(\mathbf{w}_+, i)\boldsymbol{\delta}(i) \tag{3.5.10}$$

where $f_1(\mathbf{w}_+, i)$ is the instantaneous array output power given by

$$f_1(\mathbf{w}_+, i) = \mathbf{w}_+^H(\mathbf{w}(n), i)\mathbf{x}(\ell + i)\mathbf{x}^H(\ell + i)\mathbf{w}_+(\mathbf{w}(n), i) \tag{3.5.11}$$

and $\ell$ is the time instant at which the perturbation cycle is initiated.

If the orthogonal perturbation sequence has odd symmetry, then for any $\gamma > 0$, the estimate of the gradient defined by (3.5.10) is unbiased for a given **w**(n), that is,

$$E\left[\mathbf{g}_1(\mathbf{w}(n)) \middle| \mathbf{w}(n)\right] = 2R\mathbf{w}(n) \tag{3.5.12}$$

#### *3.5.2.1 Covariance of the Gradient Estimate*

Let $V_{g_1}(\mathbf{w}(n))$ denote the covariance of the gradient estimate defined by (3.5.10). If $\{\mathbf{x}(n)\}$ is an i.i.d. Gaussian sequence, then for the time multiplex perturbation sequence defined by (3.5.8),

$$V_{g_1}(\mathbf{w}(n)) = \gamma^2\left[2L(\text{diag}(R))^2\right] + \frac{1}{\gamma^2}\left[\frac{1}{2L}\left\{\mathbf{w}^H(n)R\mathbf{w}(n)\right\}^2 I\right] \\ + 2\text{diag}\left[R\mathbf{w}(n)\mathbf{w}^H(n)R\right] + 2\mathbf{w}^H(n)R\mathbf{w}(n)\text{diag}(R) \quad (3.5.13)$$

The second and fourth terms in (3.5.13) are proportional to $\mathbf{w}^H(n)R\mathbf{w}(n)$, the quantity that the adaptive algorithm is attempting to minimize. Thus, the gradient estimate improves as the weight vector approaches the optimum. However, the first and third terms do not necessarily decrease. Interestingly, the fourth term is similar to the term in $V_{g_S}(\mathbf{w}(n))$, the covariance of the gradient estimate used in the standard algorithm. The first and second terms are penalties due to the use of perturbation for estimating the gradient. The third term is due to the mean of the gradient, which is not canceled in the single-receiver system.

### 3.5.2.2 Perturbation Noise

Although the estimated gradient is unbiased and independent of $\gamma$, the covariance of the gradient is a function of $\gamma$. Furthermore, the presence of perturbations on the weights causes an increase in the output power. This power is proportional to $\gamma^2$. An indication of the effect of the perturbation can be obtained by determining the excess output power, referred to as the perturbation noise, $\xi$ due to perturbation of weights about a nominal weight $\mathbf{w}(n)$.

For any orthogonal sequences S having a zero mean, the excess power due to perturbation about a nominal weight $\mathbf{w}(n)$ is given by

$$\xi(\gamma) = 2\gamma^2 \text{Tr}(R) \quad (3.5.14)$$

Note from (3.5.13) that $V_{g_1}(\mathbf{w}(n))$ is a convex function of $\gamma$, and the optimal value $\hat{\gamma}(\mathbf{w}(n))$ for which $V_{g_1}(\mathbf{w}(n))$ is minimum can be found. For a time multiplex perturbation sequence, the following result can be established.

Let $\hat{\gamma}(\mathbf{w}(n))$ represent the value of $\gamma(\mathbf{w}(n))$ for which $V_{g_1}(\mathbf{w}(n))$ is minimum. Then

$$\hat{\gamma}(\mathbf{w}(n)) = \left[\frac{\mathbf{w}^H(n)R\mathbf{w}(n)}{2\text{Tr}(R)}\right]^{1/2} \quad (3.5.15)$$

Let $V_{g_1}(\mathbf{w}(n))$ represent the value of $V_{g_1}(\mathbf{w}(n))$ at $\hat{\gamma}(\mathbf{w}(n))$. Then

$$\hat{V}_{g_1}(\mathbf{w}(n)) = 4\mathbf{w}^H(n)R\mathbf{w}(n)\text{diag}(R) \\ + 2\ \text{diag}\left[R\mathbf{w}(n)\mathbf{w}^H(n)R\right] \quad (3.5.16)$$

The perturbation noise when the optimal $\gamma$ is used can be obtained by substituting (3.5.15) in (3.5.14). The result is given by

$$\xi(\hat{\gamma}(n)) = \mathbf{w}^H(n)R\mathbf{w}(n) \quad (3.5.17)$$

Assuming that the gradient algorithm converges, then $\xi(\hat{\gamma}(n))$ is approximately given by

$$\xi(\hat{\gamma}(n)) \cong \hat{\mathbf{w}}^H R\hat{\mathbf{w}} \quad (3.5.18)$$

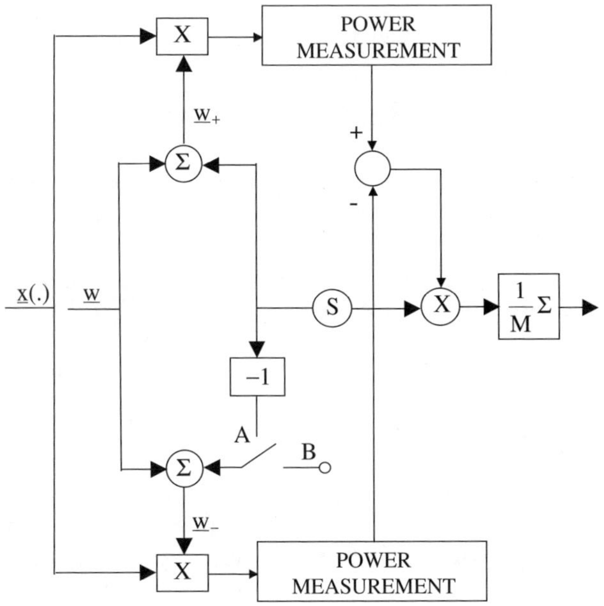


**FIGURE 3.3**
Schematic diagram of a two-receiver system.

### 3.5.3 Dual-Receiver System

In this section, a gradient estimation scheme using a processor with two receivers is described. In a two-receiver system, an estimate of the required gradient can be obtained by applying a perturbation sequence S in antiphase to the two sets of weights, and correlating the difference power from the receivers with S as shown in Figure 3.3 with switch position A. Thus, Receiver 1 has its weights perturbed according to

$$\mathbf{w}_+(\mathbf{w}(n), i) = \mathbf{w}(n) + \gamma\boldsymbol{\delta}(i), \quad 1 \le i \le M \tag{3.5.19}$$

and Receiver 2 has its weights perturbed according to

$$\mathbf{w}_-(\mathbf{w}(n), i) = \mathbf{w}(n) - \gamma\boldsymbol{\delta}(i), \quad 1 \le i \le M \tag{3.5.20}$$

Let $f_1(\mathbf{w}_+, i)$ and $f_2(\mathbf{w}_-, i)$ denote the instantaneous output power at Receivers 1 and 2, respectively. An estimate of the gradient is given by

$$\mathbf{g}_2(\mathbf{w}(n)) = \frac{1}{2\gamma M} \sum_{i=1}^{M} \left[ f_1(\mathbf{w}_+, i) - f_2(\mathbf{w}_-, i) \right] \boldsymbol{\delta}(i) \tag{3.5.21}$$

For a given weight vector $\mathbf{w}(n)$, the estimate of the gradient defined by (3.5.21) is unbiased for any orthogonal perturbation sequence S.

#### 3.5.3.1 *Dual-Receiver System with Reference Receiver*

In a two-receiver system, an estimate of the required gradient can also be obtained by using a perturbation sequence S to perturb the array weights of only one of the receivers about their nominal value $\mathbf{w}(n)$, while the other receiver has its weights fixed at $\mathbf{w}(n)$ as shown in Figure 3.3 with switch position B. Let Receiver 1 have its weights perturbed by a sequence S so that its weight vector is given by

$$\mathbf{w}_+(\mathbf{w}(n),i) = \mathbf{w}(n) + \gamma\boldsymbol{\delta}(i), \quad 1 \le i \le M \tag{3.5.22}$$

Let $f_1(\mathbf{w}_+,i)$ and $f_2(\mathbf{w},i)$ denote the output power of receivers 1 and 2, respectively. An estimate of the gradient is given by

$$g_3(\mathbf{w}(n)) = \frac{1}{\gamma M}\sum_{i=1}^{M}\left[f_1(\mathbf{w}_+,i) - f_2(\mathbf{w},i)\right]\boldsymbol{\delta}(i) \tag{3.5.23}$$

The estimate of the gradient defined by (3.5.23) is unbiased when S is an orthogonal perturbation sequence and has odd symmetry.

#### 3.5.3.2 *Covariance of Gradient*

For two-receiver systems, the following result can be established. Let $V_{g_2}(\mathbf{w}(n))$ and $V_{g_3}(\mathbf{w}(n))$ denote the covariance of the gradient estimated by (3.5.21) and (3.5.23), respectively. If $\{\mathbf{x}(n)\}$ is an i.i.d. Gaussian sequence, then for the time multiplex perturbation sequence defined by (3.5.8),

$$V_{g_2}(\mathbf{w}(n)) = 2\mathbf{w}^H(n)R\mathbf{w}(n)\mathrm{diag}(R) \tag{3.5.24}$$

and

$$V_{g_3}(\mathbf{w}(n)) = \gamma^2\left[2L(\mathrm{diag}(R))^2\right] + 2\mathbf{w}^H(n)R\mathbf{w}(n)\mathrm{diag}(R) \tag{3.5.25}$$

$V_{g_2}(\mathbf{w}(n))$ and the second term for $V_{g_3}(\mathbf{w}(n))$ are proportional to $\mathbf{w}^H(n)R\mathbf{w}(n)$, the quantity that the adaptive algorithm is attempting to minimize. Thus, the gradient estimate improves as the weight vector approaches the optimal value.

### 3.5.4 Covariance of Weights

It can be established that the weight covariance matrix is diagonalizable when the covariance of the gradient used for updating $\mathbf{w}(n)$ is $V_{g_2}(\mathbf{w}(n))$ or $V_{g_3}(\mathbf{w}(n))$. Thus, for these two cases an analysis of the weight covariance matrix is possible by developing matrix and vector difference equations using the scheme presented in Section 3.4. The results on the transient and steady-state behavior of this matrix for the two cases are presented in this section [God86].

The weight covariance matrix is not diagonalizable when the covariance of the gradient used for updating $\mathbf{w}(n)$ is $V_{g_1}(\mathbf{w}(n))$. Consequently, it is not possible to describe the transient and the steady-state behavior of the weight covariance matrix for the single-receiver system using the scheme presented in Section 3.4.

#### *3.5.4.1 Dual-Receiver System with Dual Perturbation*

Substituting $V_{g_2}(\mathbf{w}(n))$ for $V_g(\mathbf{w}(n))$ in (3.4.14), and following a procedure similar to that used in Section 3.4.5, the following matrix difference equation is derived:

$$\begin{aligned}\Sigma(n+1) &= \left(I - 4\mu\hat{\Lambda} + 4\mu^2\hat{\Lambda}^2\right)\Sigma(n)\\ &+\mu^2\frac{2}{L}\mathrm{Tr}(R)\left[\mathrm{Tr}\left(\hat{\Lambda}\Sigma(n)\right)+k_0(n)\right]\Gamma\end{aligned} \tag{3.5.26}$$

where $\Gamma$ is a diagonal matrix with its diagonal elements being the eigenvalues of P.

Let an L-dimensional vector $\boldsymbol{\vartheta}$ denote the L eigenvalues of P and an L-dimensional vector $\boldsymbol{\eta}_2(n)$ denote the L eigenvalues of the weight covariance matrix $k_{\mathbf{ww}_2}(n)$ when the covariance of the gradient used is $V_{g_2}(\mathbf{w}(n))$. Since $\mathrm{Tr}(\hat{\boldsymbol{\Lambda}}\ \Sigma(n) \equiv \hat{\boldsymbol{\lambda}}^T\boldsymbol{\eta}_2(n)$, (3.5.26) reduces to the following vector difference equation:

$$\boldsymbol{\eta}_2(n+1) = \left[I - 4\mu\hat{\Lambda} + 4\mu^2\hat{\Lambda}^2 + \mu^2\frac{2}{L}\mathrm{Tr}(R)\boldsymbol{\vartheta}\hat{\boldsymbol{\lambda}}^T\right]\boldsymbol{\eta}_2(n) + \mu^2\frac{2}{L}\mathrm{Tr}(R)k_0(n)\boldsymbol{\vartheta} \tag{3.5.27}$$

With

$$H_2 = 4\mu\hat{\Lambda} - 4\mu^2\hat{\Lambda}^2 - \frac{2}{L}\mathrm{Tr}(R)\mu^2\boldsymbol{\vartheta}\hat{\boldsymbol{\lambda}}^T \tag{3.5.28}$$

the solution of (3.5.27) is given by

$$\boldsymbol{\eta}_2(n) = (I - H_2)^n\boldsymbol{\eta}_2(0) + \mu^2\frac{2}{L}\mathrm{Tr}(R)\sum_{i=1}^{n}k_0(n-i)(I-H_2)^{i-1}\boldsymbol{\vartheta} \tag{3.5.29}$$

and $k_{\mathbf{ww}_2}(n)$ is given by

$$k_{\mathbf{ww}_2}(n) = \sum_{\ell=1}^{L}\eta_{2\ell}(n)\hat{\mathbf{Q}}_\ell\hat{\mathbf{Q}}_\ell^H \tag{3.5.30}$$

where $\boldsymbol{\eta}_2(0)$ is the vector of eigenvalues of $k_{\mathbf{ww}_2}(0)$ and $\hat{\mathbf{Q}}_\ell$, $\ell = 1, 2, \ldots, L$ are the eigenvectors of PRP. Equations (3.5.29) and (3.5.30) completely describe the transient behavior of the weight covariance matrix.

The steady-state expression for the weight covariance matrix is obtained by substituting the steady-state value of $L-1$ nonzero eigenvalues of the weight covariance matrix. The steady-state expression for these eigenvalues is given by

$$\lim_{n\to\infty}\boldsymbol{\eta}_2'(n) = \frac{\mu\dfrac{\mathrm{Tr}(R)}{2L}\hat{\mathbf{w}}^H R\hat{\mathbf{w}}}{1-\mu\dfrac{\mathrm{Tr}(R)}{2L}\displaystyle\sum_{i=1}^{L-1}\frac{1}{1-\mu\hat{\lambda}_i}}\left[\frac{1}{\hat{\lambda}_1\left(1-\mu\hat{\lambda}_1\right)},\ldots,\frac{1}{\hat{\lambda}_{L-1}\left(1-\mu\hat{\lambda}_{L-1}\right)}\right]^T \tag{3.5.31}$$

### *3.5.4.2 Dual-Receiver System with Reference Receiver*

Substituting $V_{g_3}(\mathbf{w}(n))$ for $V_g(\mathbf{w}(n))$ in (3.4.14), and following a procedure similar to that used in Section 3.4.5, the following matrix difference equation is derived:

$$
\begin{aligned}
\Sigma(n+1) = & \left(I - 4\mu\hat{\Lambda} + 4\mu^2\hat{\Lambda}^2\right)\Sigma(n) + \mu^2(2/L)\left(\gamma^2\left(\mathrm{Tr}(R)\right)^2\right. \\
& \left. + \mathrm{Tr}(R)\left(\mathrm{Tr}\left(\hat{\Lambda}\Sigma(n)\right) + k_0(n)\right)\right)\Gamma
\end{aligned}
\tag{3.5.32}
$$

Denoting the eigenvalues of the weight covariance matrix $k_{ww_3}(n)$ by an L-dimensional vector $\boldsymbol{\eta}_3(n)$ when the covariance of the gradient used is $V_{g_3}(\mathbf{w}(n))$, (3.5.32) yields the following vector difference equation:

$$
\begin{aligned}
\boldsymbol{\eta}_3(n+1) = & \left[I - 4\mu\hat{\Lambda} + 4\mu^2\hat{\Lambda}^2 + \mu^2\frac{2}{L}\mathrm{Tr}(R)\boldsymbol{\vartheta}\hat{\boldsymbol{\lambda}}^T\right]\boldsymbol{\eta}_3(n) \\
& + \mu^2\frac{2}{L}\mathrm{Tr}(R)k_0(n)\boldsymbol{\vartheta} + \mu^2\frac{2}{L}\gamma^2\left[\mathrm{Tr}(R)\right]^2\boldsymbol{\vartheta}
\end{aligned}
\tag{3.5.33}
$$

which has the solution

$$
\begin{aligned}
\boldsymbol{\eta}_3(n) = & \left(I - H_2\right)^n\boldsymbol{\eta}_3(0) \\
& + \mu^2\frac{2}{L}\mathrm{Tr}(R)\sum_{i=1}^{n}\left(k_0(n-i) + \gamma^2\mathrm{Tr}(R)\right)\left(I - H_2\right)^{i-1}\boldsymbol{\vartheta}
\end{aligned}
\tag{3.5.34}
$$

and $k_{ww_3}(n)$ is given by

$$
k_{ww_3}(n) = \sum_{\ell=1}^{L}\eta_{3\ell}(n)\hat{\mathbf{Q}}_\ell\hat{\mathbf{Q}}_\ell^H
\tag{3.5.35}
$$

where $\boldsymbol{\eta}_3(0)$ is the vector of eigenvalues of $k_{ww_3}(0)$.

The steady-state behavior is obtained by substituting the steady-state expression for $L-1$ nonzero eigenvalues given by

$$
\lim_{n\to\infty}\boldsymbol{\eta}_3'(n) = \frac{\dfrac{\mu\mathrm{Tr}(R)}{2L}\left[\hat{\mathbf{w}}^H R\hat{\mathbf{w}} + \gamma^2\mathrm{Tr}(R)\right]}{1 - \mu\dfrac{\mathrm{Tr}(R)}{2L}\displaystyle\sum_{i=1}^{L-1}\frac{1}{1-\mu\hat{\lambda}_i}}\left[\frac{1}{\hat{\lambda}_1\left(1-\mu\hat{\lambda}_1\right)}, \ldots, \frac{1}{\hat{\lambda}_{L-1}\left(1-\mu\hat{\lambda}_{L-1}\right)}\right]^T
\tag{3.5.36}
$$

It could not be established that the weight covariance matrix is diagonizable when the gradient is estimated using the single-receiver system, and thus an analysis of the behavior of the weight covariance matrix is not possible. However, some results on the misadjustment for this case are presented along with two other cases.

### 3.5.5 Misadjustment Results

In this section, exact expressions for the misadjustment are presented for the dual receiver system and bounds on the misadjustment are presented for the single-receiver case.

#### 3.5.5.1 Single-Receiver System

For a single-receiver system with

$$\gamma(\mathbf{w}(n)) = c\left[\frac{\mathbf{w}^H(n)R\mathbf{w}(n)}{2\mathrm{Tr}(R)}\right]^{1/2} \tag{3.5.37}$$

if

$$0 < \mu < \frac{1}{\frac{a}{4}\left(\frac{L-1}{L}\right)\mathrm{Tr}(R) + 1.5\hat{\lambda}_{max}} \tag{3.5.38}$$

then the misadjustment is bound by

$$b_{L1} \le M_1 \le b_{H1} \tag{3.5.39}$$

where

$$b_{L1} = \frac{\mu\left[\frac{a}{4}\left(\frac{L-1}{L}\right)\mathrm{Tr}(R) + 0.5(L-1)\hat{\mathbf{w}}^H R\hat{\mathbf{w}}\right]}{1 - \mu\frac{a}{4}\left(\frac{L-1}{L}\right)\mathrm{Tr}(R)} \tag{3.5.40}$$

$$b_{H1} = \frac{\mu\left[\frac{a}{4}\left(\frac{L-1}{L}\right)\mathrm{Tr}(R) + 0.5(L-1)\hat{\mathbf{w}}^H R\hat{\mathbf{w}}\right]}{1 - \mu\left[\frac{a}{4}\left(\frac{L-1}{L}\right)\mathrm{Tr}(R) + 1.5\hat{\lambda}_{max}\right]} \tag{3.5.41}$$

and

$$a = \left(c + \frac{1}{c}\right)^2 \tag{3.5.42}$$

Note that $c = 1$ corresponds to the optimal $\gamma$.

#### 3.5.5.2 Dual-Receiver System with Dual Perturbation

If, for a given $\mathbf{w}(n)$, the covariance of the gradient is given by

$$V_g(\mathbf{w}(n)) = V_{g_2}(\mathbf{w}(n)) \tag{3.5.43}$$

$$0 < \mu < \frac{1}{\hat{\lambda}_{max} + \mathrm{Tr}(R)/2L} \tag{3.5.44}$$

and

$$\frac{\mu\,\mathrm{Tr}(R)}{2L}\sum_{i=1}^{L-1}\frac{1}{1-\mu\hat{\lambda}_i}<1 \tag{3.5.45}$$

then the misadjustment is given by

$$M_2=\frac{\dfrac{\mu\,\mathrm{Tr}(R)}{2L}\displaystyle\sum_{i=1}^{L-1}\frac{1}{1-\mu\hat{\lambda}_i}}{1-\dfrac{\mu\,\mathrm{Tr}(R)}{2L}\displaystyle\sum_{i=1}^{L-1}\frac{1}{1-\mu\hat{\lambda}_i}} \tag{3.5.46}$$

#### 3.5.5.3 *Dual-Receiver System with Reference Receiver*

If, for a given $\mathbf{w}(n)$, the covariance of the gradient is

$$V_g(\mathbf{w}(n))=V_{g_3}(\mathbf{w}(n)) \tag{3.5.47}$$

$$0<\mu<\frac{1}{\hat{\lambda}_{max}+\mathrm{Tr}(R)/2L} \tag{3.5.48}$$

and

$$\frac{\mu\,\mathrm{Tr}(R)}{2L}\sum_{i=1}^{L-1}\frac{1}{1-\mu\hat{\lambda}_i}<1 \tag{3.5.49}$$

then the misadjustment is given by

$$M_3=\frac{\left[1+\gamma^2\dfrac{\mathrm{Tr}(R)}{\hat{\mathbf{w}}^H R\hat{\mathbf{w}}}\right]\dfrac{\mu\,\mathrm{Tr}(R)}{2L}\displaystyle\sum_{i=1}^{L-1}\frac{1}{1-\mu\hat{\lambda}_i}}{1-\dfrac{\mu\,\mathrm{Tr}(R)}{2L}\displaystyle\sum_{i=1}^{L-1}\frac{1}{1-\mu\hat{\lambda}_i}} \tag{3.5.50}$$

## 3.6 Structured Gradient Algorithm

In this section, a description and an analysis of the constrained LMS algorithm is presented when it uses a gradient estimated from an estimate of the array correlation matrix having a special structure [God89, God90, God93, God97]. The algorithm for this case is referred to as the structured gradient algorithm.

The gradient estimate given by (3.4.4) for the standard constrained LMS algorithm can be expressed as

$$\mathbf{g}_S(\mathbf{w}(n)) = 2R(n+1)\mathbf{w}(n) \tag{3.6.1}$$

where

$$R(n) = \mathbf{x}(n)\mathbf{x}^H(n) \tag{3.6.2}$$

is a noisy sample of the array correlation matrix at the nth instant of time estimated from only one array sample.

For a uniformly spaced linear array, the array correlation matrix R has the Toeplitz structure, that is,

$$R \equiv \begin{pmatrix} r_0 & r_1 & \cdots & r_{L-1} \\ r_1^* & & \ddots & \\ \vdots & \ddots & & r_1 \\ r_{L-1}^* & r_{L-2}^* & & r_0 \end{pmatrix} \tag{3.6.3}$$

where $r_i$, i = 0, 1, …, L − 1 denote the correlation between elements with lag i, defined as

$$r_i = E\left[x_m(n)x_{m+i}^*(n)\right], \quad \begin{cases} i = 0, 1, \ldots, L-1 \\ m = 1, 2, \ldots, L \end{cases} \tag{3.6.4}$$

and $x_m(n)$ denotes the signal derived from mth element at the nth time instant.

Note that not all combinations of m and i are possible in (3.6.4), as there are only L elements in the array. For i = 0, m = 1, …, L yielding L values of $r_0$. These values form the main diagonal of R in (3.6.3). For i = 1, m = 1, 2, …, L − 1 results in L − 1 values of $r_1$. These values form the second diagonal of R and so on.

The noisy sample of R used in estimating the gradient for the standard algorithm does not have the Toeplitz structure. The structured gradient algorithm exploits this structure of the array correlation matrix in estimating the gradient. It takes R(n) and estimates an array correlation matrix $\tilde{R}$ having the Toeplitz structure. The structured array correlation matrix $\tilde{R}$ is then used in gradient estimation as discussed below.

### 3.6.1 Gradient Estimate

For this algorithm, the gradient estimate is defined as follows:

$$\mathbf{g}_{st}(\mathbf{w}(n)) = 2\tilde{R}(n+1)\mathbf{w}(n) \tag{3.6.5}$$

where $\tilde{R}(n)$ is an estimate of the array correlation matrix at the nth instant of time having Toeplitz structured as in (3.6.3), and is given by

$$\tilde{R}(n) = \begin{bmatrix} \tilde{r}_0(n) & \tilde{r}_1(n) & & \tilde{r}_{L-1}(n) \\ \tilde{r}_1^*(n) & \ddots & & \\ \vdots & & \ddots & \\ \tilde{r}_{L-1}^*(n) & & & \tilde{r}_0(n) \end{bmatrix} \tag{3.6.6}$$

with

$$\tilde{r}_\ell(n) = \frac{1}{N_\ell}\sum_i x_i(n)x^*_{i+\ell}(n), \quad \ell = 0, 1, \ldots, L-1 \tag{3.6.7}$$

where $N_\ell$ denotes the number of possible combinations of elements with lag $\ell$ and summation is over all these combinations. For a linear array of equispaced elements, $N_\ell = L - l$.

Since

$$E\left[\tilde{r}_\ell(n)\right] = r_\ell, \quad \ell = 0, 1, \ldots, L-1 \tag{3.6.8}$$

it follows from (3.6.7), (3.6.6), and (3.6.3) that

$$E\left[\tilde{R}(n)\right] = R \tag{3.6.9}$$

Thus, for a given $\mathbf{w}(n)$,

$$E\left[\mathbf{g}_{st}\left(\mathbf{w}(n)\right)\right] = 2R\mathbf{w}(n) \tag{3.6.10}$$

and the gradient estimate is unbiased.

The discussion on the structured gradient algorithm presented here is for an equispaced linear array. The formulation can easily be extended to an arbitrary array.

For the equispaced linear array, each element of $\tilde{R}(n)$ is a mean value of all elements of R(n) with the same spatial correlation lags. Thus, $\tilde{r}_0(n)$ is an average of the main diagonal elements of R(n), $\tilde{r}_1(n)$ is the mean of first diagonal elements of R(n), and so on. For an array that is not an equispaced linear array, the array correlation matrix loses its Toeplitz structure, and the number of elements in R with the same spatial correlation lags is less in comparison to the equispaced case. However, there are always some elements in R with the same spatial correlation lags. Even in a completely unstructured correlation matrix, such as would be obtained from a three-element array with spacing d and 2d, the diagonal elements are always of the same correlation lag, namely lag 0.

### 3.6.2 Examples and Discussion

Examples are presented in this section to compare the performance of the structured gradient algorithm and the standard algorithm. The mean noise power for a given $\mathbf{w}(n)$ is examined as a function of the weight update iteration to see the algorithm's effectiveness in reducing noise.

Figure 3.4 to Figure 3.8 shows the plots of the mean noise power in dB for a given $\mathbf{w}(n)$ vs. the iteration number. The mean noise power for a given $\mathbf{w}(n)$ is calculated using

$$P_N\left(\mathbf{w}(n)\right) = \mathbf{w}^H(n)R\mathbf{w}(n) \tag{3.6.11}$$

A linear array of ten elements with half-wavelength spacing is assumed for these examples. The look direction is assumed to be in the broadside of the array. The power

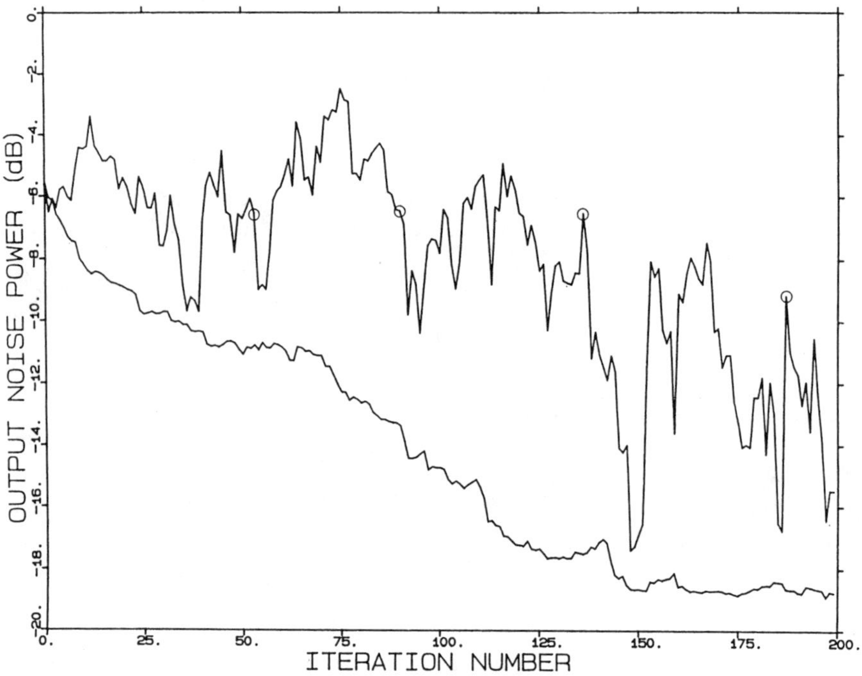


**FIGURE 3.4**
$10\log P_N(\mathbf{w}(n))$ vs. the iteration number for a 10-element linear array with one-half wavelength spacing. The curve with a solid line is for a structured gradient algorithm, and the one superimposed with a circle is for a standard LMS algorithm. Two interferences: $\theta_1 = 98°$, $p_1 = 1$, $\theta_2 = 45°$, $p_2 = 10$, $\sigma_n^2 = 0.1$, look direction angle $\theta_0 = 90°$, $p_S = 1$, $\mu_{SG} = \mu_{ST} = .0005$. (From Godara, L.C. and Gray, D.A., *J. Acoust. Soc. Am.*, 86, 1040–1046, 1989 [God89]. With permission.)

of uncorrelated noise present on each element is assumed to be equal to 0.1. Two interference sources are assumed to be present. Directions of these interferences make angles of 98° and 45° with the line of the array. The other parameters are included in figure captions. The gradient algorithm is initialized with the conventional weight. The gradient step size for the standard algorithm and the structured gradient algorithm are denoted by $\mu_{ST}$ and $\mu_{SG}$, respectively. A comparison of the two algorithms in Figure 3.4 reveals that the noise in the weights estimated by the structured gradient algorithm is much less than that estimated by the standard algorithm.

Figure 3.5 compares the two algorithms when the signal power is reduced by 20 dB compared to the scenario of Figure 3.4. Comparing Figure 3.4 and Figure 3.5, one observes that the fluctuations in the mean output noise power in Figure 3.4, where the signal power is 1, are more than in Figure 3.5 where the signal power is 0.01. Thus, the noise in the weights estimated by the standard algorithm depends on the input signal power, and increases as the signal power increases. On the other hand, the structured gradient algorithm does not appear to be sensitive to the signal power. The signal sensitivity of the two algorithms is further compared in Figure 3.6 and Figure 3.7, where the power of the second interference is increased by 10 dB. Sensitivity of the standard algorithm to the input signal level is clearly visible from the two figures. The noise fluctuation in weights estimated by standard LMS algorithm is more in Figure 3.7 where the signal power is 1.0 than that in Figure 3.6 where the signal power is .01.

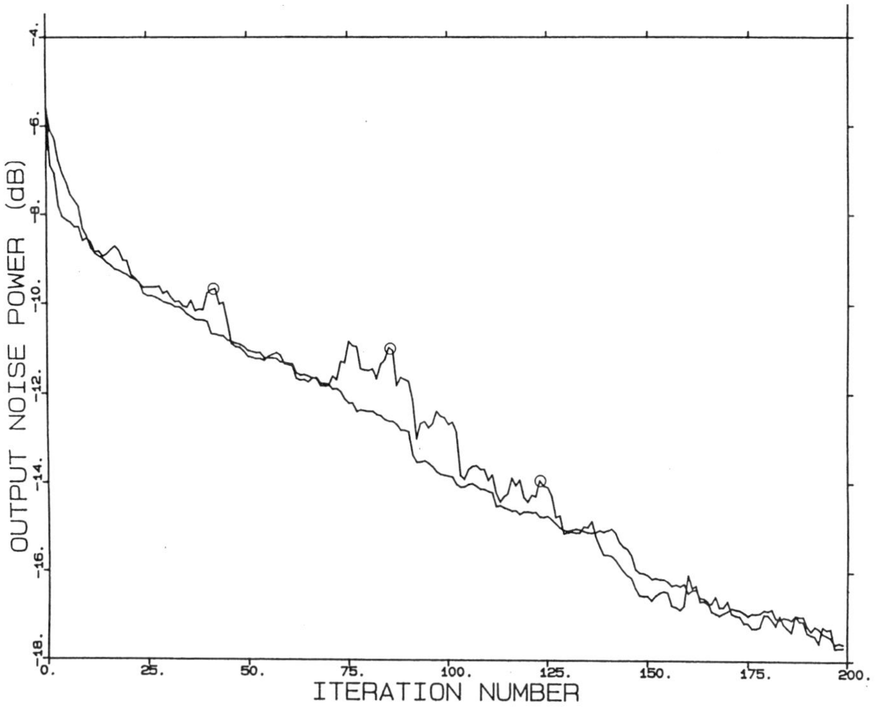


**FIGURE 3.5**
$10\log P_N(\mathbf{w}(n))$ vs. the iteration number for a 10-element linear array with one-half wavelength spacing. The curve with a solid line is for a structured gradient algorithm, and the one superimposed with a circle is for a standard LMS algorithm. Two interferences: $\theta_1 = 98°$, $p_1 = 1$, $\theta_2 = 45°$, $p_2 = 10$, $\sigma_n^2 = 0.1$, look direction angle $\theta_0 = 90°$, $p_S = .01$, $\mu_{SG} = \mu_{ST} = .0005$. (From Godara, L.C. and Gray, D.A., *J. Acoust. Soc. Am.*, 86, 1040–1046, 1989 [God89]. With permission.).

The noise in the weights estimated by the standard LMS algorithm can be reduced by using a smaller value of the gradient step size. The reduction in the step size to reduce the noise in weights means the reduction in the convergence speed of the algorithm as shown in Figure 3.8, where the step size used for the standard algorithm is one-tenth of that used in the structured gradient algorithm. For this case, the structured gradient converges faster than the standard algorithm.

It should be noted that the gradient estimate using the structured method requires more computation than the standard method. In the standard algorithm, an estimate of the gradient requires an order of L complex multiplications, whereas, for structured gradient algorithm, it requires the order of $L^2$ complex multiplications. A detailed discussion on the signal sensitivity of the LMS algorithms is presented in Section 3.14.

## 3.7 Recursive Least Mean Squares Algorithm

The recursive LMS algorithm uses all previous array samples to estimate the gradient, in comparison to the standard LMS algorithm, which uses only one array sample [God90a, God93]. In this section, the behavior of the recursive LMS algorithm is examined by deriving an expression for the covariance of the gradient.

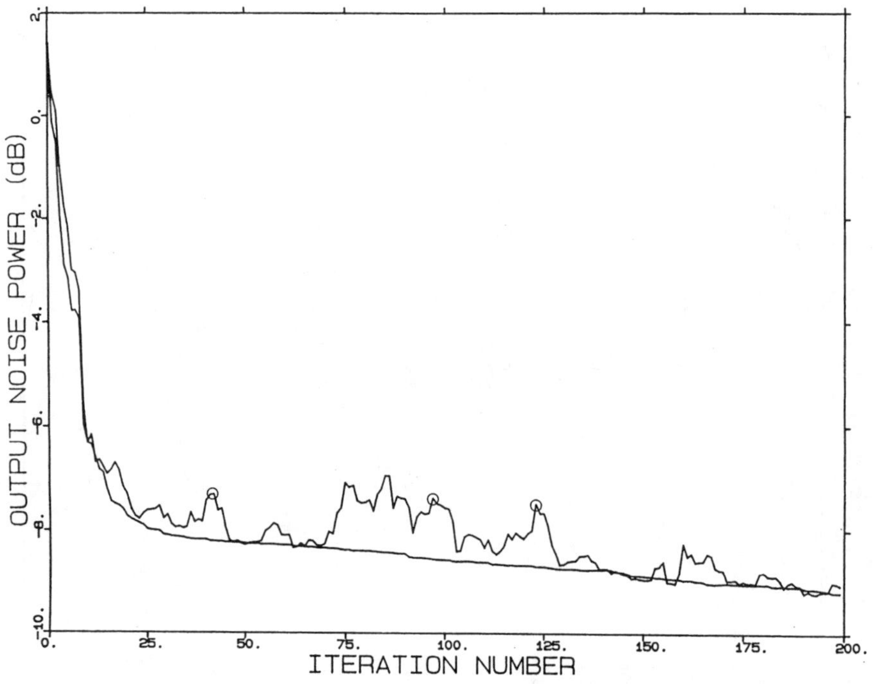


**FIGURE 3.6**
$10\log P_N(\mathbf{w}(n))$ vs. the iteration number for a 10-element linear array with one-half wavelength spacing. The curve with a solid line is for a structured gradient algorithm, and the one superimposed with a circle is for a standard LMS algorithm. Two interferences: $\theta_1 = 98°$, $p_1 = 1$, $\theta_2 = 45°$, $p_2 = 100$, $\sigma_n^2 = 0.1$, look direction angle $\theta_0 = 90°$, $p_S = .01$, $\mu_{SG} = \mu_{ST} = .00005$. (From Godara, L.C. and Gray, D.A., *J. Acoust. Soc. Am.*, 86, 1040–1046, 1989 [God89]. With permission.)

### 3.7.1 Gradient Estimate

Let $\mathbf{g}_R(\mathbf{w}(n))$ denote the estimated gradient by recursive method for a given $\mathbf{w}(n)$, defined as

$$\mathbf{g}_R(\mathbf{w}(n)) = 2\hat{R}(n+1)\mathbf{w}(n) \tag{3.7.1}$$

where

$$\hat{R}(n+1) = \frac{n\hat{R}(n) + \mathbf{x}(n+1)\mathbf{x}^H(n+1)}{n+1} \tag{3.7.2}$$

It follows from (3.7.2) that as the number of samples used in estimating the array correlation matrix increases, the matrix estimate approaches the true correlation matrix. Thus,

$$\lim_{n\to\infty} \hat{R}(n) = R \tag{3.7.3}$$

and

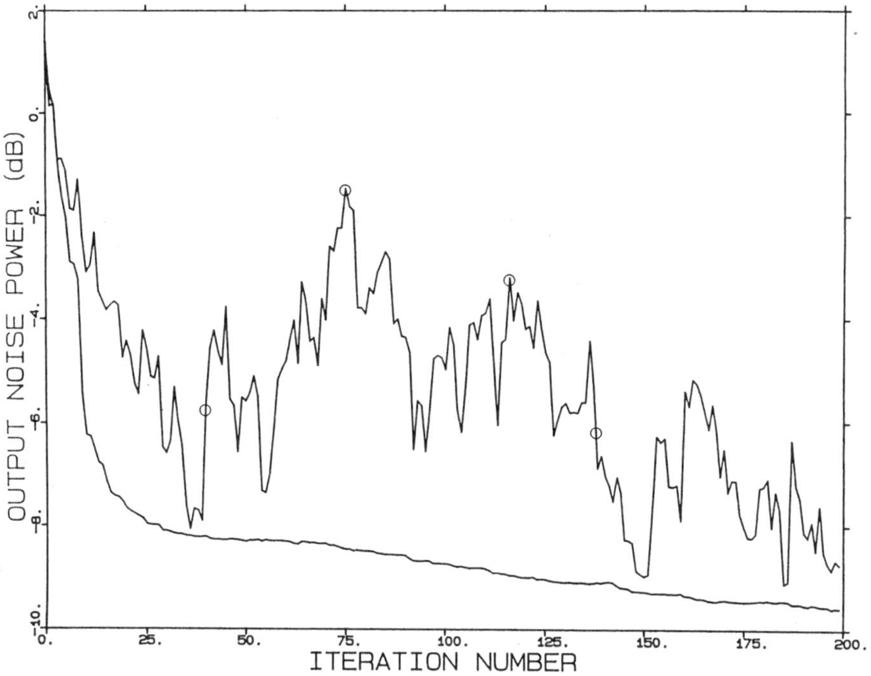


**FIGURE 3.7**
$10\log P_N(\mathbf{w}(n))$ vs. the iteration number for a 10-element linear array with one-half wavelength spacing. The curve with a solid line is for a structured gradient algorithm, and the one superimposed with a circle is for a standard LMS algorithm. Two interferences: $\theta_1 = 98°$, $p_1 = 1$, $\theta_2 = 45°$, $p_2 = 100$, $\sigma_n^2 = 0.1$, look direction angle $\theta_0 = 90°$, $p_S = 1$, $\mu_{SG} = \mu_{ST} = .00005$. (From Godara, L.C. and Gray, D.A., *J. Acoust. Soc. Am.*, 86, 1040–1046, 1989 [God89]. With permission.)

$$\lim_{n\to\infty} \mathbf{g}_R(\mathbf{w}(n)) = 2R\mathbf{w}(n) \tag{3.7.4}$$

Consequently, the gradient estimate approaches the true gradient as $n \to \infty$.

### 3.7.2 Covariance of Gradient

In this section, covariance of the gradient is established. The result is valid for large n samples, such that

$$\hat{R}(n) \cong R \tag{3.7.5}$$

It follows from (3.7.1), (3.7.2), and (3.7.5) that

$$\mathbf{g}_R(\mathbf{w}(n)) = \frac{2n}{n+1} R\mathbf{w}(n) + \frac{2}{n+1}\mathbf{x}(n+1)\mathbf{x}^H(n+1)\mathbf{w}(n) \tag{3.7.6}$$

Let $V_{g_R}(\mathbf{w}(n))$ denote the covariance of the gradient estimate defined by (3.7.1) and (3.7.2) for a given $\mathbf{w}(n)$. By definition,

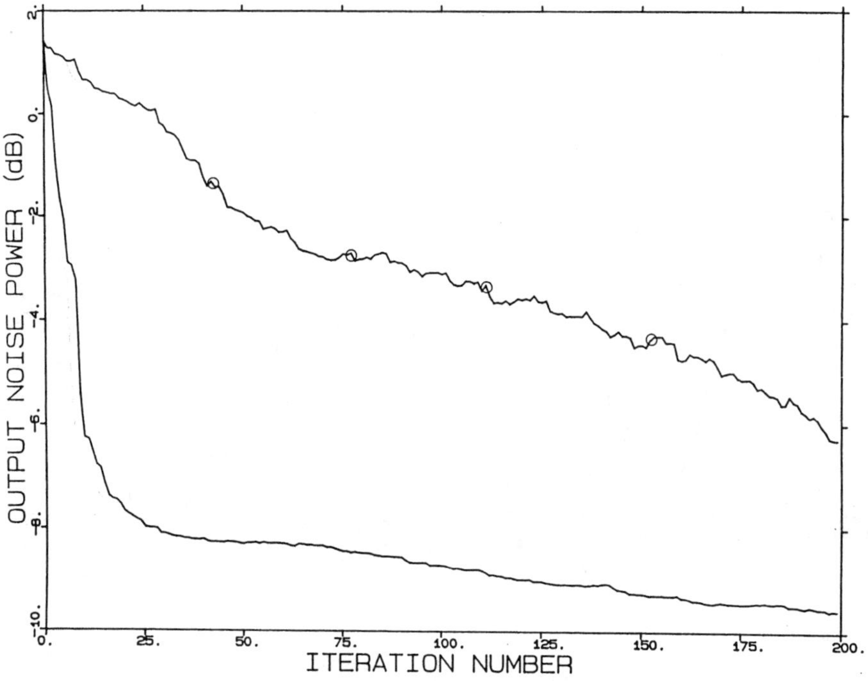


**FIGURE 3.8**
$10\log P_N(\mathbf{w}(n))$ vs. the iteration number for a 10-element linear array with one-half wavelength spacing. The curve with a solid line is for a structured gradient algorithm, and the one superimposed with a circle is for a standard LMS algorithm. Two interferences: $\theta_1 = 98°$, $p_1 = 1$, $\theta_2 = 45°$, $p_2 = 100$, $\sigma_n^2 = 0.1$, look direction angle $\theta_0 = 90°$, $p_S = 1$, $\mu_{SG} = .00005$, $\mu_{ST} = .000005$. (From Godara, L.C. and Gray, D.A., *J. Acoust. Soc. Am.*, 86, 1040–1046, 1989 [God89]. With permission.)

$$V_{g_R}(\mathbf{w}(n)) = E\left[\mathbf{g}_R(\mathbf{w}(n))\mathbf{g}_R^H(\mathbf{w}(n))\right] - \bar{\mathbf{g}}_R(\mathbf{w}(n))\bar{\mathbf{g}}_R^H(\mathbf{w}(n)) \tag{3.7.7}$$

where $\bar{\mathbf{g}}_R(\mathbf{w}(n))$ is the mean value of the gradient estimate for a given $\mathbf{w}(n)$.

It follows from (3.7.1) and (3.7.2) that

$$\bar{\mathbf{g}}_R(\mathbf{w}(n)) = 2R\mathbf{w}(n) \tag{3.7.8}$$

Thus,

$$\bar{\mathbf{g}}_R(\mathbf{w}(n))\bar{\mathbf{g}}_R^H(\mathbf{w}(n)) = 4R\mathbf{w}(n)\mathbf{w}^H(n)R \tag{3.7.9}$$

Using the following result for an i.i.d. complex Gaussian sequence $\{\mathbf{x}(k)\}$ and a Hermitian matrix A,

$$E\left[\mathbf{x}(n)\mathbf{x}^H(n)A\mathbf{x}(n)\mathbf{x}^H(n)\right] = RAR + \mathrm{Tr}(RA)R \tag{3.7.10}$$

the following is derived from (3.7.6)

$$E\left[\mathbf{g}_R(\mathbf{w}(n))\mathbf{g}_R^H(\mathbf{w}(n))\right] = 4\,R\mathbf{w}(n)\mathbf{w}^H(n)R + \frac{4}{(n+1)^2}\mathbf{w}^H(n)R\mathbf{w}(n)R \tag{3.7.11}$$

Substituting in (3.7.7) from (3.7.9) and (3.7.11),

$$V_{g_R}(\mathbf{w}(n)) = \frac{4}{(n+1)^2}\mathbf{w}^H(n)R\mathbf{w}(n)R \tag{3.7.12}$$

It follows from (3.7.12) that the covariance of the estimated gradient by the recursive method decreases as the iteration number increases and $(n + 1)^2$ times less than the covariance of the gradient estimated by the standard method. The covariance of the gradient estimated by the standard method $V_{g_S}(\mathbf{w}(n))$, is given by

$$V_{g_S}(\mathbf{w}(n)) = 4\mathbf{w}^H(n)R\mathbf{w}(n)R \tag{3.7.13}$$

### 3.7.3 Discussion

As discussed previously, the projected covariance of the gradient $PV_g(\mathbf{w}(n))P$ affects the weight covariance. Taking the projection on both sides of (3.7.12) and (3.7.13), and noting that PRP is independent of the look direction signal, one observes that the projected covariance in both the cases is proportional to the mean output power. This implies that for both the cases the projected covariance is a function of the look direction signal. This in turn makes the weight covariance at each iteration sensitive to the look direction signal. However, at the nth iteration, the weight covariance for the recursive algorithm case is less than that for the standard LMS case due to the term $(n + 1)^2$ in (3.7.12).

## 3.8 Improved Least Mean Squares Algorithm

The structured gradient algorithm exploits the structure of the array correlation matrix. However, it does not make use of the previous samples when estimating the gradient at the nth iteration. In this section, a method is presented that exploits the structure of the array correlation matrix and uses previous samples. The method is referred to as the improved method [God90a, God93].

An estimate of the gradient using the improved method is given by

$$\mathbf{g}_I(\mathbf{w}(n)) = 2\tilde{\hat{R}}(n+1)\mathbf{w}(n) \tag{3.8.1}$$

where

$$\tilde{\hat{R}}(n+1) = \frac{n\tilde{\hat{R}}(n) + \tilde{R}(n+1)}{n+1} \tag{3.8.2}$$

with $\tilde{R}(n)$ given by (3.6.6).

It can easily be shown that the gradient estimate is unbiased, that is,

$$E\left[\mathbf{g}_I(\mathbf{w}(n))\middle|\mathbf{w}(n)\right] = 2R\mathbf{w}(n) \tag{3.8.3}$$

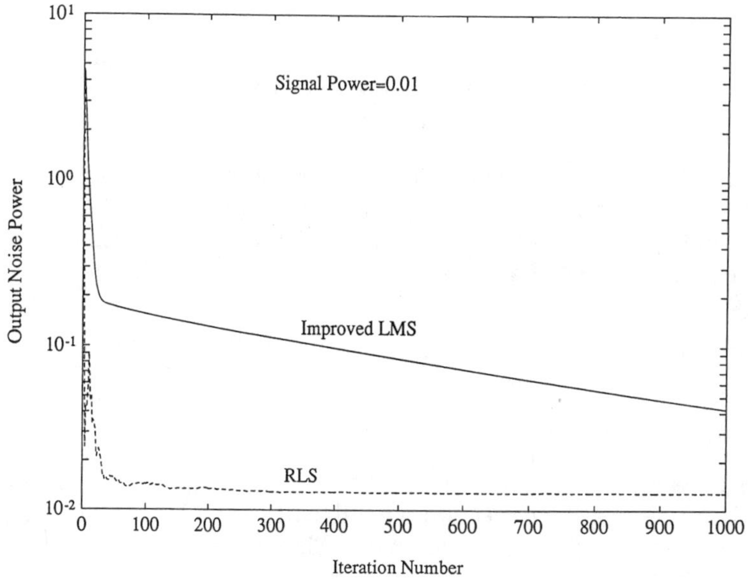


**FIGURE 3.9**
$P_N(\mathbf{w}(n))$ vs. the iteration number for a 10-element linear array with one-half wavelength spacing. Two interferences: $\theta_1 = 98°$, $p_1 = 1$, $\theta_2 = 72°$, $p_2 = 100$, $\sigma_n^2 = 0.1$, look direction angle $\theta_0 = 90°$. (From Godara, L.C., *IEEE Trans. Antennas Propagat.*, 38, 1631–1635, 1990. ©IEEE. With permission.)

The performance and the signal sensitivity of the above algorithm is now compared with a RLS algorithm that makes use of the previous samples and requires the same order of computation for computing the weights. The following form of the RLS algorithm is used for the comparison:

$$\mathbf{w}(n) = \frac{\hat{R}^{-1}(n)\mathbf{S}_0}{\mathbf{S}_0^H\hat{R}^{-1}(n)\mathbf{S}_0} \tag{3.8.4}$$

where $\hat{R}^{-1}(n)$ is updated using the Matrix Inversion Lemma and is given by (3.1.6) and (3.1.7). Note that in the absence of errors, $n \to \infty$, $\hat{R}^{-1}(n) \to R^{-1}$, and $\mathbf{w}(n) \to \hat{\mathbf{w}}$.

Figure 3.9 to Figure 3.12 compare the mean output noise power $P_N(\mathbf{w}(n))$ vs. the iteration number for various look direction signal powers when the weights $\mathbf{w}(n)$ are adjusted using the two algorithms. The mean output noise power is calculated using

$$P_N(\mathbf{w}(n)) = \mathbf{w}^H(n)R_N\mathbf{w}(n) \tag{3.8.5}$$

A linear array of ten elements with half-wavelength spacing is assumed for these examples. The variance of uncorrelated noise present on each element is assumed to be equal to 0.1. Two interference sources are assumed to be present. The first interference falls in the main lobe of the conventional array pattern and makes an angle of 98° with the line of the array. The power of this interference is taken to be 10 dB more than the uncorrelated noise power. The second interference makes an angle of 72° with the line of the array and falls in the first side-lobe of the conventional pattern. The power of this interference is 30 dB more than the uncorrelated noise power. The look direction is broadside to the array. The signal power for the four plots is varied from –10 dB below the uncorrelated power to 30 dB above the uncorrelated noise power. The gradient algorithm is initialized with the conventional weights.

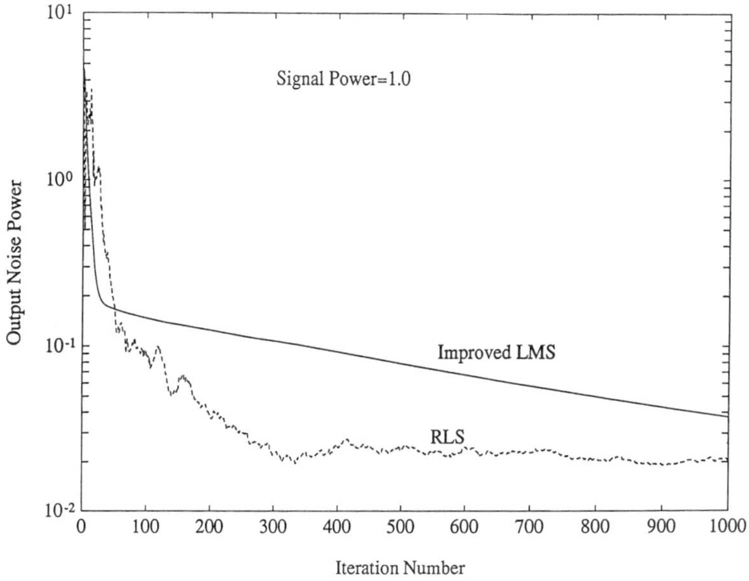


**FIGURE 3.10**
$P_N(\mathbf{w}(n))$ vs. the iteration number for a 10-element linear array with one-half wavelength spacing. Two interferences: $\theta_1 = 98°$, $p_1 = 1$, $\theta_2 = 72°$, $p_2 = 100$, $\sigma_n^2 = 0.1$, look direction angle $\theta_0 = 90°$. (From Godara, L.C., *IEEE Trans. Antennas Propagat.*, 38, 1631–1635, 1990. ©IEEE. With permission.)

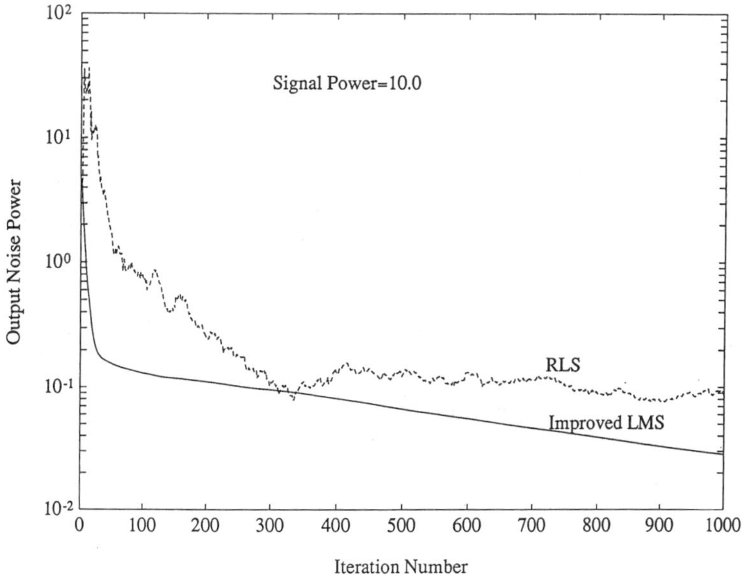


**FIGURE 3.11**
$P_N(\mathbf{w}(n))$ vs. the iteration number for a 10-element linear array with one-half wavelength spacing. Two interferences: $\theta_1 = 98°$, $p_1 = 1$, $\theta_2 = 72°$, $p_2 = 100$, $\sigma_n^2 = 0.1$, look direction angle $\theta_0 = 90°$. (From Godara, L.C., *IEEE Trans. Antennas Propagat.*, 38, 1631–1635, 1990. ©IEEE. With permission.)

For the improved LMS algorithm the gradient step size $\mu$ is taken to be equal to 0.00005 and for the RLS algorithm $\varepsilon_0$ is taken to be 0.0001. According to these figures, for a weak signal the RLS algorithm performs better than the improved algorithm. However, as the input signal power increases the output noise power of the processor using the RLS

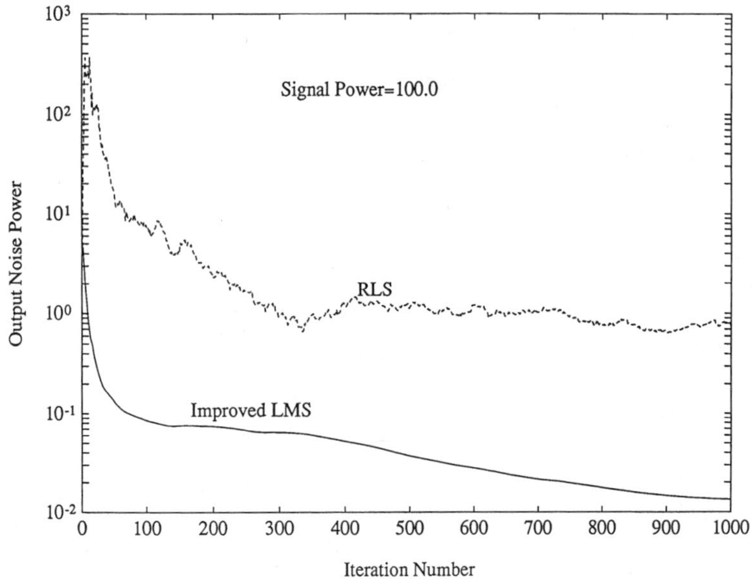


**FIGURE 3.12**
$P_N(\mathbf{w}(n))$ vs. the iteration number for a 10-element linear array with one-half wavelength spacing. Two interferences: $\theta_1 = 98°$, $p_1 = 1$, $\theta_2 = 72°$, $p_2 = 100$, $\sigma_n^2 = 0.1$, look direction angle $\theta_0 = 90°$. (From Godara, L.C., *IEEE Trans. Antennas Propagat.*, 38, 1631–1635, 1990. ©IEEE. With permission.)

algorithm increases. Thus, the RLS algorithm used in the present form is sensitive to the look direction signal. On the other hand, this is not the case for the improved LMS algorithm. Performance of the improved LMS algorithm improves as the signal power is increased, and in the presence of a strong signal it performs much better than the RLS algorithm, both in terms of convergence and the output SNR. See for example, the plots in Figure 3.12 where the input signal power is 30 dB more than the uncorrelated noise power.

Figure 3.13 compares the performance of the standard LMS algorithm, recursive LMS algorithm, and improved LMS algorithm. The noise field and array geometry used for this example are the same as those used in previous examples. The input signal power is 30 dB more than the uncorrelated noise power and the gradient step size is 0.00005. It is clear from Figure 3.13 that the output noise power of the processor at each iteration is less when the recursive algorithm and the improved algorithm are used in comparison to the output noise power using the standard algorithm. A large fluctuation in the output of the processor using the standard algorithm in comparison to the other two algorithms indicates the sensitivity of this algorithm to the look direction signal. A comparison of the recursive LMS and improved LMS show that the latter performs better, both in terms of the amount of the noise and its variation as a function of iteration number.

## 3.9 Recursive Least Squares Algorithm

The convergence speed of the LMS algorithm depends on the eigenvalues of the array correlation matrix. In an environment yielding an array correlation matrix with large eigenvalue spread the algorithm converges with a slow speed. This problem is solved

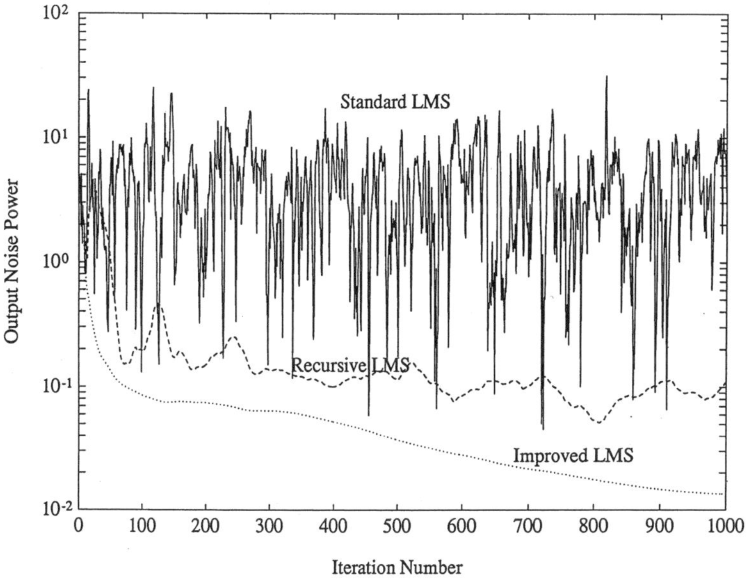


**FIGURE 3.13**
$P_N(\mathbf{w}(n))$ vs. the iteration number for a 10-element linear array with one-half wavelength spacing. Two interferences: $\theta_1 = 98°$, $p_1 = 1$, $\theta_2 = 72°$, $p_2 = 100$, $\sigma_n^2 = 0.1$, look direction angle $\theta_0 = 90°$. (From Godara, L.C., *IEEE Trans. Antennas Propagat.*, 38, 1631–1635, 1990. ©IEEE. With permission.)

with the RLS algorithm [Sch77, d'As80, God97] by replacing the gradient step size $\mu$ with a gain matrix $\hat{R}^{-1}(n)$ at the nth iteration, producing the weight update equation

$$\mathbf{w}(n) = \mathbf{w}(n-1) - \hat{R}^{-1}(n)\, \mathbf{x}(n)\varepsilon^*\left(\mathbf{w}(n-1)\right) \tag{3.9.1}$$

where $\hat{R}(n)$ is given by

$$\begin{aligned}\hat{R}(n) &= \delta_0\, \hat{R}(n-1) + \mathbf{x}(n)\, \mathbf{x}^H(n) \\ &= \sum_{k=0}^{n} \delta_0^{n-k}\, \mathbf{x}(k)\, \mathbf{x}^H(k)\end{aligned} \tag{3.9.2}$$

with $\delta_0$ denoting a real scalar less than but close to 1. The $\delta_0$ is used for exponential weighting of past data and is referred to as the forgetting factor as the update equation tends to de-emphasize the old samples. The quantity $1/(1-\delta_0)$ is normally referred to as the algorithm memory. Thus, for $\delta_0 = 0.99$ the algorithm memory is close to 100 samples. The RLS algorithm updates the required inverse of using the previous inverse and the present sample as

$$\hat{R}^{-1}(n) = \frac{1}{\delta_0}\left[\hat{R}^{-1}(n-1) - \frac{\hat{R}^{-1}(n-1)\, \mathbf{x}(n)\, \mathbf{x}^H(n)\, \hat{R}^{-1}(n-1)}{\delta_0 + \mathbf{x}^H(n)\, \hat{R}^{-1}(n-1)\, \mathbf{x}(n)}\right] \tag{3.9.3}$$

The matrix is initialized as

$$\hat{R}^{-1}(0) = \frac{1}{\varepsilon_0} I, \quad \varepsilon_0 > 0 \tag{3.9.4}$$

The RLS algorithm minimizes the cumulative square error

$$J(n) = \sum_{k=0}^{n} \delta_0^{n-k} |\varepsilon(k)|^2 \tag{3.9.5}$$

and its convergence is independent of the eigenvalues distribution of the correlation matrix.

The algorithm presented here is the exact RLS algorithm. Other forms of the RLS algorithm with improved computation efficiency are available [Fab86, Cio84]. A comparison of the convergence speed of the LMS, RLS, and some other gradient-based algorithms using quantized or clipped data indicates that RLS is the most efficient and LMS is the slowest [Gar87].

Computer simulation study of RLS, LMS, and SMI algorithms in mobile communication situations suggests that the former outperforms the latter two in flat fading channels [Fer93]. An application of the RLS algorithm for the reverse link of a cellular communication using the CDMA system is considered in [Wan94] to show an increase in channel capacity by an adaptive array.

## 3.10 Constant Modulus Algorithm

The constant modulus algorithm is gradient based [God97] and works on the premise that existing interference causes fluctuation in the amplitude of array output that otherwise has a constant modulus. It updates weights by minimizing the cost function [Chi93, God80, Tre83, Shy93]

$$J(n) = \frac{1}{2} E\left[\left(|y(n)|^2 - y_0^2\right)^2\right] \tag{3.10.1}$$

using the following equation

$$\mathbf{w}(n+1) = \mathbf{w}(n) - \mu\, \mathbf{g}(\mathbf{w}(n)) \tag{3.10.2}$$

where

$$y(n) = \mathbf{w}^H(n)\mathbf{x}(n+1) \tag{3.10.3}$$

is the array output after the nth iteration, $y_0$ is the desired amplitude in the absence of interference, and $\mathbf{g}(\mathbf{w}(n))$ denotes an estimate of the cost function gradient.

Similar to the LMS algorithm discussed previously, the constant modulus algorithm uses an estimate of the gradient by replacing the true gradient with an instant value given by

$$\mathbf{g}(\mathbf{w}(n)) = 2\varepsilon(n)\, \mathbf{x}(n+1) \tag{3.10.4}$$

where

$$\varepsilon(n) = \left(|y(n)|^2 - y_0^2\right) y(n) \quad (3.10.5)$$

The weight update equation for this case becomes

$$\mathbf{w}(n+1) = \mathbf{w}(n) - 2\mu\varepsilon(n)\,\mathbf{x}(n+1) \quad (3.10.6)$$

In appearance, this is similar to the LMS algorithm with reference signal where

$$\varepsilon(n) = r(n) - y(n) \quad (3.10.7)$$

Its application to digital, land mobile radio communication systems using TDMA to compensate for selective fading is studied in [Ohg91]. Discussions of hardware implementation of a CMA adaptive array and its BER performance for high-speed transmission in mobile communications may be found in [Ohg93a, Ohg93]. Development of CMA for beam-space array signal processing including its hardware realization has been reported in [Tan95]. The results presented in [Chi93] indicate that the beam space CMA is able to cancel interferences arriving from other than the look direction.

CMA is useful for eliminating correlated arrivals and is effective for constant modulated envelope signals such as GMSK and QPSK, which are used in digital communications. However, the algorithm is not appropriate for the CDMA system because of the required power control [Wan94]. Use of CMA to blindly separate co-channel FM signals in mobile communications has been investigated in [Par95]. A variation referred to as differential CMA reported in [Nis95] has inferior convergence characteristics compared to CMA but may be improved using direction of arrival information to make it operative in beam space.

## 3.11 Conjugate Gradient Method

An application of the conjugate gradient method [Hes52, Dan67, Sar81] to adjust the weights of an antenna array is discussed in [Cho92]. The method is generally useful for solving a set of equations of the form A**w** = **b** to obtain **w**. In this section, a brief description of the CGM is provided [God97].

For an array-processing problem, **w** denotes the array weights, A is a matrix with each of its columns denoting consecutive samples obtained from array elements, and **b** is a vector containing consecutive samples of the desired signal. Thus, a residual vector

$$\mathbf{r} = \mathbf{b} - A\mathbf{w} \quad (3.11.1)$$

denotes error between the desired signal and array output at each sample, with the sum of the squared error given by $\mathbf{r}^H\mathbf{r}$.

The method starts with an initial guess **w**(0) of the weights, obtains a residual

$$\mathbf{r}(0) = \mathbf{b} - A\mathbf{w}(0) \quad (3.11.2)$$

an initial direction vector

$$\mathbf{g}(0) = A^H \mathbf{r}(0) \tag{3.11.3}$$

and moves the weights in this direction to yield a weight update equation,

$$\mathbf{w}(n+1) = \mathbf{w}(n) - \mu(n)\,\mathbf{g}(n) \tag{3.11.4}$$

where the step size is

$$\mu(n) = \frac{\left|A^H \mathbf{r}(n)\right|^2}{\left|A^H \mathbf{g}(n)\right|^2} \tag{3.11.5}$$

The residual $\mathbf{r}(n)$ and the direction vector $\mathbf{g}(n)$ are updated using

$$\mathbf{r}(n+1) = \mathbf{r}(n) + \mu(n)\; A\,\mathbf{g}(n) \tag{3.11.6}$$

and

$$\mathbf{g}(n+1) = A^H \mathbf{r}(n+1) - \alpha(n)\mathbf{g}(n) \tag{3.11.7}$$

with

$$\alpha(n) = \frac{\left|A^H \mathbf{r}(n+1)\right|^2}{\left|A^H \mathbf{r}(n)\right|^2} \tag{3.11.8}$$

The algorithm is stopped when the residual falls below a certain predetermined level. It should be noted that the direction vector points in the direction of error surface gradient $\mathbf{r}^H(n)\mathbf{r}(n)$ at the nth iteration, which the algorithm is trying to minimize. The method converges to the error surface minimum within at most L iterations for an L-rank matrix equation, and thus provides the fastest convergence of all iterative methods [Cho91, Cho92].

Use of the conjugate gradient method to eliminate multipath fading in mobile communication situations has been studied in [Cho92] to show that the BER performance of the system using the conjugate gradient method is better than that using the RLS algorithm.

## 3.12 Neural Network Approach

In this section, a neural-network base algorithm to estimate the weights of an adaptive array system is described [God97]. For discussion on various aspects of this algorithm, referred to as Madaline rule III (MRIII), as well as other related issues, see [Wid90]. For general theory of neural networks and their applications, see [Lau90, Gel96].

The MRIII algorithm described here is applicable when the reference signal is available and minimizes the MSE between the reference signal and the modified array output, rather than the MSE between the reference signal and the array output, as is the case for other algorithms discussed previously. The array output is modified using a nonlinear mapping, such as hyperbolic tangent

$$\tanh(x) = \frac{1 - e^{-2x}}{1 + e^{-2x}} \tag{3.12.1}$$

and the weights are updated using

$$\mathbf{w}(n+1) = \mathbf{w}(n) - \mu \mathbf{g}(\mathbf{w}(n)) \tag{3.12.2}$$

where $\mu$ is the gradient step size and $\mathbf{g}(\mathbf{w}(n))$ is the instant gradient of the MSE surface with respect to the array weights $\mathbf{w}(n)$.

When the array is operating with weights $\mathbf{w}(n)$, producing the array output

$$y(n) = \mathbf{w}^H(n)\mathbf{x}(n+1) \tag{3.12.3}$$

the modified output $\tilde{y}(n)$ becomes

$$\tilde{y}(n) = \tanh(y(n)) \tag{3.12.4}$$

and the resulting error signal is given by

$$\tilde{\varepsilon}(n) = \tilde{y}(n) - r(n+1) \tag{3.12.5}$$

The instant gradient of the MSE surface with respect to the array weights $\mathbf{w}(n)$ thus becomes

$$\begin{aligned} \mathbf{g}(\mathbf{w}(n)) &= \frac{\partial(\tilde{\varepsilon}^*(n)\tilde{\varepsilon}(n))}{\partial \mathbf{w}(n)} \\ &= 2\tilde{\varepsilon}(n)\frac{\partial \tilde{\varepsilon}(n)}{\partial \mathbf{w}(n)} \\ &= 2\tilde{\varepsilon}(n)\frac{\partial \tilde{\varepsilon}(n)}{\partial y(n)}\frac{\partial y(n)}{\partial \mathbf{w}(n)} \\ &= 2\tilde{\varepsilon}(n)\frac{\partial \tilde{\varepsilon}(n)}{\partial y(n)}\mathbf{x}(n+1) \end{aligned} \tag{3.12.6}$$

Replacing $\partial\tilde{\varepsilon}(n)/\partial y(n)$ with $\Delta\tilde{\varepsilon}(n)/\Delta y$ for small $\Delta y$ in (3.12.6) results in

$$\mathbf{g}(\mathbf{w}(n)) = 2\tilde{\varepsilon}(n)\frac{\Delta\tilde{\varepsilon}(n)}{\Delta y}\mathbf{x}(n+1) \tag{3.12.7}$$

where $\Delta\tilde{\varepsilon}(n)$ denotes the change in the error output when the array output is perturbed by a small amount of $\Delta y$ and could be measured to estimate the instant gradient. The weight update equation then becomes

$$\mathbf{w}(n+1) = \mathbf{w}(n) - 2\mu\tilde{\varepsilon}(n)\frac{\Delta\tilde{\varepsilon}(n)}{\Delta y}\mathbf{x}(n+1) \tag{3.12.8}$$

The MSE surface of the error signal $\tilde{\varepsilon}(n)$ may have local minimization and thus global convergence of the MRIII algorithm is not guaranteed, which is not the case when MSE between the reference signal and the array output is minimized [Wid90]. The algorithm, however, is very robust, suitable for analog implementation, and results in fast weight updates.

The MRIII algorithm described here is suitable when the reference signal is available. A scheme to solve constrained beamforming problems using neural networks is analyzed in [Cha92], and its implementation using switched capacitor circuits is described in [Yan96]. Computer simulations and experimental results indicate the suitability of the scheme.

## 3.13 Adaptive Beam Space Processing

In this section, an adaptive algorithm to estimate the weights of the two-beam processor referred to as postbeamformer interference canceler (PIC), and discussed in Section 2.6.3 is presented and its performance is analyzed [God89a]. The analyses include the transient and steady-state behavior of the weights. The structure of the processor is shown in Figure 2.11. These results can be generalized for a general multibeam processor.

Rewrite (2.6.43) to (2.6.46) in discrete notation:

$$\psi(n) = \mathbf{V}^H\mathbf{x}(n) \tag{3.13.1}$$

$$q(n) = \mathbf{U}^H\mathbf{x}(n) \tag{3.13.2}$$

$$y(n) = \psi(n) - wq(n) \tag{3.13.3}$$

and

$$P(w) = \mathbf{V}^H R\mathbf{V} + w^* w\mathbf{U}^H R\mathbf{U} - w^*\mathbf{V}^H R\mathbf{U} - w\mathbf{U}^H R\mathbf{V} \tag{3.13.4}$$

where $\psi(n)$ denotes the output of signal beam; q(n) denotes the output of the interference beam; y(n) denotes the output of the processor; P(w) denotes the mean output of the processor for a given w; $\mathbf{V}$ and $\mathbf{U}$, respectively, denote the fixed weights of the signal-beam and the interference beam; and w is the weight applied to the interference beam output.

Let $\hat{w}$ denote the optimal weight that minimizes P(w). From (2.6.48) it is given by

$$\hat{w} = \frac{\mathbf{V}^H R\mathbf{U}}{\mathbf{U}^H R\mathbf{U}} \tag{3.13.5}$$

Define a real-time algorithm for determining the optimal weight $\hat{w}$ as

$$w(n+1) = w(n) - \mu g(w(n)) \tag{3.13.6}$$

where w(n + 1) denotes the new weight computed at the (n + 1)th iteration, μ is a positive scalar defining the step size, and g(w(n)) is an unbiased estimate of the gradient of P(w(n)) with respect to w.

It follows from (3.13.4) that the gradient of P(w(n)) with respect to w is given by

$$\nabla_w P(w(n))\Big|_{w-w(n)} = 2w(n)\mathbf{U}^H R\mathbf{U} - 2\mathbf{V}^H R\mathbf{U} \tag{3.13.7}$$

### 3.13.1 Gradient Estimate

A suitable estimate of the gradient of P(w(n)) with respect to w is given by

$$g(w(n)) = -2y(n)q^*(n) \tag{3.13.8}$$

In proposing (3.13.8), it is assumed that the gradient algorithm defined by (3.13.6) iterates at successive time instants. Thus, at time instant n + 1, the processor actually is operating with the weight w(n) computed at the previous iteration and the time instant and the array signal vector is $\mathbf{x}(n + 1)$. Note that for a given w(n), the estimate defined by (3.13.8) is unbiased, that is,

$$\begin{aligned} E\left[g(w(n))\middle|w(n)\right] &= -2E\left[y(n)q^*(n)\middle|w(n)\right] \\ &= -2E\left[\left\{\mathbf{V}^H\mathbf{x}(n) - w(n)\mathbf{U}^H\mathbf{x}(n)\right\}\mathbf{x}^H(n)\mathbf{U}\middle|w(n)\right] \\ &= -2\mathbf{V}^H R\mathbf{U} + 2w(n)\mathbf{U}^H R\mathbf{U} \end{aligned} \tag{3.13.9}$$

A particular characteristic of the gradient used in (3.13.6) that is important in determining the performance of the algorithm is the covariance. For the gradient estimate defined by (3.13.8), the following result on the convariance is established in Appendix 3.7.

Let $V_g(w(n))$ denote the covariance of the gradient estimate defined by (3.13.8) for a given w(n). If $\{\mathbf{x}(n)\}$ is an i.i.d. complex Gaussian sequence, then

$$\begin{aligned} V_g(w(n)) = 4\mathbf{U}^H R\mathbf{U}\big[&\mathbf{V}^H R\mathbf{V} + w^*(n)w(n)\mathbf{U}^H R\mathbf{U} \\ &- w(n)\mathbf{U}^H R\mathbf{V} - w^*(n)\mathbf{V}^H R\mathbf{U}\big] \end{aligned} \tag{3.13.10}$$

Note that the quantity in the square brackets is the mean output power of the PIC for a given w(n). Thus, at each iteration the covariance of the gradient estimate is proportional to the mean output power of the PIC that the adaptive algorithm defined by (3.13.6) is trying to minimize.

The convergence analysis of the algorithm defined by (3.13.6) is presented when the gradient estimate is defined by (3.13.8). In the event that $\{\mathbf{x}(n)\}$ is a sequence of i.i.d. random complex vectors, a detailed analysis of the algorithm is possible. The analysis is carried out using the approach described in Section 3.2.

### 3.13.2 Convergence of Weights

The following result on the convergence of weights is established in Appendix 3.7. For the algorithm defined by (3.13.6) and (3.13.8), if {**x**(n)} is an i.i.d. random vector sequence,

$$\left|w(0)\right| < \infty \tag{3.13.11}$$

and

$$0 < \mu < \frac{1}{\mathbf{U}^H\mathbf{R}\mathbf{U}} \tag{3.13.12}$$

then

$$\lim_{n\to\infty} E\left[w(n)\right] = \hat{w} \tag{3.13.13}$$

and the convergence of E[w(n)] to $\hat{w}$ has the time constant given by

$$\tau = -\frac{1}{\ell n\left(1 - 2\mu\mathbf{U}^H\mathbf{R}\mathbf{U}\right)} \tag{3.13.14}$$

where $\ell$n(.) denotes the natural logarithm of (.).

Note that the step size μ and the convergence time constant τ are dependent on $\mathbf{U}^H\mathbf{R}\mathbf{U}$, the average power at the output of the interference beamformer, and are independent of the output power of the signal beamformer.

### 3.13.3 Covariance of Weights

Let $K_{ww}(n)$ denote the covariance of weight w(n) at the nth iteration, that is,

$$K_{ww}(n) = E\left[\left(w(n) - \overline{w}(n)\right)\left(w(n) - \overline{w}(n)\right)^*\right] \tag{3.13.15}$$

where

$$\overline{w}(n) = E\left[w(n)\right] \tag{3.13.16}$$

The covariance of the weight $K_{ww}(n)$ satisfies the following recursive relation:

$$K_{ww}(n+1) = K_{ww}(n)\left[1 - 4\mu\mathbf{U}^H\mathbf{R}\mathbf{U} + 4\mu^2\left(\mathbf{U}^H\mathbf{R}\mathbf{U}\right)^2\right] + \mu^2 E\left[V_g\left(w(n)\right)\right] \tag{3.13.17}$$

where the expectation is taken over w. A derivation of this recursive equation is provided in Appendix 3.7.

Since the covariance of the weight at the (n + 1)th iteration depends on the covariance of the gradient estimated for a given weight at the nth iteration, it is possible to further simplify the above recursive relation for a particular method of gradient estimate. When the gradient estimate used in (3.13.6) is defined by (3.13.8), an expression for $V_g(w(n))$ is given by (3.13.10). The expression is derived with the assumption that {x(k)} is an i.i.d. complex Gaussian sequence. This assumption is necessary for the results presented throughout the remainder of this section. Taking the unconditional expectation on both sides of (3.13.10) and substituting in (3.13.17), the following difference equation for the covariance of the weight results:

$$K_{ww}(n+1) = K_{ww}(n)\left[1 - 4\mu \mathbf{U}^H \mathbf{R}\mathbf{U} + 8\mu^2 \left(\mathbf{U}^H \mathbf{R}\mathbf{U}\right)^2\right]$$
$$+ 4\mu^2 \mathrm{U}^H \mathrm{R}\mathbf{U}\left[\mathbf{V}^H \mathbf{R}\mathbf{V} + \overline{w}^*(n)\overline{w}(n)\mathbf{U}^H \mathbf{R}\mathbf{U}\right. \qquad (3.13.18)$$
$$\left. - \overline{w}^*(n)\mathrm{V}^H \mathbf{R}\mathbf{U} - \overline{w}(n)\mathbf{U}^H \mathbf{R}\mathbf{V}\right]$$

### 3.13.4 Transient Behavior of Weight Covariance

Let

$$H = 1 - 4\mu \mathbf{U}^H \mathbf{R}\mathbf{U} + 8\mu^2 \left(\mathbf{U}^H \mathbf{R}\mathbf{U}\right)^2 \qquad (3.13.19)$$

and

$$D(n) = \mathrm{U}^H \mathrm{R}\mathbf{U}\left[\mathbf{V}^H \mathbf{R}\mathbf{V} + \overline{w}^*(n)\overline{w}(n)\mathbf{U}^H \mathbf{R}\mathbf{U}\right.$$
$$\left. - \overline{w}^*(n)\mathbf{V}^H \mathbf{R}\mathbf{U} - \overline{w}(n)\mathbf{U}^H \mathbf{R}\mathbf{V}\right] \qquad (3.13.20)$$

Since

$$\lim_{n\to\infty} \overline{w}(n) = \hat{w} \qquad (3.13.21)$$

it follows from (3.13.5) and (3.13.20) that

$$\lim_{n\to\infty} D(n) = \mathbf{V}^H \mathbf{R}\mathbf{V}\mathbf{U}^H \mathbf{R}\mathbf{U} - \mathbf{V}^H \mathbf{R}\mathbf{U}\mathbf{U}^H \mathbf{R}\mathbf{V} \qquad (3.13.22)$$

From (3.13.18) to (3.13.20), it follows that

$$K_{ww}(n+1) = K_{ww}(n)H + 4\mu^2 D(n) \qquad (3.13.23)$$

which has the solution

$$K_{ww}(n) = H^n K_{ww}(0) + 4\mu^2 \sum_{i=1}^{n} D(n-i)H^{i-1} \qquad (3.13.24)$$

where $K_{ww}(0)$ is the covariance of w(0).

Since w(0) is a deterministic scalar, it follows that

$$K_{ww}(0) \equiv 0 \qquad (3.13.25)$$

and thus (3.13.24) reduces to

$$K_{ww}(n) = 4\mu^2 \sum_{i=1}^{n} D(n-i)H^{i-1} \quad (3.13.26)$$

which completely describes the transient behavior of the covariance of the weights when the gradient estimate used in (3.13.6) is defined by (3.13.8).

### 3.13.5 Steady State Behavior of Weight Covariance

Take the z transform on both sides of (3.13.23):

$$zK_{ww}(z) = K_{ww}(z)H + 4\mu^2 D(z) \quad (3.13.27)$$

where $K_{ww}(z)$ and $D(z)$ are the z transforms of $K_{ww}(n)$ and $D(n)$, respectively.

Solving for $K_{ww}(z)$, from (3.13.27),

$$K_{ww}(z) = 4\mu^2 \frac{D(z)z^{-1}}{1-Hz^{-1}} \quad (3.13.28)$$

Since $D(z)$ is stable, it follows from (3.13.28) that the stability of $K_{ww}(z)$ is guaranteed if

$$|H| < 1 \quad (3.13.29)$$

which, along with (3.13.19), implies that if

$$0 < \mu < \frac{1}{U^H R U} \quad (3.13.30)$$

then $K_{ww}(z)$ is stable. Thus, $\lim_{n\to\infty} K_{ww}(n)$ exists. This proves the existence of the limit.

To obtain the steady-state expression for the weight covariance, let

$$\hat{K}_{ww} = \lim_{n\to\infty} K_{ww}(n) \quad (3.13.31)$$

Since

$$\lim_{n\to\infty} K_{ww}(n) \equiv \lim_{n\to\infty} K_{ww}(n+1) \quad (3.13.32)$$

it follows from (3.13.19), (3.13.22), and (3.13.23) that

$$\hat{K}_{ww} = \mu \frac{V^H R V U^H R U - V^H R U U^H R V}{U^H R U\left[1 - 2\mu U^H R U\right]} \quad (3.13.33)$$

which, along with (3.13.4) and (3.13.5), leads to

$$\hat{K}_{ww} = \mu \frac{P(\hat{w})}{1 - 2\mu \mathbf{U}^H \mathbf{R} \mathbf{U}} \tag{3.13.34}$$

where $P(\hat{w})$ is the mean output power of the optimal PIC.

Thus, the steady-state weight covariance is proportional to the mean output power of the optimal PIC.

### 3.13.6 Misadjustment

In the absence of noise in the weight, the adaptive algorithm defined by (3.13.6) and (3.13.8) would converge to a steady state or optimal point on the mean output power surface. The minimum mean output power of the PIC therefore would be $P(\hat{w})$. However, the noise in the weight tends to cause the steady-state solution to vary randomly about the minimum or optimal point. This results in excess power in the output power of the PIC; the amount of excess power depends on the weight covariance.

As discussed previously, misadjustment is a dimensionless measure of the difference between the adaptive and optimal performance of a processor. It is defined as the ratio of the excess mean output power to the mean output power of the optimal PIC, that is,

$$M = \lim_{n \to \infty} \frac{E[P(w(n))] - P(\hat{w})}{P(\hat{w})} \tag{3.13.35}$$

In this section, analysis of the misadjustment is presented and an exact expression for it is derived when the gradient algorithm defined by (3.13.6) and (3.13.8) is used to estimate the weight given by (3.13.5).

Taking the expected value on both sides of (3.13.4) and using (3.13.15) and (3.13.16),

$$\begin{aligned} E[P(w(n))] = {} & \mathbf{V}^H \mathbf{R} \mathbf{V} + K_{ww}(n) \mathbf{U}^H \mathbf{R} \mathbf{U} + \overline{w}^*(n) \overline{w}(n) \mathbf{U}^H \mathbf{R} \mathbf{U} \\ & - \overline{w}^*(n) \mathbf{V}^H \mathbf{R} \mathbf{U} - \overline{w}(n) \mathbf{U}^H \mathbf{R} \mathbf{V} \end{aligned} \tag{3.13.36}$$

Taking the limit as $n \to \infty$ and subtracting $P(\hat{w})$ on both sides of (3.13.36), an expression for the steady-state excess mean output power follows:

$$\lim_{n \to \infty} E[P(w(n))] - P(\hat{w}) = \hat{K}_{ww} \mathbf{U}^H \mathbf{R} \mathbf{U} \tag{3.13.37}$$

Let $M_P$ denote the misadjustment in PIC. Equations (3.13.37), along with (3.13.35), imply that

$$M_P = \frac{\hat{K}_{ww} \mathbf{U}^H \mathbf{R} \mathbf{U}}{P(\hat{w})} \tag{3.13.38}$$

A substitution for $\hat{K}_{ww}$ from (3.13.34) in (3.13.38) leads to the following expression for the misadjustment:

$$M_P = \mu \frac{\mathbf{U}^H R \mathbf{U}}{1 - 2\mu \mathbf{U}^H R \mathbf{U}} \tag{3.13.39}$$

It follows from (3.13.39) that misadjustment in the adaptive PIC is independent of the signal in the signal channel and depends only on the mean power at the output of the interference beamformer. Furthermore, for a very small step size μ, it is proportional to this power. Thus, given this misadjustment, it is desirable that the interference beamformer weight **U** is chosen such that $\mathbf{U}^H R\mathbf{U}$ is a smaller quantity.

However, it follows from (3.13.14) that if

$$2\mu \mathbf{U}^H R \mathbf{U} \ll 1 \tag{3.13.40}$$

then

$$\tau \simeq \frac{1}{2} \mu \mathbf{U}^H R \mathbf{U} \tag{3.13.41}$$

Thus, a smaller power in the interference channel results in a longer convergence time constant, which may not be desirable.

For the range of μ that satisfies (3.13.40), the misadjustment given by (3.13.39) can be approximated as

$$M_P \simeq \mu \mathbf{U}^H R \mathbf{U} \tag{3.13.42}$$

which, along with (3.13.41) implies that the product of misadjustment and the convergence time constant is given by

$$M_P \cdot \tau = 0.5 \tag{3.13.43}$$

and is independent of array geometry and noise parameters.

### 3.13.7 Examples and Discussion

The example presented here is for a planar array of ten elements as shown in Figure 2.7. The array consists of two rings of five elements each, with half-wavelength inter-ring spacing $\mu_0$. The radius of the inner ring is 4 $\mu_0$.

A unity power signal source is assumed in the direction of the positive x-axis and an interference source is assumed in the direction of the negative x-axis. The interference power is taken to be 20 dB more than the signal power, and the uncorrelated noise power is taken to be 20 dB less than the signal power. The interference beam of the PIC is formed using

$$\mathbf{U} = P\mathbf{S}_I \tag{3.13.44}$$

and

$$P = I - \frac{\mathbf{S}_0 \mathbf{S}_0^H}{L} \tag{3.13.45}$$

where $\mathbf{S}_0$ and $\mathbf{S}_I$ are the steering vectors in the directions of the signal and interference sources, respectively, and I is the identity matrix.

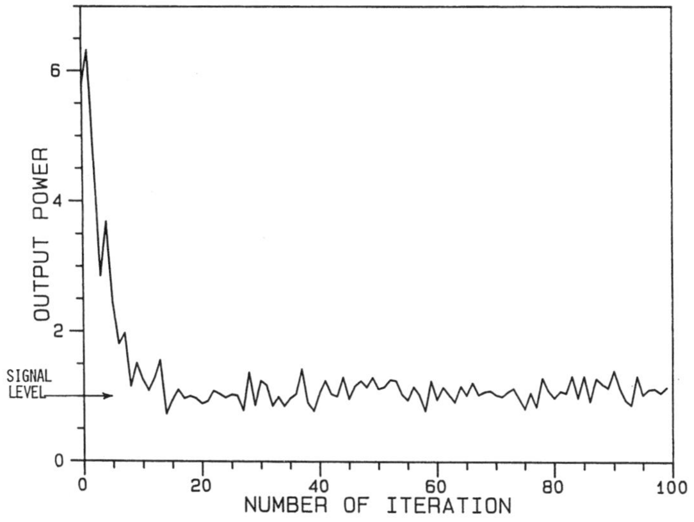


**FIGURE 3.14**
The output power averaged over 50 runs vs. the iteration number. (From Godara, L.C., *J. Acoust. Soc. Am.*, 85, 194–201, 1989 [God89a]. With permission.)

The interference beam formed using (3.13.44) and (3.13.45) ensures that the interference beam has a unity response in the interference direction and a null response in the signal direction. The signal beam is formed using the conventional weight, that is,

$$\mathbf{V} = \frac{\mathbf{S}_0^H}{L} \tag{3.13.46}$$

The algorithm is initialized with

$$w(0) = 0 \tag{3.13.47}$$

The gradient step size of $1 \times 10^{-5}$ is used, which is about one-eighth of the inverse of the estimated power of the interference beam. The power estimate at the output of the interference beam is made by averaging 100 samples. Figure 3.14 shows the PIC output power averaged over 50 runs as a function of the number of iterations. The figure shows that the output of the processor converges to the signal power in about 15 iterations. Figure 3.15 shows the norm of the weight error, that is,

$$\left[(w(n) - \hat{w})^*(w(n) - \hat{w})\right]^{1/2} \tag{3.13.48}$$

averaged over 50 runs as a function of the number of iterations. Convergence of the norm of the weight error is evident in the figure.

## 3.14 Signal Sensitivity of Constrained Least Mean Squares Algorithm

The convergence of mean weights estimated by constrained LMS algorithm to optimal weights is a function of the eigenvalues of $PR_NP$, and thus is independent of the look

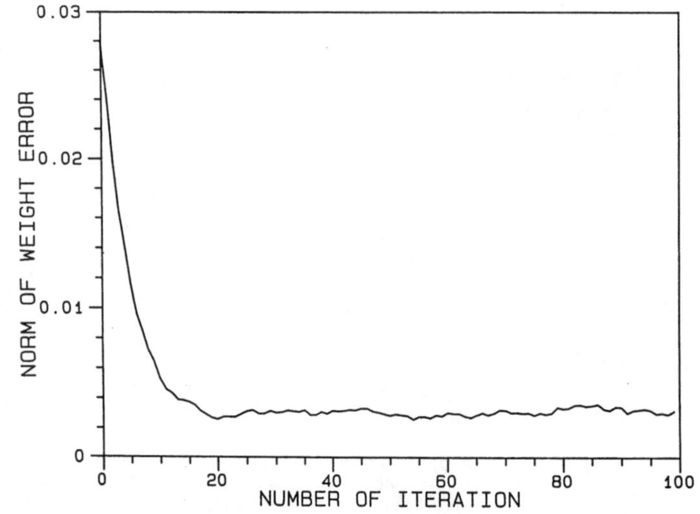


**FIGURE 3.15**
The norm of weight error averaged over 50 runs vs. the iteration number. (From Godara, L.C., *J. Acoust. Soc. Am.*, 85, 194–201, 1989 [God89a]. With permission.)

direction signal. However, this is not the case for the weight covariance matrix, which depends on the projected covariance of the gradient used for the weight update algorithm, that is, $PV_g(\mathbf{w}(n))P$. For the standard algorithm, this variance is a product of the array correlation matrix R and the mean output power $\mathbf{w}^H(n)R\mathbf{w}(n)$ at the nth instant of time. Thus, $PV_g(\mathbf{w}(n))P$, which is proportional to $\mathbf{w}^H(n)R\mathbf{w}(n)PRP$, contains a signal from the look direction, indicating that the performance of the standard LMS algorithm is not independent of the signal and that the transient behavior of weight covariance depends on it.

Results presented in Section 3.6 show that the weights estimated by the standard algorithm are sensitive to signal power in the look direction. As signal power increases, the noise in these weights tends to increase. The following, a rather heuristic argument, explains this phenomenon [God97].

Rewrite the constrained LMS algorithm as follows:

$$\mathbf{w}(n+1) = P\mathbf{w}(n) + \mathbf{S}_0/L - \mu P\mathbf{g}(\mathbf{w}(n)) \tag{3.14.1}$$

and examine the term $P\mathbf{g}(\mathbf{w}(n))$ for various estimates of the gradient. First, consider the true gradient, that is,

$$\mathbf{g}(\mathbf{w}(n)) = 2R\mathbf{w}(n) \tag{3.14.2}$$

Expressing R in the form

$$R = p_s\mathbf{S}_0\mathbf{S}_0^H + R_N \tag{3.14.3}$$

it follows that

$$P\mathbf{g}(\mathbf{w}(n)) = 2R_N\mathbf{w}(n) \tag{3.14.4}$$

Thus, the estimate of $\mathbf{w}(n+1)$ for a given $\mathbf{w}(n)$ does not depend on the signal power in the look direction when the true array correlation matrix is used in estimating the gradient.

Now consider the gradient estimate given by (3.4.4) and rewrite in the following form:

$$\mathbf{g}(\mathbf{w}(n)) = \mathbf{x}(n+1)\mathbf{x}^H(n+1)\mathbf{w}(n) \tag{3.14.5}$$

where the factor 2 has been omitted for ease of analysis.

The array signal vector $\mathbf{x}(n)$ can be expressed as

$$\mathbf{x}(n) = m_s(n)\mathbf{S}_0 + \mathbf{x}_N(n) \tag{3.14.6}$$

where $\mathbf{x}_N(n)$ is the array signal vector due to interference and uncorrelated noise only, and $m_S(n)$ is the sample of the complex modulating function of the signal.

From (3.14.5) and (3.14.6), it follows that

$$\begin{aligned}\mathbf{g}(\mathbf{w}(n)) = {} & m_s(n+1)m_s^*(n+1)\mathbf{S}_0\mathbf{S}_0^H\mathbf{w}(n) \\ & + \mathbf{x}_N(n+1)\mathbf{x}_N^H(n+1)\mathbf{w}(n) \\ & + m_s(n+1)\mathbf{S}_0\mathbf{x}_N^H(n+1)\mathbf{w}(n) \\ & + m_s^*(n+1)\mathbf{x}_N(n+1)\mathbf{S}_0^H\mathbf{w}(n)\end{aligned} \tag{3.14.7}$$

Since

$$P\mathbf{S}_0 = 0 \tag{3.14.8}$$

it follows from (3.14.7) that

$$\begin{aligned}P\mathbf{g}(\mathbf{w}(n)) = {} & P\mathbf{x}_N(n+1)\mathbf{x}_N^H(n+1)\mathbf{w}(n) \\ & + m_s^*(n+1)P\mathbf{x}_N(n+1)\mathbf{S}_0^H\mathbf{w}(n)\end{aligned} \tag{3.14.9}$$

The second term on the RHS of (3.14.9) contains $m_s^*(n + 1)$, which is a random quantity with variance equal to the look direction signal power. This makes $P\mathbf{g}(\mathbf{w}(n))$ a noisy quantity that fluctuates with the signal power and causes the $\mathbf{w}(n + 1)$ to fluctuate. The fluctuations in $\mathbf{w}(n + 1)$ increase as the signal power increases. Thus, the weights estimated by the standard algorithm are sensitive to the signal power requiring a lower step size in the presence of a strong signal for the algorithm to converge which in turn reduces its convergence speed.

This fact has been demonstrated in [Ohg93a] for a high-speed GMSK mobile communications system. The system has been implemented by mounting an array on a vehicle to measure its BER performance.

The signal sensitivity of the standard LMS algorithm is caused by the use of a sample correlation matrix in estimating the gradient, and could be reduced by using an estimate of the correlation matrix from all available samples as is done with the recursive LMS algorithm. In this case, variance of the estimated gradient is given as

$$V_{g_R}(\mathbf{w}(n)) = \frac{4}{(n+1)^2}\,\mathbf{w}^H(n)\,R\,\mathbf{w}(n)\,R \tag{3.14.10}$$

Comparing this with the variance of the standard LMS algorithm, note that the variance of the gradient estimated by the recursive method is less than that estimated by the standard method by a factor of $(n + 1)^2$. Thus, the recursive algorithm is less signal sensitive to signal power. In the limit as n increases, the signal sensitivity of the recursive LMS algorithm approaches zero.

The signal sensitivity of the LMS also can be reduced by spatial averaging instead of sample averaging, as is the case in the structured gradient algorithm. Because of spatial averaging and the fact that $m_S(n + 1)$ and $\mathbf{x}_N(n + 1)$ are not correlated, the dependence of $P\mathbf{g}_{st}(\mathbf{w}(n))$ on the signal level is substantially reduced. Thus, the weights estimated by the structured gradient algorithm are not very sensitive to the signal level in the look direction.

## 3.15 Implementation Issues

In this section, some implementation issues relating to finite precision arithmetic and real vs. complex implementation are discussed [God97].

### 3.15.1 Finite Precision Arithmetic

The convergence speed, fluctuations in array weights during adaption, and misadjustment noise are the measures of the transient and the steady-state behavior of the LMS algorithm. Theoretical performance of the algorithm and the effect of the look direction signal and gradient step size discussed in previous sections assume the existence of infinite precision, that is, the variables are allowed to take any value.

In real life, when the algorithm is implemented using digital hardware where variables can only take discrete values, other parameters affect its performance, and issues that must be considered include quantization noise as well as roundoff and truncation noise caused by finite precision arithmetic [Eva93, Ale87, Cha91, Leu91, Won91, Car84, Cio85].

First, when a b-bit quantizer is used to convert an analog signal of range $-r_{max}$ to $r_{max}$ into a digital signal, it adds quantization noise of zero mean and variance [Opp75],

$$\sigma_q^2 = \frac{-2^{2b}\, r_{max}^2}{3} \tag{3.15.1}$$

to the system. Second, the effect of finite word length of the devices where the numbers are stored causes the roundoff or truncation noise to be added to the system. This arises from the fact that when arithmetic operations are performed using these numbers, the answers are normally longer than the available word length and need to be rounded off or truncated to fit into finite word memory. Finally, all variables such as the estimated gradient, the gradient step size, and the estimated weights are only allowed to take finite values, and can be increased or decreased by a factor of 2. The combined effect of all these factors on the algorithm is a larger fluctuation in weights and a larger misadjustment than otherwise.

The misadjustment appears to be the most sensitive to the finite word length effect on weights, suggesting that the weights should be implemented using a longer word length [Ale87] and a reduction in the step size below certain levels may even cause the misadjustment to increase [Car84] which is contrary to the infinite precision case where a decrease in the step causes the misadjustment to decrease. It appears [Cio85] that the finite word-length effects are amplified in the environment, which yields smaller eigenvalues for the correlation matrix.

An important effect of the finite word length on the weight update is that when a small input does not cause the weights to move more than the least significant bit (the smallest possible increment, which depends on the number of bits used to store weights), then the algorithm stalls and weights do not change anymore [Car84], requiring a bigger step size, which in turn increases weight fluctuations.

A post-algorithm smoothing scheme suggested in [Cha91] appears to reduce weight fluctuations leading to better convergence performance. It suggests a running average of past weights. Thus, the weights are recursively updated using past weights with or without finite memory. Discussion on system design applicable to mobile satellite communications which takes into account quantization noise and other issues discussed above may be found in [Geb95].

### 3.15.2 Real vs. Complex Implementation

In some situations, the input data to the weight adaption scheme are real, and in others, the data are complex (with real and imaginary parts denoting in-phase and quadrature components). In both of these cases, the weights could be updated using the real LMS algorithm or the complex LMS algorithm. The former uses real arithmetic and real variables, and updates real weights (in-phase and quadrature component are updated separately when complex data are available), whereas the complex algorithm [Wid75] uses complex arithmetic and variables, and weights are updated as well as implemented as complex variables similar to the treatment presented in this book. For real data using complex algorithm, you need to generate the quadrature component using the Hilbert transformer or quadrature filter [Pap65], which has the following transfer functions:

$$H(f) = \begin{cases} -j & f > 0 \\ j & f < 0 \end{cases} \tag{3.15.2}$$

For a similar misadjustment, the complex algorithm converges faster than the real algorithm. More details on this topic are available in [Hor81, God86]. Some of these issues are discussed below.

#### *3.15.2.1 Quadrature Filter*

The output of the quadrature filter is related to its input by the Hilber transform. Before deriving an expression for the quadrature filter transfer function given by (3.15.2), the Hilber transform is defined and some useful properties are stated.

Let $\hat{x}(t)$ denote the Hilber transform of a real signal x(t) defined as

$$\hat{x}(t) = \frac{1}{\pi} \int_{-\infty}^{\infty} \frac{x(\tau)}{t - \tau} d\tau \tag{3.15.3}$$

The Hilber transform has the following properties. First, the Hilber transform of the Hilber transform is the negative of the original signal, that is,

$$\hat{\hat{x}}(t) = -x(t) \tag{3.15.4}$$

A signal and its Hilber transform pair form an orthogonal pair, that is,

$$\int_{-\infty}^{\infty} \hat{x}(t)x(t)dt = 0 \tag{3.15.5}$$

The Hilber transform of a constant C is zero, that is,

$$\hat{C} = 0 \tag{3.15.6}$$

The Hilber transform of cos(ωt) is sin(ωt), $\omega > 0$. If

$$x(t) = a(t)\cos\left(2\pi f_0 t + \theta\right),\ f_0 > 0 \tag{3.15.7}$$

such that the highest frequency of a(t) is less than $f_0$, then

$$\hat{x}(t) = a(t)\sin\left(2\pi f_0 t + \theta\right),\ f_0 > 0 \tag{3.15.8}$$

Now a derivation of (3.15.2) is presented. It follows from (3.15.3) that $\hat{x}(t)$ is a convolution of x(t) and $1/\pi t$, that is,

$$\hat{x}(t) = x(t) * \frac{1}{\pi t} \tag{3.15.9}$$

Thus, the Hilbert transform can be thought of as an output of a system (Hilber transformer) with an impulse response h(t) given by

$$h(t) = \frac{1}{\pi t} \tag{3.15.10}$$

Let sgn(t) denote the sign function, that is,

$$\text{sgn}(t) = \begin{cases} + & t > 0 \\ - & t < 0 \end{cases} \tag{3.15.11}$$

and F{.} denote the Fourier transform of {.}. Noting that

$$F\{\text{sgn}(t)\} = \frac{1}{j\pi t} \tag{3.15.12}$$

it follows from the duality theorem of the Fourier transform that

$$F\left\{\frac{1}{\pi t}\right\} = -j\,\text{sgn}(f) \tag{3.15.13}$$

Taking the Fourier transform on both sides of (3.15.10) and using (3.15.13), the following expression is obtained for the transfer function of the Hilber transformer, also known as the quadrature filter:

$$H(f) = -j\,\text{sgn}(f)$$
$$= \begin{cases} -j & f > 0 \\ j & f < 0 \end{cases} \tag{3.15.14}$$

In the remainder of this section, a real valued signal is denoted by $x_I(t)$ and its Hilber transform is denoted by $x_Q(t)$.

### *3.15.2.2 Analytical Signals*

A complex valued signal x(t) is said to be an analytical signal if its real and imaginary parts are related via the Hibert transform. Thus, it can be expressed as

$$x(t) = x_I(t) + jx_Q(t) \tag{3.15.15}$$

where $x_Q(t) = \hat{x}_I(t)$.

Taking the Fourier transform on both sides of (3.15.15) and using the properties of the Hilber transform, it can easily be shown that x(t) has a one-sided spectrum.

### *3.15.2.3 Beamformer Structures*

Consider the structures of two narrowband beamformers shown in Figure 3.16 and Figure 3.17. Figure 3.16 shows a real beamforming system and Figure 3.17 shows an in-phase and quadrature (IQ) or complex beamforming system. The real beamforming system has a single real valued output that can be produced by using real multiplication to achieve the weighting of the array signals. The other has a complex valued output and can be produced by using the complex multiplication to achieve the weighting of the array signals.

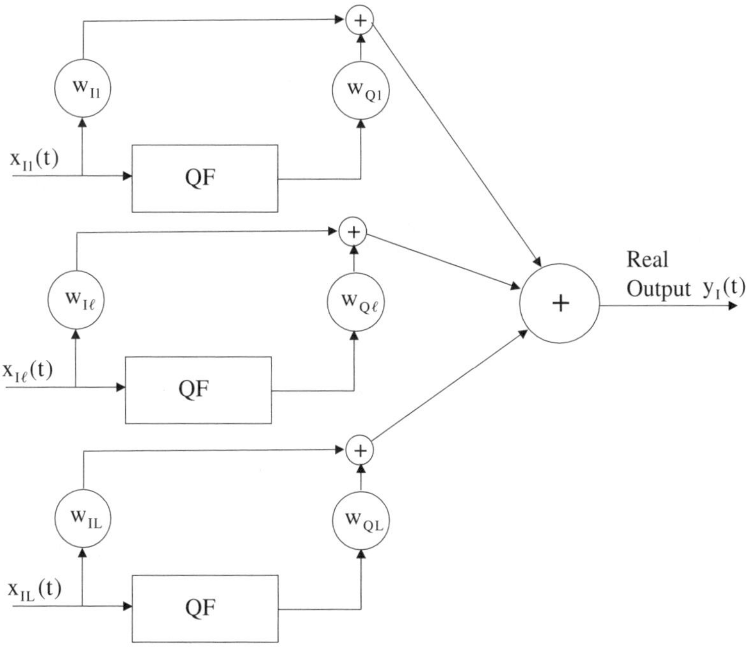


**FIGURE 3.16**
Real beamforming system.

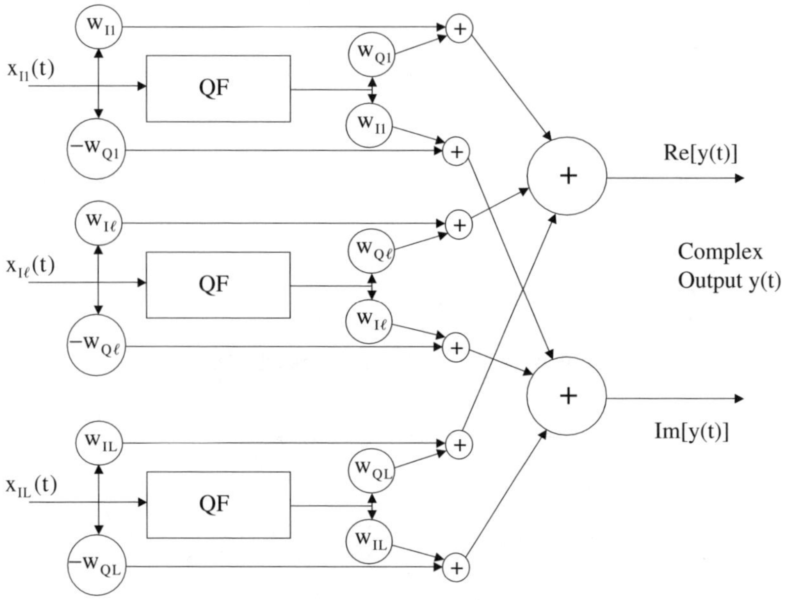


**FIGURE 3.17**
In-phase and quadrature beamforming system.

Let the L-dimensional complex vector $\mathbf{x}(t)$ denote the array signal in complex notation defined as

$$\mathbf{x}(t) = \mathbf{x}_I(t) + j\mathbf{x}_Q(t) \tag{3.15.16}$$

where the L-dimensional real vectors $\mathbf{x}_I(t)$ and $\mathbf{x}_Q(t)$ denote the in-phase and quadrature array signals, respectively.

Define the L-dimensional complex weight vector $\mathbf{w}$ as

$$\mathbf{w} = \mathbf{w}_I + j\mathbf{w}_Q \tag{3.15.17}$$

where the L-dimensional real vectors $\mathbf{w}_I$ and $\mathbf{w}_Q$ denote the weights as shown in Figure 3.16 and Figure 3.17.

Let $y_I(t)$ denote the output of the real beamforming system. It follows from Figure 3.16 that it is given by

$$\begin{aligned} y_I(t) &= \mathbf{w}_I^T\mathbf{x}_I(t) + \mathbf{w}_Q^T\mathbf{x}_Q(t) \\ &= \mathrm{Re}\left[\mathbf{w}^H\mathbf{x}(t)\right] \\ &= \frac{1}{2}\left[\mathbf{w}^H\mathbf{x}(t) + \mathbf{x}^H(t)\mathbf{w}\right] \end{aligned} \tag{3.15.18}$$

Let y(t) denote the output of the IQ beamforming system. It can easily be shown from Figure 3.17 that the output of the IQ beamforming system is given by

$$y(t) = \mathbf{w}^H\mathbf{x}(t) \tag{3.15.19}$$

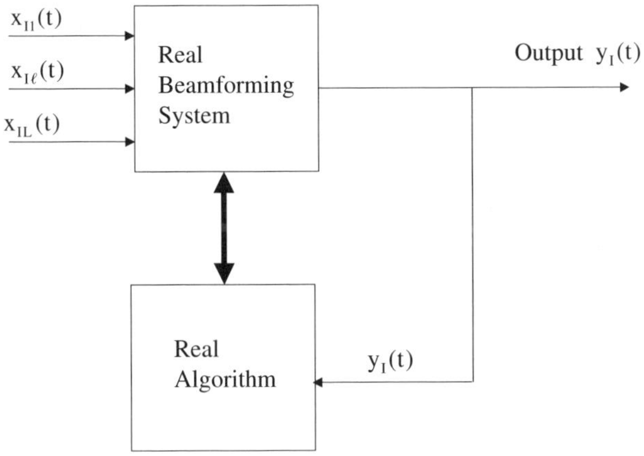


**FIGURE 3.18**
Real algorithm for real beamforming system.

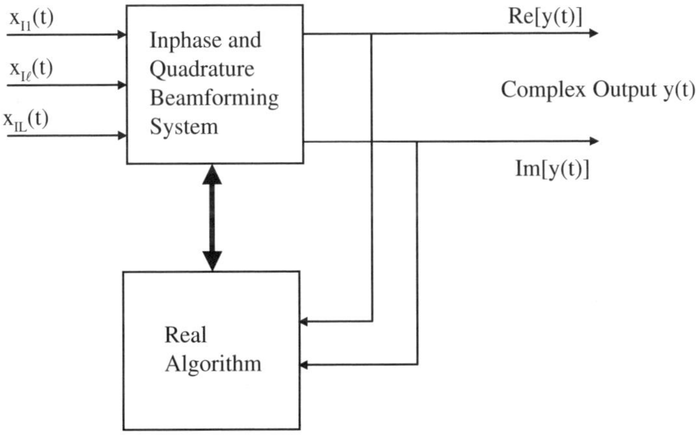


**FIGURE 3.19**
Real algorithm for IQ beamforming system.

Similarly, the beamformer structure may be developed when the reference signal is available.

Next, an implementation of the two algorithms is discussed with a view to compare the difference in convergence speed. The development presented here is for the constrained LMS algorithm. It can easily be extended for the unconstrained case.

#### *3.15.2.4 Real LMS Algorithm*

Implementation of the real LMS algorithm for the real beamforming system is shown in Figure 3.18 and for the IQ beamforming system it is shown in Figure 3.19. When all signals on the array are accessible, a suitable estimate of the required gradient of $\mathbf{w}^H R \mathbf{w}$ for $\mathbf{w} = \mathbf{w}(n)$ is

$$\mathbf{g}_R(\mathbf{w}(n)) = 4\mathbf{x}(n+1)y_I(n+1) \tag{3.15.20}$$

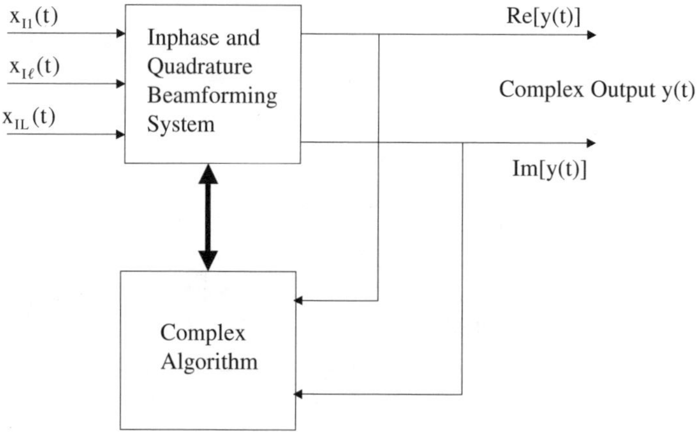


**FIGURE 3.20**
Complex algorithm for IQ beamforming system.

In the real beamforming system, $y_I(n+1)$ represents the only out of the system. In the IQ beamforming system, it is the real part of the output. In both cases, it is a real valued quantity and given by

$$y_I(n+1) = \frac{1}{2}\left[\mathbf{w}^H\mathbf{x}(n+1) + \mathbf{x}^H(n+1)\mathbf{w}(n)\right] \tag{3.15.21}$$

Note from (3.15.20) that real multiplications are used in estimating the real and imaginary parts of the complex valued quantity $\mathbf{g}_R(\mathbf{w}(n))$.

Using the result E[x(t)x(t)] = 0, it can easily be verified that the gradient given by (3.15.20) and (3.15.21) is unbiased; that is, for a given $\mathbf{w}(n)$

$$E\left[\mathbf{g}_R(\mathbf{w}(n))\right] = 2R\mathbf{w}(n) \tag{3.15.22}$$

Let $V_{g_R}(\mathbf{w}(n))$ denote the covariance of the gradient estimate given by (3.15.20) and (3.15.21). For a zero mean, stationary complex Gaussian vector process $\{\mathbf{x}(k)\}$, it is given by

$$V_{g_R}(\mathbf{w}(n)) = 4\mathbf{w}^H(n)R\mathbf{w}(n)R + 8R\mathbf{w}^H(n)\mathbf{w}(n)R \tag{3.15.23}$$

The derivation of (3.15.23) can easily be carried out following the procedure used in Section 3.4.2.

#### *3.15.2.5 Complex LMS Algorithm*

Implementation of the complex algorithm for the IQ beamforming system is shown in Figure 3.20, and for the real beamforming system in Figure 3.21. When all signals on the array are accessible, a suitable estimate of the required gradient of $\mathbf{w}^H R\mathbf{w}$ for $\mathbf{w} = \mathbf{w}(n)$ is

$$\mathbf{g}(\mathbf{w}(n)) = 2\mathbf{x}(n+1)y^*(n+1) \tag{3.15.24}$$

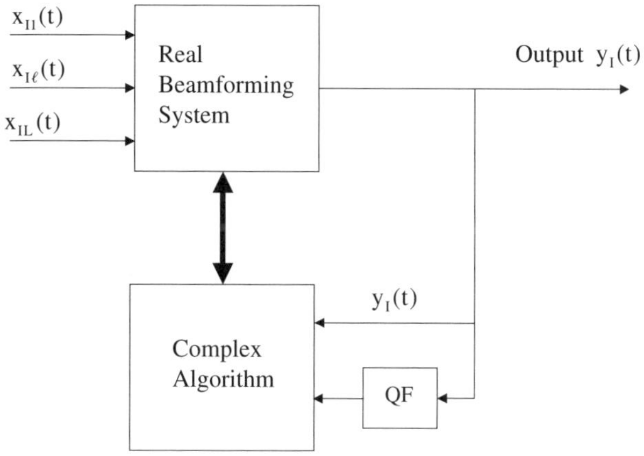


**FIGURE 3.21**
Complex algorithm for real beamforming system.

where

$$y(n+1)=\begin{cases}\text{output when the complex system is used}\\ y_I(n+1)+j\hat{y}_I(n+1)\text{ when the real system is used}\end{cases} \tag{3.15.25}$$

As $\hat{y}_I(n+1) \equiv y_Q(n+1)$, it follows from (3.15.25) that

$$y(n+1)=\mathbf{w}^H\mathbf{x}(n+1) \tag{3.15.26}$$

It follows from (3.15.24) and (3.15.25) that the gradient estimate in this case is identical to that for the standard LMS algorithm discussed in Section 3.4.1. Thus, the gradient covariance for this case is given by (3.4.6). Denoting it by $V_{g_C}(w(n))$ and rewriting (3.4.6)

$$V_{g_C}(\mathbf{w}(n))=4\mathbf{w}^H(n)R\mathbf{w}(n)R \tag{3.15.27}$$

#### *3.15.2.6 Discussion*

Comparing (3.15.23) and (3.15.27) one notes that

$$V_{g_R}(\mathbf{w}(n))=V_{g_C}(\mathbf{w}(n))+8R\mathbf{w}^H(n)\mathbf{w}(n)R \tag{3.15.28}$$

Thus, the covariance of the gradient used in the real algorithm is more than that used in the complex algorithm. The extra term $8R\mathbf{w}^H(n)\mathbf{w}(n)R$, present for the case of the real algorithm, results in more misadjustment for this case. Let $M_R$ denote the misadjustment when the gradient is given by (3.15.20) and (3.15.21). Following the procedure used in Section 3.4 it can be shown that [God86] if

$$0<\mu<\frac{1}{4\hat{\lambda}_{max}} \tag{3.15.29}$$

and

$$2\mu\sum_{i=1}^{L-1}\frac{\hat{\lambda}_i}{1-2\mu\hat{\lambda}_i}<1 \tag{3.15.30}$$

then the misadjustment is given by

$$M_R=\frac{2\mu\sum_{i=1}^{L-1}\frac{\hat{\lambda}_i}{1-2\mu\hat{\lambda}_i}}{1-2\mu\sum_{i=1}^{L-1}\frac{\hat{\lambda}_i}{1-2\mu\hat{\lambda}_i}} \tag{3.15.31}$$

The misadjustment for the complex case is given by (3.4.52). Let it be denoted by $M_C$. Comparing (3.15.30) with (3.4.52), one notes that

$$M_C(2\mu)=M_R(\mu) \tag{3.15.32}$$

Thus, it follows that misadjustment in both cases would be same if the gradient step size used in the complex case is double that used in the real case. Since for small step size the convergence time constant is inversely proportional to step size, it follows that for the same misadjustment the convergence time constant for the complex LMS algorithm is half that of the real LMS algorithm. This means that for the same misadjustment, the convergence speed of the complex LMS algorithm is twice that of the real LMS algorithm.

## Acknowledgments

Edited versions of Sections 3.4.1 to 3.4.6, 3.5, 3.6.1, 3.6.2, 3.6.3, 3.7.1, and 3.8 are reprinted from Godara, L.C., Constrained beamforming and adaptive algorithms, in *Handbook of Statistics,* Vol. 10, Bose, N.K and Rao, C.R., Eds., Copyright 1993, with permission from Elsevier. Edited versions of Sections 3.1, 3.3, 3.9 to 3.12, and 3.14 are reprinted from Godara, L.C., Application of antenna arrays to mobile communications, I. Beamforming and DOA considerations, *Proc. IEEE,* 85(8), 1195-1247, 1997. An edited version of Section 3.13 is reprinted from Godara, L.C., Analysis of transient and steady state weight covariance in adaptive postbeamformer interference canceller, *J. Acoust. Soc. Am.,* 85(1), 194-201, 1989.

## Notation and Abbreviations

| | |
|---|---|
| $\|A\|_F$ | Frobenius norm of a matrix A |
| E[.] | expectation operator |
| $E[x\|y]$ | conditional expectation for given y |
| $\ell n$ | natural logarithm |
| Tr(.) | trace of (.) |
| $(.)^H$ | Hermitian transpose of (.) |
| $(.)^T$ | transpose of (.) |
| CMA | constant modulus algorithm |

| | |
|---|---|
| IQ | in-phase and quadrature |
| LMS | least mean squares |
| MSE | mean square error |
| RHS | right-hand side |
| RLS | recursive least squares |
| SNR | signal-to-noise ratio |
| SMI | sample matrix inverse |
| TDMA | time-division multiple access |
| $\mathbf{F}$ | normalized steering vector in look direction |
| $F\{.\}$ | Fourier transform |
| $g(w(n))$ | gradient estimate for given w(n) |
| $\mathbf{g}(\mathbf{w}(n))$ | gradient estimate for given $\mathbf{w}(n)$ |
| $\mathbf{g}_1(\mathbf{w}(n))$ | gradient estimate using single-receiver system for given $\mathbf{w}(n)$ |
| $\mathbf{g}_2(\mathbf{w}(n))$ | gradient estimate using dual perturbation system for given $\mathbf{w}(n)$ |
| $\mathbf{g}_3(\mathbf{w}(n))$ | gradient estimate using reference receiver system for given $\mathbf{w}(n)$ |
| $\mathbf{g}_{st}(\mathbf{w}(n))$ | gradient estimate using structured gradient algorithm for given $\mathbf{w}(n)$ |
| $\mathbf{g}_R(\mathbf{w}(n))$ | gradient estimate using recursive LMS algorithm for given $\mathbf{w}(n)$ and gradient estimate using real LMS algorithm for given $\mathbf{w}(n)$ |
| $\mathbf{g}_I(\mathbf{w}(n))$ | gradient estimate using improved LMS algorithm for given $\mathbf{w}(n)$ |
| $\overline{\mathbf{g}}(\mathbf{w}(n))$ | mean of the gradient estimate for given $\mathbf{w}(n)$ |
| $H(f)$ | transfer functions of quadrature filter |
| $h(t)$ | impulse response |
| I | identity matrix |
| i.i.d. | independent identically distributed |
| $Im[.]$ | imaginary part of complex quantity |
| K | degree of freedom |
| $K_{ww}(n)$ | covariance of w(n) |
| $\hat{K}_{ww}$ | covariance of w(n) in limit |
| $k_{\mathbf{ww}}(n)$ | covariance matrix of $\mathbf{w}(n)$ |
| $k_{\mathbf{ww}_2}(n)$ | covariance matrix of $\mathbf{w}(n)$ in dual perturbation system |
| $k_{\mathbf{ww}_3}(n)$ | covariance matrix of $\mathbf{w}(n)$ in reference receiver system |
| $k_0(n)$ | constant denoting $\overline{\mathbf{w}}^H(n)R\overline{\mathbf{w}}(n)$ |
| L | number of elements in array |
| M | misadjustment, length of sequence S |
| $M_C$ | misadjustment in complex LMS algorithm |
| $M_P$ | misadjustment in adaptive PIC |
| $M_R$ | misadjustment in real LMS algorithm |
| $M_S$ | misadjustment in standard LMS algorithm |
| $M_U$ | misadjustment in unconstrained LMS algorithm |
| $M_1$ | misadjustment in single-receiver system |
| $M_2$ | misadjustment in dual perturbation system |
| $M_3$ | misadjustment in reference receiver system |
| $m_s(n)$ | complex modulating function of signal |
| N | number of samples |
| P | projection operator |

| | |
|---|---|
| $P(n)$ | output power at nth iteration |
| $\overline{P}(n)$ | mean output power at nth iteration |
| $P(\mathbf{w}(n))$ | mean output power for given $\mathbf{w}(n)$ |
| $P(w)$ | mean output power PIC for given w |
| $P_N(\mathbf{w}(n))$ | mean output noise power for given $\mathbf{w}(n)$ |
| $\hat{P}$ | mean output power of optimal processor |
| $p_r$ | mean power of reference signal |
| $p_s$ | signal power |
| $Q$ | matrix with columns being eigenvectors of R |
| $\hat{Q}$ | matrix with columns being eigenvectors of PRP |
| $\hat{Q}_\ell$ | eigenvector corresponding to $\hat{\lambda}_\ell$ |
| $q(n)$ | output of interference beam at nth instant of time |
| $R$ | array correlation matrix |
| $R(n)$ | estimate of R at nth instant of time using only one sample |
| $\hat{R}(n)$ | estimate of R at nth instant of time using past samples |
| $\tilde{R}(n)$ | estimate of R at nth instant of time using spatial averaging |
| $\hat{\tilde{R}}(n)$ | estimate of R using past samples and spatial averaging |
| Re[.] | real part of complex quantity |
| $R_N$ | array correlation matrix with no signal present |
| $R_{\mathbf{ww}}(n)$ | correlation matrix of $\mathbf{w}(n)$ |
| $R(N)$ | estimate of R using N samples |
| $r_i$ | correlation between elements with lag i |
| $S$ | complex vector sequence |
| $\mathbf{S}_0$ | steering vector associated with look direction |
| $\mathbf{S}_I$ | steering vector associated with interference |
| sgn(t) | sign function |
| $\mathbf{U}$ | fixed weights of interference beam |
| $\mathbf{U}_i$ | eigenvector corresponding to $\lambda_i$ of R |
| $\mathbf{V}$ | fixed weights of signal beam |
| $V_g(w(n))$ | covariance of gradient for given w(n) |
| $V_g(\mathbf{w}(n))$ | covariance of gradient for given $\mathbf{w}(n)$ |
| $V_{gC}(\mathbf{w}(n))$ | covariance of gradient in complex LMS algorithm for given $\mathbf{w}(n)$ |
| $V_{g_1}(\mathbf{w}(n))$ | covariance of gradient using single receiver system for given $\mathbf{w}(n)$ |
| $V_{g_2}(\mathbf{w}(n))$ | covariance of gradient using dual perturbation system for given $\mathbf{w}(n)$ |
| $V_{g_3}(\mathbf{w}(n))$ | covariance of gradient using reference receiver system for given $\mathbf{w}(n)$ |
| $V_{g_{st}}(\mathbf{w}(n))$ | covariance of gradient in structured LMS algorithm for given $\mathbf{w}(n)$ |
| $V_{gR}(\mathbf{w}(n))$ | covariance of gradient in recursive LMS algorithm for given $\mathbf{w}(n)$ and covariance of gradient in real LMS algorithm for given $\mathbf{w}(n)$ |
| $V_{gS}(\mathbf{w}(n))$ | covariance of gradient in standard LMS algorithm for given $\mathbf{w}(n)$ |
| $\overline{\mathbf{v}}(n)$ | mean error vector at nth iteration |
| $\hat{\mathbf{w}}$ | optimal weights of constrained processor |
| $\mathbf{w}(n)$ | array weights at nth iteration |
| $\overline{\mathbf{w}}(n)$ | mean value of $\mathbf{w}(n)$ |
| $\hat{\mathbf{w}}_{MSE}$ | optimal weights of processor with reference signal |

| | |
|---|---|
| $\hat{w}$ | optimal weight of PIC |
| $w(n)$ | PIC weight at nth iteration |
| $\overline{w}(n)$ | mean value of w(n) |
| $\mathbf{x}(n)$ | array signal vector at nth instant of time |
| $\hat{x}(t)$ | Hilber transform of x(t) |
| $x_I(t)$ | in-phase signal |
| $x_Q(t)$ | quadrature phase signal |
| $\mathbf{x}_N(n)$ | array signal vector due to interference and uncorrelated noise |
| $y(t)$ | output of IQ beamforming system |
| $y(n)$ | array output at nth instant of time |
| $y(\mathbf{w}(n))$ | array output for given $\mathbf{w}(n)$ |
| $y_I(t)$ | output of real beamforming system |
| $\mathbf{z}$ | correlation between reference signal and array signals |
| $\Gamma$ | diagonal matrix of eigenvalues of P |
| $\Lambda$ | diagonal matrix of eigenvalues of R |
| $\hat{\Lambda}$ | diagonal matrix of eigenvalues of PRP |
| $\hat{\Lambda}'$ | diagonal matrix of nonzero eigenvalues of PRP |
| $\Sigma(n)$ | diagonal matrix of eigenvalues of $k_{\mathbf{ww}}(n)$ |
| $\gamma$ | perturbation step size |
| $\hat{\gamma}(\mathbf{w}(n))$ | step size for which $V_{g1}(\mathbf{w}(n))$ is minimum |
| $\boldsymbol{\delta}(\ell)$ | L-dimensional complex vector |
| $\varepsilon(\mathbf{w}(n))$ | error between array output and reference signal for given $\mathbf{w}(n)$ |
| $\xi(\gamma)$ | perturbation noise for given $\gamma$ |
| $\xi(\mathbf{w}(n))$ | MSE for given $\mathbf{w}(n)$ |
| $\hat{\xi}$ | minimum MSE |
| $\bar{\xi}(n)$ | average value of MSE at nth iteration |
| $\mu$ | gradient step size |
| $\mu_0$ | inter-ring spacing |
| $\mu(n)$ | gradient step size at nth iteration |
| $\lambda_i$ | ith eigenvalue of R |
| $\hat{\lambda}_i$ | ith eigenvalue of PRP |
| $\lambda_{max}$ | maximum eigenvalue of R |
| $\hat{\lambda}_{max}$ | maximum eigenvalue of PRP |
| $\boldsymbol{\lambda}$ | vector of eigenvalues of R |
| $\hat{\boldsymbol{\lambda}}$ | vector of eigenvalues of PRP |
| $\hat{\boldsymbol{\lambda}}'$ | vector of nonzero eigenvalues of PRP |
| $\boldsymbol{\vartheta}$ | vector of eigenvalues of P |
| $\eta_i(n)$ | ith eigenvalue of $k_{\mathbf{ww}}(n)$ |
| $\boldsymbol{\eta}(n)$ | vector of eigenvalues of $k_{\mathbf{ww}}(n)$ |
| $\boldsymbol{\eta}_2(n)$ | vector of eigenvalues of $k_{\mathbf{ww}_2}(n)$ |
| $\boldsymbol{\eta}'(n)$ | vector of nonzero eigenvalues of $k_{\mathbf{ww}}(n)$ |
| $\psi(n)$ | output of signal beam at nth instant of time |
| $\tau$ | time constant for adaptive PIC |
| $\tau_i$ | ith time constant |

## References


Ale87 Alexander, S.T., Transient weight misadjustment properties for the finite precision LMS algorithm, *IEEE Trans. Acoust. Speech Signal Process.*, 35, 1250–1258, 1987.

Bar94 Barrett, M. and Arnott, R., Adaptive antennas for mobile communications, *Electron. Commn. Eng. J.*, 6, 203–214, 1994.

Ben92 Benesty, J. and Duhamel, P., A fast exact least mean square adaptive algorithm, *IEEE Trans. Signal Process.*, 40, 2904–2920, 1992.

Ber84 Bershad, N.J. and Qu, L.Z., LMS adaptation with correlated data: a scalar example, *IEEE Trans. Acoust. Speech Signal Process.*, 32, 695–700, 1984.

Ber85 Bershad, N.J. and Chang, Y.H., Time correlation statistics of the LMS adaptive algorithm weights, *IEEE Trans. Acoust. Speech Signal Process.*, 33, 309–312, 1985.

Ber86 Bershad, N.J. and Feintuch, P.L., A normalized frequency domain LMS adaptive algorithm, *IEEE Trans. Acoust. Speech Signal Process.*, 34, 452–461, 1986.

Ber86a Bershad, N.J., Analysis of the normalized LMS algorithm with Gaussian inputs, *IEEE Trans. Acoust. Speech Signal Process.*, 34, 793–806, 1986.

Bol87 Boland, F.B. and Foley, J.B., Stochastic convergence of the LMS algorithm in adaptive systems, *Signal Process.*, 13, 339–352, 1987.

Can80 Cantoni, A., Application of orthogonal perturbation sequences to adaptive beamforming, *IEEE Trans. Antennas Propagat.*, 28, 191–202, 1980.

Car84 Caraiscos, C. and Liu, B., A round-off error analysis of the LMS adaptive algorithm, *IEEE Trans. Acoust. Speech Signal Process.*, 32, 34–41, 1984.

Cha91 Chang, Y.H., Tzou, C.K. and Bershad, N.J., Postsmoothing for the LMS algorithm and a fixed point round-off error analysis, *IEEE Trans. Signal Process.*, 39, 959–962, 1991.

Cha92 Chang, P.R., Yang, W.H. and Chan, K.K., A neural network approach to MVDR beamforming problem, *IEEE Trans. Antennas Propagat.*, 40, 313–322, 1992.

Che90 Chen, R.Y. and Wang, C.L. On the optimum step size for the adaptive sign and LMS algorithms, *IEEE Trans. Circuits Syst.*, 37, 836–840, 1990.

Chi93 Chiba, I., Chujo, W. and Fujise, M., Beamspace constant modulus algorithm adaptive array antennas, *IEEE 8th Int. Conf. Antennas Propagat.*, 975–978, 1993.

Cho91 Choi, S., *Application of the Conjugate Gradient Method for Optimum Array Processing, Book Series on PIER (M.I.T.),* Vol. 5, Elsevier, Amsterdam, 1991, chapter 16.

Cho92 Choi, S. and Sarkar, T.K., Adaptive antenna array utilizing the conjugate gradient method for multipath mobile communication, *Signal Process.*, 29, 319–333, 1992.

Cio84 Cioffi, J.M. and Kailath, T., Fast recursive-least-square, transversal filters for adaptive filtering, *IEEE Trans. Acoust. Speech Signal Process.*, 32, 998–1005, 1984.

Cio85 Cioffi, J.M. and Werner, J.J., Effect of biases on digitally implemented data driven echo canceller, *AT&T Tech. J.*, 64, 115–138, 1985.

Cla87 Clarkson, P.M. and White, P.R., Simplified analysis of the LMS adaptive filter using a transfer function approximation, *IEEE Trans. Acoust. Speech Signal Process.*, 35, 987–993, 1987.

dAs80 d'Assumpcao, H.A., Some new signal processors for array of sensors, *IEEE Trans. Info. Theory*, 26, 441–453, 1980.

Dan67 Daniel, J.W., The conjugate gradient method for linear and nonlinear operator equations, *SIAM J. Numerical. Anal.*, 4, 10–26, 1967.

Den78 Dentino, M., McCool, J. and Widrow, B., Adaptive filtering in frequency domain, *IEEE Proc.*, 66, 1658–1659, 1978.

Eva93 Evans, J.B., Xue, P. and Liu, B., Analysis and implementation of variable step size adaptive algorithms, *IEEE Trans. Signal Process.*, 41, 2517–2535, 1993.

Ewe90 Eweda, E., Analysis and design of a signed regressor LMS algorithm for stationary and nonstationary adaptive filtering with correlated Gaussian data, *IEEE Trans. Circuits Syst.*, 37, 1367–1374, 1990.

Fab86 Fabre, P. and Gueguen, C., Improvement of the fast recursive least-squares algorithms via normalization: a comparative study, *IEEE Trans. Acoust. Speech Signal Process.*, 34, 296–308, 1986.

Fer93 Fernandez, J., Corden, I.R. and Barrett, M., Adaptive array algorithms for optimal combining in digital mobile communication systems, in Proceedings of IEEE International Conference on Antennas and Propagation, 983–986, 1993.

Feu85 Feuer, A. and Weinstein, E., Convergence analysis of LMS filters with uncorrelated Gaussian data, *IEEE Trans. Acoust. Speech Signal Process.*, 33, 222–229, 1985.

Feu93 Feuer, A. and Cristi, R., On the steady state performance of frequency domain LMS algorithms, *IEEE Trans. Signal Process.*, 41, 419–423, 1993.

Fol88 Foley, J.B. and Boland, F.M., A note on the convergence analysis of LMS adaptive filters with Gaussian data, *IEEE Trans. Acoust. Speech Signal Process.*, 36, 1087–1089, 1988.

Flo88 Florian, S. and Bershad, N.J., A weighted normalized frequency domain LMS adaptive algorithm, *IEEE Trans. Acoust. Speech Signal Process.*, 36, 1002–1007, 1988.

Fro72 Frost III, O.L., An algorithm for linearly constrained adaptive array processing, *IEEE Proc.*, 60, 926–935, 1972.

Gar86 Gardner, W.A., Comments on convergence analysis of LMS filters with uncorrelated data, *IEEE Trans. Acoust. Speech Signal Process.*, 34, 378–379, 1986.

Gar87 Gardner, W.A. and Brown III, W.A., A new algorithm for adaptive arrays, *IEEE Trans. Acoust. Speech Signal Process.*, 35, 1314–1319, 1987.

Geb95 Gebauer, T. and Gockler, H.G., Channel-individual adaptive beamforming for mobile satellite communications, *IEEE J. Selected Areas Commn.*, 13, 439–48, 1995.

Gel96 Gelenbe, E. and Barhen, J., Eds., Special issue on artificial neural network applications, *IEEE Proc.*, 78, 1996.

God80 Godard, D.N., Self-recovering equalization and carrier tracking in two-dimensional data communication systems, *IEEE Trans. Commn.*, 28, 1867–1875, 1980.

God83 Godara, L.C. and Cantoni, A., Analysis of the performance of adaptive beamforming using perturbation sequences, *IEEE Trans. Antennas Propagat.*, 31, 268–279, 1983.

God86 Godara, L.C. and Cantoni, A., Analysis of constrained LMS algorithm with application to adaptive beamforming using perturbation sequences, *IEEE Trans. Antennas Propagat.*, 34, 368–379, 1986.

God89 Godara, L.C. and Gray, D. A., A structure gradient algorithm for adaptive beamforming, *J. Acoust. Soc. Am.*, 86, 1040–1046, 1989.

God89a Godara, L.C., Analysis of transient and steady state weight covariance in adaptive postbeamformer interference canceller, *J. Acoust. Soc. Am.*, 85, 194–201, 1989.

God90 Godara, L.C., Performance analysis of structured gradient algorithm, *IEEE Trans. Antennas Propagat.*, 38, 1078–1083, 1990.

God90a Godara, L.C., Improved LMS algorithm for adaptive beamforming, *IEEE Trans. Antennas Propagat.*, 38, 1631–1635, 1990.

God93 Godara, L.C., Constrained beamforming and adaptive algorithms, in *Handbook of Statistics*, Vol. 10, Bose, N.K. and Rao, C.R., Eds., Elsevier, Amsterdam; New York, 1993.

God93a Godara, L.C., Improved LMS algorithm for adaptive beamforming, *IEEE Trans. Antennas Propagat.*, 38, 1631–1635, 1990.

God97 Godara, L.C., Application to antenna arrays to mobile communications. Part II: Beamforming and direction of arrival considerations, *IEEE Proc.*, 85, 1195–1247, 1997.

Gri69 Griffiths, L.J., A simple adaptive algorithm for real-time processing in antenna arrays, *IEEE Proc.*, 57, 1696–1704, 1969.

Har86 Harris, R.W., Chabriew, D.M. and Bishop, F.A., A variable step size (VS) adaptive filter algorithm, *IEEE Trans. Acoust. Speech Signal Process.*, 34, 309–316, 1986.

Has93 Hashemi, H., The indoor radio propagation channels, *IEEE Proc.*, 81, 943–968, 1993.

Hes52 Hestenes, M. and Stiefel, E., Method of conjugate gradients for solving linear systems, *J. Res. Nat. Bur. Standards*, 49, 409–436, 1952.

Hor79 Horowitz, L.L., et al., Controlling adaptive antenna arrays with the sample matrix inversion algorithm, *IEEE Trans. Aerosp. Electron. Syst.*, 15, 840–847, 1979.

Hor81 Horowitz, L.L. and Senne, K.D., Performance advantage of complex LMS for controlling narrowband adaptive arrays, *IEEE Trans. Circuits Syst.*, 28, 562–576, 1981.

Hud81 Hudson, J.E., *Adaptive Array Principles*, Peter Peregrins, New York; London, 1981.

Ilt85 Iltis, R.A. and Milstein, L.B., An approximate statistical analysis of the Widrow LMS algorithm with application to narrowband interference rejection, *IEEE Trans. Commn.*, 33, 121–130, 1985.

Jag90 Jaggi, S. and Martinez, A.B., Upper and lower bounds of the misadjustment in the LMS algorithm, *IEEE Trans. Acoust. Speech Signal Process.*, 38, 164–166, 1990.

Jon95 Jones, M.A. and Wickert, M.A., Direct sequence spread spectrum using directionally constrained adaptive beamforming to null interference, *IEEE J. Selected Areas Commn.*, 13, 71–79, 1995.

Ko93 Ko, C.C., A fast adaptive null-steering algorithm based on output power measurements, *IEEE Trans. Aerosp. Electron. Syst.*, 29, 717–725, 1993.

Ko93a Ko, C.C., Balabshaskar, G. and Bachl, R., Unbiased source estimation with an adaptive null steering algorithm, *Signal Process.*, 31, 283–300, 1993

Kwo92 Kwong, R.H. and Johnston, E.W., A variable step size LMS algorithm, *IEEE Trans. Signal Process.*, 40, 1633–1642, 1992.

Kwo92a Kwong, C.P., Robust design of the LMS algorithm, *IEEE Trans. Signal Process.*, 40, 2613–2616, 1992.

Lau90 Lau, C. and Widrow, B., Eds., Special issue on neural networks I, *IEEE Proc.*, 78, 1990.

Leu91 Leung, H. and Haykin, S., Error bound method and its application to the LMS algorithm, *IEEE Trans. Signal Process.*, 39, 354–358, 1991.

Lin95 Lindskog, E., Making SMI-beamforming insensitive to the sampling timing for GSM signals, in Proceedings of IEEE International Symposium on Personal, Indoor and Mobile Radio Communications, 664–668, 1995.

Mat87 Mathews, V.J. and Cho, S.H., Improved convergence analysis of stochastic gradient adaptive filters using the sign algorithm, *IEEE Trans. Acoust. Speech Signal Process.*, 35, 450–454, 1987.

Mos79 Moschner, J.L., Adaptive filters with clipped input data, Technical Report 6796–1, Information Systems Laboratory, Stanford University, Stanford, CA, June 1970.

Nag94 Nagatsuka, M., et al., Adaptive array antenna based on spatial spectral estimation using maximum entropy method, *IEICE Trans. Commn.*, E77-B, 624–633, 1994.

Nar81 Narayan, S.S. and Peterson, A.M., Frequency domain least-mean-square algorithm, *IEEE Proc.*, 69, 124–126, 1981.

Nis95 Nishimori, K., Kikuma, N. and Inagaki, N., The differential CMA adaptive array antenna using an eigen-beamspace system, *IEICE Trans. Commn.*, E78-B, 1480–1488, 1995.

Nit85 Nitzberg, R., Application of the normalized LMS algorithm to MSLC, *IEEE Trans. Aerosp. Electron. Syst.*, 21, 79–91, 1985.

Nit86 Nitzberg, R., Normalized LMS algorithm degradation due to estimation noise, *IEEE Trans. Aerosp. Electron. Syst.*, 22, 740–750, 1986.

Ohg91 Ohgane, T., Characteristics of CMA adaptive array for selective fading compensation in digital land mobile radio communications, *Electron. Commn. (Jpn.)*, 74, 43–53, 1991.

Ohg93 Ohgane, T., et al., An implementation of a CMA adaptive array for high speed GMSK transmission in mobile communications, *IEEE Trans. Vehicular Technol.*, 42, 282–288, 1993.

Ohg93a Ohgane, T., et al., BER performance of CMA adaptive array for high-speed GMSK mobile communication: a description of measurements in central Tokyo, *IEEE Trans. Vehicular Technol.*, VT-42, 484–490, 1993.

Opp75 Oppenheim, A.V. and Schafer, R.W., *Digital Signal Processing*, Prentice Hall, Englewood Cliffs, NJ, 1975.

Pap65 Papoulis, A., *Probability, Random Variables and Stochastic Processes*, McGraw-Hill, New York, 1965.

Par95 Parra, I., Xu, G. and Liu, H., Least squares projective constant modulus approach, in Proceedings of IEEE International Symposium on Personal, Indoor and Mobile Radio Communications, 673–676, 1995.

Pas96 Passerini, C., et al., Adaptive antenna arrays for reducing the delay spread in indoor radio channels, *IEE Electron. Lett.*, 32, 280–281, 1996.

Pri91 Pritzker, Z. and Feuer, A., Variable length stochastic gradient algorithm, *IEEE Trans. Signal Process.*, 39, 997–1001, 1991.

Ree62 Reed, I.S., On a moment theorem for complex Gaussian processes, *IRE Trans. Inf. Theory*, IT-8, 194–195, 1962.

Ree74 Reed, I.S., Mallett, J.D. and Brennan, L.E., Rapid convergence rate in adaptive arrays, *IEEE Trans. Aerosp. Electron. Syst.*, 10, 853–863, 1974.

Rup93 Rupp, M., The behavior of LMS and NLMS algorithms in the presence of spherically invariant processes, *IEEE Trans. Signal Process.*, 41, 1149–1160, 1993.

Sar81 Sarkar, T., Siarkiewicz, K.R. and Stratton, R.F., Survey of numerical methods for solutions of large systems of linear equations for electromagnetic field problems, *IEEE Trans. Antennas Propagat.*, 29, 847–856, 1981.

Sch77 Schultheiss, P.M., Some lessons from array processing theory, *Aspects of Signal Processing*, Part 1, Tacconi, G., Ed., Reidel, Dordrecht, 1977, pp. 309–331.

Shy93 Shynk, J.J. and Chan, C.K., Performance surfaces of the constant modulus algorithm based on a conditional Gaussian model, *IEEE Trans. Signal Process.*, 41, 1965–1969, 1993.

Slo93 Slock, D.T.M., On the convergence behavior of the LMS and the normalized LMS algorithms, *IEEE Trans. Signal Process.*, 41, 2811–2825, 1993.

Sol89 Solo, V., The limiting behavior of LMS, *IEEE Trans. Acoust. Speech Signal Process.*, 37, 1909–1922, 1989.

Sol92 Solo, V., The error variance of LMS with time-varying weights, *IEEE Trans. Signal Process.*, 40, 803–813, 1992.

Soo91 Soo, J.S. and Pang, K.K., A multiple size frequency domain adaptive filter, *IEEE Trans. Signal Process.*, 39, 115–121, 1991.

Tan95 Tanaka, T., et al., An ASIC implementation scheme to realize a beam space CMA adaptive array antenna, *IEICE Trans. Commn.*, E78-B, 1467–1473, Nov. 1995.

Tre83 Treichler, J.R. and Agee, B.G., A new approach to multipath correction of constant modulus signals, *IEEE Trans. Acoust. Speech Signal Process.*, 31, 459–472, 1983.

Van91 Van Veen, B.D., Adaptive convergence of linearly constrained beamformers based on the sample covariance matrix, *IEEE Trans. Signal Process.*, 39, 1470–1473, 1991.

Vau88 Vaughan, R.G., On optimum combining at the mobile, *IEEE Trans. Vehicular Technol.*, 37, 181–188, 1988.

Wan94 Wang, Y. and Cruz, J.R., Adaptive antenna arrays for the reverse link of CDMA cellular communication systems, *IEE Electron. Lett.*, 30, 1017–1018, 1994.

Wid66 Widrow, B., Adaptive Filters I: Fundaments, Tech. Report 6784–8, Stanford Electrical Engineering Laboratory, Stanford University, Stanford, CA, December 1966.

Wid67 Widrow, B., et al., Adaptive antenna systems, *IEEE Proc.*, 55, 2143–2158, 1967.

Wid75 Widrow, B., McCool, J.M. and Ball, M., The complex LMS algorithm, *IEEE Proc.*, 63, 719–720, 1975.

Wid76 Widrow, B. and McCool, J.M., A comparison of adaptive algorithms based on the methods of steepest descent and random search, *IEEE Trans. Antennas Propagat.*, 24, 615–637, 1976.

Wid76a Widrow, B., et al., Stationary and nonstationary learning characteristics of the LMS adaptive filter, in *Aspects of Signal Processing*, Part 1, Tacconi, G., Ed., 355–393, 1976, also in *IEEE Proc.*, 64, 1151–1162, 1976.

Wid90 Widrow, B. and Lehr, M., 30 years of adaptive neural networks: perception, madaline, and back propagation, *IEEE Proc.*, 78, 1415–1442, 1990.

Win84 Winters, J.H., Optimum combining in digital mobile radio with cochannel interference, *IEEE J. Selected Areas Commn.*, SAC–2, 528–539, 1984.

Win87 Winters, J.H., Optimum combining for indoor radio systems with multiple users, *IEEE Trans. Commn.*, 35., 1222–1230, 1987.

Win94 Winters, J.H., Salz, J. and Gitlin, R.D., The impact of antenna diversity on the capacity of wireless communication systems, *IEEE Trans. Commn.*, 42, 1740–1751, 1994.

Won91 Wong, P.W., Quantization and roundoff noises in fixed-point FIR digital filters, *IEEE Trans. Signal Process.*, 39, 1552–1563, 1991.

Yan96 Yang, W.H. and Chan, K.K., Programmable switched-capacitor neural network for MVDR beamforming, *IEEE J. Oceanic Eng.*, 21, 77–84, 1996.

Yas87 Yassa, F.F., Optimality in the choice of the convergence factor for gradient based adaptive algorithms, *IEEE Trans. Acoust. Speech Signal Process.*, 35, 48–59, 1987.

## Appendices

### Appendix 3.1

In this appendix a derivation of (3.2.90) is presented. First, the following theorem used in the derivation is established [Hor81, God86].

Theorem 3.1: Let the set of difference equations

$$D_i(n+1) = \alpha_i D_i(n) + \beta_i \left\{ \zeta(n) + \sum_{\ell=1}^{L} D_\ell(n) \right\} + c_i(n), \quad i = 1, \ldots, L \tag{3A.1}$$

be such that

$$\lim_{n\to\infty} \zeta(n) = \zeta^* \tag{3A.2}$$

$$\lim_{n\to\infty} c_i(n) = 0 \tag{3A.3}$$

$$\lim_{n\to\infty} |D_i(0)| < \infty \tag{3A.4}$$

and

all the eigenvalues of the system (3A.1) are real and positive. (3A.5)

If

$$|\alpha_{max} + \beta_{max}| < 1 \tag{3A.6}$$

and

$$\prod_{i=1}^{L} (1+\delta-\alpha_i) - \sum_{i=1}^{L} \beta_i \prod_{\ell \neq i}^{L} (1+\delta-\alpha_i) > 0, \text{ for } \delta \geq 0 \tag{3A.7}$$

then

$$\lim_{n\to\infty} \sum_{i=1}^{L} D_i(n) \tag{3A.8}$$

exists and is given by

$$\lim_{n\to\infty} \sum_{i=1}^{L} D_i(n) = \frac{\zeta^* \sum_{i=1}^{L} \frac{\beta_i}{1-\alpha_i}}{1 - \sum_{i=1}^{L} \frac{\beta_i}{1-\alpha_i}} \tag{3A.9}$$

where $\alpha_{max} + \beta_{max}$ denotes the maximum value of $\alpha_i + \beta_i$, $i = 1, \ldots, L$.

Proof of Theorem 3.1: The proof of the theorem makes use of the z-transform. If F(z) denotes the z-transform of a sequence f(n) denoted as Z{f(n)}, then z-transform of f(n + 1) is given by

$$Z\{f(n+1)\} = zF(z) - zf(0) \qquad (3A.10)$$

Taking the z-transform of the ith equation of (3A.1) and using (3A.10),

$$D_i(z) = z^{-1}\alpha_i D_i(z) + z^{-1}\beta_i\left[\zeta(z) + \sum_{\ell=1}^{L} D_\ell(z)\right] + z^{-1}c_i(z) + D_i(0) \qquad (3A.11)$$

where $D_i(0)$ denotes the value of $D_i(n)$ at n = 0.

It follows from (3A.11) that

$$\begin{aligned} D_i(z) &= \frac{z^{-1}}{1-\alpha_i z^{-1}}\beta_i\left[\zeta(z) + \sum_{\ell=1}^{L} D_\ell(z)\right] + \frac{z^{-1}c_i(z) + D_i(0)}{1-\alpha_i z^{-1}} \\ &= \frac{z^{-1}\beta_i\left[\zeta(z) + \sum_{\ell \neq i}^{L} D_\ell(z)\right] + z^{-1}c_i(z) + D_i(0)}{1-(\alpha_i + \beta_i)z^{-1}} \end{aligned} \qquad (3A.12)$$

Let

$$\upsilon(z) = \sum_{i=1}^{L} D_i(z) \qquad (3A.13)$$

It follows from the first equation of (3A.12) and (3A.13) that

$$\begin{aligned} \upsilon(z) &= z^{-1}\sum_{i=1}^{L}\frac{\beta_i}{1-\alpha_i z^{-1}}\left[\zeta(z) + \upsilon(z)\right] + \sum_{i=1}^{L}\frac{z^{-1}c_i(z) + D_i(0)}{1-\alpha_i z^{-1}} \\ &= \frac{z^{-1}\zeta(z)\sum_{i=1}^{L}\frac{\beta_i}{1-\alpha_i z^{-1}} + \sum_{i=1}^{L}\frac{z^{-1}c_i(z) + D_i(0)}{1-\alpha_i z^{-1}}}{1 - z^{-1}\sum_{i=1}^{L}\frac{\beta_i}{1-\alpha_i z^{-1}}} \end{aligned} \qquad (3A.14)$$

It can easily be shown that the characteristic equation of (3A.14) is

$$\prod_{i=1}^{L}(z-\alpha_i) - \sum_{i=1}^{L}\beta_i\prod_{\ell \neq i}^{L}(z-\alpha_\ell) = 0 \qquad (3A.15)$$

For the stability of $\upsilon(z)$, it is necessary that $D_i(z)$ is stable $\forall_i$ and that all roots of (3A.15) lie inside the unit circle.

It follows from the second equation of (3A.12) that the stability of $D_i(z)$ is guaranteed if

$$|\alpha_i + \beta_i| < 1 \quad \forall_i \tag{3A.16}$$

Equation (3A.16) follows from (3A.6). Thus, $D_i(z)$ is stable $\forall_i$.

By assumption, all eigenvalues of (3A.1) are positive and real. This implies that all the roots of (3A.15) are positive and real. Since the sign of (3A.15) is positive for large values of z, it follows that if no root of (3A.15) is to lie between z = 1 and z = ∞, then (3A.15) must be positive for z = 1 + δ for all δ ≥ 0, that is,

$$\prod_{i=1}^{L}(1+\delta-\alpha_i) - \sum_{i=1}^{L}\beta_i \prod_{\ell \neq i}^{L}(1+\delta-\alpha_\ell) > 0 \quad \text{for all } \delta \geq 0 \tag{3A.17}$$

which is true by (3A.7).

Thus, $\upsilon(z)$ is stable and $\lim_{n\to\infty} \upsilon(n) \equiv \sum_{i=1}^{L} D_i(n)$ exists, which proves the existence of (3A.8).

Equation (3A.9) is now established. From the properties of the z-transform, the final value of the sequence $\lim_{n\to\infty} f(n)$ is given by

$$\lim_{n\to\infty} f(n) = \lim_{z\to 1}(1-z^{-1})F(z) \tag{3A.18}$$

It follows from (3A.2), (3A.3), (3A.4), and (3A.18) that

$$\lim_{Z\to 1}(1-z^{-1})\zeta(z) = \zeta^* \tag{3A.19}$$

$$\lim_{Z\to 1}(1-z^{-1})c_i(z) = 0 \tag{3A.20}$$

and

$$\lim_{Z\to 1}(1-z^{-1})D_i(0) = 0 \tag{3A.21}$$

Multiplying both sides of (3A.14) by $(1 - z^{-1})$, taking the limit $z \to 1$, and using (3A.13) and (3A.19) to (3A.21),

$$\lim_{n\to\infty}\sum_{i=1}^{L} D_i(n) = \frac{\zeta^* \sum_{i=1}^{L} \frac{\beta_i}{1-\alpha_i}}{1-\sum_{i=1}^{L}\frac{\beta_i}{1-\alpha_i}} \tag{3A.22}$$

which is (3A.9). Thus, the theorem is proved.

Proof of (3.2.90): Let

$$\mathbf{D}(n) = \Lambda \boldsymbol{\eta}(n) \tag{3A.23}$$

Noting that

$$\Lambda \mathbf{1} = \boldsymbol{\lambda} \tag{3A.24}$$

and premultiplying on both sides of (3.2.73) by Λ and using (3A.23),

$$\mathbf{D}(n+1) = \{I - 4\mu\Lambda + 4\mu^2\Lambda^2 + 4\mu^2\Lambda^2\mathbf{1}\mathbf{1}^T\}\mathbf{D}(n) + 4\mu^2\xi(\overline{\mathbf{w}}(n))\Lambda^2\mathbf{1} \tag{3A.25}$$

The ith component of (3A.25) is given by

$$D_i(n+1) = \{1 - 4\mu\lambda_i + 4\mu^2\lambda_i^2\}D_i(n) + 4\mu^2\lambda_i^2\left\{\xi(\overline{\mathbf{w}}(n)) + \sum_{\ell=1}^{L} D_\ell(n)\right\} \tag{3A.26}$$

Now apply Theorem 3A.1. Equation (3A.26) satisfies the form given by (3A.1) with

$$\alpha_i = 1 - 4\mu\lambda_i + 4\mu^2\lambda_i^2 \tag{3A.27}$$

$$\beta_i = 4\mu^2\lambda_i^2 \tag{3A.28}$$

$$\zeta(n) = \xi(\overline{\mathbf{w}}(n)) \tag{3A.29}$$

and

$$c_i(n) = 0 \tag{3A.30}$$

It follows from (3A.29) that

$$\begin{aligned}\lim_{n\to\infty} \zeta(n) &= \xi(\hat{\mathbf{w}}(n)) \\ &= \hat{\xi}\end{aligned} \tag{3A.31}$$

which satisfies (3A.2). Furthermore, (3A.30) implies that (3A.3) is satisfied and $k_{ww}(0) = 0$ implies that $D_i(0) = 0$, which in turn satisfies (3A.4). Since (3A.25) is propagated by a symmetric, positive, definitive transition matrix for all values of μ, this implies that all eigenvalues of the system (3A.25) are positive and real. This satisfies condition (3A.5). The condition (3A.6) is satisfied if

$$|\alpha_i + \beta_i| < 1 \quad \forall_i \tag{3A.32}$$

or

$$|1 - 4\mu\lambda_i + 8\mu^2\lambda_i^2| < 1 \quad \forall_i \tag{3A.33}$$

This is satisfied if

$$0 < \mu < \frac{1}{2\lambda_{max}} \tag{3A.34}$$

The condition (3A.7) is now checked. Substituting for $\alpha_i$ and $\beta_i$ in (3A.7),

$$\prod_{i=1}^{L}\left\{\delta + 4\mu\lambda_i(1-\mu\lambda_i)\right\} - \sum_{i=1}^{L} 4\mu^2\lambda_i^2 \prod_{\ell \neq i}^{L}\left\{\delta + 4\mu\lambda_\ell(1-\mu\lambda_\ell)\right\} > 0, \text{ for } \delta \geq 0 \tag{3A.35}$$

It follows from (3A.34) that

$$\prod_{i=1}^{L}\left\{\delta + 4\mu\lambda_i(1-\mu\lambda_i)\right\} > 0, \text{ for } \delta \geq 0 \tag{3A.36}$$

Dividing (3A.35) by $\prod_{i=1}^{L}\left\{\delta + 4\mu\lambda_i(1-\mu\lambda_i)\right\}$, the following condition is derived:

$$\sum_{i=1}^{L} \frac{\mu\lambda_i}{\frac{\delta}{4\mu\lambda_i} + (1-\mu\lambda_i)} < 1, \text{ for } \delta \geq 0 \tag{3A.37}$$

This implies that (3A.7) is satisfied if

$$\sum_{i=1}^{L} \frac{\mu\lambda_i}{(1-\mu\lambda_i)} < 1 \tag{3A.38}$$

Thus, when (3A.34) and (3A.38) are true, all conditions of the theorem are satisfied. It follows from (3A.8) that

$$\lim_{n\to\infty} \sum_{i=1}^{L} D_i(n) \tag{3A.39}$$

exists and from (3A.9),

$$\lim_{n\to\infty} \sum_{i=1}^{L} D_i(n) = \frac{\zeta^* \sum_{i=1}^{L} \frac{\beta_i}{1-\alpha_i}}{1 - \sum_{i=1}^{L} \frac{\beta_i}{1-\alpha_i}} \tag{3A.40}$$

Noting that

$$\sum_{i=1}^{L} D_i(n) = \boldsymbol{\lambda}^T \boldsymbol{\eta}(n) \tag{3A.41}$$

(3A.39) implies that

$$\lim_{n\to\infty} \boldsymbol{\lambda}^T \boldsymbol{\eta}(n) \tag{3A.42}$$

exists.

Substituting for various quantities in (3A.40),

$$\lim_{n\to\infty} \boldsymbol{\lambda}^T \boldsymbol{\eta}(n) = \frac{\hat{\xi}\mu \sum_{i=1}^{L} \frac{\lambda_i}{1-\mu\lambda_i}}{1-\mu \sum_{i=1}^{L} \frac{\lambda_i}{1-\mu\lambda_i}} \tag{3A.43}$$

which is (3.2.90).

## Appendix 3.2

In this appendix, a derivation of (3.4.6) is presented. Rewrite (3.2.10):

$$V_g(\mathbf{w}(n)) = E\left[\mathbf{g}(\mathbf{w}(n))\mathbf{g}^H(\mathbf{w}(n))\right] - \bar{\mathbf{g}}(\mathbf{w}(n))\bar{\mathbf{g}}^H(\mathbf{w}(n)) \tag{3A.44}$$

where E[.] denotes the conditional expectation for a given $\mathbf{w}(n)$ and

$$\bar{\mathbf{g}}(\mathbf{w}(n)) = E\left[\mathbf{g}(\mathbf{w}(n))\right] \tag{3A.45}$$

It follows from (3.4.5) and (3A.45) that

$$\bar{\mathbf{g}}(\mathbf{w}(n)) = 2R\mathbf{w}(n) \tag{3A.46}$$

Thus, the second term on the RHS of (3A.44) becomes

$$\bar{\mathbf{g}}(\mathbf{w}(n))\bar{\mathbf{g}}^H(\mathbf{w}(n)) = 4R\mathbf{w}(n)\mathbf{w}^H(n)R \tag{3A.47}$$

It follows from (3.4.4) that

$$\mathbf{g}(\mathbf{w}(n))\mathbf{g}^H(\mathbf{w}(n)) = 4\mathbf{x}(n+1)\mathbf{x}^H(n+1)\mathbf{w}(n)\mathbf{w}^H(n)\mathbf{x}(n+1)\mathbf{x}^H(n+1) \tag{3A.48}$$

If $\{\mathbf{x}(k)\}$ is a complex i.i.d. Gaussian sequence, then for any Hermitian matrix A the following result holds, using (3.2.9):

$$E\left[\mathbf{x}(n)\mathbf{x}^H(n)A\mathbf{x}(n)\mathbf{x}^H(n)\right] = RAR + \mathrm{Tr}(RA)R \tag{3A.49}$$

Taking the conditional expectation on both sides of (3A.48) for a given $\mathbf{w}(n)$ and using (3A.49),

$$E\left[\mathbf{g}(\mathbf{w}(n))\mathbf{g}^H(\mathbf{w}(n))\right] = 4\mathbf{w}^H(n)R\mathbf{w}(n)R + 4R\mathbf{w}(n)\mathbf{w}^H(n)R \tag{3A.50}$$

Substituting from (3A.47) and (3A.50) in (3A.44), (3.4.6) is derived.

## Appendix 3.3

In this appendix, a derivation of (3.4.8) and (3.4.9) is presented. It is similar to results of unconstrained algorithm presented in Sections 3.2.2 and 3.2.3.

It follows from (3.4.1) and (3.4.4) that

$$\mathbf{w}(n+1) = P\left\{\mathbf{w}(n) - 2\mu\mathbf{x}(n+1)\mathbf{x}^H(n+1)\mathbf{w}(n)\right\} + \frac{\mathbf{S}_0}{L} \tag{3A.51}$$

When $\mathbf{w}(n)$ and $\mathbf{x}(n+1)$ are uncorrelated it follows by taking the unconditional expectation on both sides of (3A.51) that

$$\overline{\mathbf{w}}(n+1) = P\left\{\overline{\mathbf{w}}(n) - 2\mu R\overline{\mathbf{w}}(n)\right\} + \frac{\mathbf{S}_0}{L} \tag{3A.52}$$

where

$$\overline{\mathbf{w}}(n) = E\left[\mathbf{w}(n)\right] \tag{3A.53}$$

Define a mean error vector $\overline{\mathbf{v}}(n)$ as

$$\overline{\mathbf{v}}(n) = \overline{\mathbf{w}}(n) - \hat{\mathbf{w}} \tag{3A.54}$$

where $\hat{\mathbf{w}}$ is the optimal vector given by (2.4.21), that is,

$$\hat{\mathbf{w}} = \frac{R^{-1}\mathbf{S}_0}{\mathbf{S}_0^H R^{-1}\mathbf{S}_0} \tag{3A.55}$$

Subtracting $\hat{\mathbf{w}}$ from both sides of (3A.52) and using (3A.54), the following mean error vector update equation is derived:

$$\overline{\mathbf{v}}(n+1) = P\overline{\mathbf{v}}(n) + P\hat{\mathbf{w}} - 2\mu PR\overline{\mathbf{v}}(n) - 2\mu PR\hat{\mathbf{w}} - \hat{\mathbf{w}} + \frac{\mathbf{S}_0}{L} \tag{3A.56}$$

From (3.4.2) and the fact that

$$\hat{\mathbf{w}}^H\mathbf{S}_0 = 1 \tag{3A.57}$$

it follows that

$$P\hat{\mathbf{w}} = \hat{\mathbf{w}} - \frac{\mathbf{S}_0}{L} \tag{3A.58}$$

Similarly, (3.4.2) and (3A.55) imply that

$$\mathrm{PR}\hat{\mathbf{w}} = 0 \tag{3A.59}$$

Thus, (3A.56) becomes

$$\bar{\mathbf{v}}(n+1) = \mathrm{P}(\mathrm{I} - 2\mu\mathrm{R})\bar{\mathbf{v}}(n) \tag{3A.60}$$

where I is an identity matrix.

Since $P^2 = P$, it follows from (3A.60) that

$$\mathrm{P}\bar{\mathbf{v}}(.) = \bar{\mathbf{v}}(.) \tag{3A.61}$$

Thus, (3A.60) becomes

$$\begin{aligned}\bar{\mathbf{v}}(n+1) &= (\mathrm{I} - 2\mu\mathrm{PRP})\bar{\mathbf{v}}(n)\\ &= (\mathrm{I} - 2\mu\mathrm{PRP})^{n+1}\bar{\mathbf{v}}(0)\end{aligned} \tag{3A.62}$$

Using the properties of a norm, it follows from (3A.62) that

$$\left[1 - 2\mu\hat{\lambda}_{max}\right]^{n+1}\|\bar{\mathbf{v}}(0)\| \le \|\bar{\mathbf{v}}(n+1)\| \le \left[1 - 2\mu\hat{\lambda}_{min}\right]^{n+1}\|\bar{\mathbf{v}}(0)\| \tag{3A.63}$$

From (3.4.7),

$$0 < \mu < \frac{1}{\hat{\lambda}_{max}} \tag{3A.64}$$

Thus, $|1 - 2\mu\hat{\lambda}_{max}| < 1$ and implies that

$$\lim_{n\to\infty}\left[1 - 2\mu\hat{\lambda}_{max}\right]^{n+1} = 0 \tag{3A.65}$$

and

$$\lim_{n\to\infty}\left[1 - 2\mu\hat{\lambda}_{min}\right]^{n+1} = 0 \tag{3A.66}$$

Since $\|\bar{\mathbf{v}}(0)\| < \infty$, it follows from (3A.63) to (3A.66) that

$$\lim_{n\to\infty}\|\bar{\mathbf{v}}(n+1)\| = 0 \tag{3A.67}$$

which along with (3A.54) implies that

$$\lim_{n\to\infty} \mathrm{E}\left[\mathbf{w}(n) - \hat{\mathbf{w}}\right] = 0 \tag{3A.68}$$

This establishes

$$\lim_{n\to\infty} E[\mathbf{w}(n)] = \hat{\mathbf{w}} \tag{3A.69}$$

To obtain the convergence time constant along an eigenvector of PRP, consider

$$\bar{\mathbf{v}}(0) = \sum_{i=1}^{L} \alpha_i \hat{\mathbf{Q}}_i \tag{3A.70}$$

where $\alpha_i$, i = 1, 2, …, L are scalars and $\hat{\mathbf{Q}}_i$, i = 1, 2, …, L are eigenvectors corresponding to L eigenvalues of PRP.

From (3A.62) and (3A.70), it follows that

$$\bar{\mathbf{v}}(n+1) = (1 - 2\mu PRP)^{n+1} \sum_{i=1}^{L} \alpha_i \hat{\mathbf{Q}}_i \tag{3A.71}$$

Since eigenvectors of PRP are orthonormal, (3A.71) can be expressed as

$$\bar{\mathbf{v}}(n+1) = \sum_{i=1}^{L} \left(1 - 2\mu\hat{\lambda}_i\right)^{n+1} \alpha_i \hat{\mathbf{Q}}_i \tag{3A.72}$$

The convergence of the mean weight vector to the optimal weight vector along the ith eigenvector of PRP is therefore geometric with geometric ratio $(1 - 2\mu\hat{\lambda}_i)$. If an exponential envelope of time constant $\tau_i$ is fitted to the geometric sequence of (3A.72), then

$$\tau_i = \frac{-1}{\ell n\left(1 - 2\mu\hat{\lambda}_i\right)} \tag{3A.73}$$

where the unit of time is assumed to be one iteration. Note that if

$$2\mu\hat{\lambda}_i \ll 1 \tag{3A.74}$$

then

$$\tau_i \simeq \frac{1}{2\mu\hat{\lambda}_i} \tag{3A.75}$$

## Appendix 3.4

In this appendix, a derivation of (3.4.14) is presented [God86]. It follows from (3.4.12) and (3.4.13) that

$$k_{ww}(n) = R_{ww}(n) - \bar{\mathbf{w}}(n)\bar{\mathbf{w}}^H(n) \tag{3A.76}$$

where

$$R_{ww}(n) = E\left[\mathbf{w}(n)\mathbf{w}^H(n)\right] \tag{3A.77}$$

Taking the outer product of (3.4.1),

$$\begin{aligned}\mathbf{w}(n+1)\mathbf{w}^H(n+1) = {} & P\mathbf{w}(n)\mathbf{w}^H(n)P - \mu P\left[\mathbf{g}(\mathbf{w}(n))\mathbf{w}^H(n) + \mathbf{w}(n)\mathbf{g}^H(\mathbf{w}(n))\right]P \\ & - \mu\left[P\mathbf{g}(\mathbf{w}(n))\mathbf{F}^H + \mathbf{F}\mathbf{g}^H(\mathbf{w}(n))P\right] \\ & + \mu^2\left[P\mathbf{g}(\mathbf{w}(n))\mathbf{g}^H(\mathbf{w}(n))P\right] \\ & + \mathbf{F}\mathbf{F}^H + P\mathbf{w}(n)\mathbf{F}^H + \mathbf{F}\mathbf{w}^H(n)P\end{aligned} \tag{3A.78}$$

where $\mathbf{F} = \mathbf{S}_0/L$.

Taking the conditional expectation of both sides of (3A.78) with respect to $\mathbf{w}(n)$ and using (3.4.5),

$$\begin{aligned}E\left[\mathbf{w}(n+1)\mathbf{w}^H(n+1)\middle|\mathbf{w}(n)\right] = {} & P\mathbf{w}(n)\mathbf{w}^H(n)P \\ & - 2\mu P\left[R\mathbf{w}(n)\mathbf{w}^H + \mathbf{w}(n)\mathbf{w}^H(n)R\right]P \\ & - 2\mu\left[PR\mathbf{w}(n)\mathbf{F}^H + \mathbf{F}\mathbf{w}^H(n)RP\right] \\ & + \mu^2 PE\left[\mathbf{g}(\mathbf{w}(n))\mathbf{g}^H(\mathbf{w}(n))\middle|\mathbf{w}(n)\right]P \\ & + \mathbf{F}\mathbf{F}^H + P\mathbf{w}(n)\mathbf{F}^H + \mathbf{F}\mathbf{w}^H(n)P\end{aligned} \tag{3A.79}$$

Taking the expectation on both sides over $\mathbf{w}(n)$, (3A.79) yields

$$\begin{aligned}R_{\mathbf{ww}}(n+1) = {} & PR_{\mathbf{ww}}(n)P - 2\mu P\left[R_{\mathbf{ww}}(n)R + RR_{\mathbf{ww}}(n)\right]P \\ & - 2\mu\left[PR\overline{\mathbf{w}}(n)F^H + \mathbf{F}\overline{\mathbf{w}}^H(n)RP\right] \\ & + \mu^2 PE\left[\mathbf{g}(\mathbf{w}(n))\mathbf{g}^H(\mathbf{w}(n))\right]P + \mathbf{F}\mathbf{F}^H \\ & + P\overline{\mathbf{w}}(n)\mathbf{F}^H + \mathbf{F}\overline{\mathbf{w}}^H(n)P\end{aligned} \tag{3A.80}$$

Taking the expected value of (3.4.1),

$$\overline{\mathbf{w}}(n+1) = P\left[\overline{\mathbf{w}}(n) - \mu\overline{\mathbf{g}}(\mathbf{w}(n))\right] + \mathbf{F} \tag{3A.81}$$

It follows from (3.4.5) and (3A.81) that

$$\begin{aligned}\overline{\mathbf{w}}(n+1)\overline{\mathbf{w}}^H(n+1) = {} & P\overline{\mathbf{w}}(n)\overline{\mathbf{w}}^H(n)P - 2\mu P\left[R\overline{\mathbf{w}}(n)\overline{\mathbf{w}}^H(n) + \overline{\mathbf{w}}(n)\overline{\mathbf{w}}^H(n)R\right]P \\ & - 2\mu\left[PR\overline{\mathbf{w}}(n)\mathbf{F}^H + \mathbf{F}\overline{\mathbf{w}}^H(n)RP\right] + \mu^2 P\overline{\mathbf{g}}(\mathbf{w}(n))\overline{\mathbf{g}}^H(\mathbf{w}(n))P \\ & + \mathbf{F}\mathbf{F}^H + P\overline{\mathbf{w}}(n)\mathbf{F}^H + \mathbf{F}\overline{\mathbf{w}}^H(n)P\end{aligned} \tag{3A.82}$$

Subtracting (3A.82) from (3A.80) and using (3A.76),

$$\begin{aligned} k_{ww}(n+1) &= Pk_{ww}(n)P - 2\mu P\left[Rk_{ww}(n) + k_{ww}(n)R\right]P \\ &\quad + \mu^2 P\left\{E\left[\mathbf{g}(\mathbf{w}(n))\mathbf{g}^H(\mathbf{w}(n))\right] - E\left[\mathbf{g}(\mathbf{w}(n))\right]E\left[\mathbf{g}^H(\mathbf{w}(n))\right]\right\}P \end{aligned} \tag{3A.83}$$

By definition,

$$\begin{aligned} V_g(\mathbf{w}(n)) &= E\left[\mathbf{g}(\mathbf{w}(n))\mathbf{g}^H(\mathbf{w}(n))\middle|\mathbf{w}(n)\right] \\ &\quad - E\left[\mathbf{g}(\mathbf{w}(n))\middle|\mathbf{w}(n)\right]E\left[\mathbf{g}^H(\mathbf{w}(n))\middle|\mathbf{w}(n)\right] \end{aligned} \tag{3A.84}$$

Using (3.4.5), it follows from (3A.84) that

$$V_g(\mathbf{w}(n)) = E\left[\mathbf{g}(\mathbf{w}(n))\mathbf{g}^H(\mathbf{w}(n))\middle|\mathbf{w}(n)\right] - 4R\mathbf{w}(n)\mathbf{w}^H(n)R \tag{3A.85}$$

From (3A.85), taking the expected value over **w**,

$$E\left[V_g(\mathbf{w}(n))\right] = E\left[\mathbf{g}(\mathbf{w}(n))\mathbf{g}^H(\mathbf{w}(n))\right] - 4RR_{ww}R \tag{3A.86}$$

which implies

$$E\left[\mathbf{g}(\mathbf{w}(n))\mathbf{g}^H(\mathbf{w}(n))\right] = E\left[V_g(\mathbf{w}(n))\right] + 4RR_{ww}R \tag{3A.87}$$

From (3.4.5), taking expected value over **w**,

$$E\left[\mathbf{g}(\mathbf{w}(n))\right] = 2R\overline{\mathbf{w}}(n) \tag{3A.88}$$

The outer product of (3A.88) results in

$$E\left[\mathbf{g}(\mathbf{w}(n))\right]E\left[\mathbf{g}^H(\mathbf{w}(n))\right] = 4R\overline{\mathbf{w}}(n)\overline{\mathbf{w}}^H(n)R \tag{3A.89}$$

Subtracting (3A.89) from (3A.87) and substituting in (3A.83),

$$\begin{aligned} k_{ww}(n+1) &= Pk_{ww}(n)P - 2\mu P\left[Rk_{ww}(n) + k_{ww}(n)R\right]P \\ &\quad + \mu^2 4PRk_{ww}(n)RP + \mu^2 PE\left[V_g(\mathbf{w}(n))\right]P \end{aligned} \tag{3A.90}$$

which is (3.4.14).

## Appendix 3.5

In this appendix, a proof of the diagonalization conditions stated in Section 3.4.5 is presented [God86]. It follows directly from (3.4.1) and (3.4.12) that

$$Pk_{ww}(n)P \equiv Pk_{ww}(n) \equiv k_{ww}(n)P \equiv k_{ww}(n) \tag{3A.91}$$

Thus, (3.4.14) can be expressed as

$$\begin{aligned} k_{ww}(n+1) &= k_{ww}(n) - 2\mu\ PRP\ k_{ww}(n) - 2\mu\ k_{ww}(n)PRP \\ &\quad + \mu^2 PE\left[V_g(\mathbf{w}(n))\right]P + \mu^2 4\ PRP\ k_{ww}(n)PRP \end{aligned} \tag{3A.92}$$

Since PRP is an Hermitian matrix, a unitary matrix $\hat{Q}$ exists, such that

$$\hat{Q}^H PRP\hat{Q} = \hat{\Lambda} \tag{3A.93}$$

where $\hat{\Lambda}$ is a diagonal matrix with the diagonal elements being the eigenvalues of PRP.

Define

$$\Sigma(n) = \hat{Q}^H K_{ww}(n)\hat{Q} \tag{3A.94}$$

and

$$\Omega(n) = \hat{Q}^H PE\left[V_g(\mathbf{w}(n))\right]P\hat{Q} \tag{3A.95}$$

Pre- and postmultiplying (3A.92) by $\hat{Q}^H$ and $\hat{Q}$, respectively, and using (3A.93), (3A.94), and (3A.95),

$$\Sigma(n+1) = \left[\Sigma(n) - 2\mu\hat{\Lambda}\,\Sigma(n) - 2\mu\Sigma(n)\hat{\Lambda} + 4\mu^2\hat{\Lambda}\Sigma(n)\hat{\Lambda}\right] + \mu^2\Omega(n) \tag{3A.96}$$

In view of (3A.93), (3A.94), and (3A.95), the statement of diagonalization conditions becomes "the necessary and the sufficient condition for $\Sigma(n + 1)$, $n \geq 0$, to be a diagonal matrix is that $\Omega(n)$ is a diagonal matrix for all n." This is proved by induction.

Consider $n = 0$. Since initial weight vector $\mathbf{w}(0)$ is a known constant, it follows that

$$E\left[\mathbf{w}(0)\right] \equiv \mathbf{w}(0) \tag{3A.97}$$

It follows from (3.4.12) that

$$\begin{aligned} k_{ww}(0) &= E\left[\left(\mathbf{w}(0) - \mathbf{w}(0)\right)\left(\mathbf{w}(0) - \mathbf{w}(0)\right)^H\right] \\ &= 0 \end{aligned} \tag{3A.98}$$

Equations (3A.94) and (3A.98) imply that

$$\Sigma(0) = 0 \tag{3A.99}$$

From (3A.96) and (3A.99) one obtains

$$\Sigma(1) = \mu^2\Omega(0) \tag{3A.100}$$

It follows from (3A.100) that the necessary and sufficient condition for $\Sigma(1)$ to be a diagonal matrix is that $\Omega(0)$ is a diagonal matrix. This proves the diagonalization conditions for $n = 0$.

Consider $n = 1$. Assume that $\Sigma(1)$ is a diagonal matrix. For $n = 1$, (3A.96) becomes

$$\Sigma(2) = \left[\Sigma(1) - 2\mu\hat{\Lambda}\Sigma(1) - 2\mu\Sigma(1)\hat{\Lambda} + 4\mu^2\hat{\Lambda}\Sigma(1)\hat{\Lambda}\right] + \mu^2\Omega(1) \tag{3A.101}$$

Since $\Sigma(1)$ and $\hat{\Lambda}$ are diagonal matrices, it follows that the terms in the square bracket of (3A.101) form a diagonal matrix. Thus, it follows from (3A.101) that for $\Sigma(2)$ to be a diagonal matrix, the necessary and sufficient condition is that $\Omega(1)$ is a diagonal matrix. This proves the theorem for $n = 1$.

Finally, assume that $\Sigma(n)$ is a diagonal matrix. Since $\hat{\Lambda}$ is a diagonal matrix, the terms in the square bracket of (3A.96) form a diagonal matrix. Thus, it follows from (3A.96) that for $\Sigma(n + 1)$ to be a diagonal matrix, the necessary and sufficient condition is that $\Omega(n)$ is a diagonal matrix. This completes the steps necessary for the proof by induction.

## Appendix 3.6

In this appendix, a proof of (3.4.52) is presented [God86]. Let

$$\mathbf{D}(n) = \hat{\Lambda}'\boldsymbol{\eta}'(n) \tag{3A.102}$$

where $\boldsymbol{\eta}'(n)$ is defined by (3.4.36), and denotes the $L-1$ diagonal elements of $\hat{Q}^H \mathbf{k}_{ww}(\mathbf{n})\hat{Q}$ and $\hat{\Lambda}'$ is the diagonal matrix of $L-1$ nonzero eigenvalues of PRP.

It follows from (3.4.37), (3.4.38), and (3A.102) that

$$\mathbf{D}(n+1) = \left(I - 4\mu\hat{\Lambda}' + 4\mu^2\hat{\boldsymbol{\lambda}}'^2 + 4\mu^2\hat{\Lambda}'^2\mathbf{1}\mathbf{1}^T\right)\mathbf{D}(n) + 4\mu^2\hat{\Lambda}'^2\mathbf{1}\mathbf{k}_0(n) \tag{3A.103}$$

From (3A.103), a difference equation of the ith component of $\mathbf{D}(.)$ is given by

$$D_i(n+1) = \left(1 - 4\mu\hat{\lambda}_i' + 4\mu^2\hat{\lambda}_i'^2\right)D_i(n) + 4\mu^2\hat{\lambda}_i'^2\left(k_0(n) + \sum_{\ell=1}^{L-1} D_\ell(n)\right) \tag{3A.104}$$

With

$$\alpha_i = 1 - 4\mu\hat{\lambda}_i + 4\mu^2\hat{\lambda}_i^2 \tag{3A.105}$$

$$\beta_i = 4\mu^2\hat{\lambda}_i^2 \tag{3A.106}$$

$$\zeta(n) = k_0(n) \equiv \overline{\mathbf{w}}^H(n) R \overline{\mathbf{w}}(n) \tag{3A.107}$$

and

$$c_i(n) = 0 \tag{3A.108}$$

equation (3A.104) is similar to (3A.1) and Theorem 3A.1 can be applied, provided the conditions (3A.2) to (3A.7) are satisfied.

Since

$$\lim_{n\to\infty} \overline{\mathbf{w}}(n) = \hat{\mathbf{w}} \tag{3A.109}$$

it follows from (3A.107) that

$$\lim_{n\to\infty} \zeta(n) = \hat{\mathbf{w}}^H R \hat{\mathbf{w}} \equiv \zeta^* \tag{3A.110}$$

which is (3A.2).

Equation (3A.108) implies (3A.3), $k_{\mathbf{ww}}(n) \equiv 0$ implies that $D_i(0) = 0$, which satisfies (3A.4).

Note that (3A.103) is propagated by a symmetric, positive definite transition matrix for all values of $\mu$. This implies that all eigenvalues of the system (3A.103) are positive and real, which satisfies (3A.5). Following the argument used in (3A.32) to (3A.34), it can be shown that (3A.6) is satisfied if

$$0 < \mu < \frac{1}{2\hat{\lambda}_{max}} \tag{3A.111}$$

which is (3.4.50). Thus, (3A.6) is satisfied.

Condition (3A.7) is checked in the following. Substituting for $\alpha_i$ and $\beta_i$ in (3A.7), the condition

$$\prod_i \left[\delta + 4\mu\hat{\lambda}_i\left(1-\mu\hat{\lambda}_i\right)\right] - \sum_i 4\mu\hat{\lambda}_i^2 \prod_{\ell\neq i}\left[\delta + 4\mu\hat{\lambda}_\ell\left(1-\mu\hat{\lambda}_\ell\right)\right] > 0 \quad \text{for } \delta \geq 0 \tag{3A.112}$$

is derived. As (3A.111) implies that

$$\prod_i \left[\delta + 4\mu^2\hat{\lambda}_i\left(1-\mu\hat{\lambda}_i\right)\right] > 0 \quad \text{for } \delta \geq 0 \tag{3A.113}$$

(3A.112) becomes

$$\sum_{i=1}^{L-1} \frac{\mu\hat{\lambda}_i}{\dfrac{\delta}{4\mu\hat{\lambda}_i} + \left(1-\mu\hat{\lambda}_i\right)} < 1, \quad \delta \geq 0 \tag{3A.114}$$

This implies that (3A.7) is satisfied if

$$\sum_{i=1}^{L-1} \frac{\mu \hat{\lambda}_i}{\left(1-\mu \hat{\lambda}_i\right)} < 1 \tag{3A.115}$$

which is (3.4.51). Thus, all conditions of Theorem 3A.1 are satisfied. Thus, $\lim_{n\to\infty} \sum_{i=1}^{L-1} D_i(n)$ exists and from (3A.9) it follows that

$$\lim_{n\to\infty} \sum_{i=1}^{L-1} D_i(n) = \frac{\zeta^* \sum_{i=1}^{L-1} \frac{\beta_i}{1-\alpha_i}}{1 - \sum_{i=1}^{L-1} \frac{\beta_i}{1-\alpha_i}} \tag{3A.116}$$

It follows from (3A.102) and (3.4.47) and $\hat{\lambda}_L = 0$ ($\hat{\lambda}_{min}$ is $\hat{\lambda}_L$) that

$$\sum_{i=1}^{L-1} D_i(n) \equiv \hat{\boldsymbol{\lambda}}^T \mathbf{d}(n) \tag{3A.117}$$

From (3A.105), (3A.106), (3A.110), (3A.116), and (3A.117),

$$\lim_{n\to\infty} \hat{\boldsymbol{\lambda}}^T \mathbf{d}(n) = \frac{\hat{\mathbf{w}}^H R \hat{\mathbf{w}} \mu \sum_{i=1}^{L-1} \frac{\hat{\lambda}_i}{1-\mu\hat{\lambda}_i}}{1 - \mu \sum_{i=1}^{L-1} \frac{\hat{\lambda}_i}{1-\mu\hat{\lambda}_i}} \tag{3A.118}$$

which along with (3.4.49) implies (3.4.52).

Note that the existence of $\lim_{n\to\infty} \sum_{i=1}^{L-1} D_i(n)$, $0 < \hat{\lambda}_i < \infty \; \forall i$ and (3A.102) imply that $\lim_{n\to\infty} \boldsymbol{\eta}'(n)$ exists. This completes the derivation.

## Appendix 3.7

In this appendix, the results presented in Section 3.13 are derived. The following result, which follows from (3.2.9), is used here.

If $\{\mathbf{x}(k)\}$ is a complex i.i.d. Gaussian sequence, then for any Hermitian matrix A the following result holds:

$$E\left[\mathbf{x}(n)\mathbf{x}^H(n) A \mathbf{x}(n)\mathbf{x}^H(n)\right] = RAR + \mathrm{Tr}(RA)R \tag{3A.119}$$

where R is the array correlation matrix and Tr(.) denotes the trace.

Proof of (3.13.10): Let

$$\bar{g}(w(n)) = E\left[g(w(n)) \middle| w(n)\right] \tag{3A.120}$$

Since $V_g(w(n))$ denotes the covariance of the gradient estimate for a given w(n), it follows from the definition of covariance that

$$\begin{aligned} V_g(w(n)) &= E\left[\{g(w(n)) - \bar{g}(w(n))\}\{g^*(w(n)) - \bar{g}^*(w(n))\} \middle| w(n)\right] \\ &= E\left[g(w(n))g^*(w(n)) \middle| w(n)\right] - \bar{g}(w(n))\bar{g}^*(w(n)) \end{aligned} \tag{3A.121}$$

It follows from (3.13.3) and (3.13.8) that

$$g(w(n))g^*(w(n)) = 4\{\psi(n) - w(n)q(n)\}q^*(n)q(n)\{\psi(n) - w(n)q(n)\}^* \tag{3A.122}$$

Substituting from (3.13.1) and (3.13.2), (3A.122) leads to

$$\begin{aligned} g(w(n))g^*(w(n)) = {} & 4\mathbf{V}^H\mathbf{x}(n)\mathbf{x}^H(n)\mathbf{U}\mathbf{U}^H\mathbf{x}(n)\mathbf{x}^H(n)\mathbf{V} \\ & + 4w^*(n)w(n)\mathbf{U}^H\mathbf{x}(n)\mathbf{x}^H(n)\mathbf{U}\mathbf{U}^H\mathbf{x}(n)\mathbf{x}^H(n)\mathbf{U} \\ & - 4w(n)\mathbf{U}^H\mathbf{x}(n)\mathbf{x}^H(n)\mathbf{U}\mathbf{U}^H\mathbf{x}(n)\mathbf{x}^H(n)\mathbf{V} \\ & - 4w^*(n)\mathbf{V}^H\mathbf{x}(n)\mathbf{x}^H(n)\mathbf{U}\mathbf{U}^H\mathbf{x}(n)\mathbf{x}^H(n)\mathbf{U} \end{aligned} \tag{3A.123}$$

Taking the expectation over $\mathbf{x}(.)$ for a given w(n) on both sides of (3A.123) and using (3A.119),

$$\begin{aligned} E\left[g(w(n))g^*(w(n)) \middle| w(n)\right] = {} & 4\mathbf{V}^H\mathbf{R}\mathbf{U}\mathbf{U}^H\mathbf{R}\mathbf{V} + 4\mathbf{U}^H\mathbf{R}\mathbf{U}\mathbf{V}^H\mathbf{R}\mathbf{V} \\ & + 8w^*(n)w(n)\left(\mathbf{U}^H\mathbf{R}\mathbf{U}\right)^2 \\ & - 8w(n)\mathbf{U}^H\mathbf{R}\mathbf{U}\mathbf{U}^H\mathbf{R}\mathbf{V} \\ & - 8w^*(n)\mathbf{U}^H\mathbf{R}\mathbf{U}\mathbf{V}^H\mathbf{R}\mathbf{U} \end{aligned} \tag{3A.124}$$

Since

$$\begin{aligned} \bar{g}(w(n))\bar{g}^*(w(n)) = {} & 4w^*(n)w(n)\left(\mathbf{U}^H\mathbf{R}\mathbf{U}\right)^2 + 4\mathbf{V}^H\mathbf{R}\mathbf{U}\mathbf{U}^H\mathbf{R}\mathbf{V} \\ & - 4w(n)\mathbf{U}^H\mathbf{R}\mathbf{U}\mathbf{U}^H\mathbf{R}\mathbf{V} - 4w^*(n)\mathbf{U}^H\mathbf{R}\mathbf{U}\mathbf{V}^H\mathbf{R}\mathbf{U} \end{aligned} \tag{3A.125}$$

it follows from (3A.121), (3A.124), and (3A.125) that

$$\begin{aligned} V_g(w(n)) = {} & 4\mathbf{U}^H\mathbf{R}\mathbf{U}\left[\mathbf{V}^H\mathbf{R}\mathbf{V} + w^*(n)w(n)\mathbf{U}^H\mathbf{R}\mathbf{U} \right. \\ & \left. - w(n)\mathbf{U}^H\mathbf{R}\mathbf{V} - w^*(n)\mathbf{V}^H\mathbf{R}\mathbf{U}\right] \end{aligned} \tag{3A.126}$$

which proves the result.

Derivation of (3.13.13) and (3.13.14): Let

$$e(n) = w(n) - \hat{w} \tag{3A.127}$$

Using (3.13.6)

$$e(n+1) = e(n) - \mu g(w(n)) \tag{3A.128}$$

Taking the expected value on both sides of (3A.128),

$$E[e(n+1)] = E[e(n)] - \mu E[g(w(n))] \tag{3A.129}$$

From (3.13.9) and (3A.127), it follows that

$$E[g(w(n))|w(n)] = -2\mathbf{V}^H\mathbf{R}\mathbf{U} + 2e(n)\mathbf{U}^H\mathbf{R}\mathbf{U} + 2\hat{w}\mathbf{U}^H\mathbf{R}\mathbf{U} \tag{3A.130}$$

which along with (3.13.5), implies that

$$E[g(w(n))|w(n)] = 2e(n)\mathbf{U}^H\mathbf{R}\mathbf{U} \tag{3A.131}$$

Taking the unconditional expectation on both sides of (3A.131) and substituting in (3A.129),

$$E[e(n+1)] = E[e(n)] - 2\mu E[3(n)]\mathbf{U}^H\mathbf{R}\mathbf{U} \tag{3A.132}$$

Let

$$f(n) = E[e(n)] \tag{3A.133}$$

Then,

$$f(n+1) = (1 - 2\mu\mathbf{U}^H\mathbf{R}\mathbf{U})f(n) \tag{3A.134}$$

which implies

$$f(n+1) = (1 - 2\mu\mathbf{U}^H\mathbf{R}\mathbf{U})^{n+1} f(0) \tag{3A.135}$$

For

$$0 < \mu < \frac{1}{\mathbf{U}^H\mathbf{R}\mathbf{U}} \tag{3A.136}$$

$$|1 - 2\mu\mathbf{U}^H\mathbf{R}\mathbf{U}| < 1 \tag{3A.137}$$

Furthermore, it follows from (3.13.11), (3A.127), and (3A.133) that

$$|f(0)| < \infty \tag{3A.138}$$

Thus, it follows from (3A.135), (3A.137), and (3A.138) that

$$\lim_{n\to\infty} f(n) = 0 \tag{3A.139}$$

Equation (3A.139) along with (3A.127) and (3A.133) implies (3.13.13).

To derive (3.13.14), consider (3A.135). It follows from (3A.135) that the convergence of the mean weight to the optimum weight is geometric, with the geometric ratio $(1 - 2\mu\mathbf{U}^H R\mathbf{U})$. If an exponential envelope of time constant $\tau$ is fitted to the geometric sequence of (3A.135), then

$$\tau = -\frac{1}{\ln\left(1 - 2\mu\mathbf{U}^H R\mathbf{U}\right)} \tag{3A.140}$$

which is (3.13.14).

Derivation of (3.13.17): It follows from (3.13.15) and (3.13.16) that

$$K_{ww}(n) = R_{ww}(n) - \overline{w}(n)\overline{w}^*(n) \tag{3A.141}$$

where

$$R_{ww}(n) = E\left[w(n)w^*(n)\right] \tag{3A.142}$$

It follows from (3.13.6) that

$$\begin{aligned} w(n+1)w^*(n+1) = {} & w(n)w^*(n) + \mu^2 g(w(n))g^*(w(n)) \\ & - \mu w^*(n)g(w(n)) - \mu g^*(w(n))w(n) \end{aligned} \tag{3A.143}$$

Taking the conditional expectation with respect to w(n) on both sides of (3A.143),

$$\begin{aligned} E\left[w(n+1)w^*(n+1)\middle|w(n)\right] = {} & E\left[w(n)w^*(n)\middle|w(n)\right] \\ & + \mu^2 E\left[g(w(n))g^*(w(n))\middle|w(n)\right] \\ & - \mu E\left[w^*(n)g(w(n))\middle|w(n)\right] \\ & - \mu E\left[g^*(w(n))w(n)\middle|w(n)\right] \end{aligned} \tag{3A.144}$$

Since

$$\begin{aligned} E\left[g^*(w(n))g(w(n))\middle|w(n)\right] = {} & V_g(w(n)) \\ & + E\left[g^*(w(n))\middle|w(n)\right]E\left[g(w(n))\middle|w(n)\right] \end{aligned} \tag{3A.145}$$

and

$$E\left[g(w(n))\middle|w(n)\right] = 2w(n)\mathbf{U}^H\mathbf{R}\mathbf{U} - 2\mathbf{V}^H\mathbf{R}\mathbf{U} \tag{3A.146}$$

it follows from (3A.144) that

$$\begin{aligned} E\left[w(n+1)w^*(n+1)\middle|w(n)\right] = {} & E\left[w^*(n)w(n)\middle|w(n)\right]\left[1+4\mu^2\left(\mathbf{U}^H\mathbf{R}\mathbf{U}\right)^2 - 4\mu\mathbf{U}^H\mathbf{R}\mathbf{U}\right] \\ & +\mu^2 V_g(w(n)) + 4\mu^2\left[\mathbf{V}^H\mathbf{R}\mathbf{U}\mathbf{U}^H\mathbf{R}\mathbf{V}\right. \\ & \left. - w^*(n)\mathbf{U}^H\mathbf{R}\mathbf{U}\mathbf{V}^H\mathbf{R}\mathbf{U} - w(n)\mathbf{U}^H\mathbf{R}\mathbf{U}\mathbf{U}^H\mathbf{R}\mathbf{V}\right] \\ & + 2\mu\left[w^*(n)\mathbf{V}^H\mathbf{R}\mathbf{U} + w(n)\mathbf{U}^H\mathbf{R}\mathbf{V}\right] \end{aligned} \tag{3A.147}$$

Taking the unconditional expectation on both sides of (3A.147) and using (3A.142),

$$\begin{aligned} R_{ww}(n+1) = {} & R_{ww}(n)\left[1+4\mu^2\left(\mathbf{U}^H\mathbf{R}\mathbf{U}\right)^2 - 4\mu\mathbf{U}^H\mathbf{R}\mathbf{U}\right] \\ & +\mu^2 E\left[V_g(w(n))\right] + 4\mu^2\mathbf{V}^H\mathbf{R}\mathbf{U}\mathbf{U}^H\mathbf{R}\mathbf{V} \\ & - 4\mu^2\mathbf{U}^H\mathbf{R}\mathbf{U}\left[\overline{w}^*(n)\mathbf{V}^H\mathbf{R}\mathbf{U} + \overline{w}(n)\mathbf{U}^H\mathbf{R}\mathbf{V}\right] \\ & + 2\mu\left[\overline{w}^*(n)\mathbf{V}^H\mathbf{R}\mathbf{U} + \overline{w}(n)\mathbf{U}^H\mathbf{R}\mathbf{V}\right] \end{aligned} \tag{3A.148}$$

Taking the unconditional expectation on both sides of (3.13.6) and using (3.13.16) and (3A.146),

$$\overline{w}(n+1) = \overline{w}(n) - 2\mu\overline{w}(n)\mathbf{U}^H\mathbf{R}\mathbf{U} + 2\mu\mathbf{V}^H\mathbf{R}\mathbf{U} \tag{3A.149}$$

Taking the outer product of (3A.149),

$$\begin{aligned} \overline{w}(n+1)\overline{w}^*(n+1) = {} & \overline{w}(n)\overline{w}^*(n)\left[1+4\mu^2\left(\mathbf{U}^H\mathbf{R}\mathbf{U}\right)^2 - 4\mu\mathbf{U}^H\mathbf{R}\mathbf{U}\right] \\ & + 4\mu^2\mathbf{V}^H\mathbf{R}\mathbf{U}\mathbf{U}^H\mathbf{R}\mathbf{V} \\ & - 4\mu^2\mathbf{U}^H\mathbf{R}\mathbf{U}\left[\overline{w}^*(n)\mathbf{V}^H\mathbf{R}\mathbf{U} + \overline{w}(n)\mathbf{U}^H\mathbf{R}\mathbf{V}\right] \\ & + 2\mu\left[\overline{w}^*(n)\mathbf{V}^H\mathbf{R}\mathbf{U} + \overline{w}(n)\mathbf{U}^H\mathbf{R}\mathbf{V}\right] \end{aligned} \tag{3A.150}$$

From (3.141), (3A.148), and (3A.150), one obtains

$$\begin{aligned} K_{ww}(n+1) = {} & K_{ww}(n)\left[1+4\mu^2\left(\mathbf{U}^H\mathbf{R}\mathbf{U}\right)^2 - 4\mu\left(\mathbf{U}^H\mathbf{R}\mathbf{U}\right)\right] \\ & +\mu^2 E\left[V_g(w(n))\right] \end{aligned} \tag{3A.151}$$

which is (3.13.17).

# 4

# *Broadband Processing*

The beamformer structure of Figure 2.1 discussed earlier is for narrowband signals. As the signal bandwidth increases, beamformer performance using this structure starts to deteriorate [Rod79]. For processing broadband signals, a tap delay line (TDL) structure shown in Figure 4.1 is normally used [Rod79, May81, Voo92, Com88, Ko81, Ko87, Nun83, Yeh87, Sco83]. A lattice structure consisting of a cascade of simple lattice filters sometimes is also used [Ale87, Lin86, Iig85, Soh84], offering certain processing advantages.

Although the TDL structure with constrained optimization is the commonly used structure for broadband array signal processing, alternative methods have been proposed.

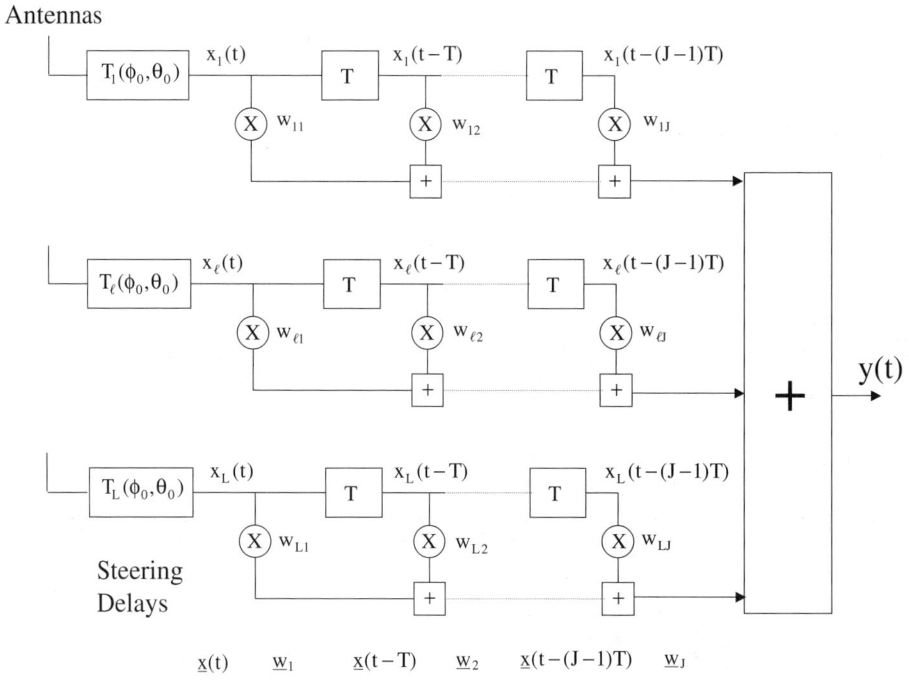


**FIGURE 4.1**
Broadband processor with tapped delay line structure.

These include adaptive nonlinear schemes, which maximizes the signal-to-noise ratio (SNR) subject to additional constraints [Win72]; a variation of the Davis beamformer [Dav67], which adapts one filter at a time to speed up convergence [Ko90]; a composite system that also utilizes a derivative of beam pattern in the feedback loop to control the weights [Tak80] to reject wideband interference; optimum filters that specify rejection response [Sim83]; master and slave processors [Hua90]; a hybrid method that uses an orthogonal transformation on data available from the TDL structure before applying weights [Che95] to improve its performance in multipath environments; the weighted Tschebysheff method [Nor94]; and the two-sided correlation transformation method [Val95].

In this chapter, details on an array processor using the TDL structure and its partitioned realization to process broadband array signals are provided, the time domain and frequency domain methods are described, and details on deriving various constraints are given [God95, God97, God99]. The treatment presented here is for solving a constrained beamforming problem, assuming that the look direction is known. It can easily be extended to the case when a reference signal is available.

## 4.1 Tapped-Delay Line Structure

In this section, a TDL structure for broadband antenna array processing is described, its frequency response and optimization are discussed, an LMS algorithm to estimate the solution of the point-constrained optimization problem is developed, and a design using minimum mean square error (MSE) between the frequency response of the processor and the desired response is presented.

### 4.1.1 Description

Figure 4.1 shows a general structure of a broadband antenna array processor consisting of L antenna elements, steering delays $T_\ell(\phi_0,\theta)$, $\ell = 1, \ldots, L$ and a delay line section of $J - 1$ delays with inter-tap delay spacing T. The steering delays $T_\ell(\phi_0,\theta_0)$, $\ell = 1, \ldots, L$ in front of each element are pure time delays and are used to steer the array in a given look direction $(\phi_0,\theta_0)$. If $\tau_\ell(\phi_0,\theta_0)$ denotes the time taken by the plane wave arriving from direction $(\phi_0,\theta_0)$, and measured from the reference point to the $\ell$th element, then the steering delay $T_\ell(\phi_0,\theta_0)$ may be selected using

$$T_\ell(\phi_0,\theta_0) = T_0 + \tau_\ell(\phi_0,\theta_0), \quad \ell = 1, \ldots, L \tag{4.1.1}$$

where $T_0$ is a bulk delay such that $T_\ell(\phi_0,\theta_0) > 0 \forall_\ell$.

If s(t) denotes the signal induced on an element present at the center of the coordinate system due to a broadband source of power density S(f) in direction $(\phi,\theta)$, then the signal induced on the $\ell$th element is given by $s(t + \tau_\ell(\phi,\theta))$, as discussed in Chapter 2.

Let $x_\ell(t)$ denote the output of the $\ell$th sensor presteered in $(\phi_0,\theta_0)$. It is given by

$$x_\ell(t) = s\big(t + \tau_\ell(\phi,\theta) - T_\ell(\phi_0,\theta_0)\big) \tag{4.1.2}$$

For a source in $(\phi_0,\theta_0)$ it becomes $x_\ell(t) = s(t - T_0)$, yielding identical wave forms after presteering delays.

The TDL structure shown in Figure 4.1 following the steering delay on each channel is a finite impulse response (FIR) filter. The coefficients of these filters are constrained to specify the frequency response in the look direction. It should be noted that these coefficients are real compared to complex weights of the narrowband processor.

It follows from Figure 4.1 that the output y(t) of the processor is given by

$$y(t) = \sum_{\ell=1}^{L} \sum_{k=1}^{J} x_\ell\left(t-(k-1)T\right) w_{\ell k} \tag{4.1.3}$$

where $w_{\ell k}$ denotes the weight on the kth tap of the $\ell$th channel. Note that the kth tap output corresponds to the output after (k – 1) delays. Thus, first tap output corresponds to the output of presteering delays and before any tapped delays section, the second tap output corresponds to the output after one delay and Jth tap output corresponds to the output after J – 1 delays.

Let **W** defined by

$$\mathbf{W}^T = \left[\mathbf{w}_1^T, \mathbf{w}_2^T, \ldots, \mathbf{w}_J^T\right] \tag{4.1.4}$$

denote LJ weights of the filter structure, with $\mathbf{w}_m$ denoting the column of L weights on the mth tap.

Define an L-dimensional vector **x**(t) to denote array signals after presteering delays, that is,

$$\mathbf{x}(t) = \left[x_1(t), x_2(t), \ldots, (x_L)\right]^T \tag{4.1.5}$$

and an LJ-dimensional vector **X**(t) to denote array signals across the TDL structure, that is,

$$\mathbf{X}^T(t) = \left[\mathbf{x}^T(t), \mathbf{x}^T(t-T), \ldots, \mathbf{x}^T\left(t-(J-1)T\right)\right] \tag{4.1.6}$$

It follows from (4.1.3) to (4.1.6) that the output y(t) of the processor in the vector notation becomes

$$y(t) = \mathbf{W}^T\mathbf{X}(t) \tag{4.1.7}$$

If **X**(t) can be modeled as a zero-mean stochastic process, then the mean output power of the processor for a given **W** is given by

$$\begin{aligned} P(\mathbf{W}) &= E\left[y^2(t)\right] \\ &= \mathbf{W}^T R \mathbf{W} \end{aligned} \tag{4.1.8}$$

where

$$R = E\left[\mathbf{X}(t)\mathbf{X}^T(t)\right] \tag{4.1.9}$$

is an LJ × LJ dimensional real matrix and denotes the array correlation matrix with its elements representing the correlation between various tap outputs. The correlation between the outputs of mth tap on the $\ell$th channel and nth tap on the kth channel is given by

$$\left(R_{m,n}\right)_{\ell,k} = E\left[x_\ell\left(t-(m-1)T\right)x_k\left(t-(n-1)T\right)\right] \tag{4.1.10}$$

Note that the L × L matrix $R_{m,n}$ denotes the correlation between the array outputs at the mth and nth taps, that is, after (m – 1) and (n – 1) delays.

Substituting from (4.1.2), it follows that

$$\left(R_{m,n}\right)_{\ell,k} = \rho\left[(m-n)T + T_\ell(\phi_0,\theta_0) - T_k(\phi_0,\theta_0) + \tau_k(\phi,\theta) - \tau_\ell(\phi,\theta)\right] \tag{4.1.11}$$

where $\rho(\tau)$ denotes the correlation function of s(t), that is,

$$\rho(\tau) = E\left[s(t)s(t+\tau)\right] \tag{4.1.12}$$

The correlation function is related to the spectrum of the signal by the inverse Fourier transform, that is,

$$\rho(\tau) = \int_{-\infty}^{\infty} S(f)\, e^{j2\pi f\tau} df \tag{4.1.13}$$

Thus, from known spectra of sources and their arrival directions, the correlation matrix may be calculated. In practice, it can also be estimated by measuring signals at the output of various taps.

For M uncorrelated directional sources, the array correlation matrix is the sum of correlation matrices due to each source, that is,

$$R = \sum_{\ell=1}^{M} R_\ell \tag{4.1.14}$$

where $R_\ell$ is the array correlation matrix due to the $\ell$th source in direction $(\phi_\ell, \theta_\ell)$.

Let $R_S$ denote the array correlation matrix due to the signal source, that is, a source in the look direction, and $R_N$ denote the array correlation matrix due to noise, that is, unwanted directional sources and other noise. The mean output signal power $P_S(\mathbf{W})$ and mean output noise power $P_N(\mathbf{W})$ for a given weight vector are, respectively, given by

$$P_S(\mathbf{W}) = \mathbf{W}^T R_S \mathbf{W} \tag{4.1.15}$$

and

$$P_N(\mathbf{W}) = \mathbf{W}^T R_N \mathbf{W} \tag{4.1.16}$$

The output SNR for given weights is

$$\begin{aligned} \mathrm{SNR}(\mathbf{W}) &= \frac{P_S(\mathbf{W})}{P_N(\mathbf{W})} \\ &= \frac{\mathbf{W}^T R_S \mathbf{W}}{\mathbf{W}^T R_N \mathbf{W}} \end{aligned} \tag{4.1.17}$$

### 4.1.2 Frequency Response

Assume that the signal induced on an element at the center of the coordinate system due to a monochromatic plane wave of frequency f can be represented in complex notation as $e^{j2\pi ft}$. Thus, the induced signal on the $\ell$th element after the steering delay due to a plane wave arriving in direction $(\phi,\theta)$ becomes $e^{j2\pi f(t+\tau_\ell(\phi,\theta)-T_\ell(\phi_0,\theta_0))}$. The frequency response $H(f,\phi,\theta)$ of the processor to a plane wave front arriving in direction $(\phi,\theta)$ is then given by

$$\begin{aligned} H(f,\phi,\theta) &= \sum_{\ell=1}^{L} e^{j2\pi f\tau_\ell(\phi,\theta)} e^{-j2\pi fT_\ell(\phi_0,\theta_0)} \sum_{k=1}^{J} \mathbf{w}_k e^{-j2\pi f(k-1)T} \\ &= \mathbf{S}^T(f,\phi,\theta)T(f)\sum_{k=1}^{J} \mathbf{w}_k e^{-j2\pi f(k-1)T} \end{aligned} \tag{4.1.18}$$

where T(f) is a diagonal matrix of steering delays given by

$$T(f) = \begin{bmatrix} e^{-j2\pi fT_1(\phi_0,\theta_0)} & & & 0 \\ & e^{-j2\pi fT_2(\phi_0,\theta_0)} & & \\ & & \ddots & \\ 0 & & & e^{-j2\pi fT_L(\phi_0,\theta_0)} \end{bmatrix} \tag{4.1.19}$$

and $\mathbf{S}(f,\phi,\theta)$ is an L-dimensional vector defined as

$$\mathbf{S}^T(f,\phi,\theta) = \left[e^{j2\pi f\tau_1(\phi,\theta)}, e^{j2\pi f\tau_2(\phi,\theta)}, \ldots, e^{j2\pi f\tau_L(\phi,\theta)}\right] \tag{4.1.20}$$

It follows from (4.1.1), (4.1.19), and (4.1.20) that

$$\mathbf{S}^T(f,\phi_0,\theta_0)T(f) = a(f)[1, 1, \ldots, 1] \tag{4.1.21}$$

where

$$a(f) = e^{-j2\pi fT_0} \tag{4.1.22}$$

In this case, the frequency response of the array steered in the look direction $(\phi_0,\theta_0)$ is given by

$$H(f,\phi_0,\theta_0) = a(f)\sum_{k=1}^{J} f_k e^{-j2\pi f(k-1)T} \tag{4.1.23}$$

where

$$f_k = \mathbf{1}^T\mathbf{w}_k, \quad k = 1, 2, \ldots, J \tag{4.1.24}$$

with **1** denoting a vector of ones.

Let **f** be a J-dimensional constraint vector defined as

$$\mathbf{f} = \left[f_1, f_2, \ldots, f_J\right]^T \tag{4.1.25}$$

and C be an LJ × J constraint matrix defined as

$$C = \begin{bmatrix} \mathbf{1} & 0 & \ldots & 0 \\ 0 & \mathbf{1} & . & . \\ . & . & . & 0 \\ 0 & \ldots & 0 & \mathbf{1} \end{bmatrix} \tag{4.1.26}$$

The J constraints defined by (4.1.24) can now be expressed as

$$C^T\mathbf{W} = \mathbf{f} \tag{4.1.27}$$

Since a(f) given by (4.1.22) corresponds to a pure time delay, the J constraints $\{f_k\}$ can be used to specify the frequency response in the direction $(\phi_0,\theta_0)$.

The processor can be forced to have a flat frequency response in the look direction by selecting **f** as follows:

$$f_i = \begin{cases} 1 & i = k_0 \\ 0 & i \neq k_0 \end{cases} \tag{4.1.28}$$

where $k_0$ is a parameter, which can itself be optimized. Frequently, $k_0$ is taken as J/2 for J, an even number, and (J + 1)/2 for J, an odd number, since for a sufficiently large J this gives close to optimum performance.

### 4.1.3 Optimization

The frequency response of an array processor in the look direction can be fixed using the J constraints in (4.1.27). The processor can minimize the non–look direction noise when weights are selected by minimizing the total mean output power such that (4.1.27) is satisfied. Thus, in situations where one is interested in finding array weights, such that the array processor minimizes the total noise and has the specified response in the look direction, the following constrained beamforming problem is considered:

$$\begin{array}{ll} \underset{\mathbf{W}}{\text{minimize}} & \mathbf{W}^T R \mathbf{W} \\ \text{subject to} & C^T\mathbf{W} = \mathbf{f} \end{array} \tag{4.1.29}$$

where **f** is a J-dimensional vector that specifies the frequency response in the look direction and C is an LJ × J constraint matrix.

Let $\hat{\mathbf{W}}$ denote the solution of the above problem. The solution is obtained by the Lagrange multipliers method [Bry69, Lue69, Pie69]. This method transforms the constrained problem into an unconstrained problem by adding the constraint function to the cost function using a J-dimensional vector of undetermined Lagrange multipliers $\boldsymbol{\lambda}$ to generate a new cost function. Let J(**W**) denote the cost function for the present problem. It is given by

$$J(\mathbf{W}) = \frac{1}{2}\mathbf{W}^T R \mathbf{W} + \boldsymbol{\lambda}^T\left(C^T\mathbf{W} - \mathbf{f}\right) \tag{4.1.30}$$

where 1/2 is added to simplify the mathematics.

Taking the gradient of (4.1.30) with respect to **W**,

$$\nabla_W J(\mathbf{W}) = R\mathbf{W} + C\boldsymbol{\lambda} \tag{4.1.31}$$

At the solution point, the cost function gradient is zero. Thus,

$$R\hat{\mathbf{W}} + C\boldsymbol{\lambda} = 0 \tag{4.1.32}$$

Assuming that the inverse of the array correlation matrix R exists, $\hat{\mathbf{W}}$ may be expressed in terms of Lagrange multipliers as

$$\hat{\mathbf{W}} = -R^{-1}C\boldsymbol{\lambda} \tag{4.1.33}$$

Since $\hat{\mathbf{W}}$ satisfies the constraint $C^T\hat{\mathbf{W}} = \mathbf{f}$, it follows from (4.1.33) that

$$-C^T R^{-1} C\boldsymbol{\lambda} = \mathbf{f} \tag{4.1.34}$$

An expression for Lagrange multipliers may be found from (4.1.34), yielding

$$\boldsymbol{\lambda} = -\left(C^T R^{-1} C\right)^{-1}\mathbf{f} \tag{4.1.35}$$

Substituting for Lagrange multipliers in (4.1.33) from (4.1.35), an expression for the optimal weights [Fro72] follows:

$$\hat{\mathbf{W}} = R^{-1}C\left(C^T R^{-1} C\right)^{-1}\mathbf{f} \tag{4.1.36}$$

Let $\hat{P}$ denote the mean output power of the processor using optimal weights, that is,

$$\hat{P} = \hat{\mathbf{W}}^T R \hat{\mathbf{W}} \tag{4.1.37}$$

Substituting for $\hat{\mathbf{W}}$ from (4.1.36),

$$\hat{P} = \mathbf{f}^T\left(C^T R^{-1} C\right)^{-1}\mathbf{f} \tag{4.1.38}$$

The point-constraint minimization problem (4.1.29) specifies J constraints on the weights such that the sum of L weights on all channels before the jth delay is equal to $f_i$. For all pass frequency responses in the look direction, all but one $f_i$, i = 1, 2, …, J are selected to be equal to zero. For i's close to J/2, $f_i$ is taken to be unity. Thus, the constraints specify that the sum of weights across the array is zero, except one near the middle of the filter that is equal to unity.

Thus, for all pass frequency responses when $f_i$, i = 1, …, J are selected as

$$f_i = \begin{cases} 1 & i = k_0 \\ 0 & i \neq k_0 \end{cases} \tag{4.1.39}$$

equation (4.1.38) becomes

$$\hat{P} = \left(C^T R^{-1} C\right)^{-1}_{k_0,k_0} \tag{4.1.40}$$

Application of broadband beamforming structures using TDL filters to mobile communications has been considered in [Win94, Des92, Ish95, Koh92] to overcome multipath fading and large delay spread in TDMA as well as in CDMA systems.

### 4.1.4 Adaptive Algorithm

A constrained LMS algorithm to estimate the optimal weights of a narrowband element space processor is discussed in Chapter 3. The corresponding algorithm to estimate the optimal weights of the broadband processor given by (4.3.36) may be developed as follows [Fro72].

Let **W**(n) denote the weights estimated at the nth iteration. At this stage, a new array sample **X**(n + 1) is available and the array output using weights **W**(n) is given by

$$y(n) = \mathbf{W}^T(n)\mathbf{X}(n+1) \tag{4.1.41}$$

For notational simplicity it is assumed that the nth iteration coincides with the nth time sample. The new weight vector **W**(n + 1) is calculated by moving in the negative direction of the cost function gradient, that is,

$$\mathbf{W}(n+1) = \mathbf{W}(n) - \mu \nabla_{\mathbf{W}} J(\mathbf{W}(n)) \tag{4.1.42}$$

where J(**W**(n)) is the cost function given by (4.1.30), with **W** replaced by **W**(n) and $\mu$ is a positive scalar. Replacing R with its noisy sample $\mathbf{X}(n+1)\mathbf{X}^T(n+1)$, it follows from (4.1.31) that

$$\begin{aligned} \nabla_{\mathbf{W}} J(\mathbf{W}(n)) &= \mathbf{X}(n+1)\mathbf{X}^T(n+1)\mathbf{W}(n) + C\boldsymbol{\lambda}(n) \\ &= y(n)\mathbf{X}(n+1) + C\boldsymbol{\lambda}(n) \end{aligned} \tag{4.1.43}$$

where $\boldsymbol{\lambda}(n)$ denotes the Lagrange multipliers at the nth iteration.

Substituting from (4.1.43) in (4.1.42),

$$\mathbf{W}(n+1) = \mathbf{W}(n) - \mu y(n)\mathbf{X}(n+1) - \mu C\boldsymbol{\lambda}(n) \tag{4.1.44}$$

Assuming that the estimated weights satisfy the constraints at each iteration, it follows from the second equation of (4.1.29) that

$$C^T \mathbf{W}(n+1) = C^T \mathbf{W}(n) = \mathbf{f} \tag{4.1.45}$$

Multiplying by $C^T$ on both sides of (4.1.44) and using (4.1.45), it follows that

$$C^T y(n)\mathbf{X}(n+1) + C^T C \boldsymbol{\lambda}(n) = 0 \tag{4.1.46}$$

Solving for $\boldsymbol{\lambda}(n)$

$$\boldsymbol{\lambda}(n) = -\left(C^T C\right)^{-1} C^T y(n)\mathbf{X}(n+1) \tag{4.1.47}$$

Substituting in (4.1.44),

$$\mathbf{W}(n+1) = \mathbf{W}(n) - \mu\, y(n) P \mathbf{X}(n+1) \tag{4.1.48}$$

where

$$P = I - C\left(C^T C\right)^{-1} C^T \tag{4.1.49}$$

is a projection operator. It follows from (4.1.45) and (4.1.49) that

$$P\mathbf{W}(n) = \mathbf{W}(n) - C\left(C^T C\right)^{-1} \mathbf{f} \tag{4.1.50}$$

Thus,

$$\mathbf{W}(n) = P\mathbf{W}(n) + C\left(C^T C\right)^{-1} \mathbf{f} \tag{4.1.51}$$

and after substitution for $\mathbf{W}(n)$, (4.1.48) becomes

$$\mathbf{W}(n+1) = P\left[\mathbf{W}(n) - \mu y(n)\mathbf{X}(n+1)\right] + \mathbf{F} \tag{4.1.52}$$

where

$$\mathbf{F} = C\left(C^T C\right)^{-1} \mathbf{f} \tag{4.1.53}$$

Thus, knowing the array weights $\mathbf{W}(n)$, array output, and array sample $\mathbf{X}(n + 1)$, the new weights $\mathbf{W}(n + 1)$ can be calculated using the constrained LMS algorithm given by (4.1.52), (4.1.53), and (4.1.49).

The algorithm is initialized at $n = 0$ using

$$\mathbf{W}(0) = \mathbf{F} \tag{4.1.54}$$

The initialization of the algorithm using weights equal to **F** is selected because it denotes the optimal weights in the presence of only white noise, that is, no directional interference. This follows from the fact that the array correlation matrix R in this case is given by

$$R = \sigma_n^2 I \tag{4.1.55}$$

Substituting in (4.1.36), it follows that

$$\begin{aligned}\hat{\mathbf{W}} &= C\left(C^T C\right)^{-1}\mathbf{f} \\ &= \mathbf{F}\end{aligned} \tag{4.1.56}$$

The convergence analysis of the algorithm may be carried out similar to that for the narrowband case discussed in Chapter 3.

A substantial amount of computation in (4.1.52) is required to compute a multiplication between an LJ-dimensional vector and matrix P. The sparse nature of matrix C allows simplification of the algorithm with reduced computation as follows.

It follows from (4.1.26) that

$$C^T C = LI \tag{4.1.57}$$

where I is an identity matrix.

Substituting in (4.1.53) and (4.1.49) yields

$$\begin{aligned}\mathbf{F} &= \frac{1}{L}C\mathbf{f} \\ &= \frac{1}{L}\begin{bmatrix} f_1\mathbf{1} & & 0 \\ & \ddots & \\ 0 & & f_J\mathbf{1}\end{bmatrix}\end{aligned} \tag{4.1.58}$$

and

$$\begin{aligned}P &= I - \frac{CC^T}{L} \\ &= I - \frac{1}{L}\begin{bmatrix} \mathbf{1}\mathbf{1}^T & & 0 \\ & \ddots & \\ 0 & & \mathbf{1}\mathbf{1}^T\end{bmatrix}\end{aligned} \tag{4.1.59}$$

From (4.1.52), (4.1.58), and (4.1.59) an update equation in $\mathbf{w}_j(n)$, $j = 0, 1, \ldots, J-1$ may be expressed as [Buc86]

$$\mathbf{w}_j(n+1) = \left[I - \frac{\mathbf{1}\mathbf{1}^T}{L}\right]\left[\mathbf{w}_j(n) - \mu y(n)\mathbf{x}(n+1-j)\right] + \frac{f_j}{L}\mathbf{1} \tag{4.1.60}$$

where $\mathbf{w}_j(n)$ denotes the L weights after the jth tap computed at the nth iteration, and $\mathbf{x}(n + 1 - j)$ denotes the array signal after the jth tap. Thus, (4.1.60) allows iterative computation of J columns of weights separately.

Noting that for an L-dimensional vector, $\mathbf{a}$

$$\mathbf{1}^T\mathbf{a} = \sum_{i=1}^{L} a_i \tag{4.1.61}$$

(4.1.60) may be implemented in summation form as [Fro72]:

$$\mathbf{w}_j(n+1) = \mathbf{w}_j(n) - \mu y(n)\mathbf{x}(n+1-j) - \frac{\mathbf{1}}{L}\sum_{\ell=1}^{L}\left[\left(\mathbf{w}_j(n)\right)_\ell - \mu y(n)x_\ell(n+1-j)\right] + \frac{f_j}{L}\mathbf{1} \tag{4.1.62}$$

where $(\mathbf{w}_j(n))_\ell$ denotes the $\ell$th component of the weight vector $\mathbf{w}_j(n)$.

### 4.1.5 Minimum Mean Square Error Design

The processor design considered in Section 4.1.3 by solving constrained optimization problems given by (4.1.29) minimizes the mean output power while maintaining a specified frequency response in the look direction. In this section, a processor design discussed in [Er85] is presented. This processor uses the TDL structure similar to that shown in Figure 4.1. The weights of the processor are estimated to minimize the MSE $\varepsilon_0$, between the frequency response of the processor in the look direction and the desired look direction response over a frequency band of interest $[f_L, f_H]$, defined as

$$\varepsilon_0 = \int_{f_L}^{f_H} \left|A(f, \phi_0, \theta_0) - H(f, \phi_0, \theta_0)\right|^2 df \tag{4.1.63}$$

where $A(f,\phi,\theta)$ denotes the desired frequency response in direction $(\phi,\theta)$. For a processor to have a flat frequency response in the look direction, it is given by

$$A(f, \phi_0, \theta_0) = \exp(j2\pi f\tau) \tag{4.1.64}$$

where $\tau$ denotes a delay parameter that may be optimized [Er85].

As the constraints on the weights are designed to minimize the deviation of the processor response from the desired response in the means squared sense, the presteering delays are not necessary. In this case, the presteering delays $T_\ell(\phi_0,\theta_0)$, $l = 1, 2, \ldots, L$ are set to zero. This is equivalent to the situation when matrix $T(f)$ is not included in the frequency response expression (4.1.18).

The processor also allows exact presteering as well as coarse presteering. For the exact presteering case, the steering delays are given by (4.1.1). This case is useful in comparing the performance of the processor using the minimum MSE design with that of the optimal processor discussed in Section 4.1.3. Coarse presteering arises when sampled signals are processed and steering delays are selected as the integer multiples of the sampling time closest to the exact delays required to steer the array in look direction.

In the treatment that follows, it is assumed that steering delays $T_\ell(\phi_0,\theta_0)$, $l = 1, 2, \ldots, L$ are included in the design and the frequency response of the processor is given by (4.1.18). However, the values of $T_\ell(\phi_0,\theta_0)$ will depend on the case under consideration, that is, no presteering, coarse presteering, or exact presteering.

#### *4.1.5.1 Derivation of Constraints*

It follows from (4.1.63) that

$$\varepsilon_0 = \sigma_0 + \mathbf{W}^T Q \mathbf{W} - 2P^T \mathbf{W} \tag{4.1.65}$$

where $\sigma_0$ is a scalar given by

$$\sigma_0 = \int_{f_L}^{f_H} A^*(f, \phi_0, \theta_0) A(f, \phi_0, \theta_0) df \tag{4.1.66}$$

$$\mathbf{W}^T Q \mathbf{W} = \int_{f_L}^{f_H} H^*(f, \phi_0, \theta_0) H(f, \phi_0, \theta_0) df \tag{4.1.67}$$

$$\mathbf{P}^T \mathbf{W} = \frac{1}{2} \int_{f_L}^{f_H} \{A^*(f, \phi_0, \theta_0) H(f, \phi_0, \theta_0) + H^*(f, \phi_0, \theta_0) A(f, \phi_0, \theta_0)\} df \tag{4.1.68}$$

Q is an LJ × LJ dimensional positive, semidefinite symmetrical matrix, and **P** is an LJ-dimensional vector.

Substituting for $H(f,\phi_0,\theta_0)$ in (4.1.67) and (4.1.68) leads to the following expressions for Q and **P** [Er85]:

$$\begin{gathered} Q_{k,\ell} = \psi\left[(\tau_i - \tau_j) + (T_j - T_i) + (n - m)T\right] \\ k = i + (m-1)L, \quad \ell = j + (n-1)L, \; i, j = 1, 2, \ldots, L, \quad m, n = 1, 2, \ldots, J \end{gathered} \tag{4.1.69}$$

where

$$\psi(\tau) = \left[f_H \operatorname{sinc}(2\pi f_H \tau) - f_L \operatorname{sinc}(2\pi f_L \tau)\right] \tag{4.1.70}$$

with

$$\operatorname{sinc}(\alpha) = \frac{\sin\alpha}{\alpha} \tag{4.1.71}$$

and

$$\mathbf{P} = \left[\mathbf{P}_1^T, \mathbf{P}_2^T, \ldots, \mathbf{P}_J^T\right]^T \tag{4.1.72}$$

where

$$\begin{gathered} [P_k]_\ell = \frac{1}{2} \int_{f_L}^{f_H} \left\{A^*(f, \phi_0, \theta_0) e^{j2\pi f(\tau_\ell - T_\ell - (k-1)T)} + A(f, \phi_0, \theta_0) e^{-j2\pi f(\tau_\ell - T_\ell - (k-1)T)}\right\} df \\ \ell = 1, 2, \ldots, L, \quad k = 1, 2, \ldots, J \end{gathered} \tag{4.1.73}$$

Let $\tilde{\mathbf{W}}$ denote an LJ-dimensional vector that minimizes $\varepsilon_0$. Thus,

$$\left.\frac{\partial \varepsilon_0}{\partial \mathbf{W}}\right|_{\mathbf{W}=\hat{\mathbf{W}}} = 0 \tag{4.1.74}$$

It follows from (4.1.65) and (4.1.74) that $\tilde{\mathbf{W}}$ satisfies

$$Q\tilde{\mathbf{W}} = \mathbf{P} \tag{4.1.75}$$

Rewrite (4.1.65) using (4.1.75) as

$$\varepsilon_0 = \left(\tilde{\mathbf{W}} - \mathbf{W}\right)^T Q\left(\tilde{\mathbf{W}} - \mathbf{W}\right) - \tilde{\mathbf{W}}^T Q \tilde{\mathbf{W}} + \sigma_0 \tag{4.1.76}$$

As the signal distortion depends on the allowed MSE between the desired look direction response and the processor response in the look direction over the frequency band of interest, the processor weights can be constrained to limit the MSE less than or equal to some threshold value $\delta_0$. Thus, an optimization problem can be formulated as discussed below.

### *4.1.5.2 Optimization*

Consider the following optimization problem:

$$\begin{aligned} &\underset{\mathbf{W}}{\text{minimize}} \quad \mathbf{W}^H R \mathbf{W} \\ &\text{subject to} \quad \varepsilon_0 \le \delta_0 \end{aligned} \tag{4.1.77}$$

Defining an error vector

$$\mathbf{V} = \tilde{\mathbf{W}} - \mathbf{W} \tag{4.1.78}$$

and using (4.1.76), the optimization problem (4.1.77) becomes

$$\begin{aligned} &\underset{\mathbf{V}}{\text{minimize}} \quad \left(\tilde{\mathbf{W}} - \mathbf{V}\right)^T R\left(\tilde{\mathbf{W}} - \mathbf{V}\right) \\ &\text{subject to} \quad \mathbf{V}^T Q \mathbf{V} \le \xi \end{aligned} \tag{4.1.79}$$

where

$$\xi = \tilde{\mathbf{W}}^T Q \tilde{\mathbf{W}} + \delta_0 - \sigma_0 \tag{4.1.80}$$

Note that (4.1.80) follows from (4.1.76), (4.1.77), and (4.1.79).

Let $\hat{\mathbf{W}}_\varepsilon$ be the solution of the optimization problem (4.1.77). It can be obtained using the Lagrange multipliers method as follows [Er85].

Let $J(\mathbf{V},\lambda)$ be the cost function defined as

$$J(\mathbf{V},\lambda) = \left(\tilde{\mathbf{W}} - \mathbf{V}\right)^T R\left(\tilde{\mathbf{W}} - \mathbf{V}\right) + \lambda\left(\mathbf{V}^T Q \mathbf{V} - \xi\right) \tag{4.1.81}$$

where $\lambda \ge 0$ is the Lagrange multiplier. As $J(\mathbf{V},\lambda)$ is a convex function of $\mathbf{V}$, the solution for any $\lambda$ is given by

$$\left.\frac{\partial J(\mathbf{V},\lambda)}{\partial \mathbf{V}}\right|_{\mathbf{V}=\hat{\mathbf{V}}(\lambda)} = 0 \tag{4.1.82}$$

Substituting from (4.1.81) it follows that

$$(R+\lambda Q)\hat{\mathbf{V}}(\lambda) = R\tilde{\mathbf{W}} \tag{4.1.83}$$

which implies

$$\hat{\mathbf{V}}^T(\lambda)R\hat{\mathbf{V}}(\lambda) + \lambda\hat{\mathbf{V}}^T(\lambda)Q\hat{\mathbf{V}}(\lambda) = \hat{\mathbf{V}}^T(\lambda)R\tilde{\mathbf{W}} \tag{4.1.84}$$

Substituting for $\mathbf{V} = \hat{\mathbf{V}}(\lambda)$ in (4.1.81) and rewriting it as

$$\begin{aligned} J\left(\hat{\mathbf{V}}(\lambda),\lambda\right) &= \tilde{\mathbf{W}}^T R\tilde{\mathbf{W}} - \tilde{\mathbf{W}}^T R\hat{\mathbf{V}}(\lambda) - \hat{\mathbf{V}}^T(\lambda)R\tilde{\mathbf{W}} \\ &\quad + \mathbf{V}^T(\lambda)R\hat{\mathbf{V}}(\lambda) + \lambda\mathbf{V}^T(\lambda)Q\hat{\mathbf{V}}(\lambda) - \lambda\xi \end{aligned} \tag{4.1.85}$$

and using (4.1.84), (4.1.85) becomes

$$J\left(\hat{\mathbf{V}}(\lambda),\lambda\right) = \tilde{\mathbf{W}}^T R\tilde{\mathbf{W}} - \tilde{\mathbf{W}}^T R\hat{\mathbf{V}}(\lambda) - \lambda\xi \tag{4.1.86}$$

It follows from (4.1.83) that

$$\hat{\mathbf{V}}(\lambda) = (R+\lambda Q)^{-1}R\tilde{\mathbf{W}} \tag{4.1.87}$$

Substituting for $\hat{\mathbf{V}}(\lambda)$ from (4.1.87) in (4.1.86),

$$\begin{aligned} \hat{J}(\lambda) &\triangleq J\left(\hat{\mathbf{V}}(\lambda),\lambda\right) \\ &= \tilde{\mathbf{W}}^T R\tilde{\mathbf{W}} - \tilde{\mathbf{W}}^T R(R+\lambda Q)^{-1}R\tilde{\mathbf{W}} - \lambda\xi \end{aligned} \tag{4.1.88}$$

It follows from the duality theorem [Lue69] that the optimum Lagrange multiplier $\hat{\lambda}$ can be obtained by maximizing $\hat{J}(\lambda)$. Thus, it follows that

$$\left.\frac{\partial \hat{J}(\lambda)}{\partial \lambda}\right|_{\lambda=\hat{\lambda}} = 0 \tag{4.1.89}$$

Substituting (4.1.88) in (4.1.89) yields

$$-\tilde{\mathbf{W}}^T R \left.\frac{\partial}{\partial \lambda}(R+\lambda Q)^{-1}\right|_{\lambda=\hat{\lambda}} R\tilde{\mathbf{W}} = \xi \tag{4.1.90}$$

To carry out the partial differentiation of $(R + \lambda Q)^{-1}$, define an invertible matrix:

$$A(\lambda) = (R + \lambda Q) \tag{4.1.91}$$

Thus,

$$A(\lambda)A^{-1}(\lambda) = I \tag{4.1.92}$$

Carrying out the partial differentiation with respect to λ results in

$$\frac{\partial A(\lambda)}{\partial \lambda} A^{-1}(\lambda) + A(\lambda)\frac{\partial A^{-1}}{\partial \lambda} = 0 \tag{4.1.93}$$

Hence,

$$\frac{\partial A^{-1}(\lambda)}{\partial \lambda} = -A^{-1}(\lambda)\frac{\partial A(\lambda)}{\partial \lambda} A^{-1}(\lambda) \tag{4.1.94}$$

Substituting for $A(\lambda)$ yields

$$\frac{\partial}{\partial \lambda}(R + \lambda Q)^{-1} = -(R + \lambda Q)^{-1} Q (R + \lambda Q)^{-1} \tag{4.1.95}$$

(4.1.90) and (4.1.95) imply that $\hat{\lambda}$ is the solution of

$$\tilde{\mathbf{W}}^T R\left(R + \hat{\lambda} Q\right)^{-1} Q \left(R + \hat{\lambda} Q\right)^{-1} R \tilde{\mathbf{W}} = \xi \tag{4.1.96}$$

(4.1.87) and (4.1.78) imply that $\hat{\mathbf{W}}_\varepsilon$, the solution of (4.1.77), is given by

$$\hat{\mathbf{W}}_\varepsilon = \tilde{\mathbf{W}} - \left(R + \hat{\lambda} Q\right)^{-1} R \tilde{\mathbf{W}} \tag{4.1.97}$$

where $\tilde{\mathbf{W}}$ satisfies (4.1.75).

See [Er85] for discussion of the processor when it has exact presteering and is designed for flat response over the entire frequency range (0,1/2T). In this case, processor performance approaches that of the TDL processor discussed in Section 4.1.3, as $\delta_0 \to 0$.

## 4.2 Partitioned Realization

The broadband processor structure shown in Figure 4.1 is sometimes referred to as an element space processor or direct form of realization compared to a beam space processor or partitioned form of realization. In the partitioned form, the processor is generally realized using two blocks as shown in Figure 4.2. The upper block forms a fixed main beam to receive the signal from the look direction and the lower block form auxiliary beams also known as secondary beams to estimate the noise (interferences and other unwanted noise) in the main beam. The lower block is designed to have no look direction

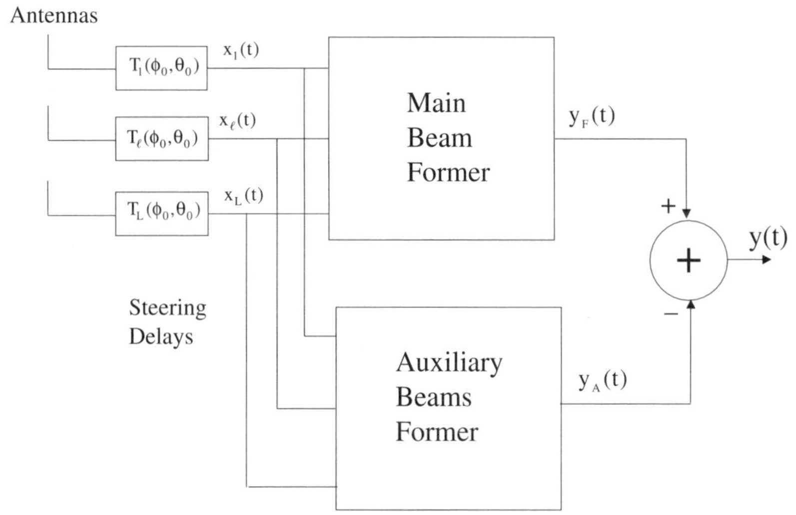


**FIGURE 4.2**
Broadband processor structure with partitioned realization.

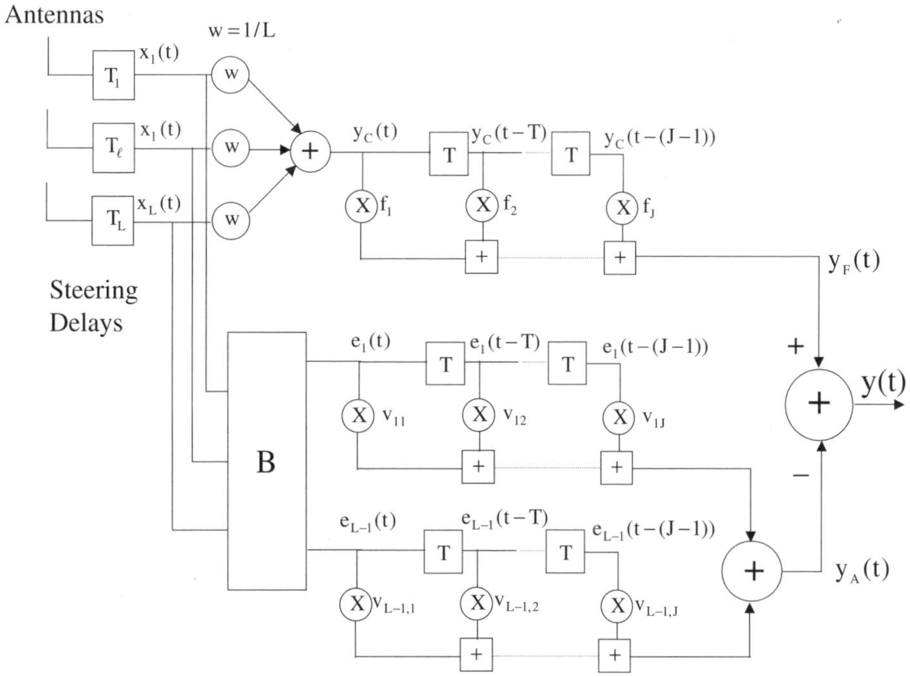


**FIGURE 4.3**
Broadband processor structure with unconstrained partitioned realization.

signal so that when its output is subtracted from the main beam it reduces the noise. The blocking of signal from the lower section may be achieved in several ways.

In one case, the array signals are processed through a signal blocking filter before processing. Signal processing in this case solves an unconstrained optimization problem. This unconstrained partitioned processor is referred to as the generalized side-lobe canceler and shown in Figure 4.3.

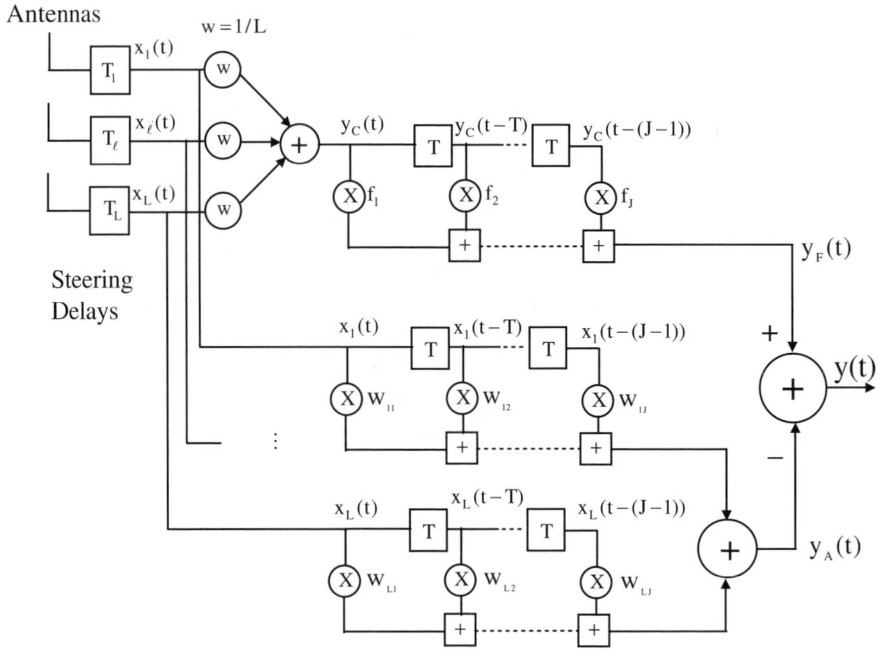


**FIGURE 4.4**
Broadband processor structure with constrained partitioned realization.

The signal in the lower section may also be blocked using constraints on its weights. In this case, the weights of the lower section are estimated by solving a constrained optimization. This form of realization is referred to as the constrained partitioned realization. Its block diagram is shown in Figure 4.4. Both forms of realization are discussed in this section.

### 4.2.1 Generalized Side-Lobe Canceler

The structure shown in Figure 4.3, also referred to as the generalized side-lobe canceler for broadband signals [Gri82], is discussed here for a point constraint, that is, the response is constrained to be unity in the look direction. Steering delays are used to align the wave form arriving from the look direction as discussed in the previous section for the element space processor. The array signals after the steering delays are passed through two sections. The upper section is designed to produce a fixed beam with a specified frequency response and the lower section consists of adjustable weights. The output of the lower section is subtracted from the output of the fixed beam to produce the processor output.

The upper section consists of a broadband conventional beam with a required frequency response obtained by selecting the coefficients $f_j$, $j = 1, \ldots, J$ of the FIR filter. Signals from all channels are equally weighted and summed to produce the output $y_C(t)$ of the conventional beam. For this realization to be equivalent to the direct form of realization, all weights need to be equal to $1/L$ and the filter coefficients $f_j$, $j = 1, \ldots, J$ need to be specified as discussed in the previous section. The output of the fixed beam is given by

$$y_F(t) = \sum_{k=0}^{J-1} f_{k+1} y_C(t - Tk) \tag{4.2.1}$$

with

$$y_C(t) = \frac{\mathbf{x}^T(t)\,\mathbf{1}}{L} \tag{4.2.2}$$

where $\mathbf{x}(t)$ denotes the array signal after presteering delays.

The fixed beam output can be expressed using the vector notation as

$$y_F(t) = \mathbf{W}_F^T \mathbf{X}(t) \tag{4.2.3}$$

where $\mathbf{X}(t)$ is an LJ-dimensional array signal vector defined by (4.1.6), $\mathbf{W}_F$ is an LJ-dimensional fixed weight given by

$$\mathbf{W}_F = C\left(C^T C\right)^{-1} \mathbf{f} \tag{4.2.4}$$

and C is the constraint matrix given by (4.1.26). Note that $\mathbf{W}_F$ is identical to $\mathbf{F}$ defined by (4.1.53).

The lower section consists of a matrix prefilter and a TDL structure. The matrix prefilter shown in the lower section is designed to block the signal arriving from the look direction. Since these signal wave forms after the steering delays are alike, the signal blocking can be achieved by selecting the matrix B such that the sum of its each row is equal to zero. For the partitioned processor to have the same degree of freedom as that of the direct form, the $L-1$ rows of the matrix B need to be linearly independent. The output $\mathbf{e}(t)$ after the matrix prefilter is an $L-1$ dimensional vector given by

$$\mathbf{e}(t) = B\mathbf{x}(t) \tag{4.2.5}$$

and can be thought of as outputs of $L-1$ beams that are then shaped by the coefficients of the FIR filter of each TDL section. Let an $L-1$ dimensional vector $\mathbf{v}_k$ denote these coefficients before the kth delay. The J vectors $\mathbf{v}_1, \mathbf{v}_2, \ldots, \mathbf{v}_J$ correspond to the J columns of weights in the tapped delay line filter in the lower section. The lower filter output is then given by

$$y_A(t) = \sum_{k=0}^{J-1} \mathbf{v}_{k+1}^T\, \mathbf{e}(t - kT) \tag{4.2.6}$$

The output may be expressed in the vector notation as

$$y_A(t) = \mathbf{V}^T \mathbf{E}(t) \tag{4.2.7}$$

where $(L-1)J$ dimensional vector $\mathbf{V}$ denotes the weights of the lower section defined as

$$\mathbf{V}^T = \left[\mathbf{v}_1^T, \mathbf{v}_2^T, \ldots, \mathbf{v}_1^T\right] \tag{4.2.8}$$

and $(L-1)J$ dimensional vector $\mathbf{E}(t)$ denotes the array signals in the lower section defined as

$$\mathbf{E}(t)^T = \left[\mathbf{e}^T(t), \mathbf{e}^T(t-T), \ldots, \mathbf{e}^T(t-(J-1)T)\right] \tag{4.2.9}$$

It follows from (4.2.3) and (4.2.7) that the array output is then given by

$$\begin{aligned} y(t) &= y_F(t) - y_A(t) \\ &= \mathbf{W}_F^T \mathbf{X}(t) - \mathbf{V}^T \mathbf{E}(t) \end{aligned} \tag{4.2.10}$$

For a given weight **V**, the mean output power of the processor is given by

$$\begin{aligned} P(\mathbf{V}) &= E\left[y^2(t)\right] \\ &= E\left[\left\{\mathbf{W}_F^T \mathbf{X}(t) - \mathbf{V}^T \mathbf{E}(t)\right\}^2\right] \\ &= \mathbf{W}_F^T R \mathbf{W}_F - \mathbf{W}_F^T R_{XE} \mathbf{V} - \mathbf{V}^T R_{XE}^T \mathbf{W}_F + \mathbf{V}^T R_{EE} \mathbf{V} \end{aligned} \tag{4.2.11}$$

where

$$R_{XE} = E\left[\mathbf{X}(t)\mathbf{E}^T(t)\right] \tag{4.2.12}$$

and

$$R_{EE} = E\left[\mathbf{E}(t)\mathbf{E}^T(t)\right] \tag{4.2.13}$$

As the array signal vectors **E**(t) and **X**(t) are related through matrix B, both matrices $R_{XE}$ and $R_{EE}$ could be rewritten in terms of R and B.

Since the response of the processor in the look direction is fixed due to the fixed beam, and the lower section contains no signal from the look direction due to the presence of the matrix prefilter, nonlook direction noise may be minimized by adjusting weights of the lower section to minimize the mean output power. Thus, the optimal weights denoted by $\hat{\mathbf{V}}$ are the solution of the following unconstrained beamforming problem:

$$\underset{\mathbf{V}}{\text{minimize}} \quad P(\mathbf{V}) \tag{4.2.14}$$

Since the mean output power surface P(**V**) is a quadratic function of **V**, the solution of the above problem can be obtained by taking the gradient of the of P(**V**) with respect to **V** and setting it equal to zero. Thus,

$$\nabla_V P(\mathbf{V})\big|_{\mathbf{V}=\hat{\mathbf{V}}} = 0 \tag{4.2.15}$$

Substituting for P(**V**) from (4.2.11),

$$R_{EE}\hat{\mathbf{V}} = R_{XE}^T \mathbf{W}_F \tag{4.2.16}$$

When the array correlation matrix R is invertible, the matrix $R_{EE}$ is invertible and (4.2.16) yields

$$\hat{\mathbf{V}} = R_{EE}^{-1} R_{XE}^T \mathbf{W}_F \tag{4.2.17}$$

It can be shown [Gri82] that when the weights in the array processors in Figure 4.1 and Figure 4.2 are optimized, the performance of the two processors is identical. The weights $\hat{\mathbf{V}}$ may be expressed using array correlation matrix as follows.

Let $\tilde{B}$ be a matrix defined as

$$\tilde{B} = \begin{bmatrix} B & & 0 \\ & \ddots & \\ 0 & & B \end{bmatrix} \tag{4.2.18}$$

It follows from (4.2.4.), (4.2.9) and (4.2.18) that

$$\begin{aligned} \mathbf{E}(t) &= \begin{bmatrix} e(t) \\ e(t+T) \\ \vdots \\ e(t-(J-1)T) \end{bmatrix} \\ &= \begin{bmatrix} B\mathbf{x}(t) \\ B\mathbf{x}(t-T) \\ \vdots \\ B\mathbf{x}(t-(J-1)T) \end{bmatrix} \\ &= \tilde{B}\mathbf{X}(t) \end{aligned} \tag{4.2.19}$$

Substituting in (4.2.12) and (4.2.13) yields

$$\begin{aligned} R_{\mathbf{XE}} &= E\left[\mathbf{X}(t)\mathbf{X}^T(t)\right]\tilde{B}^T \\ &= R\tilde{B}^T \end{aligned} \tag{4.2.20}$$

and

$$R_{\mathbf{EE}} = \tilde{B}R\tilde{B}^T \tag{4.2.21}$$

It follows from (4.2.17), (4.2.20), and (4.2.21) that

$$\hat{\mathbf{V}} = \left(\tilde{B}R\tilde{B}^T\right)^{-1}\tilde{B}R\mathbf{W}_F \tag{4.2.22}$$

Substituting in (4.2.10) from (4.2.19) and (4.2.22), the output of the processor with optimized weights becomes

$$\begin{aligned} y(t) &= \mathbf{W}_F^T\mathbf{X}(t) - \mathbf{W}_F^T R\tilde{B}^T\left(\tilde{B}R\tilde{B}^T\right)^{-1}\tilde{B}\mathbf{X}(t) \\ &= \mathbf{W}_F^T\left[I - R\tilde{B}^T\left(\tilde{B}R\tilde{B}^T\right)^{-1}\tilde{B}\right]\mathbf{X}(t) \end{aligned} \tag{4.2.23}$$

### 4.2.2 Constrained Partitioned Realization

Figure 4.4 shows a structure of the constrained partitioned realized processor [Jim77]. The main difference between the constrained processor and unconstrained processor (also referred to as the generalized side-lobe canceler in the pervious section) is that the latter uses a signal blocking matrix to stop the signal from entering the lower section and solves an unconstrained beamforming problem, whereas the constrained processor uses constraints on the weights of the lower section to eliminate the signal at the output of the lower section. Consequently, the optimization problem solved to estimate the weights of the lower section is a constrained one.

Let the LJ-dimensional vector $\mathbf{W}_F$ given by Figure 4.4 denote the weights of the fixed beam (upper section). Thus, the output of the upper section $y_F(t)$ is given by

$$y_F(t) = \mathbf{W}_F^T \mathbf{X}(t) \tag{4.2.24}$$

Let the LJ-dimensional vector $\mathbf{W}$ denote the weights of the lower section. Thus, the output of the lower section $y_A(t)$ is given by

$$y_A(t) = \mathbf{W}^T \mathbf{X}(t) \tag{4.2.25}$$

The processor output y(t) is the difference of the two outputs. Thus,

$$\begin{aligned} y(t) &= \mathbf{W}_F^T \mathbf{X}(t) - \mathbf{W}^T \mathbf{X}(t) \\ &= \left(\mathbf{W}_F - \mathbf{W}\right)^T \mathbf{X}(t) \end{aligned} \tag{4.2.26}$$

The mean output power $P(\mathbf{W})$ for given weights is given by

$$P(\mathbf{W}) = \left(\mathbf{W}_F - W\right)^T R\left(\mathbf{W}_F - \mathbf{W}\right) \tag{4.2.27}$$

The lower section is designed such that its output does not contain the look direction signal. This is achieved by selecting its weights to be the solution of the following beamforming problem:

$$\begin{aligned} &\underset{\mathbf{W}}{\text{minimize}} \quad \left(\mathbf{W}_F - \mathbf{W}\right)^T R\left(\mathbf{W}_F - \mathbf{W}\right) \\ &\text{subject to} \quad C^T \mathbf{W} = 0 \end{aligned} \tag{4.2.28}$$

It follows from (4.1.26) and the second equation of (4.2.28) that

$$\mathbf{1}^T \mathbf{w}_i = 0, \quad j = 1, 2, \ldots, J \tag{4.2.29}$$

where $\mathbf{w}_j$ denotes the weights of the jth column, that is, before the jth delay in the lower section.

Since the look direction signal wave forms on all elements after presteering delays are alike, the constraints of (4.2.29) ensure that the lower section has a null response in the look direction. Thus, the constraint in (4.2.28) achieves a null in the look direction similar to that achieved by the matrix prefilter B discussed in the previous section.

Let $\hat{\mathbf{W}}_0$ denote the solution of (4.2.28). Using the method of Lagrange multipliers discussed in Section 4.1,

$$\begin{aligned}\hat{\mathbf{W}}_0 &= \mathbf{W}_F - R^{-1}C\left(C^T R^{-1} C\right)^{-1} C^T \mathbf{W}_F \\ &= \mathbf{W}_F - \hat{\mathbf{W}}\end{aligned} \tag{4.2.30}$$

where $\hat{\mathbf{W}}$ is given by (4.1.36).

### 4.2.3 General Constrained Partitioned Realization

In this section, a processor realization in general constrained form is presented where the upper section is designed to minimize the MSE between the look direction desired response and the frequency response of the processor in the look direction over a frequency band of interest $[f_L, f_H]$, as discussed in Section 4.1.5. The lower section is designed such that its weights are constrained to yield a zero power response over the frequency band of interest to prevent signal suppression. Design details may be found in [Er86].

Let an LJ-dimensional vector $\tilde{\mathbf{W}}$ denote the weight of the upper section. These are designed using minimum MSE design and satisfy (4.1.75). The output of the upper section $y_F(t)$ is given by

$$y_F(t) = \tilde{\mathbf{W}}^T \mathbf{X}(t) \tag{4.2.31}$$

Let an LJ-dimensional vector $\mathbf{W}$ denote the weights of the lower section. Thus, the output of the lower section $y_A(t)$ is given by

$$y_A(t) = \mathbf{W}^T \mathbf{X}(t) \tag{4.2.32}$$

and the processor output y(t) is given by

$$y(t) = \left(\tilde{\mathbf{W}} - \mathbf{W}\right)^T \mathbf{X}(t) \tag{4.2.33}$$

The mean output power $P(\mathbf{W})$ for a given $\mathbf{W}$ is given by

$$P(\mathbf{W}) = \left(\tilde{\mathbf{W}} - \mathbf{W}\right)^T R\left(\tilde{\mathbf{W}} - \mathbf{W}\right) \tag{4.2.34}$$

#### *4.2.3.1 Derivation of Constraints*

Let weight vector $\mathbf{W}$ be constrained such that the power response of the lower section in the look direction is zero over the frequency band of interest, that is,

$$\int_{f_L}^{f_H} H^*\left(f, \phi_0, \theta_0\right) H\left(f, \phi_0, \theta_0\right) df = 0 \tag{4.2.35}$$

It follows from (4.2.35) and (4.1.67) that

$$\mathbf{W}^T Q \mathbf{W} = 0 \tag{4.2.36}$$

As Q is a positive semidefinite matrix, it can be factorized using its eigenvalues and eigenvectors as

$$Q = U \Lambda U^T \tag{4.2.37}$$

where $\Lambda$ is a diagonal matrix with its elements being $\lambda_i(Q)$, i = 1, 2, …, LJ, the eigenvalues of Q, such that

$$\lambda_1(Q) \geq \lambda_2(Q) \geq \cdots \geq \lambda_{LJ}(Q) \geq 0 \tag{4.2.38}$$

and U is an LJ × LJ matrix of the eigenvector of Q, that is,

$$U = \left[\mathbf{U}_1, \mathbf{U}_2, \ldots, \mathbf{U}_{LJ}\right] \tag{4.2.39}$$

where $\mathbf{U}_i$, i = 1, 2, …, LJ are the orthonormal eigenvectors of Q with the property that

$$\mathbf{U}_i^T \mathbf{U}_j = \begin{cases} 0 & i \neq j \\ 1 & i = j \end{cases} \tag{4.2.40}$$

Substituting (4.2.37) in (4.2.36)

$$\mathbf{W}^T U \Lambda U^T \mathbf{W} = 0 \tag{4.2.41}$$

Assume that Q has rank $\eta_0$. Thus, it follows from (4.2.39) and (4.2.41) that the necessary and sufficient conditions to satisfy (4.2.41) are

$$\mathbf{W}^T \mathbf{U}_i = 0, \quad i = 1, 2, \ldots, \eta_0 \tag{4.2.42}$$

Thus, the linear constraints of the form (4.2.42) can be used to ensure that the lower section has a zero power response in the look direction over the frequency range of interest. It should be noted that signal blocking in the lower section using these constraints is independent of presteering delays, that is, the processor may include exact presteering, coarse presteering, or no presteering.

#### 4.2.3.2 Optimization

Let the optimum weight vector $\hat{\mathbf{W}}$ be the solution of the following constrained beamforming problem

$$\begin{aligned} &\underset{\mathbf{W}}{\text{minimize}} \quad \left(\tilde{\mathbf{W}} - \mathbf{W}\right)^T R \left(\tilde{\mathbf{W}} - \mathbf{W}\right) \\ &\text{subject to} \qquad U_{\eta_0}^T \mathbf{W} = 0 \end{aligned} \tag{4.2.43}$$

where $U_{\eta 0}$ is the LJ × $\eta_0$ dimensional matrix given by

$$U_{\eta_0} = \left[\mathbf{U}_1, \mathbf{U}_2, \ldots, \mathbf{U}_{\eta_0}\right] \tag{4.2.44}$$

As $\mathbf{U}_i$, i = 1, 2, ..., $\eta_0$ are linearly independent, the matrix $U_{\eta 0}$ has full rank. Using the method of Lagrange multipliers,

$$\hat{\mathbf{W}} = \tilde{\mathbf{W}} - R^{-1}U_{\eta_0}\left(U_{\eta_0}^T R^{-1} U_{\eta_0}\right)^{-1} U_{\eta_0}^T \tilde{\mathbf{W}} \tag{4.2.45}$$

## 4.3 Derivative Constrained Processor

The implication of the point constraint considered in Section 4.1 is that the array pattern has a unity response in the look direction. It can be broadened using additional constraints, such as derivative constraints, along with the point constraint [Er83, Er90, Er86a, Thn93]. The derivative constraints set the derivatives of the power pattern with respect to $\phi$ and $\theta$ equal to zero in the look direction. The higher the order of derivatives, that is, first order, second order, and so on, the broader the beam in the look direction normally becomes. A broader beam is useful when the actual signal direction and known direction of the signal are not precisely the same. In such situations, the processor with the point constraint in the known direction of the signal would cancel the desired signal as if it were interference. Other directional constraints to improve the performance of the beamformer in the presence of the look direction error include multiple linear constraints [Tak85, Buc87, Gri87] and inequality constraints [Ahm83, Ahm84, Er90a, Er93].

In this section, some of these constraints are derived, a beamforming problem using these constraints is formulated, an algorithm to estimate solution of the optimization problem is presented, and the effect that choice of coordinate system origin has on the performance of an array system using derivative constraints is discussed.

Derivative constraints are derived by setting derivatives of the power response $\rho(f,\phi,\theta)$ with respect to $\phi$ and $\theta$ to zero in direction $(\phi_0,\theta_0)$. Since $H(f,\phi,\theta)$ denotes the frequency response of the processor, it follows that

$$\rho(f,\phi,\theta) = H^*(f,\phi,\theta)H(f,\phi,\theta) \tag{4.3.1}$$

The first-order derivative constraints are now derived [Er83].

### 4.3.1 First-Order Derivative Constraints

It follows from (4.3.1) that the partial derivative of the power response with respect to $\phi$ is given by

$$\frac{\partial \rho}{\partial \phi} = H^* \frac{\partial H}{\partial \phi} + \frac{\partial H^*}{\partial \phi} H \tag{4.3.2}$$

where the parameters of $\rho(f,\phi,\theta)$ and $H(f,\phi,\theta)$ are omitted for ease of notation. It follows from (4.1.18) that

$$\frac{\partial H}{\partial \phi} = \frac{\partial \mathbf{S}^T(f,\phi,\theta)}{\partial \phi} T(f) \sum_{\ell=1}^{J} \mathbf{w}_\ell e^{-j2\pi f(\ell-1)T} \tag{4.3.3}$$

Differentiating (4.1.20) with respect to $\phi$,

$$\frac{\partial \mathbf{S}^T(f,\phi,\theta)}{\partial\phi} = j2\pi f\left[\frac{\partial\tau_1(\phi,\theta)}{\partial\phi}e^{2\pi f\tau_1(\phi,\theta)},\cdots\frac{\partial\tau_L(\phi,\theta)}{\partial\phi}e^{j2\pi f\tau_L(\phi,\theta)}\right] \tag{4.3.4}$$

$$= j2\pi f\ \mathbf{S}^T(f,\phi,\theta)\Lambda_\phi(\phi,\theta)$$

where

$$\Lambda_\phi(\phi,\theta) = \begin{bmatrix} \frac{\partial\tau_1(\phi,\theta)}{\partial\phi} & & 0 \\ & \ddots & \\ 0 & & \frac{\partial\tau_L(\phi,\theta)}{\partial\phi} \end{bmatrix} \tag{4.3.5}$$

and $\tau_\ell(\phi,\theta)$ is given by (2.1.1). It can also be expressed as

$$\tau_\ell(\phi,\theta) = \frac{1}{c}\left\{(x_\ell\cos\phi + y_\ell\sin\phi)\sin\theta + z_\ell\cos\theta\right\} \tag{4.3.6}$$

where $x_\ell$, $y_\ell$, and $z_\ell$ denote the components of the $\ell$th element along the x, y, and z axis, respectively, and c denotes the speed of propagation.

Substituting (4.3.4) in (4.3.3) and noting that T(f) and $\Lambda_\phi(\phi,\theta)$ are diagonal matrices,

$$\frac{\partial H}{\partial\phi} = j2\pi f\mathbf{S}^T(f,\phi,\theta)T(f)\Lambda_\phi(\phi,\theta)\sum_{\ell=1}^{J}\mathbf{w}_\ell e^{-j2\pi f(\ell-1)T} \tag{4.3.7}$$

which implies

$$\left.\frac{\partial H}{\partial\phi}\right|_{(\phi_0,\theta_0)} = j2\pi f\mathbf{S}^T(f,\phi_0,\theta_0)T(f)\Lambda_\phi(\phi_0,\theta_0)\sum_{\ell=1}^{J}\mathbf{w}_\ell e^{-j2\pi f(\ell-1)T} \tag{4.3.8}$$

Noting from (4.1.21) that

$$S^T(f,\phi_0,\theta_0)T(f) = a(f)\mathbf{1}^T \tag{4.3.9}$$

(4.3.7) yields

$$\left.\frac{\partial H}{\partial\phi}\right|_{(\phi_0,\theta_0)} = j2\pi f a(f)\sum_{\ell=1}^{J}\mathbf{1}^T\Lambda_\phi(\phi_0,\theta_0)\mathbf{w}_\ell e^{-j2\pi f(\ell-1)T} \tag{4.3.10}$$

It follows from (4.1.23) that

$$H^*(f_1,\phi_0,\theta_0) = a^*(f)\sum_{k=1}^{J} f_k e^{j2\pi f(k-1)T} \tag{4.3.11}$$

Thus,

$$H^*\left.\frac{\partial H}{\partial \phi}\right|_{(\phi_0,\theta_0)} = j2\pi f a(f) a^*(f) \sum_{\ell=1}^{J}\sum_{k=1}^{J} f_k \mathbf{1}^T \Lambda_\phi(\phi_0,\theta_0)\mathbf{w}_\ell e^{-j2\pi f(\ell-k)T} \tag{4.3.12}$$

Noting from (4.1.22) that a(f)a*(f) = 1 and using this in (4.3.12),

$$H^*\left.\frac{\partial H}{\partial \phi}\right|_{(\phi_0,\theta_0)} = j2\pi f \sum_{\ell=1}^{J}\sum_{k=1}^{J} f_k \mathbf{1}^T \Lambda_\phi(\phi_0,\theta_0)\mathbf{w}_\ell e^{-j2\pi f(\ell-k)T} \tag{4.3.13}$$

Similarly,

$$\left.\frac{\partial H^*}{\partial \phi} H\right|_{(\phi_0,\theta_0)} = -j2\pi f \sum_{\ell=1}^{J}\sum_{k=1}^{J} f_k \mathbf{1}^T \Lambda_\phi(\phi_0,\theta_0)\mathbf{w}_\ell e^{j2\pi f(\ell-k)T} \tag{4.3.14}$$

Substituting in (4.3.2),

$$\begin{aligned}\left.\frac{\partial \rho}{\partial \phi}\right|_{(\phi_0,\theta_0)} &= 2\pi f \sum_{\ell=1}^{J}\sum_{k=1}^{J} f_k \mathbf{1}^T \Lambda_\phi(\phi_0,\theta_0)\mathbf{w}_\ell \left[je^{-j2\pi f(\ell-k)T} - je^{j2\pi f(\ell-k)T}\right] \\ &= 4\pi f \sum_{\ell=1}^{J}\sum_{k=1}^{J} f_k \mathbf{1}^T \Lambda_\phi(\phi_0,\theta_0)\mathbf{w}_\ell \sin 2\pi f(\ell-k)T\end{aligned} \tag{4.3.15}$$

Similarly,

$$\left.\frac{\partial \rho}{\partial \theta}\right|_{(\phi_0,\theta_0)} = 4\pi f \sum_{\ell=1}^{J}\sum_{k=1}^{J} f_k \mathbf{1}^T \Lambda_\theta(\phi_0,\theta_0)\mathbf{w}_\ell \sin 2\pi f(\ell-k)T \tag{4.3.16}$$

where

$$\Lambda_\theta(\phi,\theta) = \begin{bmatrix} \frac{\partial \tau_1(\phi,\theta)}{\partial \theta} & & 0 \\ & \ddots & \\ 0 & & \frac{\partial \tau_L(\phi,\theta)}{\partial \theta} \end{bmatrix} \tag{4.3.17}$$

It follows from (4.3.15) and (4.3.16), respectively, that sufficient conditions for $\left.\frac{\partial \rho}{\partial \phi}\right|_{(\phi_0,\theta_0)} = 0$ for all f > 0 are

$$\mathbf{1}^T \Lambda_\phi(\phi_0,\theta_0)\mathbf{w}_\ell = 0, \quad \ell = 1, 2, \ldots, J \tag{4.3.18}$$

and sufficient condition for $\left.\frac{\partial \rho}{\partial \theta}\right|_{(\phi_0,\theta_0)} = 0$ for all $f > 0$ are

$$\mathbf{1}^T \Lambda_\theta(\phi_0,\theta_0)\mathbf{w}_\ell = 0, \quad \ell = 1, 2, \ldots, J \tag{4.3.19}$$

Equations (4.3.18) and (4.1.19) denote 2J linear constraints on the weights of the broadband processor. These constraints are sufficient for the first-order derivatives of the power response with respect to $\phi$ and $\theta$ evaluated at $(\phi,\theta)$ to be zero. These are referred to as the first-order derivative constraints and are imposed along with the point constraint discussed previously.

Using a similar approach to the derivation of the first-order constraints presented in this section, higher-order derivative constraints may be derived by setting the higher-order derivatives of the power response with respect to $\phi$ and $\theta$ evaluated at $(\phi_0,\theta_0)$ to zero.

### 4.3.2 Second-Order Derivative Constraints

The equations describing the second-order derivative constraints follow [Er83]:

$$\mathbf{1}^T \left.\frac{\partial \Lambda_\phi(\phi,\theta)}{\partial \phi}\right|_{(\phi_0,\theta_0)} \mathbf{w}_\ell = 0, \quad \ell = 1, 2, \ldots, J \tag{4.3.20}$$

$$\mathbf{1}^T \Lambda_\phi^2(\phi_0,\theta_0)\mathbf{w}_\ell = 0, \quad \ell = 1, 2, \ldots, J \tag{4.3.21}$$

$$\mathbf{1}^T \left.\frac{\partial \Lambda_\theta(\phi,\theta)}{\partial \theta}\right|_{(\phi_0,\theta_0)} \mathbf{w}_\ell = 0, \quad \ell = 1, 2, \ldots, J \tag{4.3.22}$$

$$\mathbf{1}^T \Lambda_\theta^2(\phi_0,\theta_0)\mathbf{w}_\ell = 0, \quad \ell = 1, 2, \ldots, J \tag{4.3.23}$$

$$\mathbf{1}^T \left.\frac{\partial \Lambda_\phi(\phi,\theta)}{\partial \theta}\right|_{(\phi_0,\theta_0)} \mathbf{w}_\ell = 0, \quad \ell = 1, 2, \ldots, J \tag{4.3.24}$$

and

$$\mathbf{1}^T \Lambda_\phi(\phi_0,\theta_0)\Lambda_\theta(\phi_0,\theta_0)\mathbf{w}_\ell = 0, \quad \ell = 1, 2, \ldots, J \tag{4.3.25}$$

These equations denote 6J linear constraints that are sufficient for the second-order derivatives with respect to $\phi$ and $\theta$ evaluated at $(\phi_0,\theta_0)$ to be zero. These are imposed along with the point constraint and first-order derivative constraints.

It should be noted that these constraints depend on array geometry and are not necessarily linearly independent. In the next section, a beamforming problem with derivative constraints is considered.

### 4.3.3 Optimization with Derivative Constraints

A beamforming problem using derivative constraints may be formulated similar to the constrained beamforming problem considered previously by adding derivative constraints specified by (4.3.18) to (4.3.25) to the point constraint given by the second equation of (4.1.29).

In this case (4.1.29) becomes

$$\begin{aligned} &\underset{\mathbf{W}}{\text{minimize}} \quad \mathbf{W}^T R \mathbf{W} \\ &\text{subject to} \quad D^T \mathbf{W} = \mathbf{g} \end{aligned} \tag{4.3.26}$$

where

$$\mathbf{g}^T = \left[\mathbf{f}^T, \mathbf{0}^T, \ldots, \mathbf{0}^T\right] \tag{4.3.27}$$

and

$$D = \left[C_0 \vdots C_1 \vdots \cdots \vdots C_8\right] \tag{4.3.28}$$

with LJ × J matrices $C_0$ to $C_8$ given by

$$C_0 = \text{diag}[\mathbf{1}] \tag{4.3.29}$$

$$C_1 = \text{diag}\left[\mathbf{1}^T \Lambda_\phi(\phi_0, \theta_0)\right] \tag{4.3.30}$$

$$C_2 = \text{diag}\left[\mathbf{1}^T \Lambda_\theta(\phi_0, \theta_0)\right] \tag{4.3.31}$$

$$C_3 = \text{diag}\left[\mathbf{1}^T \left.\frac{\partial \Lambda_\phi(\phi, \theta)}{\partial \phi}\right|_{(\phi_0, \theta_0)}\right] \tag{4.3.32}$$

$$C_4 = \text{diag}\left[\mathbf{1}^T \Lambda_\phi^2(\phi_0, \theta_0)\right] \tag{4.3.33}$$

$$C_5 = \text{diag}\left[1^T \left.\frac{\partial \Lambda_\theta(\phi, \theta)}{\partial \theta}\right|_{(\phi_0, \theta_0)}\right] \tag{4.3.34}$$

$$C_6 = \text{diag}\left[\mathbf{1}^T \Lambda_\theta^2(\phi_0, \theta_0)\right] \tag{4.3.35}$$

$$C_7 = \text{diag}\left[\mathbf{1}^T \left.\frac{\partial \Lambda_\phi(\phi, \theta)}{\partial \theta}\right|_{(\phi_0, \theta_0)}\right] \tag{4.3.36}$$

and

$$C_8 = \text{diag}\left[\mathbf{1}^T \Lambda_\phi(\phi_0, \theta_0) \Lambda_\theta(\phi_0, \theta_0)\right] \tag{4.3.37}$$

The notation diag[x] in (4.3.29) to (4.3.37) is defined as

$$\mathrm{diag}[\mathbf{x}] = \begin{bmatrix} \mathbf{x} & & 0 \\ & \ddots & \\ 0 & & \mathbf{x} \end{bmatrix} \tag{4.3.38}$$

For example, in (4.3.29)

$$\mathbf{x} = \mathbf{1} \tag{4.3.39}$$

and $C_0$ is given by

$$C_0 = \begin{bmatrix} \mathbf{1} & & 0 \\ & \ddots & \\ 0 & & \mathbf{1} \end{bmatrix} \tag{4.3.40}$$

It can easily be verified from (4.3.26) to (4.3.37) that

$$C_0^T \mathbf{W} = \mathbf{f} \tag{4.3.41}$$

and

$$C_i^T \mathbf{W} = \mathbf{0}, \quad i = 1, \ldots, 8 \tag{4.3.42}$$

Equation (4.3.41) is the second equation of (4.1.29) and defines the point constraint, whereas (4.3.42) defines derivative constraints given by (4.3.18) to (4.3.25).

The optimization problem (4.3.26) is similar in form to (4.1.29). Thus, it follows from (4.1.36) that if D is of full rank, then the optimal weight $\hat{\mathbf{W}}$, the solution of (4.3.26), is given by

$$\hat{\mathbf{W}} = R^{-1} D \left( D^T R^{-1} D \right)^{-1} \mathbf{g} \tag{4.3.43}$$

The rank of D is dependent on array geometry. This is explained in the following example using a linear array [Er83].

#### 4.3.3.1 Linear Array Example

Consider a linear array along the x-axis with $x_\ell$ denoting the position of the $\ell$th element. Assume that the directional sources are in the x-y plane with the look direction making an angle $\phi_0$ with the array. In view of these assumptions, it follows that

$$\theta = 90°, \quad y_\ell = 0, \quad z_\ell = 0, \quad \ell = 1, 2, \ldots, L \tag{4.3.44}$$

These equations along with (4.3.6) imply that

$$\tau_\ell(\phi) = \frac{x_\ell \cos\phi}{c} \tag{4.3.45}$$

$$\frac{\partial \tau_\ell(\phi)}{\partial \phi} = \frac{-x_\ell \sin\phi}{c} \tag{4.3.46}$$

and

$$\frac{\partial^2 \tau_\ell(\phi)}{\partial^2 \phi} = \frac{-x_\ell \cos\phi}{c} \tag{4.3.47}$$

Now consider the constrained Equation (4.3.18) to Equation (4.3.25). Using (4.3.44) and the fact the time delay $\tau_\ell(\phi)$ is not a function of $\theta$, one notes that the constraint equations (4.3.19), (4.3.22), (4.3.23), (4.3.24), and (4.3.25) vanish. The only constraints remaining are those given by (4.3.18), (4.3.20), and (4.3.21), that is,

$$\mathbf{1}^T \Lambda_\phi(\phi_0)\mathbf{w}_\ell = 0, \quad \ell = 1, 2, \ldots, J \tag{4.3.48}$$

$$\mathbf{1}^T \frac{\partial \Lambda_\phi(\phi_0)}{\partial \phi}\bigg|_{(\phi_0)} \mathbf{w}_\ell = 0, \quad \ell = 1, 2, \ldots, J \tag{4.3.49}$$

and

$$\mathbf{1}^T \Lambda_\phi^2(\phi_0)\mathbf{w}_\ell = 0, \quad \ell = 1, 2, \ldots, J \tag{4.3.50}$$

where $\Lambda_\phi(\phi)$, $\partial\Lambda_\phi(\phi)/\partial\phi$ and $\Lambda_\phi^2(\phi)$ are diagonal matrices given by

$$\Lambda_\phi(\phi) = \begin{bmatrix} \frac{\partial \tau_1(\phi)}{\partial \phi} & & 0 \\ & \ddots & \\ 0 & & \frac{\partial \tau_L(\phi)}{\partial \phi} \end{bmatrix} \tag{4.3.51}$$

$$\frac{\partial \Lambda_\phi(\phi)}{\partial \phi} = \begin{bmatrix} \frac{\partial^2 \tau_1(\phi)}{\partial^2 \phi} & & 0 \\ & \ddots & \\ 0 & & \frac{\partial^2 \tau_L(\phi)}{\partial \phi} \end{bmatrix} \tag{4.3.52}$$

and

$$\Lambda_\phi^2(\phi) = \begin{bmatrix} \left[\frac{\partial \tau_1(\phi)}{\partial \phi}\right]^2 & & 0 \\ & \ddots & \\ 0 & & \left[\frac{\partial \tau_L(\phi)}{\partial \phi}\right]^2 \end{bmatrix} \tag{4.3.53}$$

To simplify the notation, define three L vectors $\boldsymbol{\lambda}(\phi)$, $\boldsymbol{\sigma}(\phi)$, and $\boldsymbol{\psi}(\phi)$ as

$$\boldsymbol{\lambda}(\phi) = \mathbf{1}^T \Lambda_\phi(\phi) \tag{4.3.54}$$

$$\boldsymbol{\sigma}(\phi) = \mathbf{1}^T \frac{\partial \Lambda_\phi}{\partial \phi}(\phi) \tag{4.3.55}$$

and

$$\boldsymbol{\psi}(\phi) = \mathbf{1}^T \Lambda_\phi^2(\phi) \tag{4.3.56}$$

Using (4.3.45) to (4.3.47) and (4.3.51) to (4.3.53), these become

$$\boldsymbol{\lambda}_{\phi'}(\phi_0) = -\frac{\sin\phi_0}{c}\begin{bmatrix} x_1 \\ \vdots \\ x_L \end{bmatrix} \tag{4.3.57}$$

$$\boldsymbol{\sigma}_{\phi'}(\phi_0) = -\frac{\cos\phi_0}{c}\begin{bmatrix} x_1 \\ \vdots \\ x_L \end{bmatrix} \tag{4.3.58}$$

and

$$\boldsymbol{\psi}_{\phi'}(\phi_0) = -\frac{\sin^2\phi_0}{c^2}\begin{bmatrix} x_1^2 \\ \vdots \\ x_L^2 \end{bmatrix} \tag{4.3.59}$$

The three constraint equations (4.3.48) to (4.3.50) are then given by

$$\boldsymbol{\lambda}_\phi^T(\phi_0)\mathbf{w}_\ell = 0, \quad \ell = 1, 2, \ldots, J \tag{4.3.60}$$

$$\boldsymbol{\sigma}_\phi^T(\phi_0)\mathbf{w}_\ell = 0, \quad \ell = 1, 2, \ldots, J \tag{4.3.61}$$

and

$$\boldsymbol{\psi}_\phi^T(\phi_0)\mathbf{w}_\ell = 0, \quad \ell = 1, 2, \ldots, J \tag{4.3.62}$$

Note that (4.3.60) denotes J first-order constraints equations, and (4.3.61) and (4.3.62) denote 2J second-order constraints. For a general array, there are 2J linear constraints and 6J derivative constraints as discussed previously. Thus, the constraints for a linear array are much less than those for a general array. It should be noted that these constraints are functions of the look direction.

For look direction in broadside to the array $\phi_0 = 90°$, it follows from (4.3.58) that $\boldsymbol{\sigma}_\phi(\phi_0) = \mathbf{0}$ and (4.3.61) vanish, reducing the constraints from 3J to 2J for a linear array. Similarly, for an endfire array where the look direction is parallel to the array, $\phi_0 = 0°$ or $\phi_0 = 180°$, (4.3.57) and (4.3.59), imply that both (4.3.60) and (4.3.62) vanish. Thus, for a linear array, only J second-order constraints given by (4.3.61) remain; first-order constraints have vanished.

When a beamforming problem is considered using derivative constraints, the constraints equations specifying only linearly independent constraints need to be considered. It follows from (4.3.57) and (4.3.58) that vectors $\boldsymbol{\lambda}_\phi(\phi_0)$ and $\boldsymbol{\sigma}_\phi(\phi_0)$ are not linearly independent; thus constraints (4.3.60) and (4.3.61) are not independent. Hence, only 2J constraints given by (4.3.60) and (4.3.62) need to be used in the optimization process.

For this case beamforming problem given by (4.3.26) to (4.3.37) reduce to

$$\begin{aligned} &\underset{\mathbf{W}}{\text{minimize}} \quad \mathbf{W}^T R \mathbf{W} \\ &\text{subject to} \quad D^T \mathbf{W} = \mathbf{g} \end{aligned} \tag{4.3.63}$$

where

$$\mathbf{g}^T = \left[\mathbf{f}^T, \mathbf{0}^T, \mathbf{0}^T\right] \tag{4.3.64}$$

and

$$D = \left[C_0 \vdots C_1 \vdots C_2\right] \tag{4.3.65}$$

with

$$C_0 = \begin{bmatrix} \mathbf{1} & & 0 \\ & \ddots & \\ 0 & & \mathbf{1} \end{bmatrix} \tag{4.3.66}$$

$$C_1 = \begin{bmatrix} \boldsymbol{\lambda}_\phi(\phi_0) & & 0 \\ & \ddots & \\ 0 & & \boldsymbol{\lambda}_\phi(\phi_0) \end{bmatrix} \tag{4.3.67}$$

and

$$C_2 = \begin{bmatrix} \boldsymbol{\psi}_\phi(\phi_0) & & 0 \\ & \ddots & \\ 0 & & \boldsymbol{\lambda}_\phi(\phi_0) \end{bmatrix} \tag{4.3.68}$$

For linearly independent vectors $\mathbf{1}$, $\boldsymbol{\lambda}_\phi(\phi_0)$ and $\boldsymbol{\psi}_\phi(\phi_0)$, the constraint matrix D has full rank [Er83] and the beamforming solution is given by (4.3.43).

### 4.3.4 Adaptive Algorithm

An estimate of the solution of the beamforming problem (4.3.63), which converges in mean to the optimal weights given by (4.3.43), may be made using a constrained LMS algorithm similar to that given by (4.1.52), (4.1.53), and (4.1.49). In this case, it becomes

$$\mathbf{W}(n+1) = P\left[\mathbf{W}(n) - \mu y(n)\mathbf{X}(n+1)\right] + \mathbf{G} \tag{4.3.69}$$

where the projection operator

$$P = I - D\left(D^T D\right)^{-1} D^T \tag{4.3.70}$$

and

$$\mathbf{G} = D\left(D^T D^{-1}\right)^{-1}\mathbf{g} \tag{4.3.71}$$

The algorithm is initialized at n = 0 with

$$\mathbf{W}(0) = \mathbf{G} \tag{4.3.72}$$

Note that the initial weight vector $\mathbf{W}(0)$ correspond to the optimal weight given by (4.1.43) in the presence of white noise only.

Due to the sparse nature of matrices $C_0$, $C_1$, and $C_2$, the projection operator P is sparse and allows development of a temporally decoupled update equation to estimate the J columns of L weights similar to that discussed earlier for the point constraint. For this case, the algorithm is given by [Buc86]

$$\mathbf{w}_j(n+1) = P\left[\mathbf{w}_j(n) - \mu y(n)\mathbf{x}(n+1-j) + f_j\tilde{C}\left(\tilde{C}^T\tilde{C}\right)^{-1}\mathbf{e}_1\right], \quad j = 1, 2, \ldots, J \tag{4.3.73}$$

where

$$P = I - \tilde{C}\left(\tilde{C}^T\tilde{C}\right)^{-1}\tilde{C}^T \tag{4.3.74}$$

$$\tilde{C} = \left[\mathbf{1}, \boldsymbol{\lambda}_\phi(\phi_0), \boldsymbol{\psi}_\phi(\phi_0)\right] \tag{4.3.75}$$

and

$$\mathbf{e}_1^T = [1, 0, 0] \tag{4.3.76}$$

### 4.3.5 Choice of Origin

In array system design, location of the time reference point (origin of the coordinate system) with respect to the array elements is chosen for notational convience. In most cases, it is one element of the array or array's center of gravity. These origin choices do not affect the array beam pattern or output SNR. However, this is not the case when derivative constraints are involved. The reason is that the constraint matrix D is a function of $\tau_\ell(\phi,\theta)$, which in turn depends on origin choice, as it denotes the time taken by a plane wave arriving from direction $(\phi,\theta)$ and measured from the origin to the array's $\ell$th element. This dependence of the constraint matrix D on the choice of origin affects the solution of the constrained beamforming problem. Hence, the beam pattern and the output SNR of the beamformer using optimal weights depends on the choice of origin.

The vector **G** used to initialize the adaptive algorithm is the optimal weight under white noise conditions and the output noise power is proportional to the norm of this weight, that is, $\mathbf{G}^T\mathbf{G}$. In view of this, the chosen origin should minimize $\mathbf{G}^T\mathbf{G}$ [Buc86].

It follows from (4.3.71) that

$$\mathbf{G}^T\mathbf{G} = \mathbf{g}^T\left(D^TD\right)^{-1}\mathbf{g} \tag{4.3.77}$$

The first-order and the second-order derivative constraints discussed in this section so far are sufficient to ensure that the power response derivatives evaluated at the look direction are zero. However, these constraints are not the necessary and sufficient conditions for the derivatives to be zero. In what follows is a discussion on the first-order derivative constraints for a flat-response processor. For this case, these constraints are necessary and sufficient conditions which ensure that the array beam pattern is independent of the choice of origin [Er90].

The constraint vector **f** for the case of a flat frequency response in the look direction is given by (4.1.28), that is,

$$f_i = \begin{cases} 1 & i = k_0 \\ 0 & i \neq k_0 \end{cases} \tag{4.3.78}$$

Substituting (4.3.78) in (4.3.15), it follows that

$$\left.\frac{\partial \rho}{\partial \phi}\right|_{(\phi_0,\theta_0)} = 4\pi f \sum_{\ell=1}^{J} \mathbf{1}^T \Lambda_\phi(\phi_0,\theta_0)\mathbf{w}_\ell \sin 2\pi f(\ell - k_0)T \tag{4.3.79}$$

If J is odd and $k_0 = (J+1)/2$, then (4.3.79) can be rewritten as

$$\left.\frac{\partial \rho}{\partial \phi}\right|_{(\phi_0,\theta_0)} = 4\pi f \sum_{\ell=1}^{k_0-1} \mathbf{1}^T \Lambda_\phi(\phi_0,\theta_0)\left(\mathbf{w}_{k_0+\ell} - \mathbf{w}_{k_0-\ell}\right)\sin 2\pi f\ell T \tag{4.3.80}$$

As the right hand side of (4.3.80) is a finite Fourier series, it follows that the necessary and sufficient conditions for

$$\left.\frac{\partial \rho}{\partial \theta}\right|_{(\phi_0,\theta_0)} = 0$$

for all $f > 0$ are that all series coefficients are simultaneously equal to zero, that is,

$$\mathbf{1}^T \Lambda_\phi(\phi_0,\theta_0)\left(\mathbf{w}_{k_0+\ell} - \mathbf{w}_{k_0-\ell}\right) = 0, \quad \begin{cases} \ell = 1, 2, \ldots, k_0 - 1, \\ k_0 = \dfrac{J+1}{2} \end{cases} \tag{4.3.81}$$

Similarly, the necessary and sufficient conditions for

$$\left.\frac{\partial \rho}{\partial \theta}\right|_{(\phi_0,\theta_0)} = 0$$

for all f > 0 are

$$\mathbf{1}^T \Lambda_\theta(\phi_0,\theta_0)\left(\mathbf{w}_{k_0+\ell} - \mathbf{w}_{k_0-\ell}\right) = 0, \quad \begin{cases} \ell = 1, 2, \ldots, k_0 - 1, \\ k_0 = \dfrac{J+1}{2} \end{cases} \tag{4.3.82}$$

Note that (4.3.81) and (4.3.82) denote $J - 1$ linear constraints compared to 2J linear constraints given by (4.3.18) and (4.3.19). Discussion on second-order derivative constraints may be found in [Er90], and an unconstrained partitioned realization of the processor with derivative constraints is provided by [Er86a].

## 4.4 Correlation Constrained Processor

A set of nondirectional constraints to improve the performance of a broadband array processor using a TDL structure under look direction errors is discussed in [Kik89]. These are referred to as correlation constraints, and they use known characteristics of the desired signal to estimate an LJ-dimensional correlation vector $\mathbf{r}_d$ between the desired signal and the array signal vector due to the desired signal, that is,

$$\mathbf{r}_d = E\left[s_d(t)\mathbf{X}_d(t)\right] \tag{4.4.1}$$

where $s_d(t)$ denotes the desired signal induced on the reference element, and LJ-dimensional vector $\mathbf{X}_d(t)$ denotes the array signal across the TDL structure due to the desired signal only.

The beamforming problem in this case becomes

$$\begin{aligned} &\underset{\mathbf{W}}{\text{minimize}} \quad \mathbf{W}^T R \mathbf{W} \\ &\text{subject to} \quad \mathbf{r}_d^T \mathbf{W} = \rho_0 \end{aligned} \tag{4.4.2}$$

where $\rho_0$ is a scalar constant that specifies the correlation between the desired signal and array output due to the desired signal, that is,

$$\rho_0 = \left[s_d(t)y_d(t)\right] \tag{4.4.3}$$

where

$$y_d(t) = \mathbf{W}^T \mathbf{X}_d(t) \tag{4.4.4}$$

For the desired signal with a flat spectrum over the frequency band of interest the constraint in (4.4.2) becomes $\mathbf{P}^T\mathbf{W} = 1$ [Er93]. It can easily be verified that the solution $\hat{\mathbf{W}}_C$ of the beamforming problem (4.4.2) is given by

$$\hat{\mathbf{W}}_C = R^{-1}\mathbf{r}_d\left(\mathbf{r}_d^T R^{-1}\mathbf{r}_d\right)^{-1}\rho_0 \tag{4.4.5}$$

## 4.5 Digital Beamforming

In this section, in a brief review of digital beamforming, the process of forming beams in various directions is described [God97]. First, consider the analog beamformer structure shown in Figure 4.5, where signals from all elements are weighted, delayed, and summed to form a beam. The output of the beamformer is given by

$$y(t) = \sum_{\ell=1}^{L} w_\ell x_\ell\left(t - \tau_\ell(\phi)\right) \tag{4.5.1}$$

The delay in front of each element is adjusted such that the signals induced from a given direction, where the beam needs to be pointed to, are aligned after the delays. The weights are adjusted to shape the beam.

In digital beamforming [Pri78, Dud77, Muc84, Pri79, Fan84, Mar89, Rud69, Gab84, Bra80, Syl86, DeM77], the weighted signals from various elements are sampled, stored, and summed after appropriate delays to form beams. The required delay is provided by selecting samples from different elements such that the selected samples are taken at different times. Each sample is delayed by an integer multiple of the sampling interval $\Delta$. The process is shown in Figure 4.6 for a linear array of equispaced elements where the samples of weighted signals are shown as circles. Weights on each element are not shown.

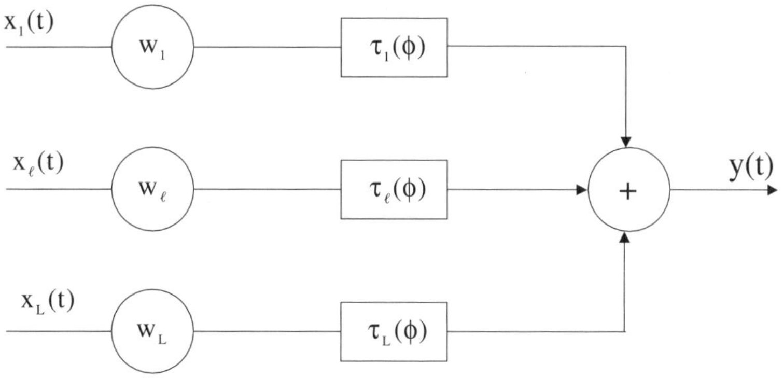


**FIGURE 4.5**
Delay and sum processor structure.

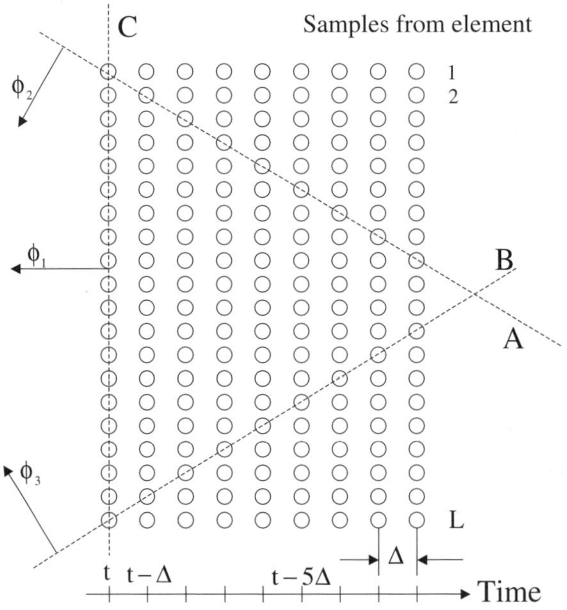


**FIGURE 4.6**
Digital beamforming process. (From Godara, L.C., Application to antenna arrays to mobile communications. Part II: Beamforming and direction of arrival considerations, *IEEE Proc.*, 85, 1195–1247, 1997. ©IEEE. With permission.)

Assume that a beam is to be formed in direction $\phi_2$. Let the direction be such that

$$\tau_\ell(\phi_2) = (\ell - 1)\Delta \tag{4.5.2}$$

Thus, the signal from the $\ell$th element needs to be delayed by $(\ell - 1)\Delta$ seconds. This may be accomplished by summing the samples on a line marked with symbol A in Figure 4.6. For this case, the samples from Element 1 are not delayed, samples from Element 2 are delayed by one sample, and so on.

Similarly, a beam may be steered in direction $\phi_3$ by summing the samples connected by the line marked with symbol B in Figure 4.6, where the signals from Lth element are not delayed, samples from element $L - 1$ are delayed by one sample, and so on. The beam formed in direction $\phi_1$, by summing the samples connected by the line marked with symbol C, does not require any delay.

It follows from the above discussion that when using this process, one can only form beams in directions that require delays equal to some integer multiple of the sampling interval, that is,

$$\tau_\ell(\phi) = k_\ell \Delta \tag{4.5.3}$$

where $k_\ell$, $\ell = 1, 2, \ldots, L$ are integers. The number of discrete directions where a beam can be exactly pointed increases with increased sampling as shown in Figure 4.7, where the sampling interval is $\Delta/2$. The figure shows that additional beams in directions $\phi_4$ and $\phi_5$ may be formed. These exact beams are normally referred to as synchronous or natural beams [Pri78], and it is possible to form a number of these beams simultaneously using a separate summing network for each beam.

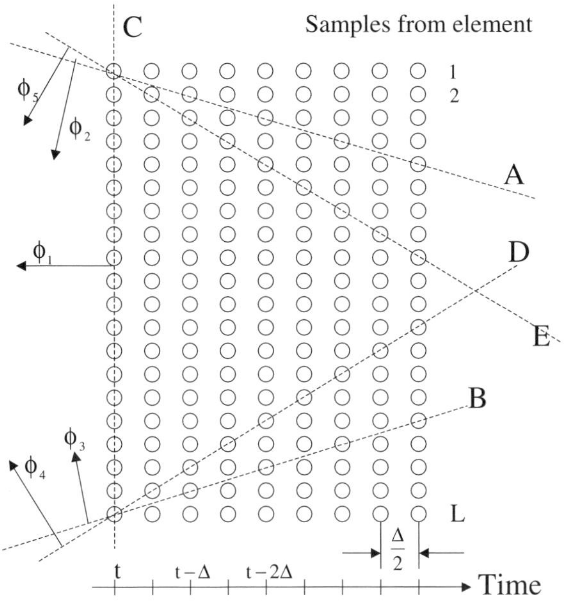


**FIGURE 4.7**
Effect of sampling on digital beamforming. (From Godara, L.C., Application to antenna arrays to mobile communications. Part II: Beamforming and direction of arrival considerations, *IEEE Proc.*, 85, 1195–1247, 1997. ©IEEE. With permission.)

The practical requirement of an adequate set of directions where simultaneous beams need to be pointed implies that the array signals be sampled at much higher rates than required by Nyquist criteria to reconstruct the wave form back from the samples [Pap75]. The high sampling rate means a large number of storage requirements along with high-speed input-output devices, analog-to-digital converters, and large bandwidth cables [Pri78].

The high sampling rate requirement may be overcome by digital interpolation [Pri78, Pri79, Syl86], which basically simulates the samples generated by high sampling rates and thus increases the effective sampling rate. The process works by sampling the array signal at a Nyquist rate or higher and padding with zeros between each sample to form a new sequence. The number of zeros padded decides the effective sampling rate. For the sampling rate to increase by L-fold, L − 1 zeros are padded to create a sequence as big as if it were created by high-speed sampling. The padded sequences then are used for digital beamforming by selecting appropriate samples as required and the beam output is passed through an FIR filter to remove the unwanted spectrum. This filter is normally referred to as an interpolation filter. The beams formed by interpolation beamformers have a slightly higher side-lobe level.

A tutorial introduction to digital interpolation beamformers is given in [Pri78], whereas some additional fundamentals of digital array processing may be found in [Dud77]. A comparison of many approaches to digital beamforming implementations is discussed in [Muc84, Mar89], who show how a real-time implementation is a trade-off between various conflicting requirements of hardware complexities, memory requirements, and system performance.

The shape of a beam, particularly its beam width, is controlled by the size of the array. Generally, a narrow beam results from a larger array. In practice, the array size is fixed and its extent is limited. A process known as extrapolation may be used [Fan84] during digital beamforming to simulate a large array extent resulting in improved beam pattern.

As the interpolation increases the effective sampling rate, the extrapolation extends the effective array length. More information on signal extrapolation schemes may be found in [Pap75, Sul91, Cad79, Son82, Jai81, Sna83].

Digital beamforming techniques for mobile satellite communications are examined in [Chu90] by studying a configuration of a digital beamforming system capable of working in transmit and receive modes. Digital beamforming for mobile satellite communications has also been reported in [Geb95, Chu90]. An introduction to digital beamforming for mobile communications may be found in [Ste87].

## 4.6 Frequency Domain Processing

A general structure of the element-space frequency domain processor is shown in Figure 4.8, where broadband signals from each element are transformed into a frequency domain using the discrete Fourier transform (DFT), and each frequency bin is processed by a narrowband processor structure. The weighted signals from all elements are summed to produce an output at each bin. The weights are selected independently by minimizing the mean output power at each frequency bin subject to steering direction constraints. Thus, the weights required for each frequency bin are selected independently and this selection may be performed in parallel, leading to faster weight update. When an adaptive algorithm such as the LMS algorithm is used for weight updating, a different step size may be used for each bin leading to faster convergence.

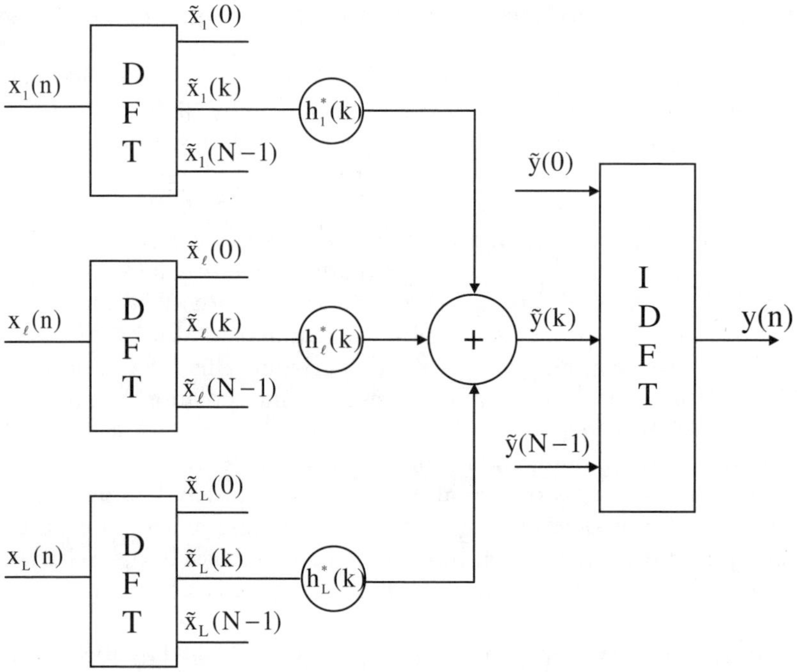


**FIGURE 4.8**
Frequency domain processor structure.

Various aspects of array signal processing in a frequency domain are reported in the literature [Hod79, Arm74, Den78, Nar81, Web84, Shy85, Flo88, Ber86, Ree85, Man82, Kum90, Cla83, Zhu90, God95, Hin81]. The optimum performance of the time domain and frequency domain processors are the same only when the signals in various frequency bins are independent. This independence assumption is mostly made in the study of frequency domain processing. When the assumption does not hold, the frequency domain processor may be suboptimal. Some of the tradeoffs and a comparison of the two processors are discussed in [Hod79, God95].

A study of the frequency domain algorithm [Web84] for coherent signals indicates that the frequency domain method is insensitive to the sampling rate, and may be able to reduce the effects of element malfunctioning on the beam pattern. A study in [Shy85] shows that due to its modular parallel structure, beam forming in the frequency domain is well suited for VLSI implementation and is less sensitive to the coefficient quantization. Computational advantages of the frequency domain method (FDM) for bearing estimation are discussed in [Ree85, Kum90, Hin81], and for correlated data are considered in [Man82, Zhu90]. A general treatment of time and frequency domain realization with a view to compare the structure of various algorithms of weight estimation in a unified manner is provided in [God95].

In this section, frequency domain processing is studied in detail using a constrained element space processor, and relationships between the time domain processor and the frequency domain processor are established [God95].

### 4.6.1 Description

Consider an L-element array immersed in a noise field consisting of uncorrelated broadband directional sources and white noise. Let s(t) be a broadband real signal, with the power spectral density S(f) induced on a reference element due to a source. The autocorrelation function

$$\rho(\tau) = E[s(t)s(t+\tau)] \tag{4.6.1}$$

is the inverse Fourier transform of S(f), that is,

$$\rho(\tau) = \int_{-\infty}^{\infty} S(f)e^{j2\pi f\tau}df \tag{4.6.2}$$

Let $x_\ell(t)$ denote the time wave form derived from the $\ell$th element after presteering. Let these wave forms be sampled at frequency $f_s$. Denoting the sampling interval by T, the sampled wave form derived from $\ell$th element becomes $x_\ell(nT)$. As the sampling period does not play any role in the treatment that follows, it has been omitted for ease of notation.

Let **x**(n) denote the L samples after presteering delays, that is,

$$\mathbf{x}(n) = [x_1,(n), x_2(n), \ldots, x_L(n)]^T \tag{4.6.3}$$

Now consider N array samples $\mathbf{x}(n - i + 1)$, i = 1, …, N, with **x**(n) denoting the most recent samples. Let these be processed by the frequency domain processor structure shown in Figure 4.8, where these are first converted into N frequency bins using discrete Fourier transforms and then processed using N narrowband processors.

Let $\tilde{y}(k)$ denote the output of the kth bin. From Figure 4.8, it follows that

$$\tilde{y}(k) = \mathbf{h}^H(k)\tilde{\mathbf{x}}(k) \tag{4.6.4}$$

where an L-dimensional complex vector $\mathbf{h}(k)$ denotes the L weights of the narrowband processor for the kth bin, that is,

$$\mathbf{h}(k) = [h_1,(k), \ldots, h_L(k)]^T \tag{4.6.5}$$

with $h_\ell(k)$ denoting the weight on the $\ell$th channel.

The L-dimensional complex vector $\tilde{\mathbf{x}}(k)$ denotes the L-frequency domain samples, that is,

$$\tilde{\mathbf{x}}(k) = [\tilde{x}_1(k), \ldots, \tilde{x}_L(k)]^T \tag{4.6.6}$$

with $\tilde{x}_\ell(k)$ denoting the frequency domain samples from the $\ell$th channel. The N frequency samples of the $\ell$th channel $\tilde{x}_\ell(k)$, k = 0, 1, …, N – 1 are related to the N time samples $x_\ell(n)$, n = 1, 2, …, N by the discrete Fourier transform [Bur85], that is,

$$\tilde{x}_\ell(k) = \sum_{i=1}^{N} x_{\ell i} e^{-j\frac{2\pi}{N}(i-1)k}, \quad k = 0, 1, \ldots, N-1 \tag{4.6.7}$$

where $x_{\ell i} \equiv x_\ell(n - i + 1)$, i = 1, 2, …, N and $x_{\ell 1} \equiv x_\ell(n)$ denotes the most recent sample.

Thus, using N array samples $\mathbf{x}(n - i + 1)$, i = 1, 2, …, N, the frequency domain processor produces N frequency domain outputs $\tilde{y}(k)$, k = 0, 1, …, N – 1. These are converted into N output time samples y(n – i + 1), i = 1, 2, …, N using the inverse DFT, that is,

$$y(n-i+1) = \frac{1}{N}\sum_{k=0}^{N-1} \tilde{y}(k) e^{j\frac{2\pi}{N}(i-1)k} \tag{4.6.8}$$

where y(n) denote the most recent output.

The most recent output corresponds to i = 1 in the LHS of (4.6.8). Thus, it follows from (4.6.8) that

$$y(n) = \frac{1}{N}\sum_{k=0}^{N-1} \tilde{y}(k) \tag{4.6.9}$$

Thus, the most recent output sample may be obtained by averaging the output of N narrowband processors without computing N-point inverse DFT. This aspect is exploited in sliding window processing, where N most recent input samples are converted into frequency domain using DFT, and the time domain output is obtained by averaging the N outputs. In this scheme, every time a new input sample arrives, a full cycle involving conversion to frequency domain using DFT, narrowband processing, and computation of output using (4.6.9) needs to be carried out.

The other processing scheme discussed previously where N input time samples are collected, converted to the frequency domain, processed using N narrowband processors, and converted back to N output time samples using the inverse DFT is referred to as block processing [Com88]. Thus, in summary, in block processing a block of N input samples is collected to be processed using narrowband processing to obtain N output time samples. On the other hand, in sliding window processing every time a new sample arrives, the complete processing cycle is invoked. The difference in the processing cycle for the two schemes is that the sliding window processing does not use the inverse DFT.

In both cases, once the N time samples are converted into N frequency domain samples, any of the narrowband processing schemes discussed in previous chapters may be used. In the next section, the relationship between the frequency domain processing discussed in this section and the time domain processing using the TDL structure discussed earlier is established.

### 4.6.2 Relationship with Tapped-Delay Line Structure Processing

Assume that the N array samples $\mathbf{x}(n-i+1)$, $i = 1, 2, \ldots, N$ are processed by two processor structures, namely, the TDL structure shown in Figure 4.1 where the processing is carried out in the time domain and frequency domain processor structure shown in Figure 4.8 where the processing is carried out in frequency domain. In the following, the conditions are derived for the two processors to produce identical outputs.

#### *4.6.2.1 Weight Relationship*

The output of the time domain processor shown in Figure 4.1 is given by

$$y(n) = \mathbf{W}^T\mathbf{X}(n) \tag{4.6.10}$$

where $\mathbf{W}$ is defined in (4.1.4) and $\mathbf{X}(n)$ is defined in (4.1.6) with t replaced by n. Rewrite (4.6.10) as

$$\begin{aligned} y(n) &= \sum_{\ell=1}^{L}\sum_{m=1}^{J} w_{\ell m} x_{\ell m}\left(n-(m-1)\right) \\ &= \sum_{\ell=1}^{L}\sum_{m=1}^{J} w_{\ell m} x_{\ell} \end{aligned} \tag{4.6.11}$$

It follows from (4.6.11) that the output at time n depends on the present input $x_\ell(n)$ and $J-1$ previous inputs, namely, $x_\ell(n-1), \ldots, x_\ell(n-J+1)$. Thus, for a given set of N samples under consideration, one is able to obtain only $N-(J+1)$ output samples, namely $y(n)$, $y(n-1), \ldots, y(n-N+J)$. This implies that for $J = N$, these samples only produce one output sample, given by

$$y(n) = \sum_{\ell=1}^{L}\sum_{m=1}^{N} w_{\ell m} x_{\ell m} \tag{4.6.12}$$

Now, consider the frequency domain processor processing the same N samples. The most recent time sample for the frequency domain processor is given by (4.6.9). For the two processors to produce identical outputs, the time samples given by (4.6.9) and (4.6.12) must be equal, that is,

$$\sum_{\ell=1}^{L}\sum_{m=1}^{N} w_{\ell m} x_{\ell m} = \frac{1}{N}\sum_{k=0}^{N-1}\tilde{y}(k) \tag{4.6.13}$$

Rewrite (4.6.4) as

$$\tilde{y}(k) = \sum_{\ell=1}^{L} h_\ell^*(k)\tilde{x}_\ell(k), \quad k = 0, 1, \ldots, N-1 \tag{4.6.14}$$

It follows from (4.6.13), (4.6.14), and (4.6.7) that

$$\begin{aligned}\sum_{\ell=1}^{L}\sum_{m=1}^{N} w_{\ell m} x_{\ell m} &= \frac{1}{N}\sum_{k=0}^{N-1}\sum_{\ell=1}^{L} h_\ell^*(k)\sum_{i=1}^{N} x_{\ell i} e^{-j\frac{2\pi}{N}(i-1)k} \\ &= \sum_{\ell=1}^{L}\sum_{i=1}^{N} x_{\ell i}\left(\frac{1}{N}\sum_{k=0}^{N-1} h_\ell^*(k) e^{-j\frac{2\pi}{N}(i-1)k}\right)\end{aligned} \tag{4.6.15}$$

The identity holds if

$$w_{\ell m} = \frac{1}{N}\sum_{k=0}^{N-1} h_\ell^*(k) e^{-j\frac{2\pi}{N}(m-1)k}, \quad \ell = 1, 2, \ldots, L, \quad m = 1, 2, \ldots, N \tag{4.6.16}$$

Thus,

$$w_{\ell m}, m = 1, 2, \ldots, N = \mathrm{DFT}\left\{\frac{h_\ell^*(k)}{N}, k = 0, 1, \ldots, N-1\right\} \tag{4.6.17}$$

It follows then that both processors produce identical outputs when the TDL structure has length equal to N and the two sets of weights are related by (4.6.16).

#### 4.6.2.2 *Matrix Relationship*

Consider the output sequence of the frequency domain structure of Figure 4.8. Assume that $M_0$ sets, each of N samples, are being processed. Let $\tilde{y}(k, m)$ denote the output of the kth frequency bin due to the mth data block. For a given $\mathbf{h}(k)$, the mean output power of the kth bin is given by

$$P(k) = \frac{1}{M_0}\sum_{m=0}^{M_0-1}\tilde{y}(k,m)\tilde{y}^*(k,m) \tag{4.6.18}$$

Following (4.6.4), the output of kth bin due to mth set is given by

$$\tilde{y}(k,m) = \mathbf{h}^H(k)\tilde{\mathbf{x}}(k,m) \tag{4.6.19}$$

This along with (4.6.18) implies that

$$P(k) = \mathbf{h}^H(k)R_f(k)\mathbf{h}(k) \tag{4.6.20}$$

where

$$R_f(k) = \frac{1}{M_0}\sum_{m=0}^{M_0-1}\tilde{\mathbf{x}}(k,m)\tilde{\mathbf{x}}^H(k,m) \tag{4.6.21}$$

is an estimate of the array correlation matrix for the kth bin.

It follows from (4.6.21) that

$$\left(R_f(k)\right)_{\ell,n} = \frac{1}{M_0}\sum_{m=0}^{M_0-1}\tilde{x}_\ell(k,m)\tilde{x}_n^*(k,m) \tag{4.6.22}$$

Since

$$\tilde{x}_\ell(k,m) = \sum_{i=1}^{N} x_{\ell i}(m)e^{-j\frac{2\pi}{N}(i-1)k}$$

it follows from (4.6.22) that

$$\left(R_f(k)\right)_{\ell,n} = \frac{1}{M_0}\sum_{m=0}^{M_0-1}\sum_{i=1}^{N} x_{\ell i}(m)e^{-j\frac{2\pi}{N}(i-1)k}\sum_{i=1}^{N} x_{ni}(m)e^{j\frac{2\pi}{N}(i-1)k} \tag{4.6.23}$$

Note that $x_\ell$ is a real variable and $\tilde{x}_\ell$ is a complex variable. Define an N-dimensional vector $\mathbf{x}_\ell(m)$ representing N samples in the tapped delay line structure on the $\ell$th channel as

$$\mathbf{x}_\ell(m) = \begin{bmatrix} x_{\ell i}(m) \\ \vdots \\ x_{\ell N}(m) \end{bmatrix}, \quad \ell = 1, 2, \ldots, L \tag{4.6.24}$$

and an N-dimensional vector $\mathbf{e}(k)$ representing N phasers at kth bin as

$$\mathbf{e}(k) = \begin{bmatrix} 1 \\ \vdots \\ e^{j\frac{2\pi}{N}nk} \\ \vdots \\ e^{j\frac{2\pi}{N}(N-1)k} \end{bmatrix} \tag{4.6.25}$$

From (4.6.23) to (4.6.25) it follows that

$$\begin{aligned}\left(R_f(k)\right)_{\ell,n} &= \frac{1}{M_0}\sum_{m=0}^{M_0-1}\mathbf{e}^H(k)\mathbf{x}_\ell(m)\mathbf{x}_n^T(m)\mathbf{e}(k)\\ &= \mathbf{e}^H(k)\left(\hat{R}_{\ell,n}\right)\mathbf{e}(k)\end{aligned} \tag{4.6.26}$$

where

$$\left(\hat{R}_{\ell,n}\right) = \frac{1}{M_0}\sum_{m=0}^{M_0-1}\mathbf{x}_\ell(m)\mathbf{x}_n^T(m) \tag{4.6.27}$$

is an $N \times N$ matrix denoting the correlation between the $\ell$th and nth elements for the tapped delay line structure, estimated from $M_0$ sets of samples, each of length N. It is an unbiased estimate for the correlation between the $\ell$th and nth elements for given $M_0$ samples. As $M_0$ increases, the estimate asymptotically approaches the true correlation. Therefore, the relationship between the frequency domain and time domain matrices holds for the true correlation matrices.

Throughout the chapter, $R_f$ and R are used to denote the frequency domain and time domain array correlation matrices, respectively, as well as their unbiased estimates. Furthermore, the correlation between the mth and nth taps is denoted by the matrix $(R_{m,n})$, and the correlation between $\ell$th and ith elements is denoted by the matrix $(\hat{R}_{\ell,i})$.

#### 4.6.2.3 *Derivation of $R_f(k)$*

Let $(R_{m,n})_{\ell,i}$ denote the correlation between $\ell$th and ith elements after mth and nth taps due to a source in direction $(\phi,\theta)$. An expression for $(R_{m,n})_{\ell,i}$ from (4.1.11) is given by

$$\left(R_{m,n}\right)_{\ell,i} = \rho\left[(m-n)T + T_\ell - T_i + \tau_i - \tau_\ell\right] \tag{4.6.28}$$

where the arguments $\phi$ and $\theta$ have been suppressed for the ease of notation.

As the correlation function is symmetrical for real signals, it follows from (4.6.2) and (4.6.28) that

$$\left(R_{m,n}\right)_{\ell,i} = \int_{-\infty}^{\infty} S(f)e^{-j2\pi fT(m-n)}e^{j2\pi f(T_i-T_\ell)}e^{j2\pi f(\tau_\ell-\tau_i)}df \tag{4.6.29}$$

Define an N-dimensional vector $\mathbf{e}(f)$ denoting N phasers at frequency f as

$$\mathbf{e}(f) = \begin{bmatrix}1\\ e^{-j2\pi fT}\\ \vdots\\ e^{-j2\pi fT(N-1)}\end{bmatrix} \tag{4.6.30}$$

It follows from (4.6.29) that the $N \times N$ matrix denoting the correlation between $\ell$th and ith elements is given by

$$\left(\hat{R}_{\ell,i}\right) = \int_{-\infty}^{\infty} S(f)\mathbf{e}(f)\mathbf{e}^H(f)e^{j2\pi f(\tau_\ell - T_\ell - \tau_i + T_i)}df \tag{4.6.31}$$

Equation (4.6.31) along with (4.6.26) implies that

$$\begin{aligned}\left(R_f(k)\right)_{\ell,i} &= \mathbf{e}^H(k)\left(\hat{R}_{\ell,i}\right)\mathbf{e}(k)\\ &= \int_{-\infty}^{\infty} S(f)a(f,k)e^{j2\pi f(\tau_\ell - T_\ell - \tau_i + T_i)}\end{aligned} \tag{4.6.32}$$

where

$$a(f,k) = \mathbf{e}^H(k)\mathbf{e}(f)\mathbf{e}^H(f)\mathbf{e}(k) \tag{4.6.33}$$

Substituting for $\mathbf{e}(f)$ and $\mathbf{e}(k)$, (4.6.33) becomes

$$a(f,k) = \frac{\sin^2 \pi\left(k + \frac{f}{f_s}N\right)}{\sin^2 \frac{\pi}{N}\left(k + \frac{f}{f_s}N\right)} \tag{4.6.34}$$

Using steering vector notation, one obtains from (4.6.32) the following compact expression for $R_f(k)$:

$$R_f(k) = \int_{-\infty}^{\infty} S(f)a(f,k)\tilde{\mathbf{S}}(f,\phi,\theta)\tilde{\mathbf{S}}^H(f,\phi,\theta)df \tag{4.6.35}$$

where $\tilde{\mathbf{S}}(f,\phi,\theta)$ denotes the steering vector in $(\phi,\theta)$ direction for an array presteered in $(\phi_0,\theta_0)$.

#### *4.6.2.4 Array with Presteering Delays*

Noting that the steering vector in $(\phi_0,\theta_0)$ direction for an array presteered in $(\phi_0,\theta_0)$ is identical to $\mathbf{1}$, it follows from (4.6.33) and (4.6.35) that the matrix $R_f(k)$ due to a source in a presteered direction is given by

$$R_f(k) = \alpha(k)\mathbf{1}\mathbf{1}^T \tag{4.6.36}$$

where

$$\alpha(k) = \mathbf{e}^H(k)\left[\int_{-\infty}^{\infty} S(f)\mathbf{e}(f)\mathbf{e}^H(f)df\right]\mathbf{e}(k) \tag{4.6.37}$$

The matrix in the square brackets on the right side of (4.6.37) is a spectrum-dependent quantity. Let it be denoted by A. Its (m,n)th element $A_{m,n}$ is given by

$$A_{m,n} = \int_{-\infty}^{\infty} S(f) e^{-j2\pi f(m-n)T} df \tag{4.6.38}$$

$A_{m,n}$ can be evaluated for a specific spectrum using (4.6.38). For example, for a brick-wall type of spectrum given by

$$S(f) = \begin{cases} a_0 & f_L < |f| < f_H \\ 0 & \text{otherwise} \end{cases} \tag{4.6.39}$$

it becomes

$$A_{m,n} = 2a_0 \left[ \frac{\sin 2\pi f_H (m-n)}{2\pi(m-n)} - \frac{\sin 2\pi f_L (m-n)}{2\pi(m-n)} \right] \tag{4.6.40}$$

where $f_H$ and $f_L$ are assumed to be normalized with respect to the sampling frequency.

#### 4.6.2.5 *Array without Presteering Delays*

For this case, the steering delays $T_i = 0$, $i = 1, 2, \ldots, L$. Thus, it follows from (4.6.32) that

$$R_f(k) = \int_{-\infty}^{\infty} S(f) a(f,k) \mathbf{S}(f,\phi,\theta) \mathbf{S}^H(f,\phi,\theta) df \tag{4.6.41}$$

where $\mathbf{S}(f,\phi,\theta))$ denotes the steering vector in $(\phi,\theta)$ direction for an array without presteering. Note that this matrix in general is not equal to a matrix that depends on the energy from the kth bin only, namely

$$\tilde{R}_f(k) = \int_{k\Delta f}^{(k+1)\Delta f} S(f) \mathbf{S}(f,\phi,\theta) \mathbf{S}^H(f,\phi,\theta) df \tag{4.6.42}$$

with $\Delta f = 1/N$ denoting the bandwidth of a frequency bin.

#### 4.6.2.6 *Discussion and Comments*

The results presented here show that when a broadband correlation matrix is transformed into narrowband matrices, these matrices depend on the spectrum of the signal beyond the bandwidth of their particular frequency bins, which is controlled by the parameter $a(f,k)$ given by (4.6.34). Figure 4.9 and Figure 4.10 show how this parameter behaves as a function of the frequency for $N = 10$ and $N = 100$, respectively. The plots are for $k = 0$, and show the normalized value of the parameter with respect to its maximum value $N^2$.

### 4.6.3 Transformation of Constraints

As discussed in Section 4.3, the weights of the broadband element space processor using TDL are subjected to various constraints to make the processor robust against various uncertainties. In this section, some of these constraints are transformed for narrowband processors operating in the frequency domain.

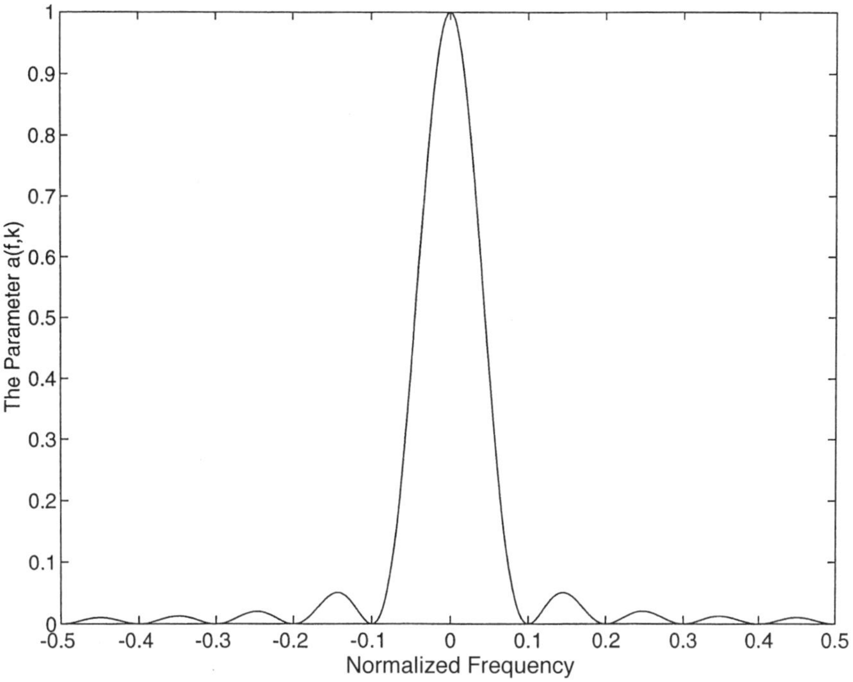


**FIGURE 4.9**
The parameter a(f,k) defined by (4.6.33), normalized with respect to its maximum value, vs. frequency for N = 10 and k = 0. (From Godara, L.C., Application of the fast Fourier transform to broadband beamforming, *J. Acoust. Soc. Am.*, 98, 230–240, 1995. With permission.)

#### 4.6.3.1 *Point Constraints*

Assume that the weights of the TDL are constrained in the look direction, such that

$$\sum_{\ell=1}^{L} w_{\ell m} = f_m, \quad m = 1, 2, \ldots, N \tag{4.6.43}$$

where $f_m$, m = 1, 2, …, N specifies the frequency response of the processor in the look direction as discussed in Section 4.1.2. Note that (4.6.43) is obtained by rewriting (4.1.24) with J replaced by N.

Summing on both sides of (4.6.16) over $\ell$,

$$\sum_{\ell=1}^{L} w_{\ell m} = \frac{1}{N}\sum_{k=0}^{N-1}\sum_{\ell=1}^{L} h_\ell^*(k) e^{-j\frac{2\pi}{N}(m-1)k}, \quad m = 1, 2, \ldots, N \tag{4.6.44}$$

This along with (4.6.43) implies that

$$f_m = \frac{1}{N}\sum_{k=0}^{N-1}\sum_{\ell=1}^{L} h_\ell^*(k) e^{-j\frac{2\pi}{N}(m-1)k}, \quad m = 1, 2, \ldots, N \tag{4.6.45}$$

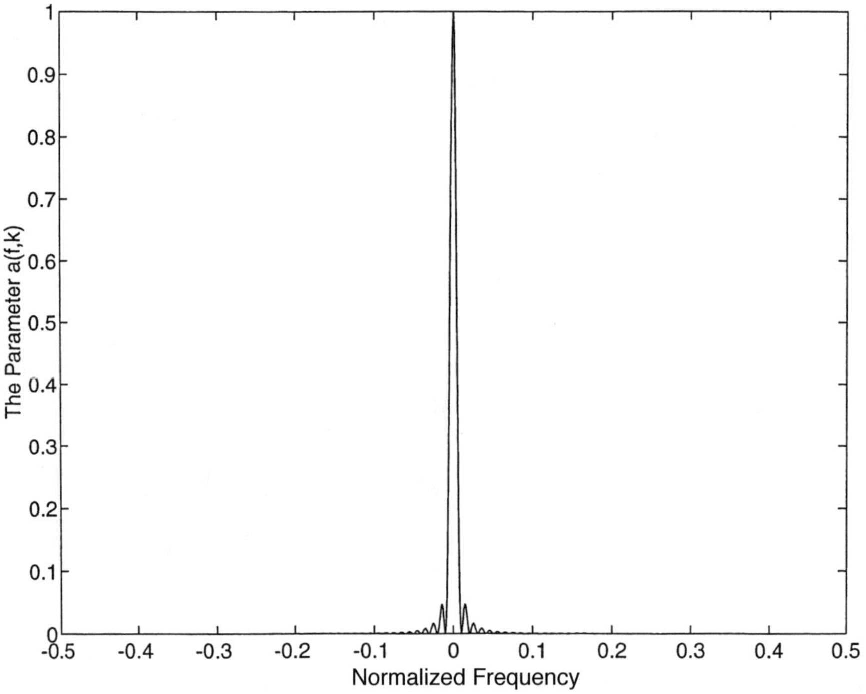


**FIGURE 4.10**
The parameter a(f,k) defined by (4.6.33), normalized with respect to its maximum value, vs. frequency for N = 100 and k = 0. (From Godara, L.C., Application of the fast Fourier transform to broadband beamforming, *J. Acoust. Soc. Am.*, 98, 230–240, 1995. With permission.)

Taking the inverse DFT on both sides, after rearrangements

$$\sum_{\ell=1}^{L} h_\ell^*(k) = \sum_{m=1}^{L} f_m e^{j\frac{2\pi}{N}(m-1)k}, \quad k = 0, 1, \ldots, N-1 \tag{4.6.46}$$

Thus, the equivalent constraints on the weights of the kth bin processor are given by

$$h^H(k)\mathbf{1} = \tilde{f}_k, \quad k = 0, 1, 2, \ldots, N-1 \tag{4.6.47}$$

where $\tilde{f}_k$ specifies the constraint on the weights of the kth bin processor. It follows from (4.6.46) that

$$\tilde{f}_k = \sum_{m=1}^{N} f_m e^{j\frac{2\pi}{N}(m-1)k} \tag{4.6.48}$$

Thus, $\tilde{f}_k$, k = 0, 1, 2, …, N – 1 are the coefficients of inverse DFT of $Nf_m$, m = 1, 2, …, N.

#### *4.6.3.2 Derivative Constraints*

The derivative constraints for the broadband processor are discussed in detail in Section 4.3. These are imposed alongside the point constraints to broaden the beamwidth, which helps to overcome the pointing errors. First-order constraints are given by (4.3.18) and (4.3.19). Rewriting,

$$\mathbf{1}^T \Lambda_\phi(\phi_0, \theta_0)\mathbf{w}_m = 0, \quad m = 1, 2, \ldots, N \tag{4.6.49}$$

and

$$\mathbf{1}^T \Lambda_\theta(\phi_0, \theta_0)\mathbf{w}_m = 0, \quad m = 1, 2, \ldots, N \tag{4.6.50}$$

where $\Lambda_\phi(\phi,\theta)$ and $\Lambda_\theta(\phi,\theta)$ are diagonal matrices given by

$$\Lambda_\phi(\phi,\theta) = \begin{bmatrix} \dfrac{\partial \tau_1(\phi,\theta)}{\partial \phi} & & 0 \\ & \ddots & \\ 0 & & \dfrac{\partial \tau_L(\phi,\theta)}{\partial \phi} \end{bmatrix} \tag{4.6.51}$$

and

$$\Lambda_\theta(\phi,\theta) = \begin{bmatrix} \dfrac{\partial \tau_1(\phi,\theta)}{\partial \theta} & & 0 \\ & \ddots & \\ 0 & & \dfrac{\partial \tau_L(\phi,\theta)}{\partial \theta} \end{bmatrix} \tag{4.6.52}$$

Rewrite (4.6.16) in vector notation as

$$\mathbf{w}_m = \frac{1}{N}\sum_{k=0}^{N-1} \mathbf{h}^*(k) e^{-j\frac{2\pi}{N}(m-1)k}, \quad m = 1, 2, \ldots, N \tag{4.6.53}$$

Substituting in (4.6.49) and (4.6.50),

$$\frac{1}{N}\sum_{k=0}^{N-1} \mathbf{1}^T \Lambda_\phi(\phi_0, \theta_0)\mathbf{h}^*(k) e^{-j\frac{2\pi}{N}(m-1)k} = 0, \quad m = 1, 2, \ldots, N \tag{4.6.54}$$

and

$$\frac{1}{N}\sum_{k=0}^{N-1} \mathbf{1}^T \Lambda_\theta(\phi_0, \theta_0)\mathbf{h}^*(k) e^{-j\frac{2\pi}{N}(m-1)k} = 0, \quad m = 1, 2, \ldots, N \tag{4.6.55}$$

Taking the inverse DFT on both sides of (4.6.54) and (4.6.55), the following equivalent constraints on the narrowband weights result:

$$\mathbf{1}^T \Lambda_\phi(\phi_0, \theta_0)\mathbf{h}^*(k) = 0, \quad k = 0, 1, \ldots, N-1 \tag{4.6.56}$$

and

$$\mathbf{1}^T \Lambda_\theta(\phi_0, \theta_0)\mathbf{h}^*(k) = 0, \quad k = 0, 1, \ldots, N-1 \tag{4.6.57}$$

Alternatively, these may be expressed as

$$\mathbf{h}^H(k)\Lambda_\phi(\phi_0, \theta_0)\mathbf{1} = 0, \quad k = 0, 1, \ldots, N-1 \tag{4.6.58}$$

and

$$\mathbf{h}^H(k)\Lambda_\theta(\phi_0, \theta_0)\mathbf{1} = 0, \quad k = 0, 1, \ldots, N-1 \tag{4.6.59}$$

Following a similar procedure, the second-order derivative constraints for the weights of the broadband processor given by (4.3.20) to (4.3.25) can be transformed for the weights of the narrowband processors. These are given by

$$\mathbf{h}^H(k)\left.\frac{\partial \Lambda_\phi(\phi, \theta)}{\partial \phi}\right|_{(\phi_0, \theta_0)} \mathbf{1} = 0, \quad k = 0, 1, \ldots, N-1 \tag{4.6.60}$$

$$\mathbf{h}^H(k)\Lambda_\phi^2(\phi_0, \theta_0)\mathbf{1} = 0, \quad k = 0, 1, \ldots, N-1 \tag{4.6.61}$$

$$\mathbf{h}^H(k)\left.\frac{\partial \Lambda_\theta(\phi, \theta)}{\partial \theta}\right|_{(\phi_0, \theta_0)} \mathbf{1} = 0, \quad k = 0, 1, \ldots, N-1 \tag{4.6.62}$$

$$\mathbf{h}^H(k)\Lambda_\theta^2(\phi_0, \theta_0)\mathbf{1} = 0, \quad k = 0, 1, \ldots, N-1 \tag{4.6.63}$$

$$\mathbf{h}^H(k)\left.\frac{\partial \Lambda_\phi(\phi, \theta)}{\partial \theta}\right|_{(\phi_0, \theta_0)} \mathbf{1} = 0, \quad k = 0, 1, \ldots, N-1 \tag{4.6.64}$$

and

$$\mathbf{h}^H(k)\Lambda_\phi(\phi_0, \theta_0)\Lambda_\theta(\phi_0, \theta_0)\mathbf{1} = 0, \quad k = 0, 1, \ldots, N-1 \tag{4.6.65}$$

## 4.7 Broadband Processing Using Discrete Fourier Transform Method

In the previous section, an FDM to process broadband signals was discussed in which broadband time domain data are transformed into narrowband frequency domain data using DFT, and are then processed using narrowband processing schemes. The processed signals are transformed into broadband time domain signals using the inverse DFT. Thus, the implementation is done using narrowband processors operating at different frequency bins.

In contrast, in the time domain method (TDM) discussed in Section 4.1.3, the processor is implemented in a time domain using a TDL structure, as shown in Figure 4.1. The weights of the broadband processor are obtained by solving the constrained beamforming problem when the look direction information is available.

In this section, the DFT method for estimating the weights of the broadband processor using a TDL structure of Figure 4.1 is discussed, and the performance of the broadband processor using the DFT method is compared with that using the time domain method [God99]. The method is discussed by considering the beamforming problem with the point constraint. In this case the TDM solves the following beamforming problem:

$$\begin{aligned} &\underset{\mathbf{W}}{\text{minimize}} \quad \mathbf{W}^T R \mathbf{W} \\ &\text{subject to} \quad C^T \mathbf{W} = \mathbf{f} \end{aligned} \tag{4.7.1}$$

where C is the constraint matrix defined in (4.1.26) and **f** is a J-dimensional vector selected to specify the frequency response in the look direction. The weights $\hat{\mathbf{W}}$ estimated by the TDM are the solution of (4.7.1), and are given by

$$\hat{\mathbf{W}} = R^{-1}C\left(C^T R^{-1} C\right)^{-1} C^T \mathbf{f} \tag{4.7.2}$$

The DFT method estimates the weights of the broadband processor of Figure 4.1 in two steps. First, it estimates the weights of narrowband processors by minimizing the mean output power of each frequency bin, and then uses the relations developed in the last section between the time domain and frequency domain structures for identical outputs to transform these into the required weights. It also maintains the same frequency response in the look direction as is done by the TDM using the appropriate constraints developed in the last section. Figure 4.11 shows a schematic diagram of the DFT method.

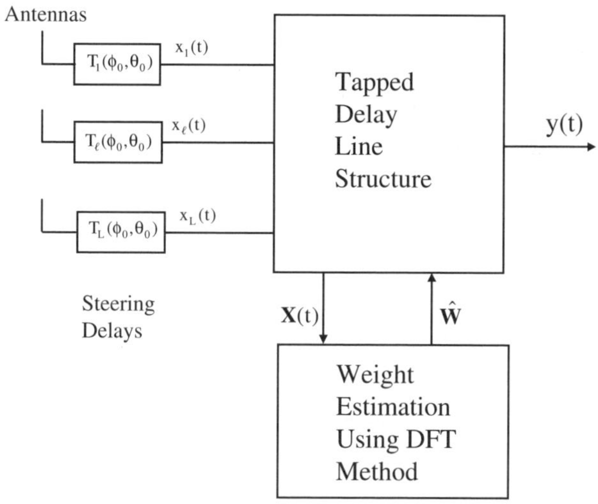


**FIGURE 4.11**
Schematic dagram of broadband processor using DFT methods.

The similarity between the TDM and DFT methods is that both estimate the weights of the TDL structure of Figure 4.1. The main difference between the two is that this method minimizes the mean output power of each frequency bin, rather than minimizing the mean output power of the processor, as is done by the TDM. This implies that if the sum of the mean output powers from all frequency bins is not equal to the mean output power of the processor shown in Figure 4.1, then the realized processor using the DFT method does not maximize the mean output SNR in the absence of errors, as is case with the processor using the TDM to estimate the weights. However, this method offers the potential for a large amount of computational savings for real-time applications due to its parallel nature of implementation as discussed later in this section.

As the DFT method minimizes the mean output power of each frequency bin and then uses the relations between the time domain and the frequency domain structures for the identical outputs, the performance of the realized processor in the absence of implementation errors is the same as the processor implemented in the frequency domain. However, there are important differences.

The main difference between the DFT method and the FDM is that this method uses the optimized weights of the narrowband processors operating at different frequency bins to estimate the optimal weights of the time domain broadband processor. The processor is implemented in the time domain and the received signal flows in the time domain structure without encountering the delay associated with the frequency domain implementation. This may be important for some applications.

As broadband processor performance using the DFT method to estimate the weights when implemented in the time domain is identical to that implemented in the frequency domain, this fact presents a framework for comparing the performance of time domain and frequency domain implementations under identical conditions.

### 4.7.1 Weight Estimation

The DFT method uses the following procedure to estimate the weights of the time-domain broadband-constrained processor using a TDL structure of length J.

1. Estimate narrowband array correlation matrices $R_f(k)$, $k = 0, \ldots, J-1$ using

$$R_f(k)_{\ell i} = \mathbf{e}^H(k)\left(\hat{R}_{\ell,i}\right)\mathbf{e}(k), \quad \ell, i = 1, \ldots, L \tag{4.7.3}$$

where

$$\mathbf{e}(k) = \left[1, e^{j\frac{2\pi}{J}k}, \ldots, e^{j\frac{2\pi}{J}(J-1)k}\right]^T \tag{4.7.4}$$

and $(\hat{R}_{\ell,i})$ is a $J \times J$ matrix denoting the correlation between samples from $\ell$th and ith elements given by (4.6.27).

2. Estimate $\hat{\mathbf{h}}(k)$, $k = 0, \ldots, (J-1)$ using

$$\hat{\mathbf{h}}(k) = \frac{R_f^{-1}(k)\mathbf{1}\tilde{f}_k^*}{\mathbf{1}^T R_f^{-1}(k)\mathbf{1}} \tag{4.7.5}$$

which are the solutions of the following narrowband beamforming problems:

$$\begin{aligned} &\underset{\mathbf{h}(k)}{\text{minimize}} \quad \mathbf{h}^H(k)R_f(k)\mathbf{h}(k) \\ &\text{subject to} \quad \mathbf{h}^H(k)\mathbf{1} = \tilde{f}_k \end{aligned} \qquad k = 0, \dots, J-1 \tag{4.7.6}$$

where

$$\tilde{f}_k = \sum_{m=1}^{J} f_m e^{j\frac{2\pi}{J}(m-1)k}, \quad k = 0, \dots, J-1 \tag{4.7.7}$$

Equation (4.7.7) ensures that the required frequency response in the desired direction is maintained. It should be noted that due to the symmetry property of the Fourier transform, one only needs to estimate $\hat{J}$ narrowband weights $\hat{\mathbf{h}}(k)$, $k = 0, \dots, (\hat{J} - 1)$, where

$$\hat{J} = \begin{cases} \dfrac{J+1}{2}, & \text{when J is odd} \\ \dfrac{J}{2}, & \text{when J is even} \end{cases} \tag{4.7.8}$$

3. Estimate the weights of the time domain structure of Figure 4.1 using

$$\hat{w}_{m\ell} = \frac{1}{J}\sum_{k=0}^{J-1} \hat{h}_\ell^*(k) e^{-j\frac{2\pi}{J}(m-1)k}, \quad m = 1, \dots, J, \quad \ell = 1, \dots, L \tag{4.7.9}$$

The block diagram shown in Figure 4.12 summarizes the method to estimate the weights of the broadband processor using the proposed technique.

### 4.7.2 Performance Comparison

In this section, examples are presented to compare the output SNR of the processor using the weights estimated by the DFT method and the TDM. The weights for the TDM are computed using (4.7.2), whereas for the DFT method, they are computed using (4.7.3) to (4.7.9). Both methods use actual $LJ \times LJ$ dimensional array correlation matrix R, and produce LJ weights of the TDL structure.

A linear array of equispaced elements is used in the presence of one interference source. The element spacing is measured in wavelengths of the desired signal at the highest frequency. The signal bandwidth is expressed in terms of the normalized frequency with respect to sampling frequency. The sampling frequency is taken to be equal to twice the highest frequency of the desired signal. Thus, the normalized highest frequency of the desired signal is identical to 0.5. All directional sources are assumed to be of the brick-wall type spectrum. The directional sources considered for the study are assumed to be of two bandwidths, referred to as the large bandwidth and the small bandwidth. The normalized frequency band for the large bandwidth is from [0.15, 0.5], whereas for the small bandwidth it is [0.45, 0.5]. The desired signal of unit power is assumed to be present broadside to the array.

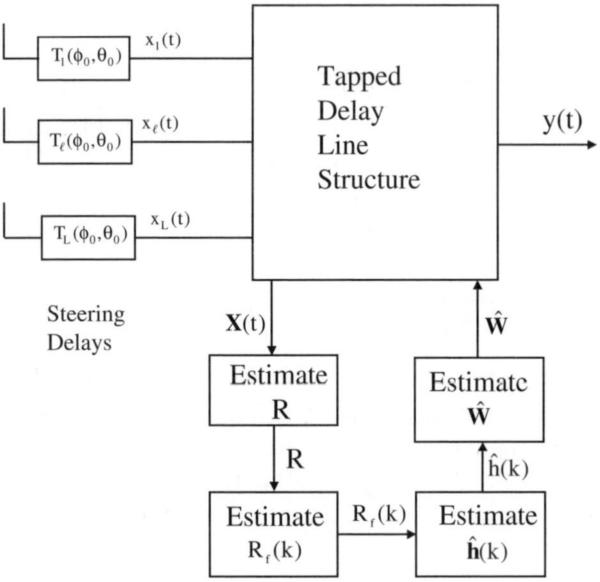


**FIGURE 4.12**
Block diagram of DFT method.

The output SNR is computed using

$$\mathrm{SNR} = \frac{\hat{\mathbf{W}}^H R_S \hat{\mathbf{W}}}{\hat{\mathbf{W}}^H R_N \hat{\mathbf{W}}}$$

with $R_S$ denoting the actual array correlation matrix due to signal only, and $R_N$ denoting the actual array correlation matrix due to interference and background noise only. The SNR is plotted as a function of the angle of the interference by varying it from 0° to 180°. The array is constrained to have the all-pass response in the desired signal direction by selecting

$$f_i = \begin{cases} 1 & i = \dfrac{J+1}{2} \\ 0 & \text{otherwise} \end{cases} \tag{4.7.10}$$

where the filter-length parameter J is assumed to be an odd integer.

The performance comparison is carried out by varying the length of the filter, number of elements in the array, the signal bandwidth, and interference-to-background-noise ratio to see how various parameters affect the result.

### 4.7.2.1 *Effect of Filter Length*

In order to compare the performance of the two methods for a different number of taps, a five-element array is used in the presence of a directional interference of power 10 dB above the signal level and the white noise power 10 dB below the signal level. Figure 4.13 shows $SNR_T$(dB) – $SNR_D$(dB), or equivalently, $10\log_{10}(SNR_T/SNR_D)$, as a function of interference angle for various filter lengths with $SNR_T$ and $SNR_D$, respectively, denoting the output SNR of the processor using the TDM and DFT methods.

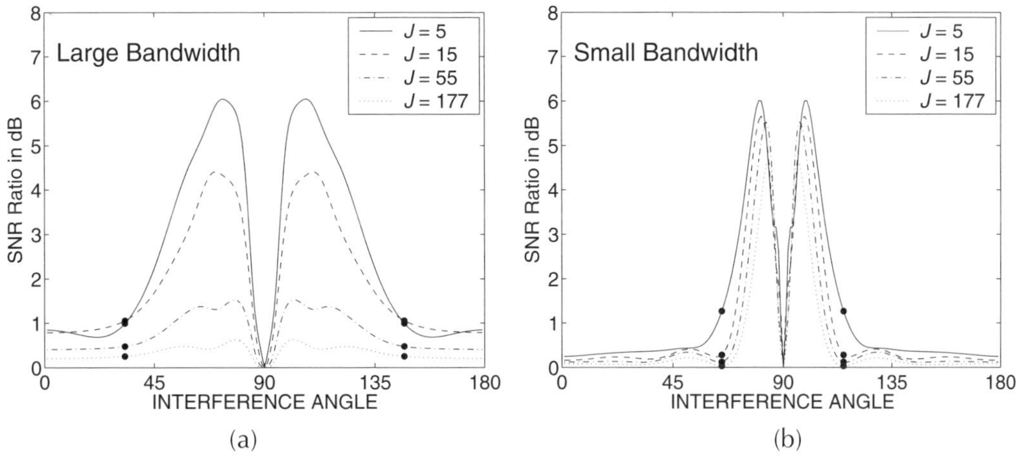


**FIGURE 4.13**
10 log($SNR_T/SNR_D$) vs. interference angle with number of elements = 5, signal power = 1.0, interference power = 10.0, and white noise power = 0.1. (a) Large bandwith. (b) Small bandwidth. (From Godara, L.C. and Jahromi, M.R.S., *IEEE Trans. Signal Process.*, 47, 2386–2395, 1999. ©IEEE. With permission.)

The two bullets on each curve indicate the beamwidth of the antenna array used. An expression for the beamwidth of the main lobe for the narrowband array with a large number of elements is given by [Col85]

$$\text{Beamwidth} = 2\arcsin\left(\frac{\lambda_0}{\text{Ld}}\right)$$

where $\lambda_0$ is the wavelength of the narrowband signal, and d is the element spacing in meters. Beamwidth for the broadband arrays has been taken to be the average of the two beamwidths computed at the lowest and the highest frequencies of the signal.

Figure 4.13 shows that the difference between the two SNRs is smaller when the interference is outside the main lobe than the case when it is within the main lobe, except when the interference is close to the look direction, in which case the processor generally is not used for its interference canceling capability and thus the situation is of no practical significance.

Above certain values of filter length, the results for the two bandwidths are different. For a small bandwidth signal, the difference between the two SNRs is very small when the interference is outside the main lobe, whereas it is reasonably high when it is within the main lobe, except when the interference is close to the look direction.

In the case of large bandwidth signals, the difference between the two output SNRs does not become as small as that for the small bandwidth case when the interference is outside the main lobe. The difference is more sensitive to the filter length above certain values for the large bandwidth case compared to the small bandwidth case and decreases as the filter length is increased.

#### 4.7.2.2 *Effect of Number of Elements in Array*

For this example, the array element numbers are varied to study their effect on the performance difference of the two methods. Figure 4.14 shows the difference in the two SNRs for the both bandwidth sources. When the interference is outside the main lobe, an increase in the number of elements causes a decrease in the difference between the SNRs

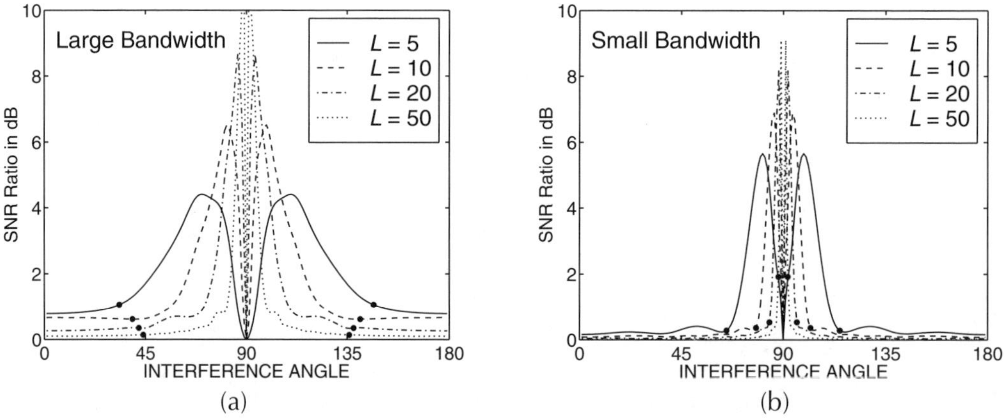


**FIGURE 4.14**
10 log($SNR_T/SNR_D$) vs. interference angle with J = 15, signal power = 1.0, interference power = 10.0, and white noise power = 0.1. (a) Large bandwith. (b) Small bandwidth. (From Godara, L.C. and Jahromi, M.R.S., *IEEE Trans. Signal Process.*, 47, 2386–2395, 1999. ©IEEE. With permission.)

obtained by the two methods. This implies that increasing the number of elements improves the output SNR of the DFT method more than that of the TDM. Thus, when the interference is away from the look direction, the output SNR achievable by the DFT method approaches that of the TDM as the number of elements are increased. It should be noted that an increase in the number of elements in the array causes a decrease in the array beamwidth. Thus, as the number of elements in the array increases, the sector outside the main lobe increases. When an interference is present in this sector, the difference in the two SNRs is small.

Figure 4.14 also shows that the maximum value of the difference between the SNRs of the two methods increases as the number of elements in the array is increased. Furthermore, the direction of interference where the maximum difference between the SNRs occurs moves closer to the look direction as the number of elements is increased. Thus, it means that as the number of elements is increased, the interference canceling capability of the DFT method decreases relative to the TDM when the interference is close to the look direction.

This is a very interesting result. It says that the interference-canceling capability of the DFT method decreases, as the interference is very close to the look direction. In practice, a situation in which interference is close to the look direction rarely occurs, and even if it did, the interference-canceling capability of a processor is low for all practical purposes. However, in the presence of the look direction error, situations do occur when the desired source is not in the look direction and a processor treats it as interference. Extra precautions are necessary to overcome such situations. It appears from these results that the DFT method provides this beam-broadening capability naturally. This aspect of the DFT method is further explored in a later section to show that it is robust against look direction errors.

#### *4.7.2.3 Effect of Interference Power*

Figure 4.15 shows the difference in SNRs achievable by the TDM and DFT methods for various interference power levels at a given background noise. This figure shows that the performance of the processor using the DFT method deteriorates relative to the one using the TDM as interference power increases. This is true for small as well as large bandwidth signals. However, the deterioration is comparatively low when the interference is outside the main lobe. For the small bandwidth case, it is hardly noticeable.

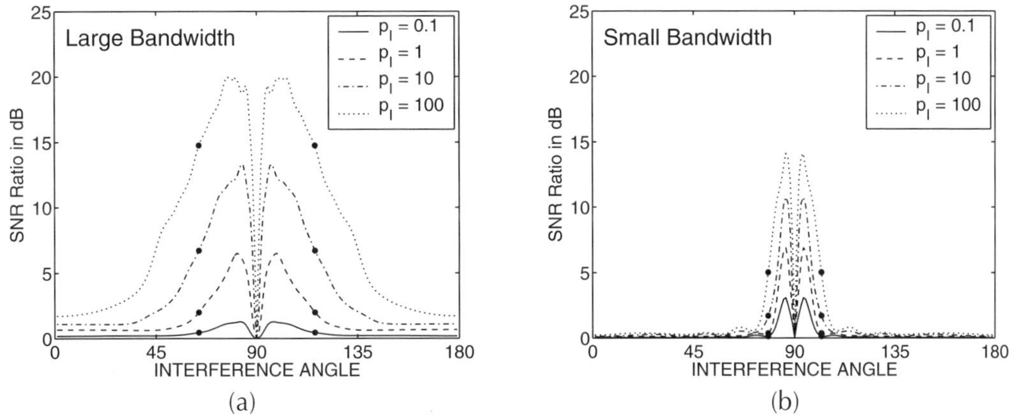


**FIGURE 4.15**
10 log($SNR_T/SNR_D$) vs. interference angle with number of elements = 10, J = 15, signal power = 1.0, and white noise power = 0.01. (a) Large bandwith. (b) Small bandwidth. (From Godara, L.C. and Jahromi, M.R.S., *IEEE Trans. Signal Process.*, 47, 2386–2395, 1999. ©IEEE. With permission.)

### 4.7.3 Computational Requirement Comparison

In this section, examples are presented to compare the two methods based on their computational requirements to estimate weights of the TDL filter once the time domain array correlation matrix has been computed. The computation count reflects the floating-point operations required for weight estimation. Denoting the computation count for the TDM and the DFT method by $O_T$ and $O_D$, respectively, one obtains from (4.7.2) and (4.7.3) to (4.7.9) that

$$O_T = 2J^3\left(L^3 + L^2 + 2L + 1\right) + 2LJ^2 \tag{4.7.11}$$

and

$$O_D = 4J^3L^2 + 4J^2L^2 + 4JL^3 + 8J^2L + 4JL^2 + 10JL \tag{4.7.12}$$

It should be noted that no allowance has been made in either of the methods for any special matrix structure that might be used to reduce computation count.

Figure 4.16 shows the ratio of the floating-point operation for the TDM to the DFT method, $O_T/O_D$, as a function of the filter length for a varying number of elements. The TDM requires more computation than the DFT method, and a reduction of the order of 50 is possible using an array of 100 elements with a tapped delay line filter of length 100. It should be noted that an increase in filter length does not increase the computational savings as much as that achievable by increasing the number of elements. This is also evident from approximations of $O_T$ and $O_D$ for large J and L. Approximating (4.7.11) and (4.7.12) for large J and L lead to

$$O_T \approx 2J^3L^3 \tag{4.7.13}$$

and

$$O_D \approx 4J^3L^2 \tag{4.7.14}$$

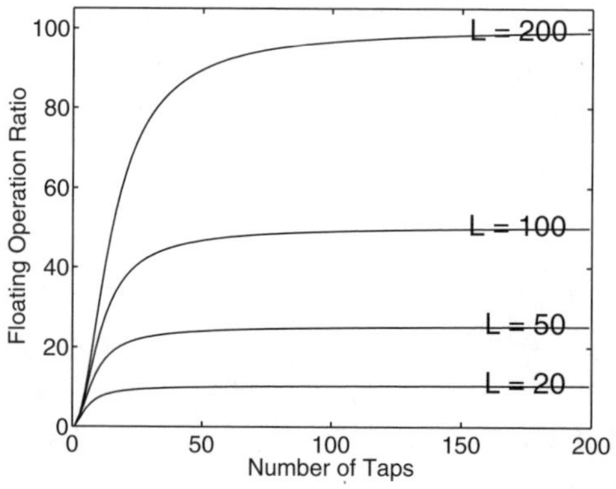


**FIGURE 4.16**
Ratio of the required floating point operations using the time domain method to DFT method ($O_T/O_D$) vs. number of taps (J). (From Godara, L.C. and Jahromi, M.R.S., *IEEE Trans. Signal Process.*, 47, 2386–2395, 1999. ©IEEE. With permission.)

It follows from (4.7.13) and (4.7.14) that

$$\frac{O_T}{O_D} \approx \frac{L}{2} \tag{4.7.15}$$

Thus, it follows that the reduction of the order of L/2 is possible using the DFT method.

### 4.7.4 Schemes to Reduce Computation

In this section, a number of schemes are discussed to reduce the computational requirements for weight estimation using the DFT method.

#### *4.7.4.1 Limited Number of Bins Processing*

The DFT method basically divides the entire spectrum into a number of frequency bins and processes signals in each bin. The weights at each bin are selected by minimizing the mean output power of each bin subject to constraints. In practice, the processing of all bins is not necessary, as the desired signal only covers a part of the spectrum, and thus one is only interested in canceling the interference that overlaps the signal bandwidth. Hence, one only needs to select weights by minimizing the mean output power of those bins that are in the vicinity of the signal bandwidth. The weights for bins outside this range may be selected to provide the maximum SNR under no directional sources. The conventional processor maximizes the output SNR in the absence of a directional source environment. Thus, selecting the weights of the antenna array is done by solving the optimal beamforming problem for those bins in the vicinity of the signal bandwidth and using equal weighting for other bins.

Since the equal weighting process does not require any computation, processing a limited number of bins reduces the computation load substantially, depending on the signal bandwidth. Computer analyses have shown that good results are obtained by processing two extra bins, one on each side of the signal bandwidth. Let $\tilde{J}$ denote the number of bins in the vicinity of the signal bandwidth that need to be processed by solving the optimization problem. Thus, $\tilde{J}$ is given by

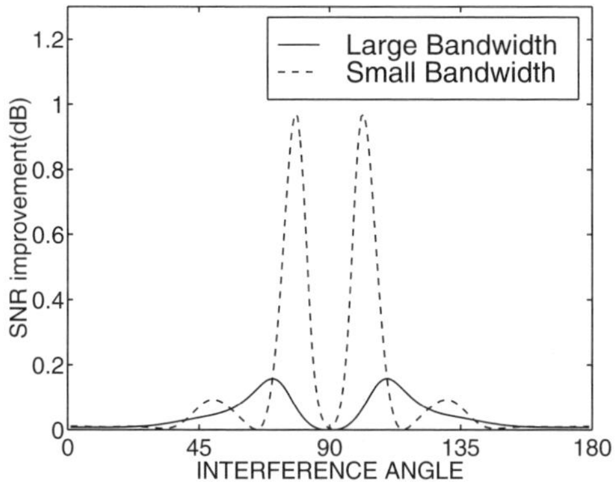


**FIGURE 4.17**
SNR improvement using the bin elimination method compared to the DFT method vs. interference angle with number of elements = 5, J = 101, signal power = 1.0, interference power = 10.0, and white noise power = 0.1. (From Godara, L.C. and Jahromi, M.R.S., *IEEE Trans. Signal Process.*, 47, 2386–2395, 1999. ©IEEE. With permission.)

$$\tilde{J} = \begin{cases} \lceil (J+1)B_S \rceil + 2 & J \text{ is odd} \\ \lceil JB_S \rceil + 2, & J \text{ is even} \end{cases} \tag{4.7.16}$$

where $\lceil x \rceil$ denotes an integer greater than or equal to x, and $B_S$ denotes the normalized signal bandwidth, that is,

$$B_S = (f_H - f_L)/f_S \tag{4.7.17}$$

In order to illustrate the computational efficiency and performance improvement provided by this scheme, consider the parameters of Figure 4.17. For this case, the bin elimination scheme requires the processing of eight bins for the small bandwidth signal and 38 bins for the large bandwidth case, compared with 51 bins by the normal DFT method. The floating-point operations required to process these bins reduce to 16% and 75% of the normal DFT method for the two cases, respectively.

Figure 4.17 shows the improvement in output SNR using this method compared with the normal DFT method, which processes all the bins. The SNR improvement is evident for all interference directions. Thus, the processor using this method not only requires less computation time but also attains higher output SNR compared to the normal DFT method.

#### *4.7.4.2 Parallel Processing Schemes*

It is possible to increase the computation speed of the FDM by carrying out many computations in parallel. Hardware complexity and, thus, system cost, is expected to increase as more and more parallel processing is carried out to increase processing speed. Thus, there is a tradeoff between speed, which is vital in real-time operations, and system hardware cost. In this section, selected schemes are discussed, and their computational requirements are compared with the TDM.

##### *4.7.4.2.1 Parallel Processing Scheme 1*

A block diagram showing the steps involved in this scheme to estimate the weights is shown in Figure 4.18. The scheme processes all frequency bins in parallel. The number of bins $\hat{J}$ required to be processed for a J-tap filter is given by (4.7.8).

$$\begin{array}{c}R\\\downarrow\end{array}$$

```
k=1        k=j         k=ĵ
```

$$R_f(k)_{\ell i} = \mathbf{e}^H(k)\left(\hat{R}_{\ell i}\right)\mathbf{e}(k)$$
$$\ell, i = 1, \cdots, L$$

$$\hat{\mathbf{h}}(k) = \frac{R_f^{-1}(k)\mathbf{1}\tilde{f}_k^*}{\mathbf{1}^T R_f^{-1}(k)\mathbf{1}}$$

$$\hat{w}_{m\ell} = \frac{1}{J}\sum_{k=0}^{J-1}\hat{h}_\ell^*(k)e^{-j\frac{2\pi}{J}(m-1)k},$$
$$m = 1, \cdots, J,\ \ell = 1, \cdots, L$$

$$\downarrow \hat{W}$$

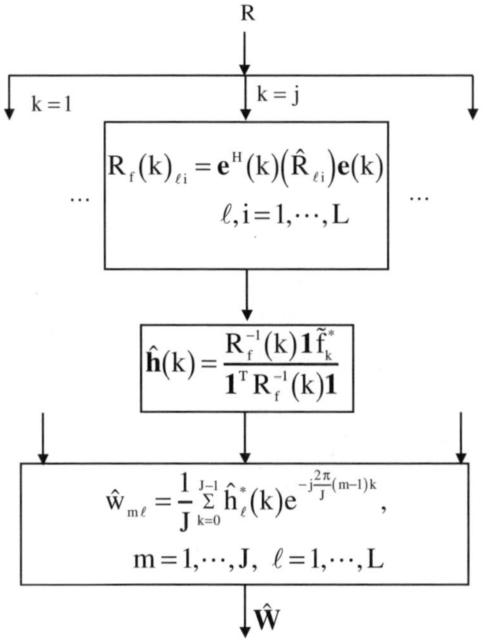

**FIGURE 4.18**
Block diagram of Parallel Processing Scheme Number 1. (From Godara, L.C. and Jahromi, M.R.S., *IEEE Trans. Signal Process.*, 47, 2386–2395, 1999. ©IEEE. With permission.)

It should be noted that when weights are estimated for real-time operations, the time taken by the processor is an important measure of its performance, and the parallel processing scheme minimizes this time. Let $O_{D1}$ denote the computation count that reflects this fact, and which represents the time taken to estimate the weights rather than to measure total computation requirements. Then, the number of floating-point operations $O_{D1}$ required to estimate the weights is given by

$$O_{D1} = 8J^2L^2 + 8JL^2 + 8L^3 + 8J^2L + 8L^2 + 20L \qquad (4.6.18)$$

#### 4.7.4.2.2 Parallel Processing Scheme 2

This scheme not only processes all frequency bins in parallel but carries out matrix multiplications in parallel. Computation of each element of matrix $R_f(k)$ requires the following operation:

$$R_f(k)_{\ell i} = \mathbf{e}^H(k)\left(\hat{R}_{\ell,i}\right)\mathbf{e}(k), \quad \ell, i = 1, \ldots, L \qquad (4.7.19)$$

The scheme carries out multiplication of $\mathbf{e}(k)$ with each column of $(\hat{R}_{\ell,i})$ in parallel to reduce computation time from J vector multiplications to l vector multiplication. The resulting vector is then multiplied with $\mathbf{e}^H(k)$. The total time to compute each element of $R_f(k)$ reduces from J + 1 complex vector multiplications to two complex vector multiplications. A block diagram of the scheme is shown in Figure 4.19.

Let the number of floating-point operations required to estimate weights with this scheme be denoted by $O_{D2}$. The solution, then, is

$$O_{D2} = 16L^2J + 8L^3 + 8J^2L + 8L^2 + 20L \qquad (4.7.20)$$

[Figure 4.19 diagram:

R

k = 1, k = j, k = ĵ

$$z_j(k) = \sum_{m=1}^{J} (\hat{R}_{\ell i})_{jm} e_m(k)$$

$z_1(k)$ ... $z_j(k)$

$$R_f(k)_{\ell i} = \mathbf{e}^H(k)\mathbf{z}(k)$$

$$\hat{\mathbf{h}}(k) = \frac{R_f^{-1}(k)\mathbf{1}\tilde{f}_k^*}{\mathbf{1}^T R_f^{-1}(k)\mathbf{1}}$$

$$\hat{w}_{m\ell} = \frac{1}{J}\sum_{k=0}^{J-1} \hat{h}_\ell^*(k) e^{-j\frac{2\pi}{J}(m-1)k},$$
$m = 1,\cdots,J, \quad \ell = 1,\cdots,L$

$\hat{\mathbf{w}}$]

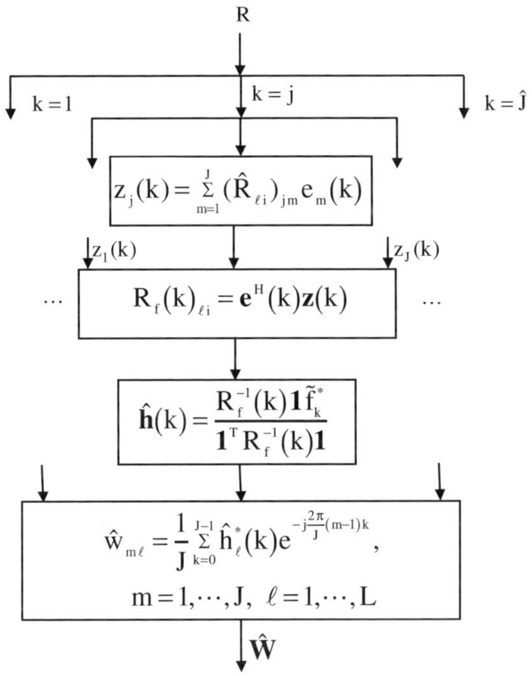

**FIGURE 4.19**
Block diagram of Parallel Processing Scheme Number 2. (From Godara, L.C. and Jahromi, M.R.S., *IEEE Trans. Signal Process.*, 47, 2386–2395, 1999. ©IEEE. With permission.)

### 4.7.4.2.3 Parallel Processing Scheme 3

The FDM requires estimation of $L^2$ elements of matrix $R_f(k)$. This scheme estimates these elements in parallel, as shown in Figure 4.20. Thus, by computing $R_f(k)_{\ell i}$, $\ell, i = 1, ..., L$ in parallel, it saves time of the order of $L^2$ in the matrix estimation. Let the floating-point operations required to estimate weights using this scheme be denoted by $O_{D3}$. Then

$$O_{D3} = 8J^2 + 8J + 8L^3 + 8J^2L + 8L^2 + 20L \qquad (4.7.21)$$

It should be noted that this scheme incorporates the processing of all frequency bins in parallel but does not carry out the multiplications of $\mathbf{e}(k)$ with $(\hat{R}_{\ell i})$ in parallel, as suggested by Scheme 2. However, it is possible to carry out these operations in parallel by combining all of the above schemes to get the maximum speed for real-time operations. The floating-point operations required to estimate the weights using the combined scheme are given by the following expression:

$$O_{DC} = 16J + 8L^3 + 8J^2L + 8L^2 + 20L \qquad (4.7.22)$$

Figure 4.21 compares the ratios of floating-point operations required to estimate the optimal weights using the TDM to the FDM using various parallel processing schemes. Figure 4.21(a) shows the results for an array with 100 elements as a function of filter length.

Figure 4.21(b) shows the floating-point operations ratio as a function of the number of elements using 100 taps. The successive parallel processing schemes require less processing time, and thus, a substantial increase in computation speed is possible using them. Using a 100-element array with a filter length of 100 taps, a 50-fold computational savings is

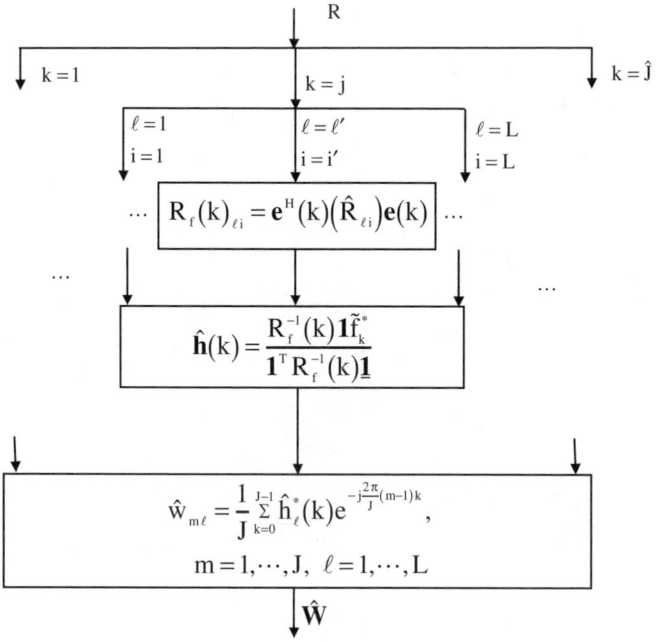


**FIGURE 4.20**
Block diagram of Parallel Processing Scheme Number 3. (From Godara, L.C. and Jahromi, M.R.S., *IEEE Trans. Signal Process.*, 47, 2386–2395, 1999. ©IEEE. With permission.)

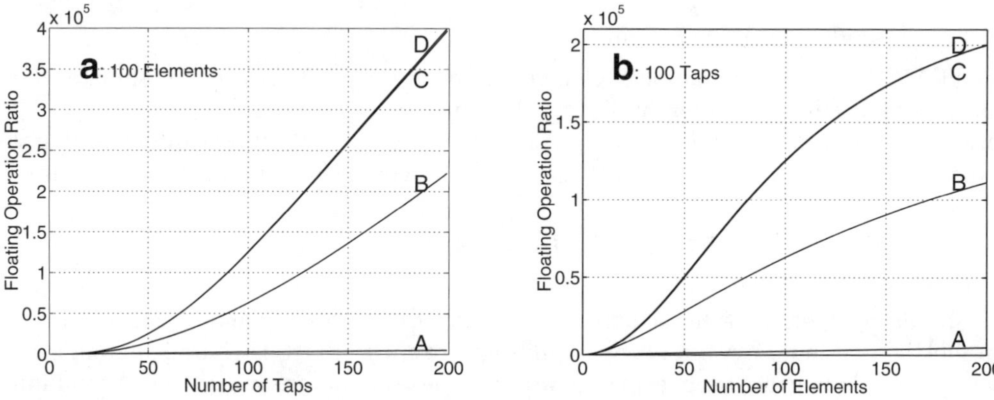


**FIGURE 4.21**
Ratio of the required floating point operations using the time domain method to DFT method ($O_T/O_D$) (a) vs. number of taps for 100 elements, (b) vs. number of elements for 100 taps. Curve A: Parallel Processing Scheme Number 1; Curve B: Parallel Processing Scheme Number 2; Curve C: Parallel Processing Scheme Number 3; Curve D: combination of all parallel processing schemes. (From Godara, L.C. and Jahromi, M.R.S., *IEEE Trans. Signal Process.*, 47, 2386–2395, 1999. ©IEEE. With permission.)

possible without any parallel processing, and 125,620-fold using all parallel processing schemes, compared to the TDM.

It should be noted that the schemes discussed in this section to increase processing speed tend to do so by increasing system complexity. The limited bin-processing scheme not only reduces the computation requirements of the DFT method but also has a potential to improve its performance without increasing system complexity, as is the case with parallel processing schemes.

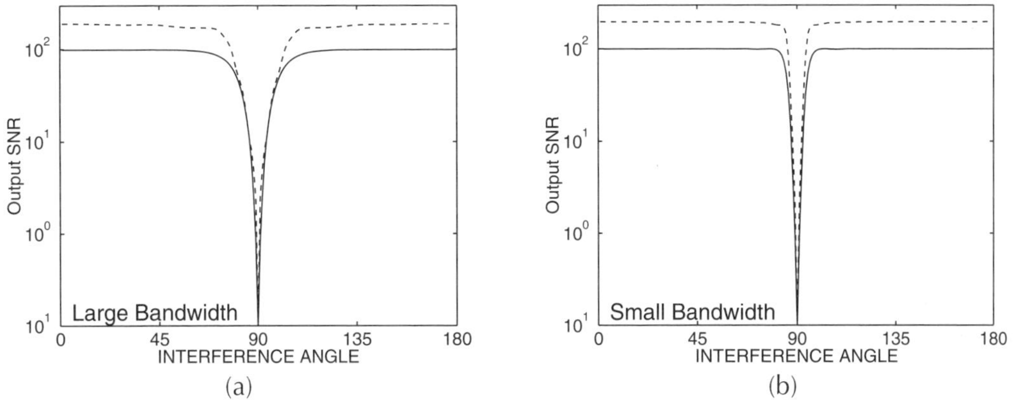


**FIGURE 4.22**
Output SNR vs. interference angle with number of taps = 17, signal power = 1.0, interference power = 10.0, and white noise power = 0.1. Solid line depicts the result using the time domain method with 10 elements and dotted line depicts the result using the DFT method with 20 elements. (a) Large bandwidth. (b) Small bandwidth. (From Godara, L.C. and Jahromi, M.R.S., *IEEE Trans. Signal Process.*, 47, 2386–2395, 1999. ©IEEE. With permission.)

### 4.7.5 Discussion

It follows from the results presented so far that the DFT method is computationally more efficient than the TDM, the output SNR of the beamformer is lower when the weights are estimated by the DFT method compared to the case when the weights are estimated by the TDM, and the interference-canceling capability of the DFT method decreases more than the TDM when the interference approaches the look direction. Some of these issues are reexamined in this section with the view to show that by appropriate choice of filter length and number of elements in the array, it is possible to achieve a higher output SNR with less processing time using the DFT method than the TDM. The DFT method is also robust against look direction errors.

#### *4.7.5.1 Higher SNR with Less Processing Time*

It is possible to obtain better SNR using the DFT method by increasing the number of elements or filter length such that the required processing time remains less than when using the TDM. Two examples are presented to demonstrate this fact.

In the first example, an array with 20 elements uses the DFT method and an array with 10 elements uses the TDM to estimate the weights. Performance of the two methods is compared in Figure 4.22, where results are displayed for both small and large bandwidth cases. The figure shows that the DFT method yields better performance than the TDM. For this case, computational savings of 14% are possible without using any parallel processing, and when using combined parallel processing, 99%.

The second example uses a five-element array and a filter of 17 taps for the TDM and 177 taps for the DFT method. Results for both bandwidth sources displayed in Figure 4.23 indicate that the DFT method performance is almost equal to that of the TDM. The DFT method for this case requires about 21% less computation time than the TDM. It should be noted that the computational savings have been achieved by using parallel processing, which increases hardware cost. Reduction in hardware cost could be achieved by using the bin elimination method, which reduces 89 parallel stages to 11 stages for the small bandwidth case and to 65 parallel stages for the large bandwidth case.

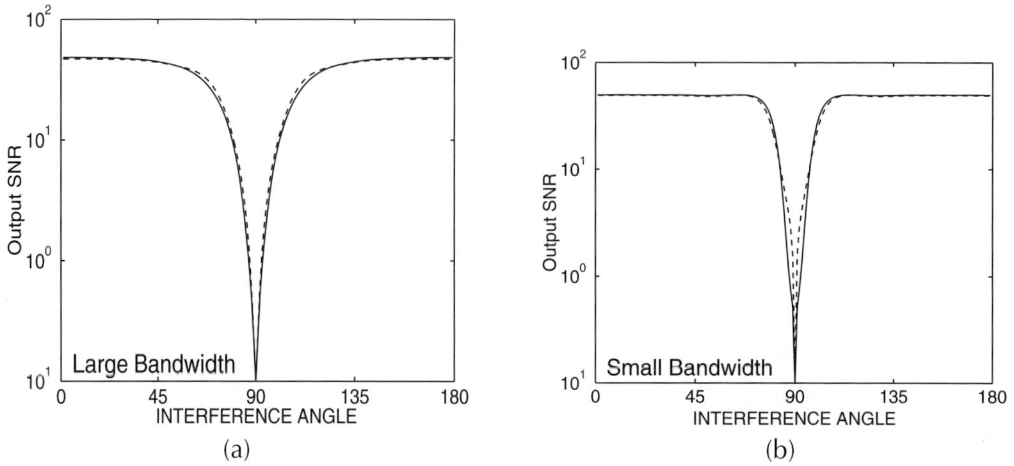


**FIGURE 4.23**
Output SNR vs. interference angle with number of elements = 5, signal power = 1.0, interference power = 10.0, and white noise power = 0.1. Solid line depicts the result using the time domain method with 17 taps and dotted line depicts the result using the DFT method with 177 taps. (a) Large bandwidth. (b) Small bandwidth. (From Godara, L.C. and Jahromi, M.R.S., *IEEE Trans. Signal Process.*, 47, 2386–2395, 1999. ©IEEE. With permission.)

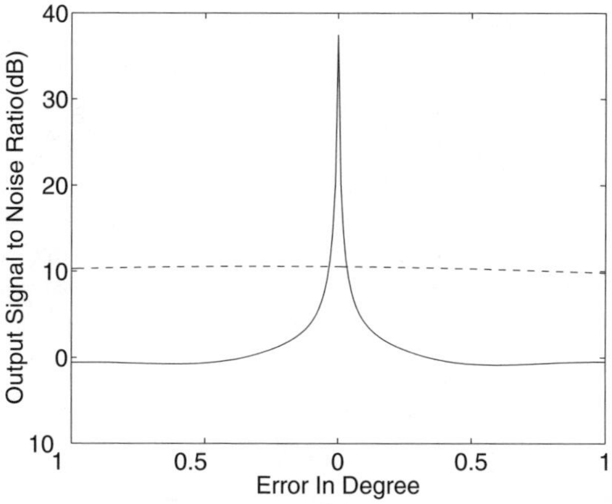


**FIGURE 4.24**
Output SNR vs. the look direction error with number of elements = 10, number of taps = 15, bandwidth [0.15,0.5], signal power = 1.0, interference power = 100.0, interference direction = 75 degrees with the line of the array, and white noise power = 0.001. Solid line depicts the result using the time domain method with 17 taps and dotted line depicts the result using the DFT method with 177 taps. (From Godara, L.C. and Jahromi, M.R.S., *IEEE Trans. Signal Process.*, 47, 2386–2395, 1999. ©IEEE. With permission.)

### *4.7.5.2 Robustness of DFT Method*

Processor performance is compared when weights are estimated using the two methods in the presence of the look direction error (LDE). It is assumed that the actual signal direction is different from the look direction. The weights in both cases are constrained in the look direction.

Figure 4.24 shows the output SNR of the processor using the two methods as a function of look direction error. The error is measured relative to the look direction and is assumed positive in the counterclockwise direction. Thus, errors 1° and −1° mean that the signal direction is, respectively, 91° and 89° relative to the line of the array.

Figure 4.24 shows that the DFT method is robust against the look direction error in the presence of a single interference. The SNR of the processor using the DFT is about 10 dB more than the one using the TDM in the presence of look direction error of less than 0.5°. Although computer simulation shows the robustness of the DFT method against pointing error, a theoretical explanation does not seem to exist.

## 4.8 Performance

Performance of broadband arrays as a function of the number of various parameters such as the number of taps, tap spacing, array geometry, array aperture, and signal bandwidth has been considered in the literature [May81, Voo92, Com88, Ko81, Ko87, Nun83, Yeh87, Sco83] to understand their influence on the behavior of arrays. An analysis [May81] of broadband array using eigenvalues of the array correlation matrix indicates that the product of the array aperture and fractional bandwidth (FBW) of the signal is an important parameter of the broadband array in determining its performance. The FBW is defined as the ratio of the bandwidth to the center frequency of the signal. The number of taps required on each element depends on this parameter as well as on the shape of the array, with more taps needed for a complex shape. A study [Voo92, Com88] of the SNR as a function of inter-tap spacing indicates that there is a range of spacing that yields close to maximum attainable SNR and depends on the FBW of the signal. This range includes quarter wavelength spacing at the center frequency $f_0$. The quarter wavelength spacing produces a 90° phase shift at $f_0$ and is equal to $1/4f_0$. By measuring the tap spacing as a multiple of this delay, the inter-tap spacing with the multiple around 1/FBW yields close to the highest attainable SNR. With the multiple between 1/FBW to 4/FBW, a larger number of taps for an equivalent performance is necessary.

A study of the jamming rejection capability [Ko81] and tracking performance of the array in nonstationary environment [Ko87] also indicates that when tap spacing is measured in terms of the signal's center frequency, the best performance is achieved when the spacing is $1/4f_0$. For this tap spacing, the array correlation matrix has less eigenvalue spread, which is the reason for this performance. The eigenvalue spread of a matrix indicates the range of values that its eigenvalues take. A bigger ratio of the largest eigenvalue to the smallest eigenvalue indicates a larger spread.

The TDL filter tends to increase the degrees of freedom of the array that may be traded against the number of elements such that an array with L elements is able to suppress more than $L-1$ directional interferences provided that their center frequencies are not the same and fall within the FBW of the signal [Yeh87].

## Acknowledgments

Edited versions of Sections 4.5 and 4.8 are reprinted from Godara, L.C., Application of antenna arrays to mobile communications, I. Beamforming and DOA considerations, *Proc. IEEE.*, 85(8), 1195-1247, 1997. An edited version of Section 4.6 is reprinted from Godara, L.C., Application of the fast Fourier transform to broadband beamforming, *J. Acoust. Soc. Am.*, 98(1), 230-240, 1995.

## Notation and Abbreviations

| | |
|---|---|
| $(\phi,\theta)$ | direction in three-dimensional coordinate system, Figure 2.2 |
| $(\phi)$ | direction with respect to line array |
| $(\phi_0,\theta_0)$ | look direction |
| $\Delta f$ | bandwidth of frequency bin |
| $[f_L,f_H]$ | frequency band of interest |
| a(f,k) | scalar defined in (4.6.33) |
| DFT | discrete Fourier transform |
| FDM | frequency domain method |
| FIR | finite impulse response |
| LDE | look direction error |
| MMSE | minimum mean square error |
| MSE | mean square error |
| TDL | tapped delay line |
| TDM | time domain method |
| $A(f,\phi,\theta)$ | desired frequency response in direction $(\phi,\theta)$ |
| B | matrix prefilter |
| $\tilde{B}$ | matrix with B as diagonal elements |
| $B_S$ | normalized signal bandwidth |
| C | $LJ \times J$ dimensional constraint matrix |
| $C_k$ | constraint matrix |
| D | constraint matrix |
| diag[**x**] | matrix with **x** as diagonal elements |
| **E**(t) | column of matrix prefilter outputs across TDL structure |
| **e**(t) | column of L – 1 outputs of matrix prefilter |
| **e**(k) | column of N phasers at kth bin defined in (4.6.25) |
| **e**(f) | column of N phasers at frequency f defined in (4.6.30) |
| **F** | optimal weights with point constraints, only white noise present |
| **f** | J-dimensional constraint vector |
| $f_k$ | kth component of **f** |
| $\tilde{f}_k$ | kth coefficient of inverse DFT of $Nf_m$, m = 1, 2, …, N |
| $f_s$ | sampling frequency |
| **G** | optimal weights with directional constraints, only white noise present |
| **g** | constraint vector |
| H, $H(f,\phi,\theta)$ | frequency response of TDL processor in direction $(\phi,\theta)$ |
| **h**(k) | L weights of narrowband processor for kth bin |
| $h_\ell(k)$ | weight on $\ell$th channel for kth bin |
| J | number of taps in tapped delay line filter |
| $\tilde{J}$ | number of bins that need processing in bin elimination method |

| | |
|---|---|
| $\hat{J}$ | number of bins that need processing due to DFT properties |
| $J(\mathbf{W})$ | cost function |
| $J(V,\lambda)$ | cost function |
| L | number of elements |
| M | number of directional sources |
| $M_0$ | number of data sets of N samples |
| N | Number of samples processed by frequency domain method |
| $O_T$ | Number of floating-point operations using TDM |
| $O_D$ | Number of floating-point operations using DFT method |
| P | projection operator |
| $P(\mathbf{W})$ | mean out power of a processor for given **W** |
| $P_S(\mathbf{W})$ | mean output signal power for given **W** |
| $P_N(\mathbf{W})$ | mean output noise power for given weight |
| P(k) | mean out power of narrowband processor for kth bin |
| $\hat{P}$ | mean output power of TDL processor using optimal weights |
| **P** | LJ-dimensional column vector defined by (4.1.68) |
| Q | LJ × LJ matrix defined by (4.1.67) |
| R | array correlation matrix |
| $R_S$ | array correlation matrix due to signal source |
| $R_N$ | array correlation matrix due to noise |
| $R_\ell$ | array correlation matrix due to $\ell$th source in direction $(\phi_\ell,\theta_\ell)$ |
| $(R_{m,n})$ | matrix denoting correlation after (m − 1) and (n − 1) delays |
| $(\hat{R}_{\ell,i})$ | matrix denoting correlation between $\ell$th and ith elements |
| $R_f(k)$ | array correlation matrix in frequency domain for kth bin |
| $\tilde{R}_f(k)$ | array correlation matrix using energy from kth bin only |
| $R_{\mathbf{XE}}$ | matrix of correlation between **X**(t) and **E**(t) |
| $R_{\mathbf{EE}}$ | matrix of correlation between **E**(t) and **E**(t) |
| $\mathbf{r}_d$ | correlation between desired signal and array signal vector |
| S(f) | power spectral density of s(t) |
| $\mathbf{S}(f,\phi,\theta)$ | steering vector at frequency f in direction $(\phi,\theta)$ |
| $\tilde{\mathbf{S}}(f,\phi,\theta)$ | steering vector in $(\phi,\theta)$ direction for array presteered in $(\phi_0,\theta_0)$ |
| s(t) | signal induced on reference element |
| SNR | signal-to-noise ratio |
| SNR(**W**) | SNR for given **W** |
| $SNR_T$ | SNR using TDM |
| $SNR_D$ | SNR using DFT method |
| T | inter-tap spacing, sampling interval |
| T(f) | diagonal matrix of steering delays |
| $T_\ell(\phi_0,\theta_0)$ | steering delay on $\ell$th element |
| $T_0$ | bulk delay to make $T_\ell(\phi_0,\theta_0)$ a positive quantity |
| U | LJ × LJ matrix of the eigenvector of Q |

| | |
|---|---|
| $U_{\eta 0}$ | matrix of eigenvectors associated with $\eta_0$ nonzero eigenvalues of Q |
| $\mathbf{U}_i$ | eigenvector associated with ith eigenvalue of Q |
| $\mathbf{V}$ | error vector, column of (L – 1)J weights of TDL structure |
| $\hat{\mathbf{V}}$ | (L – 1)J dimensional optimal weights of TDL structure |
| $\mathbf{v}_k$ | column of L – 1 weights on the kth tap of TDL structure |
| $\mathbf{W}$ | column of LJ weights of TDL structure |
| $\mathbf{W}_F$ | column of LJ fixed weight |
| $\hat{\mathbf{W}}$ | optimal weights of TDL processor |
| $\hat{\mathbf{W}}_0$ | optimal weights of constrained partioned processor |
| $\tilde{\mathbf{W}}$ | weight vector which minimizes $\varepsilon_0$ |
| $\mathbf{W}(n)$ | weights estimated at the nth iteration |
| $\mathbf{w}_m$ | column of L weights on the mth tap of TDL structure |
| $w_{\ell k}$ | weight on the kth tap of the $\ell$th channel |
| $\mathbf{X}(t)$ | column of array signals across the TDL structure |
| $\mathbf{X}(n)$ | array signals at nth instant of time |
| $\mathbf{x}(t)$ | column of array signals after presteering delays |
| $\tilde{\mathbf{x}}(k)$ | column of frequency domain array signals for kth bin |
| $\tilde{\mathbf{x}}(k,m)$ | array signals for kth bin from mth data set |
| $x_\ell(t)$ | output of $\ell$th sensor presteered in $(\phi_0,\theta_0)$ |
| $x_{\ell i}$ | output of $\ell$th sensor before ith tap |
| $x_{\ell i}(m)$ | output of $\ell$th sensor before ith tap from mth data set |
| $\mathbf{x}_\ell(m)$ | N outputs of $\ell$th sensor across TDL filter from mth data set |
| $x_\ell$ | components of the $\ell$th element along x-axis |
| $\tilde{x}_\ell(k)$ | output of $\ell$th sensor for kth bin |
| $\tilde{x}_\ell(k,m)$ | output of $\ell$th sensor from mth data set for kth bin |
| $y(t)$ | output of processor |
| $y_\ell$ | components of $\ell$th element along y-axis |
| $y(n)$ | output at nth instant of time |
| $\tilde{y}(k)$ | output of processor at kth bin |
| $\tilde{y}(k,m)$ | output of processor at kth bin from mth data set |
| $y_A(t)$ | output of auxiliary beams |
| $y_F(t)$ | output of fixed beam |
| $z_\ell$ | components of $\ell$th element along z-axis |
| $\Lambda$ | diagonal matrix with elements being eigenvalues of Q |
| $\Lambda_\phi(\phi,\theta)$ | diagonal matrix defined by (4.3.5) |
| $\Lambda_\theta(\phi,\theta)$ | diagonal matrix defined by (4.3.17) |
| $\Lambda_\phi(\phi)$ | diagonal matrix defined by (4.3.51) |
| $\Delta$ | sampling interval |
| $\eta_0$ | rank of Q |
| $\delta_0$ | threshold value |
| $\sigma_0$ | normalizing constant |

| | |
|---|---|
| $\boldsymbol{\sigma}(\phi)$ | vector defined by (4.3.55) |
| $\boldsymbol{\psi}(\phi)$ | vector defined by (4.3.56) |
| $\varepsilon_0$ | MSE between $A(f,\phi_0,\theta_0)$ and $H(f,\phi_0,\theta_0)$ |
| $\lambda$ | Lagrange multiplier |
| $\lambda_i(Q)$ | ith eigenvalue of Q |
| $\boldsymbol{\lambda}$ | J-dimensional vector of undetermined Lagrange multipliers |
| $\boldsymbol{\lambda}(n)$ | Lagrange multipliers at nth iteration |
| $\boldsymbol{\lambda}(\phi)$ | vector defined by (4.3.54) |
| $\rho(\tau)$ | correlation function of s(t) |
| $\rho$, $\rho(f,\phi,\theta)$ | power response of TDL processor in direction $(\phi,\theta)$ |
| $\rho_0$ | correlation between desired signal and array output |
| $\tau_\ell(\phi,\theta)$ | delay faced by signal from source in $(\phi,\theta)$ on $\ell$th element |
| $\tau_\ell(\phi)$ | delay faced by signal from source in $(\phi)$ on $\ell$th element |
| $\tau$ | delay parameter |

## References


Ahm83 Ahmed, K.M. and Evans, R.J., Broadband adaptive array processing, *IEE Proc.*, 130, Pt. F, 433–440, 1983.

Ahm84 Ahmed, K.M. and Evans, R.J., An adaptive array processor with robustness and broadband capabilities, *IEEE Trans. Antennas Propagat.*, 32, 944–950, 1984.

Ale87 Alexandrou, D., Boundary reverberation rejection via constrained adaptive beamforming, *J. Acoust. Soc. Am.*, 82, 1274–1290, 1987.

Arm74 Armijo, L., Daniel, W. and Labuda, W.M., Applications of the FFT to antenna array beamforming, in Proceedings of IEEE Electronics and Aerospace Systems Convention, Washington, DC, pp. 381–383, 1974.

Ber86 Bershad, N.J. and Feintuch, P.L., A normalized frequency domain LMS adaptive algorithm, *IEEE Trans. Acoust. Speech Signal Process.*, 34, 452–461, 1986.

Bra80 Brady, J.J., A serial phase shift beamformer using charge transfer devices, *J. Acoust. Soc. Am.*, 68, 504–506, 1980.

Bry69 Bryson, Jr., A.E. and Ho, Y.C., *Applied Optimal Control*, Blaisdell, Waltham, MA, 1969.

Buc86 Buckley, K.M and Griffiths, L.J., An adaptive generalized sidelobe canceller with derivative constraints, *IEEE Trans. Antennas Propagat.*, 34, 311–319, 1986.

Buc87 Buckley, K.M., Spatial/spectral filtering with linearly constrained minimum variance beamformers, *IEEE Trans. Acoust. Speech Signal Process.*, 35, 249–266, 1987.

Bur85 Burrus, C.S. and Parks, T.W., *DFT/FFT and Convolution Algorithms: Theory and Implementation*, Wiley, New York, 1985.

Cad79 Cadzow, J.A., An extrapolation procedure for band-limited signals, *IEEE Trans., Acoust. Speech Signal Process.*, 27, 4–12, 1979.

Che95 Chern, S.J. and Sung, C.Y., The hybrid Frost's beamforming algorithm for multiple jammers suppression, *Signal Process.*, 43, 113–132, 1995.

Chu90 Chujo, W. and Yasukawa, K., Design study of digital beam forming antenna applicable to mobile satellite communications, *IEEE Antennas and Propagation Symposium Digest*, 400–403, 1990.

Cla83 Clark, G.A., Parker, S.R. and Mitra, S.K., A unified approach to time-and frequency-domain realization of FIR adaptive digital filters, *IEEE Trans. Acoust. Speech Signal Process.*, 31, 1073–1083, 1983.

Col85 Colin, R.E., *Antenna and Radio Wave Propagation*, McGraw-Hill, New York, 1985, chapter 3.

Com88 Compton, Jr., R.T., The bandwidth performance of a two element adaptive array with tapped delay line processing, *IEEE Trans. Antennas Propagat.*, 36, 5–14, 1988.

DeM77 DeMuth, G.J., Frequency domain beamforming techniques, in *Proceedings of IEEE International Conference on Acoustics, Speech and Signal Processing*, 713–715, 1977.

Dav67 Davies, D.E.N., Independent angular steering of each zero of the directional pattern for a linear array, *IEEE Trans. Antennas Propagat. (Commn.)*, 15, 296–298, 1967.

Den78 Dentino, M., McCool, J. and Widrow, B., Adaptive filtering in frequency domain, *IEEE Proc.* 66, 1658–1659, 1978.

Des92 Despins, C.L.B., Falconer, D.D. and Mahmoud, S.A., Compound strategies of coding, equalization and space diversity for wideband TDMA indoor wireless channels, *IEEE Trans. Vehicular Technol.*, 41, 369–379, 1992.

Dud77 Dudgeon, D.E., Fundamentals of digital array processing, *IEEE Proc.*, 65, 898–904, 1977.

Er83 Er, M.H. and Cantoni, A., Derivative constraints for broadband element space antenna array processors, *IEEE Trans. Acoust. Speech Signal Process.*, 31, 1378–1393, 1983.

Er85 Er, M.H. and Cantoni, A., A new approach to the design of broadband element space antenna array processors, *IEEE J. Oceanic Eng.*, 10, 231–240, 1985.

Er86 Er, M.H. and Cantoni, A., A new set of linear constraints for broadband time domain element space processors, *IEEE Trans. Antennas Propagat.*, 34, 320–329, 1986

Er86a Er, M.H. and Cantoni, A., An unconstrained portioned realization for derivative constrained broadband antenna array processors, *IEEE Trans. Acoust. Speech Signal Process.*, 34, 1376–1379, 1986.

Er90 Er, M.H. and Ng, B.P., On derivative constrained broadband beamforming, *IEEE Trans. Acoust. Speech Signal Process.*, 38, 551–552, 1990.

Er90a Er, M.H., An alternative implementation of quadratically constrained broadband beamformers, *Signal Process.*, 38, 17–23, 1990.

Er93 Er, M.H., On the limiting solution of quadratically constrained broadband beamformers, *IEEE Trans. Signal Process.*, 41, 418–419, 1993.

Fan84 Fan, H., El-Masry, E.I. and Jenkins, W.K., Resolution enhancement of digital beamforming, *IEEE Trans. Acoust. Speech Signal Process.*, 32, 1041–1052, 1984.

Flo88 Florian, S. and Bershad, N.J., A weighted normalized frequency domain LMS adaptive algorithm, *IEEE Trans. Acoust. Speech Signal Process.*, 36, 1002–1007, 1988.

Fro72 Frost, III, O.L., An algorithm for linearly constrained adaptive array processing, *IEEE Proc.*, 60, 926–935, 1972.

Gab84 Gabel, R.A. and Kurth, R.R., Hybrid time-delay/phase-shift digital beamforming for uniform collinear array, *J. Acoust. Soc. Am.*, 75, 1837–1847, 1984.

Geb95 Gebauer, T. and Gockler, H.G., Channel-individual adaptive beamforming for mobile satellite communications, *IEEE J. Selected Areas Commn.*, 13, 439–448, 1995.

God95 Godara, L.C., Application of the fast Fourier transform to broadband beamforming, *J. Acoust. Soc. Am.*, 98, 230–240, 1995.

God97 Godara, L.C., Application to antenna arrays to mobile communications. Part II: Beamforming and direction of arrival considerations, *IEEE Proc.*, 85, 1195–1247, 1997.

God99 Godara, L.C. and Jahromi, M.R.S., Limitations and capabilities of frequency domain constrained beamforming schemes, *IEEE Trans. Signal Process.*, 47, 2386–2395, 1999.

Gri82 Griffiths, L.J. and Jim, C.W., An alternative approach to linearly constrained adaptive beamforming, *IEEE Trans. Antennas Propagat.*, 30, 27–34, 1982.

Gri87 Griffiths, L.J. and Buckley, K.M., Quiescent pattern control in linearly constrained adaptive arrays, *IEEE Trans. Acoust. Speech Signal Process.*, 35, 917–926, 1987.

Hin81 Hinich, M.J., Frequency-wave number array processing, *J. Acoust. Soc. Am.*, 69, 732–737, 1981.

Hod79 Hodgkiss, W.S., Adaptive array processing: time vs. frequency domain, *International Conference on Acoustics, Speech, and Signal Processing*, 282–284, 1979.

Hua90 Huang, K.C., Chang, S.H. and Chen, Y.H., An alternative structure for adaptive broadband beamforming with imperfect arrays, *J. Acoust. Soc. Am.*, 87, 1218–1226, 1990.

Iig85 Iiguni, Y., Sakai, H. and Tokumaru, H., Convergence properties of simplified gradient adaptive lattice algorithms, *IEEE Trans. Acoust. Speech Signal Process.*, 33, 1427–1434, 1985.

Ish95 Ishii, N. and Kohno, R., Spatial and temporal equalization based on an adaptive tapped-delay-line array antenna, *IEICE Trans. Commn.*, E78-B, 1162–1169, 1995.

Jai81 Jain, A.K. and Ranganath, S., Extrapolation algorithms for discrete signals with application in spectral estimation, *IEEE Trans. Acoust. Speech Signal Process.*, 29, 830–845, 1981.

Jim77 Jim, C.W., A comparison of two LMS constrained optimal array structures, *IEEE Proc.*, 65, 1730–1731, 1977.

Kik89 Kikuma, N. and Takao, K. Broadband and robust adaptive antenna under correlation constraints, *IEE Proc.*, 136, Pt. H, 85–89, 1989.

Ko81 Ko, C.C., Jamming rejection capability of broadband Frost power inversion array, *IEE Proc.* 128, Pt. F, 140–151, 1981.

Ko87 Ko, C.C., Tracking performance of a broadband tapped delay line adaptive array using the LMS algorithm, *IEE Proc.*, 134, Pt. F, 295–302, 1987.

Ko90 Ko, C.C., Fast null steering algorithm for broadband power inversion array, *IEE Proc.*, 137, Pt. F, 377–383, 1990.

Koh92 Kohno, R., Wang, H. and Imai, H., Adaptive array antenna combined with tapped delay line using processing gain for spread spectrum CDMA systems, paper presented at IEEE International Symposium on Personal, Indoor and Mobile Radio Communications, Boston, 1992.

Kum90 Kumaresan, R., On a frequency domain analog of Prony's method, *IEEE Trans. Acoust. Speech Signal Process.*, 38, 168–170, 1990.

Lin86 Ling, F., Manolakis, D. and Proakis, J.G., Numerically robust least-squares lattice ladder algorithms with direct updating of the reflection coefficients, *IEEE Trans. Acoust. Speech Signal Process.*, 34, 837–845, 1986.

Lue69 Luenberger, D.G., Optimization by Vector Space Method, Wiley, New York, 1969.

Man82 Mansour, D. and Gray, Jr., A.H., Unconstrained frequency-domain adaptive filter, *IEEE Trans. Acoust. Speech Signal Process.*, 30, 168–170, 1982.

Mar89 Maranda, B., Efficient digital beamforming in the frequency domain, *J. Acoust. Soc. Am.*, 86, 1813–1819, 1989.

May81 Mayhan, J.T., Simmons, A.J. and Cummings, W.C., Wide-band adaptive nulling using tapped delay lines, *IEEE Trans. Antennas Propagat.*, 29, 923–936, 1981.

Muc84 Mucci, R.A., A comparison of efficient beamforming algorithms, *IEEE Trans., Acoust. Speech Signal Process.*, 32, 548–558, 1984.

Nar81 Narayan, S.S. and Peterson, A.M., Frequency domain least-mean-square algorithm, *IEEE Proc.*, 69, 124–126, 1981.

Nor94 Nordebo, S., Claesson, I. and Nordhom, S., Weighted Tschebysheff approximation for the design of broadband beamformers using quadratic programming, *IEEE Signal Process. Lett.*, 1, 103–105, 1994.

Nun83 Nunn, D., Performance assessments of a time domain adaptive processor in a broadband environment, *IEE Proc.* 130, Parts F and H, 139–145, 1983.

Pap75 Papoulis, A., A new algorithm in spectral analysis and band-limited extrapolation, *IEEE Trans. Circuits Syst.*, 22, 735–742, 1975.

Pie69 Pierre, D.A., *Optimization Theory with Applications*, Wiley, New York, 1969.

Pri78 Pridham, R.G. and Mucci, R.A., A novel approach to digital beamforming, *J. Acoust. Soc. Am.*, 63, 425–434, 1978.

Pri79 Pridham, R.G. and Mucci, R.A., Digital interpolation beamforming for low-pass and bandpass signals, *IEEE Proc.*, 67, 904–919, 1979.

Ree85 Reed, F.A., Feintuch, P.L. and Bershad, N.J., The application of the frequency domain LMS adaptive filter to split array bearing estimation with a sinusoidal signal, *IEEE Trans. Acoust. Speech Signal Process.*, 33, 61–69, 1985.

Rod79 Rodgers, W.E. and Compton, Jr., R.T., Adaptive array bandwidth with tapped delay line processing, *IEEE Trans. Aerosp. Electron. Syst.*, 15, 21–28, 1979

Rud69 Rudnick, P., Digital beamforming in the frequency domain, *J. Acoust. Soc. Am.*, 46, 1089–1090, 1969.

Sco83 Scott, K.K., Transversal filter techniques for adaptive array applications, *IEE Proc.*, 130, Pts. F and H, 29–35, 1983.

Shy85 Shynk, J.J. and Gooch, R.P., Frequency-domain adaptive pole-zero filtering, *IEEE Proc.*, 73, 1526–1528, 1985.

Sim83 Simaan, M., Optimum array filters for array data signal processing, *IEEE Trans. Acoust. Speech Signal Process.*, 31, 1006–10015, 1983.

Sna83 Snaz, J.L.C. and Huang, T.H., Discrete and continuous band-limited signal extrapolation, *IEEE Trans., Acoust. Speech Signal Process.*, 31, 1276–1285, 1983.

Soh84 Sohie, G.R.L. and Sibul, L.H., Stochastic convergence properties of the adaptive gradient lattice, *IEEE Trans. Acoust. Speech Signal Process.*, 32, 102–107, 1984.

Son82 Sonnenschein, A. and Dickinson, B.W., On a recent extrapolation procedure for band-limited signals, *IEEE Trans. Circuits Syst.*, 29, 116–117, 1982.

Ste87 Steyskal, H., Digital beamforming antenna, an introduction, *Microwave J.*, Jan. 1987, pp. 107–124.

Syl86 Sylva, P.D., Menard, P. and Roy, D., A reconfigurable real-time interpolation beamformer, *IEEE J. Oceanic Eng.*, 11, 123–126, 1986.

Sul91 Sullivan, B.J., Effect of sampling rate on the conjugate gradient method applied to signal extrapolation, *IEEE Trans. Signal Process.*, 39, 1235–1238, 1991.

Tak85 Takao, K. and Ishizaki, T., Constraints of the output power minimization adaptive array for broadband desired signal, *Trans. Inst. Electron. Commn. Eng. (Jpn.)*, J68-B, 411–418, 1985.

Tak80 Takao, K. and Komiyama, K., An adaptive antenna for rejection of wideband interference, *IEEE Trans. Aerosp. Electron. Syst.*, 16, 452–459, 1980.

Thn93 Thng, I., Cantoni, A. and Leung, Y.H., Derivative constrained optimum broadband antenna arrays, *IEEE Trans. Signal Process.*, 41, 2376–2388, 1993.

Val95 Valaee, S. and Kabal, P., Wideband array processing using a two-sided correlation transformation, *IEEE Trans. Signal Process.*, 43, 160–172, 1995.

Van91 Van Veen, B.D., Minimum variance beamforming with soft response constraints, *IEEE Trans. Signal Process.*, 39, 1964–1972, 1991.

Voo92 Vook, E.W. and Compton Jr., R.T., Bandwidth performance of linear arrays with tapped delay line processing, *IEEE Trans. Aerosp. Electron. Syst.*, 28, 901–908, 1992.

Web84 Weber, M.E. and Heisler, R., A frequency-domain beamforming algorithm for wideband, coherent signal processing, *J. Acoust. Soc. Am.*, 76, 1132–1144, 1984.

Win72 Winkler, L.P. and Schwartz, M., Adaptive nonlinear optimization of the signal-to-noise ratio of an array subject to a constraint, *J. Acoust. Soc. Am.*, 52, 39–51, 1972.

Win94 Winters, J.H., Salz, J. and Gitlin, R.D., The impact of antenna diversity on the capacity of wireless communication systems, *IEEE Trans. Commn.*, 42, 1740–51, 1994.

Yeh87 Yeh, C.C., Hong, Y.J. and Ucci, D.R., Use of tapped delay line adaptive array to increase the number of degrees of freedom for interference suppression, *IEEE Trans. Aerosp. Electron. Syst.*, 23, 809–813, 1987.

Zhu90 Zhu, J.X. and Wang, H., Adaptive beamforming for correlated signal and interference: A frequency domain smoothing approach, *IEEE Trans. Acoust. Speech Signal Process.*, 38, 193–195, 1990.

# 5

# *Correlated Fields*



Interference canceling capabilities of the optimal antenna array processors discussed in previous chapters assumed implicitly or explicitly that the desired signals arriving from the look direction and the nonlook directional interferences are not correlated. Correlation between the desired signal and unwanted interference exists in situations of multipath arrivals and deliberate jamming, and affects the performance of antenna array processors as discussed in [Wid82, God90, Tak87, Sha85, Han86, Han88, Lut86, Red87, Zol88, Ali92, Qia95, Cho87, Tak86, Wil88, Han92].

The two directional signals are said to be fully correlated or coherent when one is the delayed and scaled version of the other. For two sinusoidal signals, this amounts to the fixed-phase relation between the two. The coherence between two signals normally arises from deliberate jamming using so-called smart jammers, whereas the multipath signals normally result in partial correlation. The study of antenna array systems presented in this chapter includes general correlated fields with coherence as a special case. Unless otherwise explicitly stated, only two directional sources are assumed to be present to facilitate the derivation of analytical expressions for the performance measure of antenna array processors.

Correlation between the desired signal and an interference limits the applicability of various weight estimation schemes. For example, when the weights are estimated by minimizing the mean output power subject to the look direction constraint, the processor cancels the desired signal while maintaining the constraint. The reason this happens is that the processor, while minimizing the mean output power, adjusts the phase of the correlated interference induced on each antenna such that the power of the sum of the signal and the interference that is correlated with the signal is minimized, causing the signal cancelation. This is consistent with the design that the processor minimizes the output power. The optimal weights design is based on the assumption that the signal is not correlated with interference.

The correlation $\delta_{xy}(f)$ between two broadband signals x(t) and y(t) is defined in terms of their power spectrum [Car87]:

$$\delta_{xy}(f) = \frac{G_{xy}(f)}{\sqrt{G_{xx}(f)G_{yy}(f)}} \tag{5.1}$$

with $G_{xy}(f)$ denoting the cross-power spectrum. It is related to the cross-correlation function

$$\rho_{xy}(\tau) = E[x(t)y(t+\tau)] \tag{5.2}$$

by the inverse Fourier transform

$$G_{xy}(f) = \int_{-\infty}^{\infty} \rho_{xy}(\tau)e^{j2\pi f\tau}d\tau \tag{5.3}$$

This chapter shows that the correlation between the desired signal and the unwanted interference severely degrades the performance of antenna array systems, and techniques are presented to improve their performance by decorrelating the directional sources. Both narrowband and broadband arrays are discussed.

## 5.1 Correlated Signal Model

Consider an array of L omnidirectional elements immersed in the far field of two sinusoidal sources. One source is a signal source and the second is interference. Let $p_S$ and $p_I$ represent the powers of the signal source and the interference, respectively; and let $\sigma_n^2$ denote the variance of the random noise component on each element with the temporal narrowband spectrum and spatially white spectrum.

Let an L dimensional vector **x**(t) represent the L wave forms derived from L elements of the array, and let a complex scalar δ, which lies within the unit circle, represent the correlation coefficient between the two sources. Assuming the center of the coordinate system as the time reference, the vector **x**(t) can be expressed as

$$\mathbf{x}(t) = \sqrt{p_S}\, m_S(t)\mathbf{S}_0 + \sqrt{p_I}\left(\delta^* m_S(t) + \sqrt{1-|\delta|^2}\, m_I(t)\right)\mathbf{S}_1 + \mathbf{n}(t) \tag{5.1.1}$$

where $\mathbf{S}_0$ and $\mathbf{S}_1$ are steering vectors in the signal direction and in the interference direction, respectively; * denotes the complex conjugate; $\mathbf{n}(t)$ represents the random noise component; and $m_S(t)$ and $m_I(t)$ are zero-mean, unit-variance, complex low-pass processes associated, respectively, with the signal source and the interference source. It is assumed that $m_S(t)$, $m_I(t)$, and $n_\ell(t)$ are mutually uncorrelated with $n_\ell(t)$ denoting the $\ell$th component of $\mathbf{n}(t)$.

The value of the complex scalar $\delta$ decides the correlated field under consideration. When $\delta$ lies on the unit circle, that is $|\delta| = 1$, the two sources are coherent and their fixed-phase difference is given by $\delta_p$, the phase of $\delta$. On the other hand, when it lies inside the unit circle with $|\delta| < 1$, the two sources are partially correlated and $\delta = 0$ corresponds to the uncorrelated field case. For the uncorrelated field, (5.1.1) becomes

$$\mathbf{x}(t) = \sqrt{p_S}\, m_s(t)\mathbf{S}_0 + \sqrt{p_I}\, m_I(t)\mathbf{S}_1 + \mathbf{n}(t) \tag{5.1.2}$$

Equation (5.1.2) is identical to (2.1.15) with $M = 2$, $m_1 = \sqrt{p_S}\, m_s(t)$, $m_2 = \sqrt{p_I}\, m_I(t)$, $\mathbf{S}_1 = \mathbf{S}_0$, and $\mathbf{S}_2 = \mathbf{S}_1$.

From (5.1.1) it follows that the array correlation matrix R can be expressed as

$$\begin{aligned} R &= E\left[\mathbf{x}(t)\mathbf{x}^H(t)\right] \\ &= ASA^H + \sigma_n^2 I \end{aligned} \tag{5.1.3}$$

where $L \times 2$ dimensional matrix

$$A = \left[\mathbf{S}_0, \mathbf{S}_I\right] \tag{5.1.4}$$

and $2 \times 2$ dimensional source correlation matrix

$$S = \begin{bmatrix} p_S & \sqrt{p_S p_I}\,\delta \\ \sqrt{p_S p_I}\,\delta^* & p_I \end{bmatrix} \tag{5.1.5}$$

Note that (5.1.4) is identical to (2.1.27) with $M = 2$. However, the source correlation matrix S for the correlated case given by (5.1.5) differs from the uncorrelated case given by (2.1.28) due to the presence of off-diagonal terms containing the correlation coefficient.

The above equations show how the correlation between the two sources affects R. It follows from these expressions that when two sources are uncorrelated, that is $|\delta| = 0$, S is a diagonal matrix guaranteeing R to be positive definite (assuming A is of full rank, which requires that steering vectors corresponding to all directional sources are linearly independent [God81]). The presence of correlation affects the rank of S and thus of R. In the presence of correlation, the matrix R becomes ill conditioned and may not be invertible, making it difficult for estimation of the weights of the optimal beamformer, which relies on existence of the inverse of R. Thus, a beamforming scheme, which is optimal in the absence of correlated arrival, is not able to cancel a correlated interference.

In the next section, the behavior of the constrained element space processor (ESP) discussed in Section 2.4 is analyzed.

## 5.2 Optimal Element Space Processor

Consider the narrowband ESP shown in Figure 2.1. The output y(t) and the mean output power P(**w**) of the processor for the given weights **w** are given by

$$y(t) = \mathbf{w}^H \mathbf{x}(t) \tag{5.2.1}$$

and

$$P(\mathbf{w}) = \mathbf{w}^H R \mathbf{w} \tag{5.2.2}$$

Let $\hat{\mathbf{w}}$ represent the L weights of the processor that minimizes the mean output power subject to unity constraint in the look direction, that is,

$$\begin{aligned} &\underset{\mathbf{w}}{\text{minimize}} \quad \mathbf{w}^H R \mathbf{w} \\ &\text{subject to} \quad \mathbf{w}^H \mathbf{S}_0 = 1 \end{aligned} \tag{5.2.3}$$

The processor with these weights is referred to as the optimal processor in Chapter 2. An expression for the mean output power of the optimal processor in the presence of correlated arrival is derived [Red87] below. Substituting for R from (5.1.3) to (5.1.5) in (5.2.2) it follows that

$$\begin{aligned} P(\mathbf{w}) &= \left[\mathbf{w}^H \mathbf{S}_0, \ \mathbf{w}^H \mathbf{S}_I\right] \begin{bmatrix} p_S & \sqrt{p_S p_I}\,\delta \\ \sqrt{p_S p_I}\,\delta^* & p_I \end{bmatrix} \begin{bmatrix} \mathbf{S}_0^H \mathbf{w} \\ \mathbf{S}_I^H \mathbf{w} \end{bmatrix} + \sigma_n^2 \mathbf{w}^H \mathbf{w} \\ &= p_S \mathbf{w}^H \mathbf{S}_0 \mathbf{S}_0^H \mathbf{w} + p_I \mathbf{w}^H \mathbf{S}_I \mathbf{S}_I^H \mathbf{w} + \sqrt{p_S p_I}\,\delta\, \mathbf{w}^H \mathbf{S}_0 \mathbf{S}_I^H \mathbf{w} \\ &\quad + \sqrt{p_S p_I}\,\delta^*\, \mathbf{w}^H \mathbf{S}_I \mathbf{S}_0^H \mathbf{w} + \sigma_n^2 \mathbf{w}^H \mathbf{w} \end{aligned} \tag{5.2.4}$$

To solve beamforming problem (5.2.3) using the Lagrange multiplier method define a cost function,

$$J(\mathbf{w}) = \frac{1}{2} P(\mathbf{w}) + \lambda\left(\mathbf{w}^H \mathbf{S}_0 - 1\right) \tag{5.2.5}$$

where $\lambda$ is the Lagrange multiplier. The solution $\hat{\mathbf{w}}$ is obtained by setting the partial differentiation of the cost function with respect to **w** equal to zero. Thus,

$$\left.\frac{\partial J(\mathbf{w})}{\partial \mathbf{w}}\right|_{\mathbf{w}=\hat{\mathbf{w}}} = 0 \tag{5.2.6}$$

Substituting for J(**w**) in (5.2.6) and using (5.2.4),

$$\lambda \mathbf{S}_0 = -\left\{p_S \mathbf{S}_0 \mathbf{S}_0^H + p_I \mathbf{S}_I \mathbf{S}_I^H + \sqrt{p_S p_I}\, \delta \mathbf{S}_0 \mathbf{S}_I^H + \sqrt{p_S p_I}\, \delta^* \mathbf{S}_I \mathbf{S}_0^H + \sigma_n^2\right\} \hat{\mathbf{w}} \tag{5.2.7}$$

Define

$$\beta = \frac{\mathbf{S}_0^H \mathbf{S}_I}{L} \tag{5.2.8}$$

and note that

$$\hat{\mathbf{w}}^H \mathbf{S}_0 = 1 \tag{5.2.9}$$

$$\mathbf{S}_0^H \mathbf{S}_0 = L \tag{5.2.10}$$

and

$$\mathbf{S}_I^H \mathbf{S}_I = L \tag{5.2.11}$$

Premultiplying (5.2.7) by $\mathbf{S}_0^H$, and using (5.2.8) to (5.2.11) one obtains

$$\lambda = -p_S - \mathbf{S}_I^H \hat{\mathbf{w}}\left(p_I \beta + \sqrt{p_S p_I}\, \delta\right) - \sqrt{p_S p_I}\, \delta^* \beta - \frac{\sigma_n^2}{L} \tag{5.2.12}$$

Substituting for $\lambda$ in (5.2.7), premultiplying it with $\mathbf{S}_I^H$, using (5.2.8) to (5.2.11), and solving for $\hat{\mathbf{w}}^H \mathbf{S}_I$, one obtains the optimized processor response in the interference direction, that is,

$$\hat{\mathbf{w}}^H \mathbf{S}_I = \frac{\sigma_n^2 \beta - \sqrt{p_S p_I}\, \delta L \rho}{\sigma_n^2 + p_I L \rho} \tag{5.2.13}$$

where

$$\rho = 1 - \frac{\mathbf{S}_0^H \mathbf{S}_I \mathbf{S}_I^H \mathbf{S}_0}{L^2} \tag{5.2.14}$$

Note that $\rho$ is also given by

$$\rho = 1 - \beta \beta^* \tag{5.2.15}$$

Equation (5.2.13) describes the response of the beamformer in the interference direction. Consider $\sigma_n^2 = 0$. Equation (5.2.13) for this case becomes

$$\hat{\mathbf{w}}^H \mathbf{S}_I = \sqrt{\frac{p_S}{p_I}}\, \delta \tag{5.2.16}$$

It follows from (5.2.16) that the response of the optimal processor in the absence of white noise is not zero, as is the case for the uncorrelated sources case. Thus, it does not cancel the interference when it is correlated with the desired signal.

The implication of (5.2.16) is that even though the response of the processor in the look direction is unity ($\hat{\mathbf{w}}^H \mathbf{S}_I = 1$), the processor using weights $\hat{\mathbf{w}}$ suppresses the look direction signal. This aspect is now examined by deriving an expression for the mean output power of the optimal processor.

It follows from (5.2.4) and the fact that an expression for $\hat{\mathbf{w}}^H \mathbf{S}_I$ has been derived previously, that to evaluate $P(\hat{\mathbf{w}})$, only an expression for $\hat{\mathbf{w}}^H \hat{\mathbf{w}}$ is necessary. This can be obtained by premultiplying (5.2.7) with $\hat{\mathbf{w}}^H$ and using (5.2.8) to (5.2.11). Thus,

$$\sigma_n^2 \hat{\mathbf{w}}^H \hat{\mathbf{w}} = -\lambda - p_S - p_I \hat{\mathbf{w}}^H \mathbf{S}_I \mathbf{S}_I^H \hat{\mathbf{w}} - \sqrt{p_S p_I}\, \delta \mathbf{S}_I^H \hat{\mathbf{w}} + \sqrt{p_S p_I}\, \delta^* \hat{\mathbf{w}}^H \mathbf{S}_I \tag{5.2.17}$$

where $\lambda$ is given by (5.2.12) and $\hat{\mathbf{w}}^H \mathbf{S}_I$ is given by (5.2.13). Substituting for $\sigma_n^2 \hat{\mathbf{w}}^H \hat{\mathbf{w}}$ from (5.2.17), $\lambda$ from (5.2.12), and $\hat{\mathbf{w}}^H \mathbf{S}_I$ from (5.2.13) in (5.2.4),

$$\begin{aligned} \hat{P} &= \hat{\mathbf{w}}^H R \hat{\mathbf{w}} \\ &= p_S + \frac{\sigma_n^2}{L} + \frac{-p_S p_I |\delta|^2 L\rho + (1-\rho) p_I \sigma_n^2 + \sqrt{p_S p_I}\left(\beta\delta^* + \beta^* \delta\right)\sigma_n^2}{p_I L\rho + \sigma_n^2} \end{aligned} \tag{5.2.18}$$

Now consider $\sigma_n^2 = 0$. For this case

$$\hat{P} = p_S\left(1 - |\delta|^2\right) \tag{5.2.19}$$

It follows from (5.2.19) that the mean output power of the processor decreases as the magnitude of the correlation constant increases and reduces to zero for coherent sources. The processor in this case completely cancels the desired signal.

In the next section, the optimized postbeamformer interference canceler (PIC) processor is studied in a correlated field environment. It is shown that the performance of the optimized PIC processor is identical to that of the optimal ESP [God90]. Using this fact, a derivation of the output signal-to-noise ratio (SNR) of the optimal ESP is presented in Section 5.4.

## 5.3 Optimized Postbeamformer Interference Canceler Processor

The narrowband PIC processor structure is shown in Figure 2.6.3. Section 2.6.3 shows that the mean output power of the processor for a given weight w is given by

$$P(w) = \mathbf{V}^H R \mathbf{V} + w^* w \mathbf{U}^H R \mathbf{U} - w^* \mathbf{V}^H R \mathbf{U} - w \mathbf{U}^H R \mathbf{V} \tag{5.3.1}$$

where L dimensional complex vectors $\mathbf{V}$ and $\mathbf{U}$, respectively, denote the fixed weights of the signal beam and the interference beam, and a complex scalar w denotes the adjustable weight. The optimal weight $\hat{w}$, which minimizes the mean output power P(w), is given by (2.6.48), that is,

$$\hat{w} = \frac{\mathbf{V}^H R \mathbf{V}}{\mathbf{U}^H R \mathbf{U}} \tag{5.3.2}$$

Assume that the signal beam is formed using the conventional beam forming weights, that is,

$$\mathbf{V} = \frac{1}{L}\mathbf{S}_0 \tag{5.3.3}$$

and that the interference beam is selected as follows:

$$\mathbf{U} = P\mathbf{S}_I \tag{5.3.4}$$

where P is a projection matrix given by

$$P = I - \frac{\mathbf{S}_0 \mathbf{S}_0^H}{L} \tag{5.3.5}$$

Equation (5.3.3) ensures that the signal beam response in the look direction is unity. The interference beam selected using (5.3.4) and (5.3.5) has a unity response in the interference direction and has a null in the look direction. This form of interference beam has been selected to facilitate the derivation of the output SNR for the optimized ESP.

Next, an expression for $P(\hat{w})$, the mean output power of the optimized PIC, is derived, and it is shown that $P(\hat{w})$ is equal to $\hat{P}$, the mean output power of the optimal ESP in the presence of correlated sources. It follows from (5.3.1) and (5.3.2) that the mean output power $P(\hat{w})$ of the optimal PIC is given by

$$P(\hat{w}) = \mathbf{V}^H R \mathbf{V} - \frac{\mathbf{U}^H R \mathbf{V} \mathbf{V}^H R \mathbf{U}}{\mathbf{U}^H R \mathbf{U}} \tag{5.3.6}$$

Substituting for **V** and **U** from (5.3.3) and (5.3.4) in (5.3.6) results in an expression for $P(\hat{w})$. This is achieved by evaluating $\mathbf{V}^H R\mathbf{V}$, $\mathbf{V}^H R\mathbf{U}$, $\mathbf{U}^H R\mathbf{V}$, and $\mathbf{U}^H R\mathbf{U}$, and substituting in (5.3.6). It follows from (5.3.3), (5.3.4), and (5.3.5) that

$$\begin{aligned} \mathbf{V}^H \mathbf{U} &= \frac{\mathbf{S}_0^H P \mathbf{S}_I}{L} \\ &= 0 \end{aligned} \tag{5.3.7}$$

$$\begin{aligned} \mathbf{U}^H \mathbf{U} &= \mathbf{S}_I^H P \mathbf{S}_I \\ &= L\rho \end{aligned} \tag{5.3.8}$$

These, along with (5.1.3), imply that

$$\mathbf{V}^H R \mathbf{U} = \mathbf{V}^H A S A^H \mathbf{U} \tag{5.3.9}$$

$$\mathbf{V}^H R \mathbf{V} = \mathbf{V}^H A S A^H \mathbf{V} + \frac{\sigma_n^2}{L} \tag{5.3.10}$$

and

$$\mathbf{U}^H R\mathbf{U} = \mathbf{U}^H ASA^H\mathbf{U} + L\rho\sigma_n^2 \tag{5.3.11}$$

From (5.3.3), (5.1.4), and (5.2.8),

$$V^H A = [1 \quad \beta] \tag{5.3.12}$$

Similarly,

$$\begin{aligned} A^H\mathbf{U} &- \begin{bmatrix} \mathbf{S}_0^H P\mathbf{S}_I \\ \mathbf{S}_I^H P\mathbf{S}_I \end{bmatrix} \\ &= \begin{bmatrix} 0 \\ L\rho \end{bmatrix} \end{aligned} \tag{5.3.13}$$

Thus,

$$\begin{aligned} \mathbf{V}^H R\mathbf{U} &= [1 \quad \beta] \begin{bmatrix} p_S & \sqrt{p_S p_I}\,\delta \\ \sqrt{p_S p_I}\,\delta^* & p_I \end{bmatrix} \begin{bmatrix} 0 \\ L\rho \end{bmatrix} \\ &= \sqrt{p_S p_I}\,\delta L\rho + p_I \beta L\rho \end{aligned} \tag{5.3.14}$$

$$\begin{aligned} \mathbf{V}^H R\mathbf{U} &= [1 \quad \beta] \begin{bmatrix} p_S & \sqrt{p_S p_I}\,\delta \\ \sqrt{p_S p_I}\,\delta^* & p_I \end{bmatrix} \begin{bmatrix} 1 \\ \beta^* \end{bmatrix} + \frac{\sigma_n^2}{L} \\ &= p_S + (1-\rho)p_I + \sqrt{p_S p_I}\left(\beta\delta^* + \beta^*\delta\right) + \frac{\sigma_n^2}{L} \end{aligned} \tag{5.3.15}$$

and

$$\begin{aligned} \mathbf{U}^H R\mathbf{U} &= [0 \quad L\rho] \begin{bmatrix} p_S & \sqrt{p_S p_I}\,\delta \\ \sqrt{p_S p_I}\,\delta^* & p_I \end{bmatrix} \begin{bmatrix} 0 \\ L\rho \end{bmatrix} + L\rho\sigma_n^2 \\ &= L^2\rho^2 p_I + L\rho\sigma_n^2 \end{aligned} \tag{5.3.16}$$

It follows from (5.3.14) along with (5.2.15) and (5.3.16) that

$$\frac{\mathbf{U}^H R\mathbf{V}\mathbf{V}^H R\mathbf{U}}{\mathbf{U}^H R\mathbf{U}} = \frac{p_S p_I |\delta|^2 L^2\rho + p_I^2(1-\rho)\rho L^2 + \sqrt{p_S p_I}\,L^2 p_I \rho\left(\beta\delta^* + \beta^*\delta\right)}{L^2\rho p_I + L\sigma_n^2} \tag{5.3.17}$$

Subtracting (5.3.17) from (5.3.15), the following expression for $P(\hat{w})$ results:

$$P(\hat{w}) = p_S + \frac{\sigma_n^2}{L} + \frac{-p_S p_I |\delta|^2 L\rho + (1-\rho)p_I\sigma_n^2 + \sqrt{p_S p_I}\left(\beta\delta^* + \beta^*\delta\right)\sigma_n^2}{p_I L\rho + \sigma_n^2} \tag{5.3.18}$$

Comparing (5.2.18) and (5.3.18), shows that the mean output powers of the optimal ESP and the optimal PIC with **V** and **U** selected using (5.3.3) and (5.3.5), respectively, are the same. Thus, in the presence of the correlated sources, the two processors perform identically.

In the next section, an expression for the output SNR of the two processors is derived [God90]. As the performance of the two processors is the same, the PIC processor is used for derivation of results.

## 5.4 Signal-to-Noise Ratio Performance

For the ease of analysis, rewrite (5.1.1) by regrouping terms containing $m_S(t)$ as follows:

$$\mathbf{x}(t) = m_S(t)\left(\sqrt{p_S}\,\mathbf{S}_0 + \delta^* \sqrt{p_I}\,\mathbf{S}_I\right) + m_I(t)\sqrt{p_I}\sqrt{1-|\delta|^2}\,\mathbf{S}_I + n(t) \tag{5.4.1}$$

From (5.1.3) and (5.4.1),

$$R = \left[p_S \mathbf{S}_0 \mathbf{S}_0^H + |\delta|^2 p_I \mathbf{S}_I \mathbf{S}_I^H + \sqrt{p_S p_I}\left(\delta \mathbf{S}_0 \mathbf{S}_I^H + \delta^* \mathbf{S}_I \mathbf{S}_0^H\right)\right] + p_I\left(1-|\delta|^2\right)\mathbf{S}_I \mathbf{S}_I^H + \sigma_n^2 I \tag{5.4.2}$$

It should be noted that the array correlation matrix is composed of three terms. The first term in square brackets is contributed by the signal source. Let it be denoted by $R_S$. The second and the third terms on the RHS of (5.4.2) are contributions due to the interference source (the component that is uncorrelated with the signal source) and the random noise. Let these be denoted by $R_I$ and $R_n$, respectively.

It follows from (5.3.1) that the output signal power $P_S(\hat{w})$, residual interference power $P_I(\hat{w})$, and the output uncorrelated noise power $P_n(\hat{w})$ of the optimal PIC, respectively are given by

$$P_S(\hat{w}) = \mathbf{V}^H R_S \mathbf{V} + \hat{w}^* \hat{w} \mathbf{U}^H R_S \mathbf{U} - \hat{w}^* \mathbf{V}^H R_S \mathbf{U} - \hat{w} \mathbf{U}^H R_S \mathbf{V} \tag{5.4.3}$$

$$P_I(\hat{w}) = \mathbf{V}^H R_I \mathbf{V} + \hat{w}^* \hat{w} \mathbf{U}^H R_I \mathbf{U} - \hat{w}^* \mathbf{V}^H R_I \mathbf{U} - \hat{w} \mathbf{U}^H R_I \mathbf{V} \tag{5.4.4}$$

and

$$P_n(\hat{w}) = \mathbf{V}^H R_n \mathbf{V} + \hat{w}^* \hat{w} \mathbf{U}^H R_n \mathbf{U} - \hat{w}^* \mathbf{V}^H R_n \mathbf{U} - \hat{w} \mathbf{U}^H R_n \mathbf{V} \tag{5.4.5}$$

Let $P_N(\hat{w})$ denote the total noise at the optimal PIC output. This consists of output uncorrelated noise power and residual interference power, that is,

$$P_N(\hat{w}) = P_I(\hat{w}) + P_n(\hat{w}) \tag{5.4.6}$$

First, consider $P_S(\hat{w})$ and evaluate various terms on the RHS of (5.4.3). It follows from (5.3.2), (5.3.14), and (5.3.16) that

$$\hat{w} = \frac{\sqrt{p_S p_I}\,\delta + p_I \beta}{L \rho p_I + \sigma_n^2} \tag{5.4.7}$$

and

$$\hat{w}^*\hat{w} = \frac{p_S p_I |\delta|^2 + p_I^2(1-\rho) + p_I\sqrt{p_S p_I}(\delta^*\beta + \delta\beta^*)}{(L\rho p_I + \sigma_n^2)} \tag{5.4.8}$$

Substituting for $R_S$ from the first term on the RHS of (5.4.2) and using (5.3.3) to (5.3.5),

$$\mathbf{V}^H R_S \mathbf{V} = p_S + |\delta|^2 p_I(1-\rho) + \sqrt{p_S p_I}(\delta\beta^* + \delta^*\beta) \tag{5.4.9}$$

$$\mathbf{U}^H R_S \mathbf{U} = |\delta|^2 p_I L^2 \rho^2 \tag{5.4.10}$$

and

$$\mathbf{U}^H R_S \mathbf{V} = |\delta|^2 p_I L\rho\beta^* + \sqrt{p_S p_I}\,\delta^* L\rho \tag{5.4.11}$$

It follows from (5.4.7) and (5.4.11) that

$$\hat{w}\mathbf{U}^H R_S \mathbf{V} = \frac{L\rho p_I\left[p_S|\delta|^2 + p_I(1-\rho)|\delta|^2 + \sqrt{p_S p_I}|\delta|^2\beta^*\delta + \sqrt{p_S p_I}\,\delta^*\beta\right]}{L\rho p_I + \sigma_n^2} \tag{5.4.12}$$

Substituting for $R_S$ from (5.4.8) to (5.4.12) in (5.4.3),

$$\begin{aligned}
P_S(\hat{w}) &= p_S + |\delta|^2 p_I(1-\rho) + \sqrt{p_S p_I}(\delta\beta^* + \delta^*\beta) \\
&\quad + \frac{p_S p_I^2|\delta|^4 L^2\rho^2 + p_I^3(1-\rho)|\delta|^2 L^2\rho^2 + p_I^2|\delta|^2 L^2\rho^2\sqrt{p_S p_I}(\delta\beta^* + \delta^*\beta)}{(L\rho p_I + \sigma_n^2)^2} \\
&\quad - \frac{2L\rho p_I p_S|\delta|^2 + 2L\rho p_I^2(1-\rho)|\delta|^2}{L\rho p_I + \sigma_n^2} \\
&\quad - \frac{L\rho p_I\sqrt{p_S p_I}|\delta|^2(\delta\beta^* + \delta^*\beta) + L\rho p_I\sqrt{p_S p_I}|\delta|^2(\delta\beta^* + \delta^*\beta)}{L\rho p_I + \sigma_n^2} \qquad (5.4.13) \\
&= p_S\left[1 + \left(\frac{p_I|\delta|^2 L\rho}{L\rho p_I + \sigma_n^2}\right)^2 - \frac{2L\rho p_I|\delta|^2}{L\rho p_I + \sigma_n^2}\right] \\
&\quad + p_I(1-\rho)|\delta|^2\left[1 + \frac{p_I^2 L^2\rho^2}{(L\rho p_I + \sigma_n^2)^2} - \frac{2L\rho p_I}{L\rho p_I + \sigma_n^2}\right] \\
&\quad + \sqrt{p_S p_I}(\delta\beta^* + \delta^*\beta)\left[1 + \frac{p_I^2|\delta|^2 L^2\rho^2}{(L\rho p_I + \sigma_n^2)^2} - \frac{L\rho p_I|\delta|^2 + L\rho p_I}{L\rho p_I + \sigma_n^2}\right]
\end{aligned}$$

After manipulation, (5.4.13) leads to

$$P_S(\hat{w}) = p_S\left[1 - \frac{|\delta|^2 L\rho p_I}{L\rho p_I + \sigma_n^2}\right]^2 + p_I(1-\rho)|\delta|^2\left[\frac{\sigma_n^2}{L\rho p_I + \sigma_n^2}\right]^2$$
$$+ \sqrt{p_S p_I}\,(\delta\beta^* + \delta^*\beta)\frac{\sigma_n^2}{L\rho p_I + \sigma_n^2}\left(1 - \frac{L\rho p_I|\delta|^2}{L\rho p_I + \sigma_n^2}\right) \quad (5.4.14)$$

Next, an expression for the residual interference at the output of the optimal PIC is derived. Substituting from (5.3.3), (5.3.4), and the second term on the RHS of (5.4.2) for $R_I$ in (5.4.4), results in

$$P_I(\hat{w}) = P_I\left(1 - |\delta|^2\right)\left[1 - \rho + \hat{w}^*\hat{w}L^2\rho^2 - \hat{w}^*\beta L\rho - \hat{w}\beta^* L\rho\right] \quad (5.4.15)$$

which along with (5.4.7) and (5.4.8) implies that

$$P_I(\hat{w}) = p_I\left(1 - |\delta|^2\right)\left[1 - \rho + \frac{p_S p_I|\delta|^2 L^2\rho^2 + p_I^2(1-\rho)L^2\rho^2}{\left(L\rho p_I + \sigma_n^2\right)^2}\right.$$
$$+ \frac{p_I\sqrt{p_S p_I}\,L^2\rho^2(\delta^*\beta + \delta\beta^*)}{\left(L\rho p_I + \sigma_n^2\right)^2}$$
$$\left. - \frac{L\rho\sqrt{p_S p_I}\,(\delta^*\beta + \delta\beta^*)}{L\rho p_I + \sigma_n^2} + \frac{2L\rho p_I(1-\rho)}{L\rho p_I + \sigma_n^2}\right] \quad (5.4.16)$$
$$= p_I\left(1 - |\delta|^2\right)\left[(1-\rho)\left(1 + \left(\frac{p_I L\rho}{L\rho p_I + \sigma_n^2}\right)^2 - \frac{2L\rho p_I}{L\rho p_I + \sigma_n^2}\right)\right.$$
$$\left. + \frac{p_S p_I|\delta|^2 L^2\rho^2}{\left(L\rho p_I + \sigma_n^2\right)^2} + \frac{L\rho\sqrt{p_S p_I}\,(\delta^*\beta + \delta\beta^*)}{L\rho p_I + \sigma_n^2}\left(\frac{L\rho p_I}{L\rho p_I + \sigma_n^2} - 1\right)\right]$$

After manipulation, (5.4.16) leads to

$$P_I(\hat{w}) = p_I\left(1 - |\delta|^2\right)\left[\frac{\sigma_n^2\left\{(1-\rho)\sigma_n^2 - L\rho\sqrt{p_S p_I}\,(\delta^*\beta + \delta\beta^*)\right\} + p_S p_I|\delta|^2 L^2\rho^2}{\left(L\rho p_I + \sigma_n^2\right)^2}\right] \quad (5.4.17)$$

Similarly, an expression for the uncorrelated noise power at the optimal PIC output is given by

$$P_n(\hat{w}) = \frac{\sigma_n^2}{L} + L\rho p_I \sigma_n^2 \frac{p_S|\delta|^2 + p_I(1-\rho) + \sqrt{p_S p_I}\,(\delta^*\beta + \delta^*\beta)}{\left(L\rho p_I + \sigma_n^2\right)} \quad (5.4.18)$$

Substituting from (5.4.17) and (5.4.18) in (5.4.6), after manipulation

$$P_N(\hat{w}) = \frac{\sigma_n^2}{L}\frac{Lp_I + \sigma_n^2}{L\rho p_I + \sigma_n^2} + \frac{|\delta|^2 p_S L\rho p_I}{L\rho p_I + \sigma_n^2}\left(1 - \frac{|\delta|^2 L p_I \rho}{L\rho p_I + \sigma_n^2}\right)$$
$$+ \frac{|\delta|^2\sqrt{p_S p_I} L\rho p_I(\delta^*\beta + \delta\beta^*) - |\delta|^2 p_I(1-\rho)\sigma_n^2}{(L\rho p_I + \sigma_n^2)}\sigma_n^2 \tag{5.4.19}$$

It can easily be verified that for $|\delta| = 0$, (5.4.19) reduces to (2.7.72).

Let $SNR(\hat{w})$ denote the output SNR of the optimal PIC defined as

$$SNR(\hat{w}) = \frac{P_S(\hat{w})}{P_N(\hat{w})} \tag{5.4.20}$$

Substituting for $P_S(\hat{w})$ and $P_N(\hat{w})$ from (5.4.14) and (5.4.19) in (5.4.20),

$$SNR(\hat{w}) = \frac{p_S\left(1+\gamma-|\delta|^2\right)^2 + p_I(1-\rho)|\delta|^2\gamma^2 + \sqrt{p_S p_I}(\delta^*\beta + \beta^*\delta)\gamma\left(1+\gamma-|\delta|^2\right)}{\frac{\sigma_n^2}{L}\left(\frac{1}{\rho}+\gamma\right)(1+\gamma) + |\delta|^2\left[p_S\left(1+\gamma-|\delta|^2\right) + \gamma\sqrt{p_S p_I}(\delta^*\beta + \beta^*\delta) - p_I(1-\rho)\gamma^2\right]} \tag{5.4.21}$$

where

$$\gamma = \frac{\sigma_n^2}{L\rho p_I} \tag{5.4.22}$$

As discussed in the previous section, the optimal PIC and ESP behave identically in the absence of errors. Thus, the expression for the output SNR given by (5.4.21) is true for both processors. Let it be denoted by SNRO. In the following, some special cases are considered.

### 5.4.1 Zero Uncorrelated Noise

Zero uncorrelated noise corresponds to $\sigma_n^2 = 0$. From (5.4.22) it follows that for this case, $\gamma = 0$, which along with (5.4.21) implies that

$$SNRO = \frac{1-|\delta|^2}{|\delta|^2} \tag{5.4.23}$$

Thus, in the absence of uncorrelated noise, the output SNR is independent of the array geometry and noise environment. It only depends on the magnitude of the correlation coefficient and is independent of its phase. It should be noted here that $\gamma = 0$ indirectly assumes that $p_I$ and $\rho$ are not identical to zero.

### 5.4.2 Strong Interference and Large Number of Elements

Now, consider a case when there is a strong interference source in the presence of nonzero uncorrelated noise, and the array consists of a large number of elements, such that

$$\gamma \to 0 \tag{5.4.24}$$

It follows from (5.4.21) and (5.4.24) that

$$\text{SNRO} = \frac{\left(1-|\delta|^2\right)^2}{\dfrac{\sigma_n^2}{L\rho p_S} + |\delta|^2\left(1-|\delta|^2\right)} \tag{5.4.25}$$

Thus, the output SNR in this case is less than that for the zero uncorrelated noise case and decreases as the uncorrelated noise power increases.

### 5.4.3 Coherent Sources

This corresponds to $|\delta| = 1$. Substituting $|\delta| = 1$ in (5.4.21), the following expression results for the output SNR when the signal source and the interference source are fully correlated:

$$\text{SNRO} = \frac{\gamma^2\left(p_S + p_I(1-\rho) + \sqrt{p_S p_I}\,\Omega\right)}{\dfrac{\sigma_n^2}{L}\left(\dfrac{1}{\rho} + \gamma\right)(1+\gamma) + p_S\gamma - p_I(1-\rho)\gamma^2 + \gamma\sqrt{p_S p_I}\,\Omega} \tag{5.4.26}$$

where

$$\Omega = \beta e^{-j\delta_P} + \beta^* e^{j\delta_P} \tag{5.4.27}$$

with $\delta_p$ denoting the phase of the correlation coefficient.

It follows from (5.2.15) that

$$\beta = \sqrt{1-\rho}\, e^{j\beta_P} \tag{5.4.28}$$

where $\beta_p$ denoting the phase of $\beta$. Thus, (5.4.27) implies

$$\Omega = 2\sqrt{1-\rho}\cos\left(\beta_p - \delta_p\right) \tag{5.4.29}$$

Substituting for $\gamma$ and $\Omega$ in (5.6.26) leads to the following expression for SNRO for fully correlated sources:

$$\text{SNRO} = \frac{A\dfrac{\sigma_n^2}{L}}{\left(\dfrac{\sigma_n^2}{L}\right)^2 + B\dfrac{\sigma_n^2}{L} + C} \tag{5.4.30}$$

with

$$A = p_S + p_I(1-\rho) + \sqrt{p_S p_I}(1-\rho)2\cos(\beta_p - \delta_p) \quad (5.4.31)$$

$$B = 2p_I\rho \quad (5.4.32)$$

$$C = p_I^2 + p_S p_I \rho + 2p_I^{3/2}\rho\sqrt{p_S}(1-\rho)\cos(\beta_p - \delta_p) \quad (5.4.32)$$

It follows from (5.4.30) that for fully correlated sources, the output SNR (1) increases as $\sigma_n^2/L$ increases for low values of $\sigma_n^2/L$, and (2) decreases as $\sigma_n^2/L$ increases for high values of $\sigma_n^2/L$. Furthermore, the output SNR attains the maximum value

$$SNRO_{max} = \frac{A}{2\sqrt{C} + B} \quad (5.4.34)$$

when

$$\sigma_n^2 = L\sqrt{C} \quad (5.4.35)$$

### 5.4.4 Examples and Discussion

In this section, some examples are presented to understand the effect of correlation on the output SNR. For the results presented in Figure 5.1 to Figure 5.4, a linear array of one-half wavelength spacing is used. The signal source of unity power is assumed broadside to the array. Interference direction is measured relative to the line of the array. The correlation phase is measured at the center of the coordinate system, which is at an end element of the array. The correlation phase is assumed to be equal to 45°.

Figure 5.1 shows the output SNR as a function of the uncorrelated noise power for various values of $|\delta|$. The curve with the solid line is for fully correlated sources and agrees with the results presented in the previous section. One observes from the figure that as $|\delta|$ increases, the output SNR (1) decreases for low values of uncorrelated noise, and (2) increases for high values of uncorrelated noise. The reason for the increase in the output SNR as $|\delta|$ increases in the presence of high uncorrelated noise power is that, for this scenario, the optimal processor tends to behave as the conventional processor. The response of the conventional processor in the direction of the interference is fixed, and thus the processor does not minimize the output power by canceling the desired signal.

Figure 5.2 to Figure 5.4 show the output SNR as a function of $|\delta|$ for various values of $\sigma_n^2$. Figure 5.2 is for an array with four elements whereas Figure 5.3 is for an array with ten elements. Comparing Figure 5.2 and Figure 5.3, it is apparent that for a given noise field, an increase in the number of elements in an array causes the output SNR to increase in the absence of correlation but has a reverse effect when the sources are fully correlated. Note the difference in the scales for the two figures.

Figure 5.4 shows the results for an array with ten elements when interference is in direction 65°. A comparison between Figure 5.3 and Figure 5.4 reveals that the effect of correlation on the output SNR is more when the interference is far from the look direction. For the scenario of Figure 5.4, the output SNR decreases as $|\delta|$ increases at all values of $\sigma_n^2$. However, the reduction at higher values of $\sigma_n^2$ is much less than at lower values of $\sigma_n^2$.

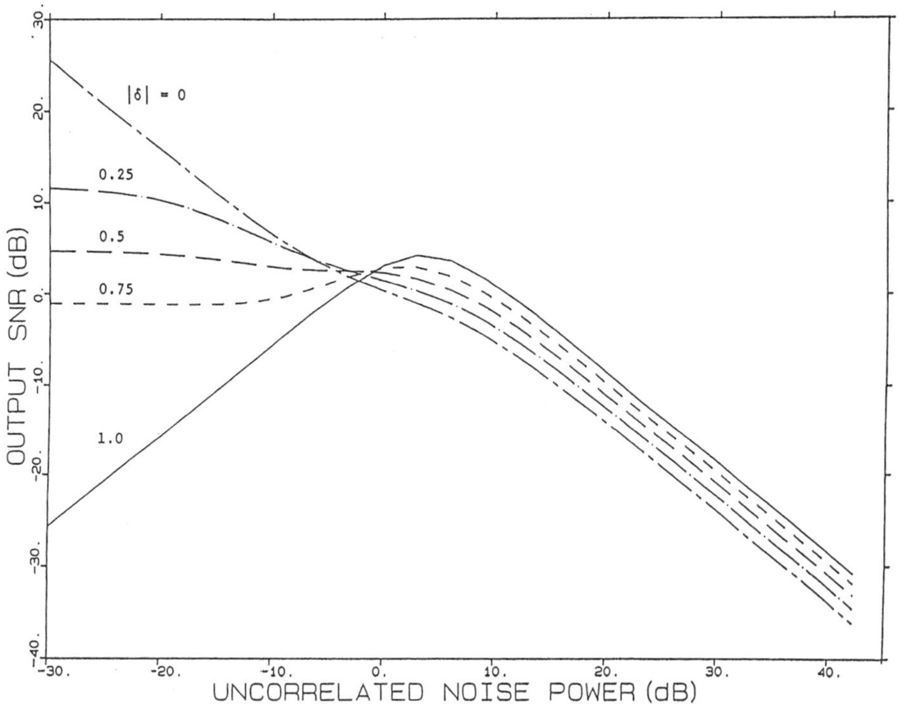


**FIGURE 5.1**
Output SNR vs. the uncorrelated noise power for a four element linear array with one-half wavelength spacing for various values of $|\delta|$, $p_I = 1$, $\theta_I = 85°$, $p_S = 1$, $\theta_0 = 90°$, $\delta_p = 45°$. (From Godara, L.C., *IEEE Trans. Acoust. Speech Signal Process.*, 38, 1–15, 1990. ©IEEE. With permission.)

## 5.5 Methods to Alleviate Correlation Effects

Many beamforming schemes have been devised to cancel an interference source that is correlated with the signal. In principle, these work by restoring the rank of R. In this section, some of these are briefly reviewed [God97].

In some earlier work [Wid82, Gab80], a mechanical movement of the array perpendicular to the look direction was suggested to reduce the signal cancelation effect by the correlated interference. The scheme generally known as the spatial dither algorithm works on the principle that as the movement is perpendicular to the look direction, the signal induced in the array is not affected, whereas the interference that arrives from a direction different from that of the signal gets modulated with this motion. This causes a reduction in interference, as noted in [Cho87] where the dither algorithm is further developed such that a mechanical movement is not required.

The spatial smoothing scheme [Eva81] uses a notion of spatial averaging by subdividing the array into smaller subarrays, and estimates the array correlation matrix by averaging the correlation matrices estimated from each such subarray. The use of spatial smoothing for beamforming is discussed in [Sha85, Red87] showing that the use of this method reduces effective correlation between the interference and desired signal resulting in reduced signal cancelation caused by optimal beamforming. Details on spatial smoothing are provided in Section 5.6.

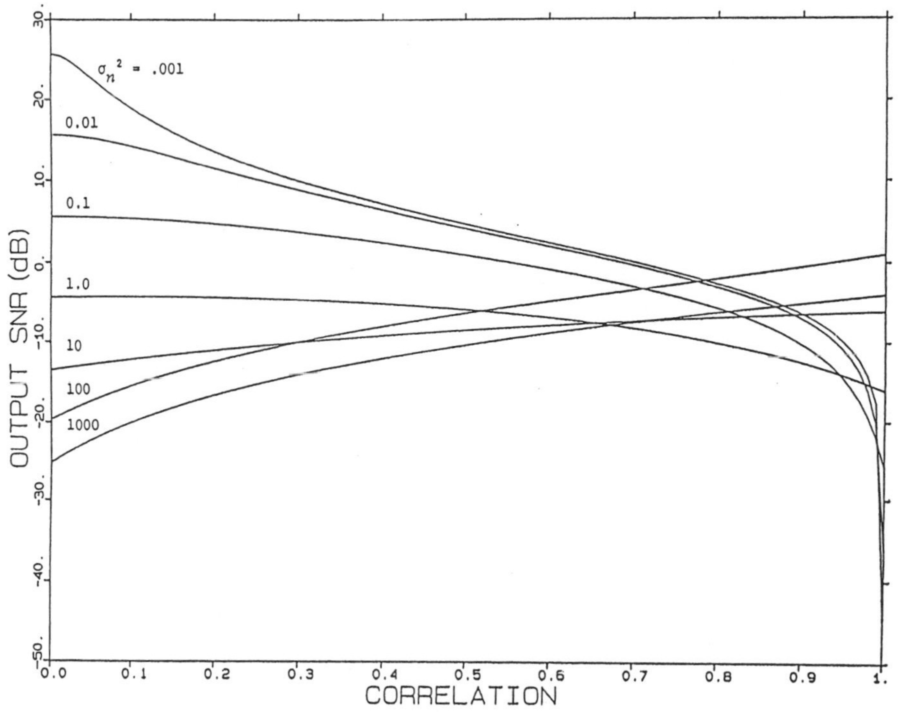


**FIGURE 5.2**
Output SNR vs. the magnitude of the correlation coefficient for a four-element linear array with one-half wavelength spacing for various values of $\sigma_n^2$, $p_I = 100$, $\theta_I = 85°$, $p_S = 1$, $\theta_0 = 90°$, $\delta_p = 45°$. (From Godara, L.C., *IEEE Trans. Acoust. Speech Signal Process.*, 38, 1–15, 1990. ©IEEE. With permission.)

The spatial smoothing method uses uniform averaging of all matrices obtained from various subarrays, that is, each matrix is weighted equally. This results in an estimate of the matrix that is not as good as the one that could have been obtained from given subarray matrices. Ideally in the absence of correlation, the array correlation matrix for a uniformly spaced linear array has a Toeplitz structure, that is, elements of the matrix along each diagonal are equal, and the estimated matrix by the spatial smoothing scheme is not the closest to the Toeplitz matrix. An estimated matrix that is closest to a Toeplitz matrix is obtained by a spatial averaging technique [Tak87, Lim90]. This technique weighs each subarray matrix differently and then optimize the weights such that it minimizes the mean square error between the weighted matrix and a Toeplitz matrix. When this matrix is used to estimate the weights of the beamformer, the resulting system reduces more interference than that given by the uniform weighted matrix estimate.

It should be noted that the number of rows and columns in the estimated matrix is equal to the number of elements in the subarray and not equal to the number of elements in the full array. Thus, the weights estimated by this matrix could only be applied to one of the subarrays. Consequently, not all array elements are used for beamforming. This reduces the array aperture and its degrees of freedom. For an environment consisting of M − 1 direction interferences and the desired signal, the subarray size should be at least M + 1 and the number of subarrays should be at least M(M − 1) + 1 [Tak87].

A scheme that does not reduce the degrees of freedom of the array is described in [God90]. It decorrelates the sources by structuring the correlation matrix as the Toeplitz type by averaging along each diagonal, and uses the resulting matrix to estimate the weights of the full array. An adaptive algorithm to estimate the weights of an array based

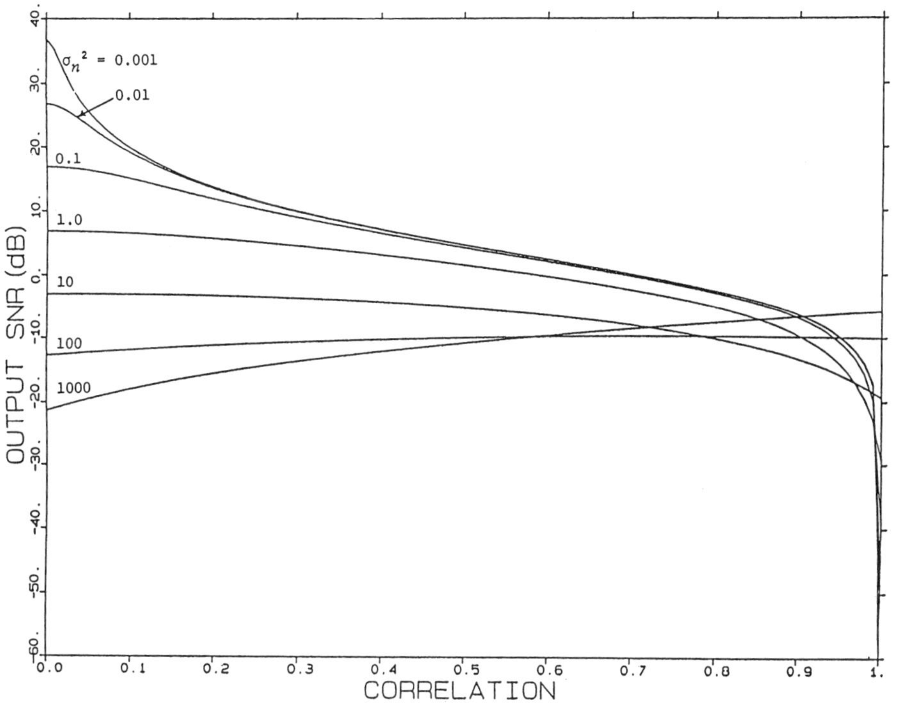


**FIGURE 5.3**
Output SNR vs. the magnitude of the correlation coefficient for a ten-element linear array with one-half wavelength spacing for various values of $\sigma_n^2$, $p_I = 100$, $\theta_I = 85°$, $p_S = 1$, $\theta_0 = 90°$, $\delta_p = 45°$. (From Godara, L.C., *IEEE Trans. Acoust. Speech Signal Process.*, 38, 1–15, 1990. ©IEEE. With permission.)

on this principle is presented in [God91], and the concept is extended to broadband beamforming in [God92]. Details are provided in Section 5.7 and Section 5.8.

A beamforming scheme [Wid82] based on master and slave concepts cancels the correlated arrival by the use of two channels. In one channel, the look direction signal is blocked, and then weights are estimated by solving the constrained beamforming problem. These weights are then used on the second channel. As the signal is not present at the time of weight estimation, the beamformer does not cancel the signal. However, the process only works for one correlated interference. It is extended for the multiple correlated interference case in [Lut86] where an array of 2M – 1 elements is required to cancel M – 1 interferences.

Other schemes that require some knowledge of the interference, such as direction or the correlation matrix due to interference only, are discussed in [Han86, Han88, Qia95, Wil88, Han92]. Many of the schemes discussed above improve the array performance in the presence of correlated arrivals by treating the correlated components as interferences and canceling them by forming nulls in their directions using beamforming techniques. These methods do not utilize the correlated components as is done in the diversity-combining techniques discussed in Chapter 7. In diversity combining, various components are added in a way to improve the signal level.

The RAKE receiver [Vau88, Tur80, Pri58, Faw64] achieves this increase in signal level for a CDMA system by using a number of demodulators operating in parallel to track each component employing the user code for that signal. The signal delay is identified by sliding the code sequence to obtain the maximum correlation with the received component. The signals are added at the baseband after appropriate delay and amplitude scaling. The receiver, however, does not cancel unwanted interference by shaping the beam pattern.

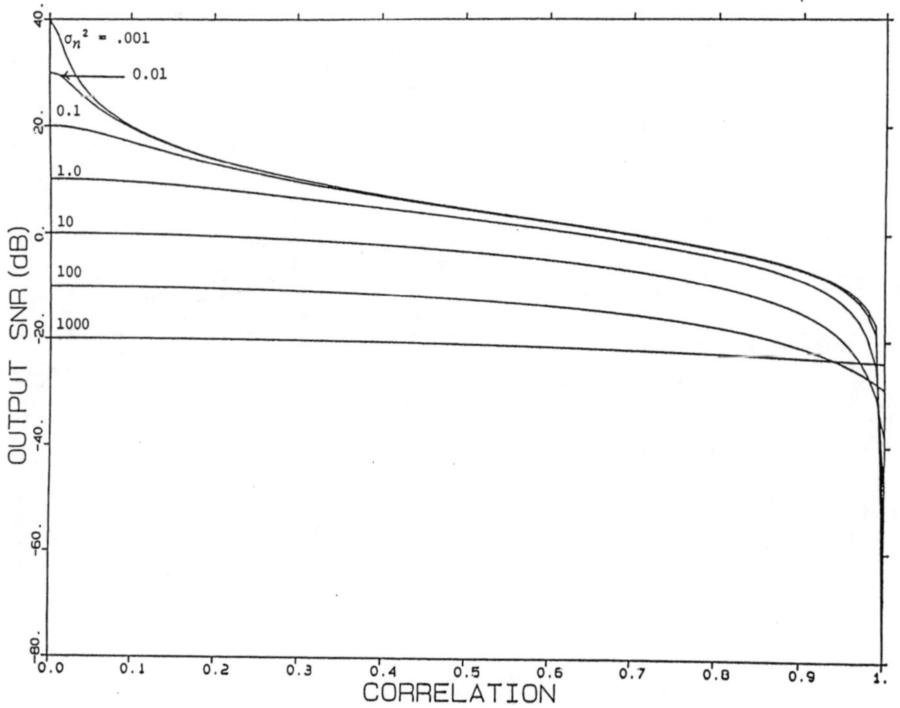


**FIGURE 5.4**
Output SNR vs. the magnitude of the correlation coefficient for a ten-element linear array with one-half wavelength spacing for various values of $\sigma_n^2$, $p_I = 100$, $\theta_I = 65°$, $p_S = 1$, $\theta_0 = 90°$, $\delta_p = 45°$. (From Godara, L.C., *IEEE Trans. Acoust. Speech Signal Process.*, 38, 1–15, 1990. ©IEEE. With permission.)

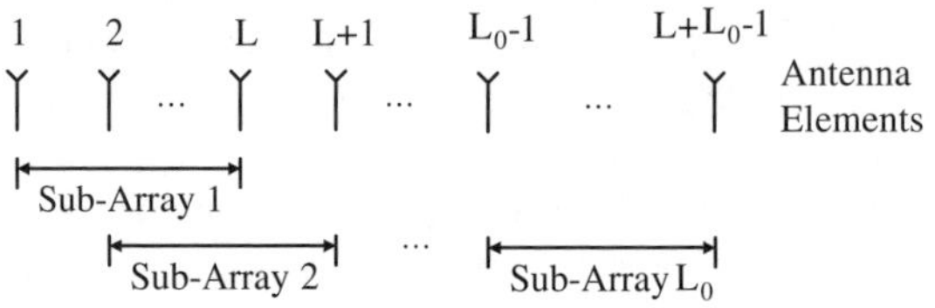


**FIGURE 5.5**
Construction of subarrays.

## 5.6 Spatial Smoothing Method

The spatial smoothing method, also known as the subarray averaging method, estimates the weights of an L-element antenna array system using an augmented array of more than L elements, and is suitable for a linear array of equispaced elements. The signals induced on these extra elements are only used to restore the rank of the array correlation matrix to be used in weight estimation. These signals are not used to produce the array output.

The method divides the array in $L_0$ subarrays of size L such that the first subarray consists of Element 1 to Element L, the second one consists of Element 2 to Element L + 1, and so on as shown in Figure 5.5.

Let the L dimensional vectors $\mathbf{x}_1(t)$, $\mathbf{x}_2(t)$, …, $\mathbf{x}_{L_0}(t)$ denote the array signal vectors of $L_0$ subarrays, that is,

$$\mathbf{x}_1(t) = \left[x_1(t), x_2(t), \ldots, x_L(t)\right]^T \tag{5.6.1}$$

$$\mathbf{x}_2(t) = \left[x_2(t), x_3(t), \ldots, x_{L+1}(t)\right]^T \tag{5.6.2}$$

$$\mathbf{x}_{L_0}(t) = \left[x_{L_0}(t), x_{L_0+1}(t), \ldots, x_{L+L_0-1}(t)\right]^T \tag{5.6.3}$$

where $x_k(t)$ denotes the signal induced on the kth element of the augmented array (full array).

Let $R_k$ denote the array correlation matrix of the kth subarray, that is,

$$R_k = E\left[\mathbf{x}_k(t)\mathbf{x}_k^H(t)\right] \tag{5.6.4}$$

Define the spatially smoothed correlation matrix $\bar{R}$ by averaging $R_k$, $k = 1, 2, \ldots, L_0$, that is,

$$\bar{R} = \frac{1}{L_0}\sum_{k=1}^{L_0} R_k \tag{5.6.5}$$

and use this to estimate the weights of the array system. It follows from (5.6.3) that to form $L_0$ subarrays of size L, one needs $L + L_0 - 1$ elements.

As shown in [Sha85], the matrix $\bar{R}$ has full-rank iff $L_0 \geq L - 1$. Thus, to estimate an $L \times L$ dimensional full-rank spatially smoothed correlation matrix to estimate weights of an L-element array system, at least $2(L - 1)$ elements are necessary.

### 5.6.1 Decorrelation Analysis

In this section, an analysis is presented that shows the decorrelation effect of the spatial smoothing method [Red87]. It follows from (2.1.15) that the array signal vector $\mathbf{x}(t)$ due to M directional sources and white noise can be expressed in the matrix notation as

$$\mathbf{x}(t) = A\mathbf{s}(t) + \mathbf{n}(t) \tag{5.6.6}$$

where the M dimensional vector $\mathbf{s}(t)$ is defined as

$$\mathbf{s}(t) = \left[m_1(t), m_2(t), \ldots, m_M(t)\right]^T \tag{5.6.7}$$

with $m_k(t)$ denoting the modulating function of the kth source and

$$A = \left[\mathbf{S}(\theta_1), \ldots, \mathbf{S}(\theta_M)\right] \tag{5.6.8}$$

with $\mathbf{S}(\theta_k)$ denoting the steering vector associated with the kth source in direction $\theta_k$, that is,

$$\mathbf{S}(\theta_k) = \left[e^{j2\pi f_0 \tau_1(\theta_k)}, \ldots, e^{j2\pi f_0 \tau_L(\theta_k)}\right] \tag{5.6.9}$$

with

$$\tau_\ell(\theta_k) = (\ell - 1)\frac{d}{c}\cos\theta_k \tag{5.6.10}$$

Now, consider the array signal vector for the kth subarray. Following (5.6.6) to (5.6.10), it can be expressed as

$$\mathbf{x}_k(t) = A_k\mathbf{s}(t) + \mathbf{n}_k(t) \tag{5.6.11}$$

where $\mathbf{n}_k(t)$ denotes the random noise vector received by the kth subarray,

$$A_k = \left[\mathbf{S}_k(\theta_1), \mathbf{S}_k(\theta_2), \ldots, \mathbf{S}_k(\theta_M)\right]^T \tag{5.6.12}$$

and

$$\mathbf{S}_k(\theta) = \left[e^{j2\pi f_0 \tau_1^k(\theta)}, e^{j2\pi f_0 \tau_2^k(\theta)}, \ldots, e^{j2\pi f_0 \tau_L^k(\theta)}\right]^T \tag{5.6.13}$$

with $\tau_\ell^k(\theta)$ denoting the propagation delay from the origin to the $\ell$th element in the kth subarray. As the kth subarray is comprised of elements k to k + (L – 1), it follows that

$$\tau_1^k(\theta) = (k-1)\frac{d}{c}\cos\theta \tag{5.6.14}$$

$$\tau_2^k(\theta) = k\frac{d}{c}\cos\theta \tag{5.6.15}$$

and

$$\tau_L^k(\theta) = (k+L-2)\frac{d}{c}\cos\theta \tag{5.6.16}$$

Thus, (5.6.13) becomes

$$\mathbf{S}_k(\theta) = \begin{bmatrix} 1 \\ e^{j2\pi\hat{d}\cos\theta} \\ \vdots \\ e^{j2\pi\hat{d}(L-1)\cos\theta} \end{bmatrix} e^{j2\pi\hat{d}(k-1)\cos\theta} \tag{5.6.17}$$

where $\hat{d} = d/\lambda$ and $\lambda$ denotes the wavelength corresponding to $f_0$.

Substituting (5.6.17) in (5.6.12) and using (5.6.8),

$$A_k = A\Phi^{k-1} \tag{5.6.18}$$

where $\Phi$ is an $M \times M$ diagonal matrix with

$$\Phi_{m,m} = e^{j2\pi\hat{d}\cos\theta_m}, \qquad m = 1, 2, \ldots, M \tag{5.6.19}$$

It follows from (5.6.11) and (5.6.18) that

$$\mathbf{x}_k = A\Phi^{k-1}\mathbf{s}(t) + \mathbf{n}_k(t) \tag{5.6.20}$$

Equations (5.6.4) and (5.6.20) imply that

$$R_k = A\Phi^{k-1}S\left(\Phi^{k-1}\right)^H A^H + \sigma^2 I \tag{5.6.21}$$

where S is the source covariance matrix defined as

$$S = E\left[\mathbf{s}(t)\mathbf{s}^H(t)\right] \tag{5.6.22}$$

For uncorrelated sources, S is given by (2.1.28). For correlated sources, $S_{i,j}$ denotes the correlation between ith and jth sources. For the correlated source model presented in Section 5.1, $S_{i,j}$ is given by (5.1.5).

The following expression for the spatially smooth correlation matrix results after substituting for $R_k$ from (5.6.21) in (5.6.5):

$$\bar{R} = A\bar{S}A^H + \sigma^2 I \tag{5.6.23}$$

where

$$\bar{S} = \frac{1}{L_0}\sum_{k=1}^{L_0} \Phi^{k-1}S\left(\Phi^{k-1}\right)^H \tag{5.6.24}$$

denotes the smoothed sources covariance matrix.

To understand the effect of spatial smoothing on the correlation between difference sources, consider $\bar{S}_{ij}$. It follows from (5.6.24) that

$$\bar{S}_{i,j} = \frac{S_{i,j}}{L_0}\sum_{k=1}^{L_0} \Phi_{i,i}^{k-1}\left(\Phi_{j,j}^{k-1}\right)^* \tag{5.6.25}$$

Equation (5.6.25) along with (5.6.19) imply that

$$\bar{S}_{i,j} = \begin{cases} S_{i,j} & i = j \\ \dfrac{S_{i,j}}{L_0}\displaystyle\sum_{k=1}^{L_0} e^{j2\pi\hat{d}\psi_{ij}(k-1)} & i \neq j \end{cases} \tag{5.6.26}$$

where

$$\psi_{ij} = \cos\theta_i - \cos\theta_j \tag{5.6.27}$$

Using

$$1 + a + a^2 + \ldots, a^{N-1} = \frac{1 - a^N}{1 - a} \tag{5.6.28}$$

$i \neq j$ terms in (5.6.26) simplifies to

$$\bar{S}_{i,j} = \frac{S_{i,j}}{L_0} \frac{\sin \pi \hat{d} \psi_{ij} L_0}{\sin \pi \hat{d} \psi_{ij}} e^{j2\pi \hat{d} \psi_{ij}(L_0 - 1)} \tag{5.6.29}$$

which reduces as $L_0$ increases and goes to zero in the limit.

Thus, the sources progressively get decorrelated as the number of subarrays is increased, and the rate of decorrelation depends on the element spacing and the source directions.

### 5.6.2 Adaptive Algorithm

In this section, updating the weights of an array processor from available array samples using the spatial smoothing method is discussed [Sha85]. Assume that N samples of the kth subarray signal vectors $\mathbf{x}_k(n)$, n = 1, 2, …, N are available. It follows then from (5.6.4) and (5.6.5) that an estimate of the spatially smoothed correlation matrix is given by

$$\begin{aligned} \hat{\bar{R}}_N &= \frac{1}{L_0} \sum_{k=1}^{L_0} \frac{1}{N} \sum_{n=1}^{N} \mathbf{x}_k(n) \mathbf{x}_k^H(n) \\ &= \frac{1}{NL_0} \sum_{n=1}^{N} \sum_{k=1}^{L_0} \mathbf{x}_k(n) \mathbf{x}_k^H(n) \end{aligned} \tag{5.6.30}$$

Using the next sample $\mathbf{x}_k(N + 1)$, the matrix $\hat{\bar{R}}_{N+1}$ becomes

$$\begin{aligned} \hat{\bar{R}}_{N+1} &= \frac{1}{(N+1)L_0} \sum_{n=1}^{N+1} \sum_{k=1}^{L_0} \mathbf{x}_k(n) \mathbf{x}_k^H(n) \\ &= \frac{1}{(N+1)L_0} \sum_{n=1}^{N} \sum_{k=1}^{L_0} \mathbf{x}_k(n) \mathbf{x}_k^H(n) + \frac{1}{(N+1)L_0} \sum_{k=1}^{L_0} \mathbf{x}_k(N+1) \mathbf{x}_k^H(N+1) \\ &= \frac{N\hat{\bar{R}}_N}{N+1} + \frac{1}{(N+1)L_0} \sum_{k=1}^{L_0} \mathbf{x}_k(N+1) \mathbf{x}_k^H(N+1) \end{aligned} \tag{5.6.31}$$

Thus, using (5.6.31), the spatially smoothed correlation matrix can be updated as new samples arrive and the new matrix can be used to update the weights. For example, this matrix can be used to estimate the power surface gradient, and the gradient-based adaptive algorithm discussed previously can be employed to update the weights of an array processor.

Equation (5.6.31) can also be employed to update the inverse of the correlation matrix by making successive use of the Matrix Inverse Lemma and the weights can be estimated using the sample inversion algorithm as discussed in Chapter 3.

## 5.7 Structured Beamforming Method

In this section, the use of a structured correlation matrix for estimating optimal weights is discussed, and the decorrelation effect of this technique on the correlated environment is examined [God90]. For ease of analysis, only two sources are assumed to be correlated. The presence of other uncorrelated sources does not alter the analysis.

For the linear array of equispaced receivers immersed in a homogeneous noise field, the array correlation matrix has a Toeplitz structure, that is, the entries along any diagonal are equal. In the presence of correlated sources, the array correlation matrix does not have this structure. The technique proposed here uses an estimate of the array correlation matrix constrained to have this structure. This constraint is implemented by averaging the unconstrained array correlation matrix along the diagonals. Let this matrix, referred to as the structured correlation matrix, be denoted by $\tilde{R}$. The entries along the mth diagonal of $\tilde{R}$ are given by

$$\tilde{R}_m = \frac{1}{L-m}\sum_{\ell=1}^{L-m} R_{\ell,\ell+m} \qquad m = 0, 1, \ldots, L-1 \tag{5.7.1}$$

Using the structured correlation matrix, the following expression is obtained for the weights of the optimal ESP with unity constraint in the look direction:

$$\tilde{\mathbf{w}} = \frac{\tilde{R}^{-1}\mathbf{S}_0}{\mathbf{S}_0^H\tilde{R}^{-1}\mathbf{S}_0} \tag{5.7.2}$$

The mean output power of the processor for a given $\tilde{\mathbf{w}}$ is given by

$$P(\tilde{\mathbf{w}}) = \tilde{\mathbf{w}}^H R \tilde{\mathbf{w}} \tag{5.7.3}$$

It should be noted that the use of the structured correlation matrix has been made only in the feedback loop to calculate the weights of the processor and not in estimating the output power. The structured correlation matrix can be used in obtaining the weights of the optimal PIC processor by replacing R by $\tilde{R}$ in (5.3.2).

### 5.7.1 Decorrelation Analysis

The decorrelation effect of this method, referred to as the structured method, is examined [God90] in this section. Rewrite (5.4.2) in the following form:

$$R = \left[p_S\mathbf{S}_0\mathbf{S}_0^H + p_I\mathbf{S}_I\mathbf{S}_I^H + \sigma_n^2 I\right] + \sqrt{p_S p_I}\left[\delta\mathbf{S}_0\mathbf{S}_I^H + \delta^*\mathbf{S}_I\mathbf{S}_0^H\right] \tag{5.7.4}$$

The term in the first set of square brackets on the RHS of (5.7.4) is not a function of the correlation coefficient, has a Toeplitz structure, and is not affected by averaging along the diagonals. The term in the second set of square bracket depends on δ and does not have a Toeplitz structure. Thus, it is sufficient to examine the effect of averaging along the diagonals on this term. Let this term be denoted by Q, that is,

$$Q = \delta \mathbf{S}_0 \mathbf{S}_I^H + \delta^* \mathbf{S}_I \mathbf{S}_0^H \tag{5.7.5}$$

For an equispaced linear array, the $\ell$th component of a steering vector associated with $\theta$ is given by

$$(\mathbf{S})_\ell = e^{j2\pi\hat{d}(\ell-1)\cos\theta} \tag{5.7.6}$$

where $\hat{d}$ is the spacing between the elements measured in wavelengths, and $\theta$ is the direction of a source relative to the line of the array.

In writing (5.7.6), Element 1 is taken as the time reference. Consider the $(\ell,k)$th element of $\mathbf{S}_0\mathbf{S}_I$

$$\left(\mathbf{S}_0\mathbf{S}_I^H\right)_{\ell,k} = \exp\left\{j2\pi\hat{d}\left[(\ell-1)\cos\theta_0 - (k-1)\cos\theta_I\right]\right\} \tag{5.7.7}$$

with $\theta_0$ and $\theta_1$ respectively denoting the direction of the signal and the interference relative to the line of the array.

Let

$$\phi = \cos\theta_0 - \cos\theta_I \tag{5.7.8}$$

Then the mth diagonal of $\mathbf{S}_0\mathbf{S}_I^H$ is given by

$$\left(\mathbf{S}_0\mathbf{S}_I^H\right)_{\ell,\ell+m} = \exp\left(-j2\pi\hat{d}m\cos\theta_I\right)\exp\left[j2\pi\hat{d}(\ell-1)\phi\right] \qquad \ell = 1, 2, \ldots, L-m \tag{5.7.9}$$

Let $q_m$ denote the average of the mth diagonal. It follows then from (5.7.9) that

$$q_m = \exp\left(-j2\pi\hat{d}m\cos\theta_I\right)\frac{1}{L-m}\sum_{\ell=1}^{L-m}\exp\left[j2\pi\hat{d}(\ell-1)\phi\right] \tag{5.7.10}$$

Using the identity

$$1 + a + a^2 + \ldots + a^{N-1} = \frac{1-a^N}{1-a} \tag{5.7.11}$$

from (5.7.10),

$$q_m = \exp\left(-j2\pi\hat{d}m\cos\theta_I\right)\frac{1}{L-m}\frac{1-\exp\left(j2\pi\hat{d}(L-m)\phi\right)}{1-\exp\left(j2\pi\hat{d}\phi\right)} \tag{5.7.12}$$

is obtained, which, after manipulation, leads to

$$q_m = \exp\left(j\pi\hat{d}(L-1)\phi\right)\exp\left\{-j\pi\hat{d}m\left(\cos\theta_1 + \cos\theta_0\right)\right\}\frac{\sin\pi(L-m)\hat{d}\phi}{(L-m)\sin\hat{d}\phi} \tag{5.7.13}$$

Now consider $\mathbf{S}_1\mathbf{S}_0^H$. Let $q_m^1$ denote the average of the mth diagonal of this matrix, that is,

$$q_m^1 = \exp\left(-j2\pi\hat{d}m\cos\theta_0\right)\frac{1}{L-m}\sum_{\ell=1}^{L-m}\exp\left(-j2\pi\hat{d}(\ell-1)\phi\right) \tag{5.7.14}$$

which, after manipulation, leads to

$$q_m^1 = \exp\left(-j\pi\hat{d}(L-1)\phi\right)\exp\left(-j\pi\hat{d}m\left(\cos\theta_I + \cos\theta_0\right)\right)\frac{\sin\pi(L-m)\hat{d}\phi}{(L-m)\sin\pi\hat{d}\phi} \tag{5.7.15}$$

It follows from (5.7.5), (5.7.10), and (5.4.14) that the entries along the mth diagonal of the structured matrix $\tilde{Q}$, the term in the second set of square brackets in (5.7.4), is given by

$$\tilde{Q}_m = \delta q_m + \delta^* q_m^1 \tag{5.7.16}$$

which, along with (5.7.13) and (5.7.15), imply that

$$\begin{aligned}\tilde{Q}_m &= \exp\left\{-j\pi\hat{d}m\left(\cos\theta_I + \cos\theta_0\right)\right\}\frac{\sin\pi(L-m)\hat{d}\phi}{(L-m)\sin\pi\hat{d}\phi} \\ &\quad \left(\delta\exp\left\{j\pi\hat{d}(L-1)\phi\right\} + \delta^*\exp\left\{-j\pi\hat{d}(L-1)\phi\right\}\right)\end{aligned} \tag{5.7.17}$$

If $|\delta|$ and $\delta_p$, respectively, denote the magnitude and phase of the correlation coefficient measured at the reference point, Element 1 in the present case (5.7.17) reduces to

$$\tilde{Q}_m = \exp\left\{-j\pi\hat{d}m\left(\cos\theta_I + \cos\theta_0\right)\right\}\frac{2|\delta|\sin\pi(L-m)\hat{d}\phi}{(L-m)\sin\pi\hat{d}\phi}\cos\Psi_p \tag{5.7.18}$$

where

$$\Psi_p = \delta_p + \pi\hat{d}(L-1)\phi \tag{5.7.19}$$

is the phase of the correlation coefficient measured at the center of the array.

Equation (5.7.18) describes the mth diagonal of the component of the structured correlation matrix that depends on the correlation coefficient. From (5.7.18), the following observations can be made. For

$$\Psi_p = (2n+1)\frac{\pi}{2}, \quad n = 0, \pm 1, \ldots, \tag{5.7.20}$$

$\cos\Psi_p = 0$ and thus $\tilde{Q}_m$, $m = 0, 1, \ldots, L-1$ reduce to zero. Thus, for these values of the correlation phase, the two sources are completely decorrelated. This result is independent of the magnitude of the correlation coefficient.

For a given element spacing and source direction, the magnitudes of $\tilde{Q}_m$, m = 0, 1, …, L − 1 behave like a well-known function (sin x)/x, with the zeros given by

$$(L-m)\hat{d}\phi = n, \quad n = 0, \pm 1, \ldots, \tag{5.7.21}$$

As the number of elements in the array increases, the magnitude of $\tilde{Q}_m$ decreases. The greatest reduction occurs in the elements of the principal diagonal correspond to m = 0. The magnitude of $\tilde{Q}_0$ is given by

$$\frac{2|\delta|\sin \pi L\hat{d}\phi}{L\sin \hat{d}\phi}\cos \Psi_p \tag{5.7.22}$$

As m increases, the effect of increased elements in the array declines. The last diagonal of $\tilde{Q}$ that consists of only one element, $\tilde{Q}_{L,L}$, is not affected.

#### 5.7.1.1 Examples and Discussion

For these examples, a linear array with one-half wavelength spacing is used. A unity power signal source is assumed to be present broadside to the array. Unless otherwise specified, the correlation phase is measured at an end element of the array.

Figure 5.6 compares the power patterns of the conventional beamformer, optimal beamformer, and structured beamformer. Eight interferences are assumed in the directions of

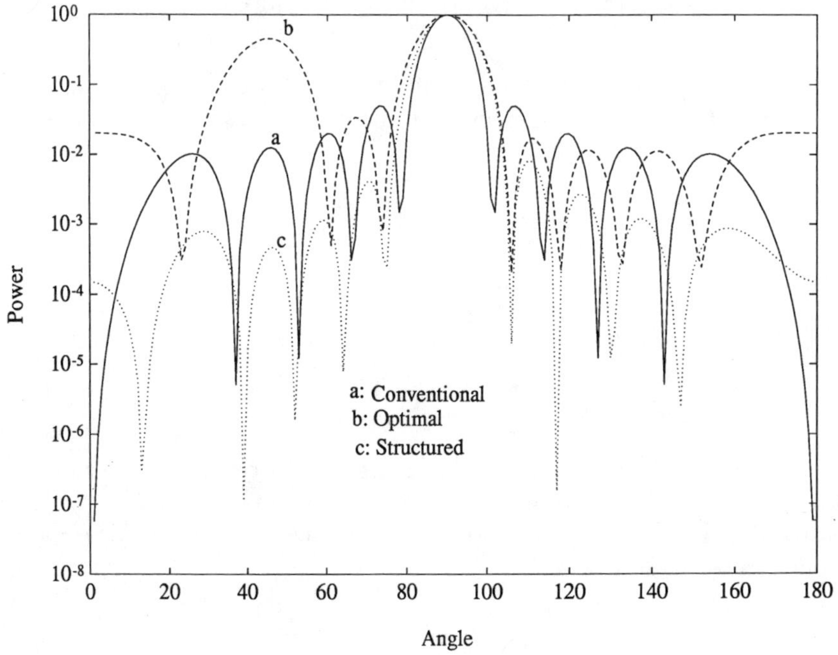


**FIGURE 5.6**
Power pattern of an element space processor using conventional, optimal, and structured beamforming methods using a ten-element linear array with one-half wavelength spacing in the presence of eight directional interferences in directions 25°, 45°, 60°, 108°, 120°, 135°, and 155°, each with unity power. Look direction is 90°, $\sigma_n^2 = 4$ sources in 45° and 90° are correlated with $|\delta| = 1$, $\delta_p = 45°$. (From Godara, L.C., *IEEE Trans. Acoust. Speech Signal Process.*, 38, 1–15, 1990. ©IEEE. With permission.)

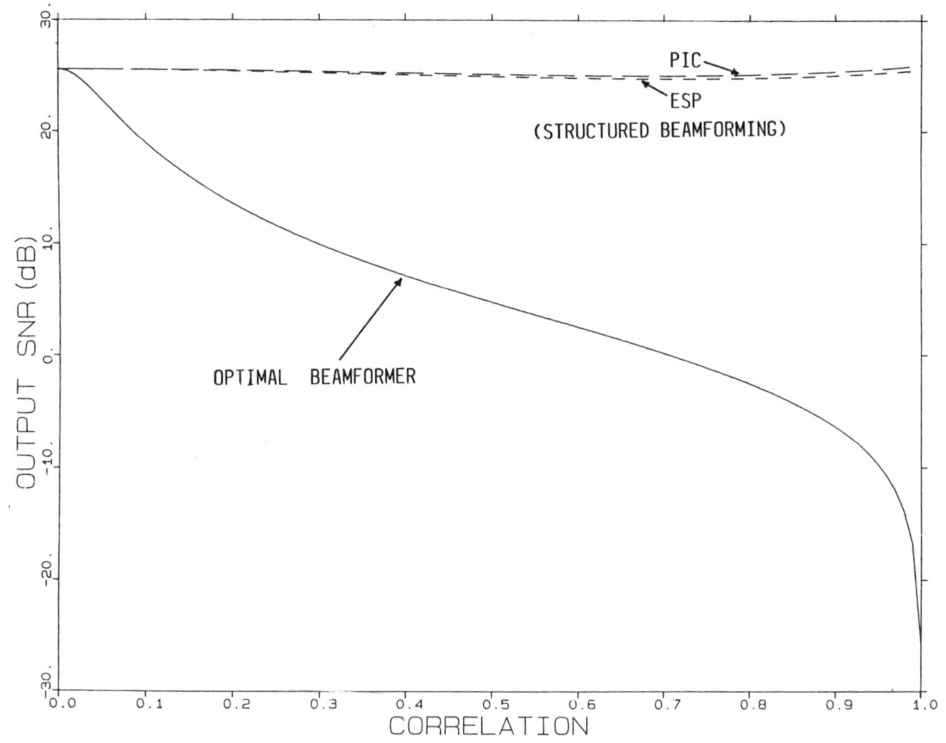


**FIGURE 5.7**
Output SNR of the element space processor and the PIC processor using structured beamforming method vs. the magnitude of the correlation coefficient for a four-element linear array with one-half wavelength spacing in the presence of one directional interference of unity power in direction 85°. Correlation phase is 90° measured at the center of the array. $\sigma_n^2 = 0.001$. (From Godara, L.C., *IEEE Trans. Acoust. Speech Signal Process.*, 38, 1–15, 1990. ©IEEE. With permission.)

the side-lobes of the conventional pattern. The interference in the 45° direction is correlated with the signal source. It is clear from the figure that even in the presence of correlated arrivals, a ten-element array using the structured method is capable of nulling eight direction sources while maintaining a specified response in the look direction. As expected, the increased response of the optimal processor in the direction of the correlated interference is clearly visible.

Figure 5.7 shows the output SNRs of the PIC and the ESP using the structured method and compares the result to that of the optimal beamformer. The phase of the correlation coefficient measured at the center of the array is assumed to be 90°. The magnitude of the correlation has almost no effect on the output SNRs of the two processors when the structured method is used. This agrees with the analysis presented in the previous section. The output SNR of the optimal beamformer reduces to about –25 dB when the two sources are fully correlated.

### 5.7.2 Structured Gradient Algorithm

The structured gradient algorithm uses the structured array correlation matrix to estimate the required gradient to update the weights, and is discussed in detail in Section 3.6. In this section, an analysis of this algorithm is focused on its use in updating the weights in the presence of correlated arrivals. The analysis is presented for an equispaced linear array in the presence of two correlated sources [God91].

Let an L-dimensional vector $\mathbf{w}(n + 1)$ denote the weights of the ESP updated by the structured gradient algorithm, that is,

$$\mathbf{w}(n+1) = P\{\mathbf{w}(n) - \mu \mathbf{g}_{st}(\mathbf{w}(n))\} + \frac{\mathbf{S}_0}{L} \tag{5.7.23}$$

where P is the projection operator given by (3.4.2), and $\mathbf{g}_{st}(\mathbf{w}(n))$ denotes the gradient estimate defined by (3.6.5) to (3.6.7). It follows from (3.6.5) to (3.6.7) that

$$E\left[\mathbf{g}_{st}(\mathbf{w}(n))\middle|\mathbf{w}(n)\right] = 2\tilde{R}\mathbf{w}(n) \tag{5.7.24}$$

where $\tilde{R}$ is given by (5.7.1).

#### *5.7.2.1 Gradient Comparison*

Let $\mathbf{g}(\mathbf{w}(n))$ denote the gradient of the mean output power for a given $\mathbf{w}(n)$ when the sources are not correlated, that is,

$$\mathbf{g}(\mathbf{w}(n)) = 2R_0\mathbf{w}(n) \tag{5.7.25}$$

where $R_0$ denotes the array correlation matrix when sources are not correlated, that is, $\delta = 0$. It follows from (5.7.4) and (5.7.5) that

$$R = R_0 + \sqrt{p_S p_I}\, Q \tag{5.7.26}$$

Let an L-dimensional error vector $\mathbf{e}(n)$ denote the difference between the true gradient used in the standard LMS algorithm in the absence of the correlated field given by (5.7.25), and the expected value of the gradient used by the structured gradient algorithm in the presence of correlated field, that is,

$$\mathbf{e}(n) = \mathbf{g}(\mathbf{w}(n)) - E\left[\mathbf{g}_{st}(\mathbf{w}(n))\middle|\mathbf{w}(n)\right] \tag{5.7.27}$$

The normalized norm of the error vector approaches zero in the limit as $L \to \infty$, that is,

$$\lim_{L\to\infty}\sqrt{\frac{\mathbf{e}^H(n)\mathbf{e}(n)}{L}} = 0 \tag{5.7.28}$$

Equation (5.7.28) is now established. It follows from (5.7.24), (5.7.25), and (5.7.27) that

$$\mathbf{e}(n) = 2R_0\mathbf{w}(n) - 2\tilde{R}\mathbf{w}(n) \tag{5.7.29}$$

Since $\tilde{R}$ is obtained from R by averaging along the diagonals, it follows from (5.7.4) and (5.7.5) that

$$\tilde{R}_0 = R_0 + \sqrt{p_S p_I}\,\tilde{Q} \tag{5.7.30}$$

where $\tilde{Q}$ is a matrix having a Toeplitz structure with the entries along the mth diagonal given by

$$\tilde{Q}_m = \frac{1}{L-m}\sum_{\ell=1}^{L-m} Q_{\ell,\ell+m} \qquad m = 0, 1, \ldots, L-1 \tag{5.7.31}$$

Note that the matrix $R_0$ denotes the array correlation matrix of an equispaced linear array immersed in an uncorrelated noise field and thus has a Toeplitz structure. Hence, it is not affected by averaging along the diagonals.

It follows from (5.7.29) and (5.7.30) that

$$\mathbf{e}(n) = -2\sqrt{p_S p_I}\,\tilde{Q}\mathbf{w}(n) \tag{5.7.32}$$

Taking the dot product, dividing by L, and taking the limit on both sides,

$$\lim_{L\to\infty}\frac{\mathbf{e}^H(n)\mathbf{e}(n)}{L} = 4p_S p_I \mathbf{w}^H(n) C \mathbf{w}(n) \tag{5.7.33}$$

where

$$C = \lim_{L\to\infty}\frac{\tilde{Q}^H\tilde{Q}}{L} \tag{5.7.34}$$

Consider the matrix C. Its $(\ell,n)$th element is given by

$$C_{\ell,n} = \lim_{L\to\infty}\frac{1}{L}\sum_{k=1}^{L}\tilde{Q}_{\ell,k}\tilde{Q}_{k,n} \tag{5.7.35}$$

It follows from (5.7.18) that the $(\ell,k)$th element of $\tilde{Q}$ is given by

$$\tilde{Q}_{\ell,k} = \alpha_0 \exp\{-j\beta_0(k-\ell)\}\frac{\sin \pi\hat{d}\phi(L-|k-\ell|)}{(L-|k-\ell|)} \tag{5.7.36}$$

where

$$\alpha_0 = \frac{2|\delta|\cos\psi_P}{\sin\hat{d}\phi} \tag{5.7.37}$$

and

$$\beta_0 = \pi\hat{d}(\cos\theta_I + \cos\theta_0) \tag{5.7.38}$$

From (5.7.35) and (5.7.36), it follows that

$$C_{\ell,n} = \alpha_0^2 \exp\{-j\beta_0(n-\ell)\}\lim_{L\to\infty}\frac{1}{L}\sum_{k=1}^{L}\frac{\sin \pi\hat{d}\phi(L-|k-\ell|)}{(L-|k-\ell|)}\frac{\sin\left(\pi\hat{d}\phi(L-|n-k|)\right)}{(L-|n-k|)} \tag{5.7.39}$$

Since $-1 \le \sin x \le 1 \forall x$ and $1/(L - |n-k|) \le 1$ for $1 \le n \le L$, $1 \le k \le L$, it follows that

$$-1 \le \frac{\sin\left(\pi \hat{d}\phi\left(L - |k-\ell|\right)\right)\sin\left(\pi \hat{d}\phi\left(L - |n-k|\right)\right)}{\left(L - |n-k|\right)} \le 1 \tag{5.7.40}$$

Thus,

$$-\alpha_0^2\alpha_k \le C_{\ell,n}\exp\{j\beta_0(n-\ell)\} \le \alpha_k\alpha_0^2 \tag{5.7.41}$$

where

$$\alpha_k = \lim_{L\to\infty}\frac{1}{L}\sum_{k=1}^{L}\frac{1}{L-|k-\ell|} \tag{5.7.42}$$

Since

$$\begin{aligned}\sum_{k=1}^{L}\frac{1}{L-|k-\ell|} &= \sum_{k=1}^{\ell}\frac{1}{L+k-\ell} + \sum_{k=\ell+1}^{L}\frac{1}{L-k+\ell}\\ &= \left(\frac{1}{L-\ell+1}+\cdots+\frac{1}{L}\right)+\left(\frac{1}{L-1}+\frac{1}{L-2}+\cdots+\frac{1}{\ell}\right)\\ &\le \left(1+\frac{1}{2}+\cdots+\frac{1}{L}\right)+\left(\frac{1}{L}+\frac{1}{L-1}+\cdots+1\right)\\ &\le \sum_{k=1}^{L}\frac{2}{k}\end{aligned} \tag{5.7.43}$$

From (5.7.42) and (5.7.43), it follows that

$$\begin{aligned}\alpha_k &\le \lim_{L\to\infty}\sum_{k=1}^{L}\frac{2}{Lk}\\ &= 0\end{aligned} \tag{5.7.44}$$

Along with (5.7.39), (5.7.44) implies that

$$C_{\ell,n} = 0 \forall \ell \text{ and } \forall n \tag{5.7.45}$$

Thus, $C \equiv 0$ and it follows from (5.7.33) that

$$\lim_{L\to\infty}\frac{e^H(n)e(n)}{L} = 0 \tag{5.7.46}$$

This implies (5.7.28).

Thus, the normalized error vector norm approaches zero in the limit as $L \to \infty$. The error vector is the difference between the true gradient used in the standard LMS algorithm in the presence of the uncorrelated noise field and the expected value of the gradient used by the structured gradient algorithm in the presence of correlated arrivals. Since the use of the true gradient in the standard LMS algorithm leads the estimated weight vector to the optimal weight vector $\hat{\mathbf{w}}$, and the processor using $\hat{\mathbf{w}}$ minimizes the total noise when the noise field is not correlated, it follows that by using the gradient estimated by the structured method, the mean value of the estimated weight would approach to $\hat{\mathbf{w}}$ for an infinitely large array. Thus, the processor in the presence of correlated arrivals would have the same antenna pattern (in the mean sense) as it has in the presence of the uncorrelated noise field. Thereby, the correlated jammer would be canceled.

### *5.7.2.2 Weight Vector Comparison*

In this section, a comparison is made between the normalized error between the expected values of the weights estimated by the standard method when the noise field is not correlated, and by the structured method when the noise field is correlated. Let $\mathbf{w}(n)$ denote the weights estimated by the standard LMS algorithm in the absence of correlation, that is, when $\delta = 0$ and $\tilde{\mathbf{w}}(n)$ denotes the weights estimated by structured method. It is assumed for the purpose of the comparison that at the nth iteration, both methods have the same weight vector, that is, $\tilde{\mathbf{w}}(n) = \mathbf{w}(n)$.

Let

$$\hat{\mathbf{e}}(n+1) = E\left[\tilde{\mathbf{w}}(n+1)\right] - E\left[\mathbf{w}(n+1)\right] \tag{5.7.47}$$

Now it is shown that

$$\lim_{L\to\infty} \sqrt{\frac{\hat{\mathbf{e}}^H(n+1)\hat{\mathbf{e}}(n+1)}{L}} = 0 \tag{5.7.48}$$

It follows from (5.7.24) and (5.7.26) that

$$E\left[\mathbf{g}_{st}\left(\mathbf{w}(n)\middle|\mathrm{w}(n)\right)\right] = 2R_0\mathbf{w}(n) + 2\sqrt{p_S p_I}\,\tilde{Q}\mathbf{w}(n) \tag{5.7.49}$$

Denoting the gradient estimate of (3.4.4) by $\mathbf{g}_s(\mathbf{w}(n))$, from (3.4.4) one obtains

$$E\left[\mathbf{g}_s\left(\mathbf{w}(n)\middle|\mathrm{w}(n)\right)\right] = 2R_0\mathbf{w}(n) \tag{5.7.50}$$

Taking the expected value on both sides of (3.4.1) and (5.7.23), using (5.7.47), (5.7.49), (5.7.50), and the fact that both the weight vectors are identical at the nth iteration $\{\tilde{\mathbf{w}}(n) = \mathbf{w}(n)\}$,

$$\hat{\mathbf{e}}(n+1) = 2\mu\sqrt{p_S p_I}\,P\tilde{Q}\overline{\mathbf{w}}(n) \tag{5.7.51}$$

where

$$\overline{\mathbf{w}}(n) = E\left[\mathbf{w}(n)\right] \tag{5.7.52}$$

Thus,

$$\lim_{L\to\infty}\sqrt{\frac{\hat{\mathbf{e}}^H(n+1)\hat{\mathbf{e}}(n+1)}{L}} = 4\mu^2 p_S p_I \lim_{L\to\infty}\overline{\mathbf{w}}^H(n)\frac{\tilde{Q}^H P\tilde{Q}}{L}\overline{\mathbf{w}}(n) \tag{5.7.53}$$

Now consider $\lim_{L\to\infty}\tilde{Q}^H P\tilde{Q}/L$. It follows from (3.4.2) that

$$\lim_{L\to\infty}\frac{\tilde{Q}^H P\tilde{Q}}{L} = \lim_{L\to\infty}\frac{\tilde{Q}^H\tilde{Q}}{L} - \lim_{L\to\infty}\frac{\tilde{Q}^H\mathbf{S}_0\mathbf{S}_0^H\tilde{Q}}{L^2} \tag{5.7.54}$$

It follows from (5.7.34) and (5.7.45) that

$$\lim_{L\to\infty}\frac{\tilde{Q}^H\tilde{Q}}{L} = 0 \tag{5.7.55}$$

Since $\tilde{Q}^H = \tilde{Q}$ and $(\mathbf{S}_0)_m = \exp(j2\pi\hat{d}(m-1)\cos\theta_0)$, it follows from (5.7.36)

$$\begin{aligned}\left(\tilde{Q}^H\mathbf{S}_0\mathbf{S}_0^H\tilde{Q}\right)_{\ell,k} &= \sum_{n=1}^{L}\sum_{m=1}^{L}\tilde{Q}_{\ell,m}\left(\mathbf{S}_0\right)_m\left(\mathbf{S}_0\right)_n^*\tilde{Q}_{n,k} \\ &= \sum_{n=1}^{L}\sum_{m=1}^{L}\alpha_0^2\exp\{-j\beta_0(m-\ell+k-n)\}\exp\left\{j2\pi\hat{d}(m-n)\cos\theta_0\right\} \\ &\quad \frac{\sin\pi\hat{d}\phi(L-|m-\ell|)}{L-|m-\ell|}\frac{\sin\pi\hat{d}\phi(L-|k-n|)}{L-|k-n|}\end{aligned} \tag{5.7.56}$$

Substituting for $\beta_0$ from (5.7.38) and rearranging,

$$\begin{aligned}\left(\tilde{Q}^H\mathbf{S}_0\mathbf{S}_0^H\tilde{Q}\right)_{\ell,k} &= \alpha_0^2\exp\{-j\beta_0(k-\ell)\} \\ &\quad \sum_{m=1}^{L}\exp\left\{-j\pi\hat{d}(\cos\theta_I-\cos\theta_0)m\right\}\frac{\sin\pi\hat{d}\phi(L-|m-\ell|)}{L-|m-\ell|} \\ &\quad \sum_{n=1}^{L}\exp\left\{j\pi\hat{d}(\cos\theta_I-\cos\theta_0)n\right\}\frac{\sin\pi\hat{d}\phi(L-|n-k|)}{L-|n-k|}\end{aligned} \tag{5.7.57}$$

Since $-1 \le \sin x \le 1\forall x$, it follows that

$$\begin{aligned}\exp\{j\beta_0(k-\ell)\}\left(\tilde{Q}^H\mathbf{S}_0\mathbf{S}_0^H\tilde{Q}\right)_{\ell,k} &\le \alpha_0^2\sum_{m=1}^{L}\exp\left\{-j\pi\hat{d}(\cos\theta_I-\cos\theta_0)m\right\} \\ &\quad \sum_{n=1}^{L}\exp\left\{j\pi\hat{d}(\cos\theta_I-\cos\theta_0)n\right\}\end{aligned} \tag{5.7.58}$$

and

$$\exp\{j\beta_0(k-\ell)\}\left(\tilde{Q}^H \mathbf{S}_0 \mathbf{S}_0^H \tilde{Q}\right)_{\ell,k} \geq -\alpha_0^2 \sum_{m=1}^{L} \exp\left\{-j\pi\hat{d}\left(\cos\theta_I - \cos\theta_0\right)m\right\} \sum_{n=1}^{L} \exp\left\{j\pi\hat{d}\left(\cos\theta_I - \cos\theta_0\right)n\right\} \tag{5.7.59}$$

Using

$$a + a^2 + a^3 + \cdots + a^N = a\frac{\left(1-a^N\right)}{(1-a)} \tag{5.7.60}$$

to sum the series on the RHS of (5.7.58) and (5.7.59), dividing by $L^2$ on both sides, and taking the limit as $L \to \infty$,

$$0 \leq \lim_{L\to\infty} \frac{\exp\{j\beta_0(k-\ell)\}\left(\tilde{Q}^H \mathbf{S}_0 \mathbf{S}_0^H \tilde{Q}\right)_{\ell,k}}{L^2} \leq 0 \tag{5.7.61}$$

This implies that

$$\lim_{L\to\infty} \frac{\tilde{Q}^H \mathbf{S}_0 \mathbf{S}_0^H \tilde{Q}}{L^2} = 0 \tag{5.7.62}$$

From (5.7.54), (5.7.55) and (5.7.62) it follows that

$$\lim_{L\to\infty} \frac{\tilde{Q}^H P \tilde{Q}}{L} = 0 \tag{5.7.63}$$

This along with (5.7.53) establishes (5.4.48).

The implication of this result is that when the array has an infinitely large number of elements, the structured LMS algorithm in the presence of correlated arrivals yields the same weight vector in the mean sense as estimated by the standard LMS algorithm when the sources are not correlated.

The structured gradient algorithm analyzed above only uses a snapshot available at the (n +1)st iteration of the weight update to estimate the gradient. The improved LMS algorithm discussed in Section 3.8 makes use of all available samples to estimate a gradient required to update the array weights. Results similar to those given by (5.7.28) and (5.7.48) for the structured gradient algorithm may also be established for the improved LMS algorithm following a procedure similar to that discussed above.

Now some numerical examples are presented to compare the performance of the structured gradient algorithm and the improved algorithm with that of the standard LMS algorithm in the presence of correlated field.

#### *5.7.2.3 Examples and Discussion*

Figure 5.8 and Figure 5.9 compare the mean output power $P(\mathbf{w}(n))$ and the output SNR vs. the iteration number when the weights are adjusted using the three algorithms. A

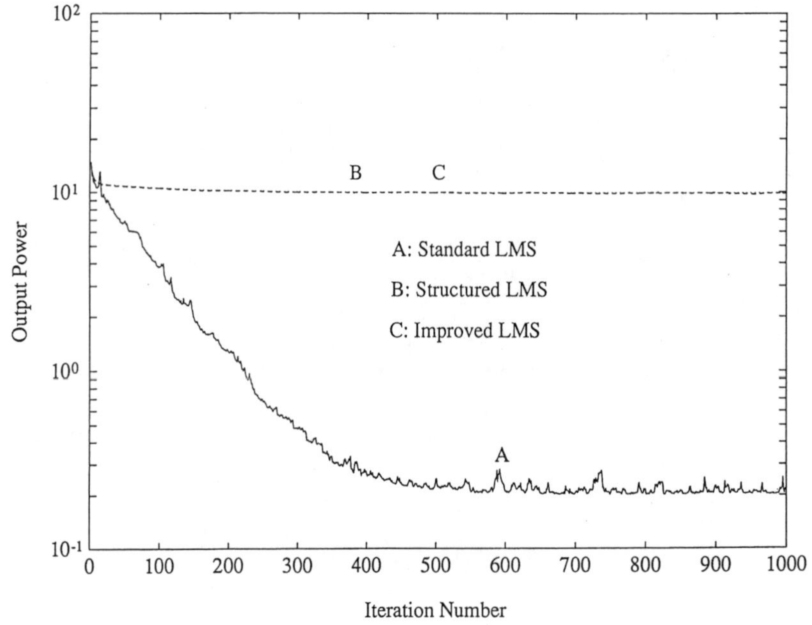


**FIGURE 5.8**
Output power vs. the iteration number for a ten-element linear array with one-half wavelength spacing in the presence of two directional interferences with direction = 65°, power = 1, $|\delta| = 0.99$, $\delta_p = 45°$ and direction = 72°, power = 100. Look direction is 90° with signal power = 1. $\sigma_n^2 = 0.01$. (From Godara, L.C., *J. Acoust. Soc. Am.*, 89, 1730–1736, 1991. With permission.)

linear array of ten elements with one-half wavelength spacing is assumed for these examples. That variance of uncorrelated noise present on each element is assumed to be 0.01. Two interference sources are assumed to be present. The first interference makes an angle of 65° with the line of the array, and is correlated with the signal source present broadside to the array. The magnitude of correlation is taken to be equal to 0.99, and the correlation phase measured at the reference point (one of the side elements of the array) is equal to 45°. The second interference makes an angle of 72° with the line of the array, and is not correlated with the signal source. The power of the signal source, as well as of the correlated interference, is 20 dB above the white noise power. The power of the second interference is 40 dB above the white noise power. All algorithms are initialized with the conventional weights, that is,

$$\mathbf{w}(0) = \frac{\mathbf{S}_0}{L} \tag{5.7.64}$$

and the gradient step size μ in each case is taken to be equal to 0.00005.

The mean output power for a given $\mathbf{w}(n)$ is calculated using

$$P(\mathbf{w}(n)) = \mathbf{w}^H(n) R \mathbf{w}(n) \tag{5.7.65}$$

and the output SNR is calculated using

$$\text{SNR} = \frac{P_S(\mathbf{w}(n))}{P_N(\mathbf{w}(n))} \tag{5.7.66}$$

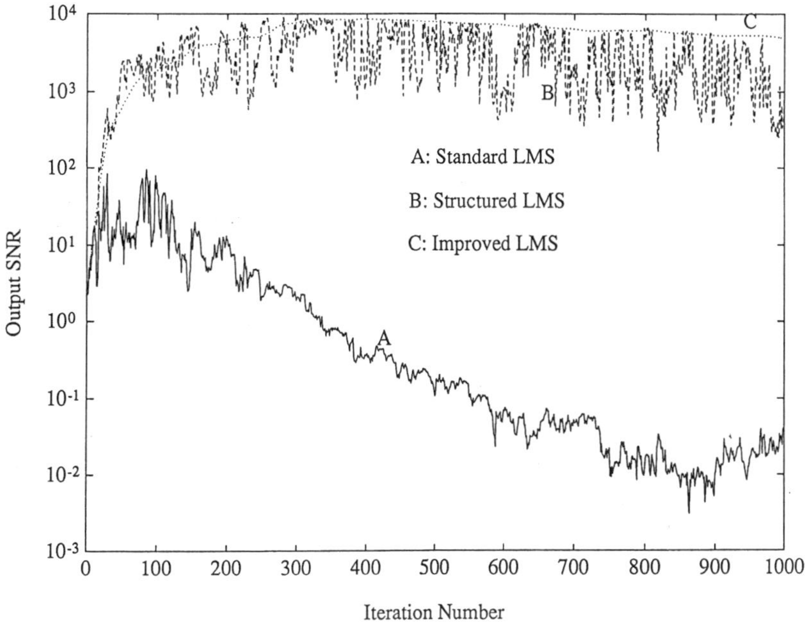


**FIGURE 5.9**
Output SNR vs. the iteration number for a ten-element linear array with one-half wavelength spacing in the presence of two directional interferences with direction = 65°, power = 1, $|\delta| = 0.99$, $\delta_p = 45°$ and direction = 72°, power = 100. Look direction is 90° with signal power = 1. $\sigma_n^2 = 0.01$. (From Godara, L.C., *J. Acoust. Soc. Am.*, 89, 1730–1736, 1991. With permission.)

with $P_S(\mathbf{w}(n))$ and $P_N(\mathbf{w}(n))$, respectively, denoting the mean output signal power and the total mean output noise power for a given $\mathbf{w}(n)$. $P_S(\mathbf{w}(n))$ is calculated using

$$P_S(\mathbf{w}(n)) = \mathbf{w}^H(n)R_S\mathbf{w}(n) \tag{5.7.67}$$

where $R_S$ is the correlation matrix due to signal only, that is,

$$R_S = E[\mathbf{x}_S(t)\mathbf{x}_S^H(t)] \tag{5.7.68}$$

and

$$\mathbf{x}_S(t) = m_S(t)\left(\sqrt{p_S}\mathbf{S}_0 + \delta^* \sqrt{p_I}\mathbf{S}_I\right) \tag{5.7.69}$$

Note that $\mathbf{x}_S(t)$ is the array signal vector contributed by the signal source.

The mean output noise power is calculated using

$$P_N(\mathbf{w}(n)) = \mathbf{w}^H(n)R_N\mathbf{w}(n)$$

where $R_N$ is the noise only array correlation matrix. It is given by

$$R_N = E[\mathbf{x}_N(t)\mathbf{x}_N^H(t)] \tag{5.7.70}$$

with

$$\mathbf{x}_N(t) = m_I(t)\sqrt{1-|\delta|^2}\,\mathbf{S}_I + m_I'(t)\mathbf{S}_I' + \mathbf{n}(t) \tag{5.7.71}$$

where $m_I'$ and $S_I'$ characterize the second interference.

Figure 5.8 shows that the mean output power of the processor is close to the input signal power when the structured LMS algorithm and the improved LMS algorithm are used to update the weights. Thus, in the presence of the correlated arrivals the processor is able to cancel the correlated directional interference without canceling the desired signal. This agrees with the theoretical results presented previously. The processor using the standard LMS algorithm to update the weights cancels the desired signal and the mean output power falls below the level of the input signal power, as expected.

Figure 5.9 compares the output SNR for the three algorithms. The output SNR achievable by the processor using the structured method and the improved method is much higher than by the standard LMS case. There are lots of fluctuations in the output SNR curve of the structured method compared to that of the improved method. The reason for these fluctuations is that the structured method uses only one sample to estimate the gradient in comparison to all available samples used by the improved method. Thus, the correlated interference can be canceled and the close proximity to the convergence point is quickly attained by using the improved method to update the weights of an adaptive beamformer. It should be noted here that though the theoretical results are presented to show the performance of these algorithms for an infinitely large array, the example presented for a ten-element array demonstrates the correlated jammer cancelation capability of these algorithms for an array that is not so large.

## 5.8 Correlated Broadband Sources

In this section, an array processor using the tapped delay line (TDL) structure of Figure 4.1 is considered in the presence of correlated broadband directional sources. The structured beamforming method is proposed to cancel the correlated interferences using a linear array of equispaced elements, and its performance is analyzed [God92].

The array correlation matrix for an equispaced linear array using a TDL filter has a special structure in the presence of uncorrelated directional sources, and the correlated field destroys this structure. This structure is examined in the next section.

### 5.8.1 Structure of Array Correlation Matrix

Consider a linear array of L equispaced elements immersed in a homogeneous and uncorrelated noise field. For ease of analysis, assume that the array is aligned with the positive x-axis and that one of the elements is situated at the origin. Let the origin of the coordinate system be taken as the time reference. Thus, the time taken by a plane wave arriving from direction $\theta$ and measured from Element $\ell$ to the origin is given by

$$\tau_\ell(\theta) = \frac{d(\ell-1)\cos\theta}{c} \tag{5.8.1}$$

where d is the spacing between the elements and c is the speed of propagation. It is assumed that the spacing is less than a half-wavelength at all frequencies of interest. Let an L-dimensional vector **x**(t) denote the sensor output after presteering delays $T_\ell(\theta_0)$, $\ell = 1, 2, \ldots, L$. These delays are selected such that the L output waveforms of the presteered sensors due to a broadband source in the look direction are identical. As discussed previously, an array may be presteered in direction $\theta_0$ using

$$T_\ell(\theta_0) = T_0 + \tau_\ell(\theta_0), \quad \ell = 1, 2, \ldots, L \tag{5.8.2}$$

where $T_0$ is a bulk delay, such that

$$T_\ell(\theta_0) \geq 0, \quad \forall \ell \tag{5.8.3}$$

Let an LJ dimensional vector **X**(t) defined by (4.1.6) denote the array signals across the TDL structure and $R_0$ denote the array correlation matrix in the absence of source correlation. Let $R_0(m,n)$ denote the (m,n)th block of $R_0$ given by

$$R_0(m,n) = E\left[\mathbf{x}(t-(m-1)T)\mathbf{x}^T(t-(n-1)T)\right] \tag{5.8.4}$$

It follows from (4.1.11) that $[R_0(m,n)]_{\ell,k}$ due to a source in direction $\theta$ is given by

$$\begin{aligned}\left[R_0(m,n)\right]_{\ell,k} &= \rho\left((m-n)T + T_\ell(\theta_0) - T_k(\theta_0) + \tau_k(\theta) - \tau_\ell(\theta)\right)\\ \ell,k &= 1, 2, \ldots, L\end{aligned} \tag{5.8.5}$$

where $\rho(\tau)$ denotes the correlation function defined by (4.1.12).

Let $[R_0(m,n)]_{\ell,\ell+k}$, $k = 0, 1, \ldots, L-1$ denote the kth diagonal of the matrix $R_0(m,n)$. Thus, (5.8.5) can be expressed as

$$\begin{aligned}\left[R_0(m,n)\right]_{\ell,\ell+k} &= \rho\left((m-n)T + T_\ell - T_{\ell+k} + \tau_{\ell+k} - \tau_\ell\right)\\ k = 0, 1, \ldots, L-1, &\quad \ell = 1, 2, \ldots, L-k\end{aligned} \tag{5.8.6}$$

where the parameters $\theta_0$ and $\theta$ are omitted for the ease of notation. It follows from (5.8.1) and (5.8.2) that

$$T_\ell - \tau_\ell = T_0 + \frac{d}{c}(\ell - 1)(\cos\theta_0 - \cos\theta) \tag{5.8.7}$$

and

$$T_{\ell+k} - \tau_{\ell+k} = T_0 + \frac{d}{c}(\ell + k - 1)(\cos\theta_0 - \cos\theta) \tag{5.8.8}$$

Equations (5.8.7) and (5.8.8) imply that

$$T_\ell - \tau_\ell - \left(T_{\ell+k} - T_{\ell+k}\right) = -\frac{d}{c}k(\cos\theta_0 - \cos\theta) \tag{5.8.9}$$

Substituting from (5.8.9) in (5.8.6),

$$\left[R_0(m,n)\right]_{\ell,\ell+k} = \rho\left((m-n)T - \frac{d}{c}k(\cos\theta_0 - \cos\theta)\right)$$
$$k = 0, 1, \ldots, L-1, \quad \ell = 1, 2, \ldots, L-k \tag{5.8.10}$$

As seen on the RHS of (5.8.10), the correlation function parameter only depends on k and not on $\ell$. Thus, it follows that all $L-k$ elements of the kth diagonal of the matrix $R_0(m,n)$ are the same. Hence, each $L \times L$ block of the array correlation matrix $R_0(m,n)$, $m,n = 1, 2, \ldots, J$ has the Toeplitz structure in the absence of correlation. The existence of correlation between the directional sources destroys this structure. An array-processing method is discussed later in the chapter to restore this structure in the array correlation matrix before using it to estimate the weights of the TDL structure.

### 5.8.2 Correlated Field Model

Without any loss of generality, assume that there are two correlated broadband directional sources. One source is a signal source and the other source is interference. Let $p_S$ and $p_I$ represent the powers of the signal source and the interference source, respectively. Let $\theta_0$ and $\theta_I$ denote the directions of the two sources, respectively. Assume that the interference contains a component of the desired signal such that the output of a sensor present at the center of the coordinate system, assumed to be the time reference, can be expressed as

$$x(t) = \sqrt{p_S}\, m_S(t) + \sqrt{p_I}\left[\alpha m_S(t - T_c) + \sqrt{(1-\alpha^2)}\, m_I(t)\right] + n(t) \tag{5.8.11}$$

where $m_S(t)$ and $m_I(t)$ are zero-mean unit variance, low-pass processes associated with the signal source and the interference source, respectively; $n(t)$ is the random noise component with a zero mean and variance equals $\sigma_n^2$; $\alpha$ is a positive real scalar denoting the magnitude of correlation, and $T_c$ is a real scalar denoting the time delay for the correlated field. For two coherent sources, the magnitude of correlation equals 1. It is assumed that $m_S(t)$, $m_I(t)$, and $n(t)$ are mutually uncorrelated. The autocorrelation functions of $m_S(t)$ and $m_I(t)$ are denoted by $\rho_S(\tau)$ and $\rho_I(\tau)$, respectively.

It should be noted that although the following analysis is for a specialized model of (5.8.11) emphasizing a multipath application, the results are equally valid for a more general correlated field model of the type

$$x(t) = \sqrt{p_S}\, m_S(t) + \sqrt{p_I}\left[\alpha m_2(t) + \sqrt{(1-\alpha^2)}\, m_I(t)\right] + n(t) \tag{5.8.12}$$

with the cross-correlation

$$\rho(\tau) = E\left[m_S(t) m_2(t-\tau)\right] \tag{5.8.13}$$

that is assumed to be band limited.

It follows from (5.8.11) that the output of the mth tap on $\ell$th sensor, presteered in the $\theta_0$ direction, is given by

$$\begin{aligned} x_\ell(t-(m-1)T) = \sqrt{p_S}\, m_S(t-(m-1)T-T_0) \\ +\sqrt{p_I}\{\alpha m_S(t-T_c+\tau_\ell-T_\ell-(m-1)T) \\ +\sqrt{1-\alpha^2}\, m_I(t+\tau_\ell-T_\ell-(m-1)T)\} \\ +n_\ell(t-(m-1)T-T_\ell) \end{aligned} \tag{5.8.14}$$

Thus,

$$\begin{aligned} [R(m,n)]_{\ell,k} &= E[x_\ell(t-(m-1)T)x_k(t-(n-1)T)] \\ &= p_S\rho_S[(m-n)T] + p_I\{\alpha^2\rho_S[(m-n)T+\tau_k-\tau_\ell+T_\ell-T_k] \\ &\quad + (1-\alpha^2)\rho_I[(m-n)T+\tau_k-\tau_\ell+T_\ell-T_k]\} \\ &\quad + \sqrt{p_S p_I}\,\alpha\{\rho_S[\tau_k-T_k+(m-n)T-T_c+T_0] \\ &\quad + \rho_S[-\tau_\ell+T_\ell+(m-n)T+T_c-T_0]\} + \sigma_n^2\delta(\ell-k)\delta(m-n) \end{aligned} \tag{5.8.15}$$

where

$$\delta(i) = \begin{cases} 0 & i \neq 0 \\ 1 & i = 0 \end{cases} \tag{5.8.16}$$

It follows from (5.8.15) that the array correlation matrix R in the presence of correlated sources can be expressed as

$$R = [R_S + R_I^1 + \sigma_n^2 I] + Q \tag{5.8.17}$$

where $R_S$ denotes the array correlation matrix due to the signal source in the look direction, $R_I^1$ denotes the array correlation matrix due to the interference with the effective autocorrelation given by

$$\rho_I(\tau) = \alpha^2\rho_S(\tau) + (1-\alpha^2)\rho_I(\tau) \tag{5.8.18}$$

and Q denotes the array correlation matrix due to the cross-correlation between the signal source and the interference. Expressions for $R_S$, $R_I^1$, and Q are given by the first term, second term, and third term, respectively, on the RHS of (5.8.15).

The quantity inside the square brackets on the RHS of (5.8.17) represents the total array correlation matrix due to the uncorrelated noise field. Thus,

$$R = R_0 + Q \tag{5.8.19}$$

### 5.8.3 Structured Beamforming Method

For a linear array of equispaced elements immersed in a homogeneous and uncorrelated noise field, the array correlation matrix has a block Toeplitz structure; that is, each block of an L × L dimensional correlation matrix arising from the correlation of sensor vectors at any two taps has the Toeplitz structure. The correlation between two directional sources destroys this structure. The structured beamforming method described here uses an estimate of the array correlation matrix with the constraint that the estimated matrix has the block Toeplitz structure. Let $\tilde{R}$ denote the array correlation matrix estimated with this structure. The weights of the beamformer estimated with the structured method are calculated using the following expression:

$$\tilde{W} = \hat{R}^{-1}C\left(C^T\hat{R}^{-1}C\right)^{-1}\mathbf{f} \tag{5.8.20}$$

where

$$\hat{R} = \tilde{R} + \hat{\beta}I \tag{5.8.21}$$

$\hat{\beta}$ is a positive scalar selected such that $\hat{R}$ is positive definite, **f** is given by (4.1.25), and $\tilde{R}$ is the structured correlation matrix.

An estimate of the structured correlation matrix is made by averaging each block of the L × L matrix along its diagonals. Let $\tilde{R}(m,n)$ denote the (m,n)th block of the averaged matrix. The entries along the kth diagonal of $\tilde{R}(m,n)$ are given by

$$\tilde{R}(m,n) = \frac{1}{L-k}\sum_{\ell=1}^{L-k}\left[R(m,n)\right]_{\ell,\ell+k}, \quad k = 0, 1, \ldots, L-1 \tag{5.8.22}$$

### 5.8.4 Decorrelation Analysis

In this section, an analysis is presented to show the decorrelation effect of the structured method when the block correlation matrix is estimated by averaging along the diagonals. It follows from (5.8.17) and (5.8.19) that the matrix $R_0$ is not affected by the above method since it has the block Toeplitz structure. Thus, it is sufficient to examine the effect of averaging along the diagonals on matrix Q. It follows from (5.8.15) that

$$\begin{aligned}\left[Q(m,n)\right]_{\ell,k} &= \rho_S\left[\tau_k - T_k + (m-n)T - T_c + T_0\right] \\ &\quad + \rho_S\left[-\tau_\ell - T_\ell + (m-n)T + T_c - T_0\right]\end{aligned} \tag{5.8.23}$$

where the constant $\alpha\sqrt{p_S p_I}$ has been suppressed for ease of analysis.

$Q_k(m,n)$, the kth diagonal of Q(m,n) is given by

$$\begin{aligned}Q_k(m,n) &= \rho_S\left[\tau_{\ell+k} - T_{\ell+k} + (m-n)T - T_c + T_0\right] \\ &\quad + \rho_S\left[-\tau_\ell + T_\ell + (m-n)T + T_c - T_0\right]\end{aligned} \tag{5.8.24}$$

Since

$$\tau_\ell = \frac{d}{c}(\ell-1)\cos\theta_I \tag{5.8.25}$$

$$T_\ell = T_0 + \frac{d}{c}(\ell - 1)\cos\theta_S \tag{5.8.26}$$

$$\tau_{\ell+k} = \frac{d}{c}(\ell + k - 1)\cos\theta_I \tag{5.8.27}$$

and

$$T_{\ell+k} = T_0 + \frac{d}{c}(\ell + k - 1)\cos\theta_S \tag{5.8.28}$$

it follows from the fact that $\rho_S(\tau) = \rho_S(-\tau)$ and (5.8.24) that

$$\begin{aligned} Q_k(m,n) = \rho_S&\left[(\ell + k - 1)\psi + \eta - T_c\right] \\ &+ \rho_S\left([\ell - 1]\psi - \eta - T_c\right) \end{aligned} \tag{5.8.29}$$

where

$$\psi = \frac{d}{c}\left(\cos\theta_I - \cos\theta_0\right) \tag{5.8.30}$$

and

$$\eta = (m - n)T \tag{5.8.31}$$

Thus, from (5.8.22), it follows that

$$\begin{aligned} \tilde{Q}_k(m,n) = &\frac{1}{L-k}\sum_{\ell=1}^{L-k}\rho_S\left[(\ell + k - 1)\psi + \eta - T_c\right] \\ &+ \frac{1}{L-k}\sum_{\ell=1}^{L-k}\rho_S\left[(\ell - 1)\psi - \eta - T_c\right] \end{aligned} \tag{5.8.32}$$

Substituting from (4.1.13) in (5.8.32) $\tilde{Q}_k(m,n)$,

$$\begin{aligned} \tilde{Q}_k(m,n) &= \frac{1}{L-k}\int_{-\infty}^{\infty} S(f)\sum_{\ell=1}^{L-k} e^{j2\pi f\{(\ell+k-1)\psi+\eta-T_c\}}df \\ &+ \frac{1}{L-k}\int_{-\infty}^{\infty} S(f)\sum_{\ell=1}^{L-k} e^{j2\pi f\{(\ell-1)\psi-\eta-T_c\}}df \\ &= \int_{-\infty}^{\infty} S(f)e^{-j2\pi f T_c}\left(e^{-j2\pi\eta} + e^{j2\pi f(k\psi+\eta)}\right)\frac{1}{L-k}\sum_{\ell=1}^{L-k} e^{j2\pi f(\ell-1)\psi}df \end{aligned} \tag{5.8.33}$$

where S(f) denotes the power spectral density of the desired signal.

Using

$$1 + a + a^2 + \cdots + a^{N-1} = \frac{1 - a^N}{1 - a} \tag{5.8.34}$$

in (5.8.33),

$$\begin{aligned}\tilde{Q}_k(m,n) &= \int_{-\infty}^{\infty} S(f) e^{-j2\pi f T_c} \frac{e^{-j2\pi f \eta} + e^{j2\pi f(k\psi+\eta)}}{L-k} \frac{1 - e^{j2\pi f(L-k)\psi}}{1 - e^{j2\pi f \psi}} df \\ &= \int_{-\infty}^{\infty} \left[ S(f) \exp\left\{ -j2\pi f \left( T_c - \frac{L-1}{2}\psi \right) \right\} \right. \\ &\quad \left. \frac{2\cos \pi f(k\psi + 2\eta) \sin \pi f(L-k)\psi}{(L-k)\sin \pi f \psi} df \right]\end{aligned} \tag{5.8.35}$$

Using

$$S(f) = S(-f) \tag{5.8.36}$$

in (5.8.35),

$$\tilde{Q}_k(m,n) = \int_0^{\infty} \frac{4S(f)}{L-k} \frac{\cos 2\pi f \psi_c \cos \pi f(k\psi + 2\eta) \sin \pi f(L-k)\psi}{\sin \pi f \psi} df \tag{5.8.37}$$

where

$$\psi_c = T_c - \frac{L-1}{2}\frac{d}{c}(\cos\theta_I - \cos\theta_0) \tag{5.8.38}$$

is the correlation delay time measured at the center of the array.

The following result is true for a signal source of finite bandwidth. The result is proved later in the discussion.

If S(f) = 0 outside the frequency range of interest $[f_L, f_H]$ and bounded over this finite range, then

$$\lim_{L\to\infty} \frac{1}{L} \sum_{i,j=1}^{L} \left| \tilde{Q}(m,n)_{i,j} \right|^2 = 0 \tag{5.8.39}$$

where $\tilde{Q}(m,n)$ is the (m,n)th block of the structured correlation matrix arising from the correlated source. The above result states that the Hilbert–Schmidt norm of a matrix goes to zero as $L \to \infty$.

The Hilbert–Schmidt norm of a matrix A satisfies the following axiom [Gra77]:

$$\|A\| = 0, \text{ iff } A = 0, \text{ the all-zero matrix.} \tag{5.8.40}$$

Thus, it follows that $\tilde{Q}(m,n)$ is an all-zero matrix for an infinitely large array; hence, $\tilde{Q}$ is an all-zero matrix in the limit. This along with (5.8.17) implies that in the limit,

$$\tilde{R} = R_0 \tag{5.8.41}$$

Thus, for an infinitely large array, the effect of correlation is completely canceled using the structured beamforming method.

Although the results presented in this section hold for a large array, numerical examples are presented later to show that the method presented here performs satisfactorily for a relatively small array.

The result given by (5.8.39) is now proved. Let

$$G(f,k) = \frac{4S(f)\cos 2\pi f\psi_c \cos \pi f(k\psi + 2\eta)\sin(L-k)\psi}{\sin \pi f\psi} \tag{5.8.42}$$

First, it is shown that $|G(f,k)|$ is bounded over the frequency range of interest $[f_L, f_H]$ for every k. It follows from (5.8.30) that

$$f\psi = \frac{d}{\lambda}(\cos\theta_I - \cos\theta_0) \tag{5.8.43}$$

It is assumed that inter-element spacing is less than one-half wavelength at all frequencies, that is,

$$\frac{d}{\lambda} < \frac{1}{2} \tag{5.8.44}$$

Equations (5.8.43) and (5.8.44) imply that

$$-1 < f\psi < 1 \tag{5.8.45}$$

From

$$0 \le \theta_I \le \pi \tag{5.8.46}$$

$$0 \le \theta_0 \le \pi \tag{5.8.47}$$

and

$$\theta_I \ne \theta_0 \tag{5.8.48}$$

it follows that

$$\cos\theta_I - \cos\theta_0 \ne 0 \tag{5.8.49}$$

and thus

$$f\psi \ne 0 \tag{5.8.50}$$

From (5.8.45) and (5.8.50),

$$\sin \pi f \psi \neq 0 \tag{5.8.51}$$

As

$$|S(f)| < \infty \tag{5.8.52}$$

for the frequency range of interest, it follows from (5.8.42) and (5.8.52) that

$$|G(f,k)| < \infty \quad \forall k \tag{5.8.53}$$

Now an outline of the proof of the result is presented. Since

$$\sum_{i,j=1}^{L} \left|\tilde{Q}(m,n)_{i,j}\right|^2 = L\left|\tilde{Q}_0(m,n)\right|^2 + \sum_{k=1}^{L-1} (L-k)\left|\tilde{Q}_k(m,n)\right|^2 \tag{5.8.54}$$

we need to show that

$$\lim_{L\to\infty}\left|\tilde{Q}_0(m,n)\right|^2 + \lim_{L\to\infty}\frac{1}{L}\sum_{k=1}^{L-1} 2(L-k)\left|\tilde{Q}_k(m,n)\right|^2 = 0 \tag{5.8.55}$$

Consider the first term. From (5.8.37) and (5.8.42), it follows that

$$\begin{aligned}\lim_{L\to\infty}\left|\tilde{Q}_0(m,n)\right|^2 &= \lim_{L\to\infty}\left|\int_{f_L}^{f_H} \frac{1}{L} G(f,0)\right|^2 \\ &\le \lim_{L\to\infty}\frac{1}{L^2}\int_{f_L}^{f_H} \left|G(f,0)\right|^2 df\end{aligned} \tag{5.8.56}$$

As $|G(f,0)|$ is bounded, the integration yields a finite value. Thus,

$$\lim_{L\to\infty}\left|\tilde{Q}_0(m,n)\right|^2 = 0 \tag{5.8.57}$$

Now, consider the second term of (5.8.55). From (5.8.37) it follows that

$$\text{second term} = \lim_{L\to\infty}\frac{2}{L}\sum_{k=1}^{L-1}\frac{1}{L-k}\left|\int_{f_L}^{f_H} G(f,k)df\right|^2 \tag{5.8.58}$$

Since $G(f,k)$ is finite for every k, the integral exists. Let it be denoted by $V_0(k)$. Let $V_0$ be such that

$$V_0(k) \le V_0, \quad k = 1, 2, \ldots, L-1 \tag{5.8.59}$$

From (5.8.58) and (5.8.59), it follows that

$$\begin{aligned}\text{second term} &= \lim_{L\to\infty}\frac{2}{L}\sum_{k=1}^{L-1}\frac{V_0^2(k)}{L-k}\\ &\le 2V_0^2\lim_{L\to\infty}\frac{1}{L}\left(1+\frac{1}{2}+\cdots+\frac{1}{L-1}\right)\\ &=0\end{aligned}\tag{5.8.60}$$

From (5.8.57) and (5.8.60), it follows that (5.8.55) is true. This completes the proof.

#### *5.8.4.1 Examples and Discussion*

Figure 5.10 shows power patterns for an eight-element linear array in the presence of six directional broadband sources using three beamforming methods. All sources are assumed to have the brick-wall type of spectrum with normalized cutoff frequencies of 0.45 and 0.5. The power of each source is 20 dB above the power of white noise present on each element of the array. Five interferences are assumed to be in the far field of the array and are in directions of 22°, 50°, 68°, 112°, and 130° relative to the line of the array, and coincide with the side-lobes of the conventional array pattern. The signal source is to the array broadside. The interference in the direction of 50° is fully correlated with the signal source and delayed by 45° at the maximum frequency. The phase delay is specified at the origin of the coordinates system with array situated along the x axis. The spacing between the elements of the array is taken to be one-half wavelength at the maximum frequency. The delay line filter has nine taps (J = 9) with one sample delay between taps. The parameter $\hat{\beta}$ of (5.8.21) is taken to be equal to 8. The vector **f** is selected as follows.

$$f_i=\begin{cases}1 & i=5\\ 0 & \text{otherwise}\end{cases}\qquad i=1,2,\ldots,9$$

Figure 5.10 compares the power patterns of the conventional, optimal, and structured beamformers. The figure shows that the power pattern of the optimal beamformer has an increased response in the direction of the correlated jammer, and this increased response is responsible for the cancelation of the look direction signal. The power pattern of the structured beamformer shown in plot C has its response about −48 dB in the direction of the correlated jammer and has clearly suppressed it. The SNR measured at the output of the array using the conventional, optimal, and structured beamformers is 45, 1, and 527, respectively.

Figure 5.11 compares the Hilbert–Schmidt norm of the structured as well as the unstructured block of the array correlation matrix as a function of the number of elements in the array. The L × L dimensional block of the array correlation matrix considered corresponds to m = 1 and n = 1. Two sources are considered for the example. The look direction signal is broadside to the array and the correlated interference is in the direction of 50° relative to the line of the array. The other parameters are the same as in Figure 5.10. As seen in the figure, the norm of the structured correlation matrix decreases as the number of the elements in the array increases. On the other hand, the norm for the unstructured matrix increases.

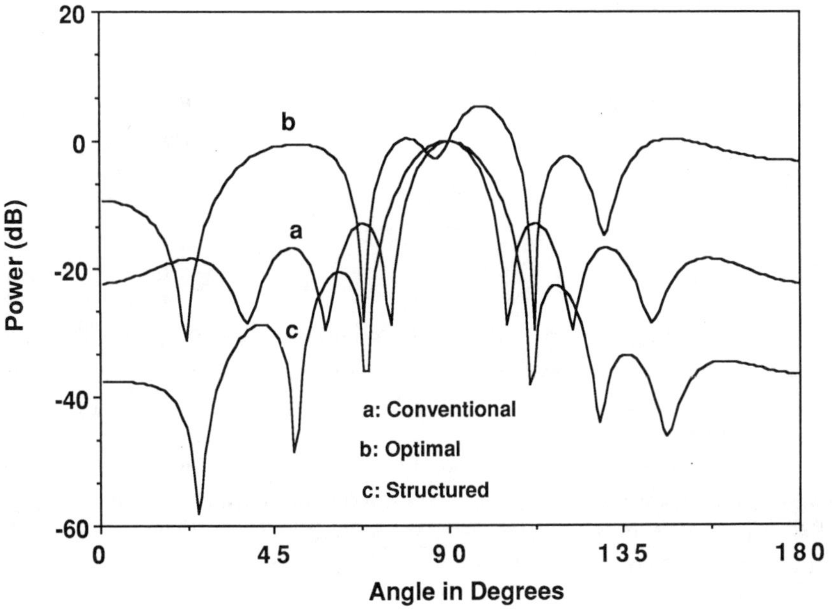


**FIGURE 5.10**
Power patterns of an element space processor using conventional, optimal, and structured beamforming methods using an eight-element linear array with one-half wavelength spacing at the maximum frequency in the presence of six directional broadband interferences in directions 22°, 50°, 68°, 90°, 112°, and 130°, each with unity power and frequency range (0.45, 0.5). Look direction is 90°, $\sigma_n^2 = 0.01$; sources in 50° and 90° are correlated with correlation phase delay of 45° at the maximum frequency measured at the origin. (From Godara, L.C., *J. Acoust. Soc. Am.*, 92, 2702–2708, 1992. With permission.)

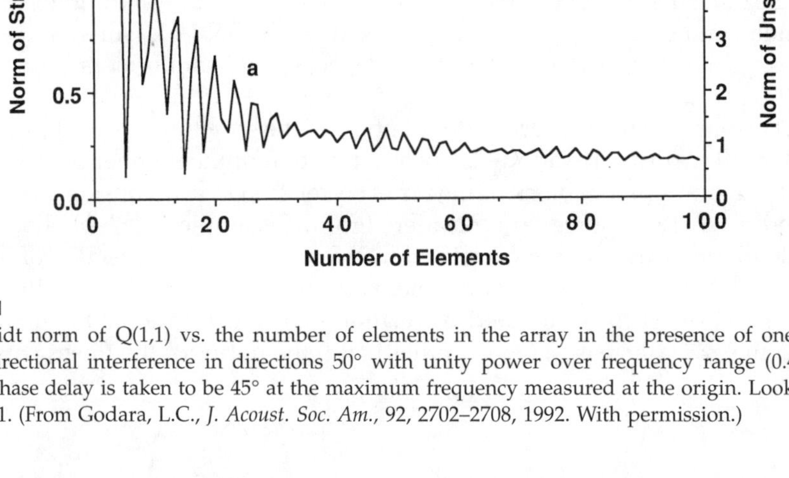

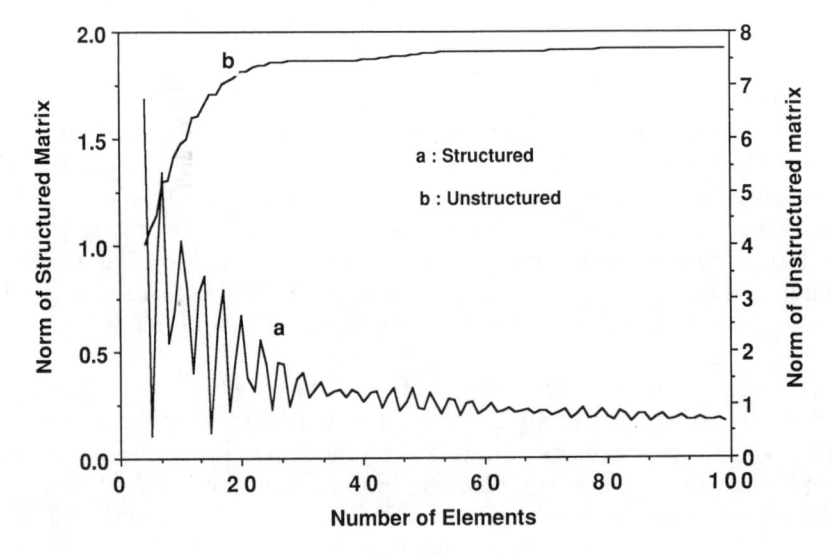

**FIGURE 5.11**
Hilber-Schmidt norm of Q(1,1) vs. the number of elements in the array in the presence of one broadband correlated directional interference in directions 50° with unity power over frequency range (0.45, 0.5). The correlation phase delay is taken to be 45° at the maximum frequency measured at the origin. Look direction is 90°, $\sigma_n^2 = 0.01$. (From Godara, L.C., *J. Acoust. Soc. Am.*, 92, 2702–2708, 1992. With permission.)

## Acknolwedgments

Edited versions of Sections 5.3, 5.4, and 5.7 are reprinted from Godara, L.C., Beamforming in the presence of correlated arrivals using structured correlation matrix, *IEEE Trans. Acoust. Speech Signal Process.*, 38(1), 1-15, 1990. An edited version of Section 5.5 is reprinted from Godara, L.C., Application of antenna arrays to mobile communications, I. Beamforming and DOA considerations, *Proc. IEEE.*, 85(8), 1195-1247, 1997. An edited version of Section 5.8 is reprinted from Godara, L.C., Beamforming in the presence of broadband correlated arrivals, *J. Acoust. Soc. Am.*, 92(5), 2702-2708, 1992.

## Notation and Abbreviations

| | |
|---|---|
| ESP | element space processor |
| PIC | postbeamformer interference canceler |
| SNR | signal-to-noise ratio |
| $SNR(\hat{w})$ | output SNR of optimal PIC |
| SNRO | SNR of optimal ESP and optimal PIC |
| TDL | tapped delay line |
| $A$ | matrix of steering vectors |
| $A_k$ | matrix of steering vectors for kth subarray |
| $d$ | spacing between elements |
| $\hat{d}$ | spacing between elements measured in wavelengths |
| $\mathbf{e}(n)$ | difference between $\mathbf{g}(\mathbf{w}(n))$ and $E[\mathbf{g}_{st}(\mathbf{w}(n))\mathbf{w}(n)]$ |
| $\hat{\mathbf{e}}(n+1)$ | difference between $E[\tilde{\mathbf{w}}(n+1)]$ and $E[\mathbf{w}(n+1)]$ |
| $\mathbf{f}$ | constraint vector |
| $G_{xy}(f)$ | cross-power spectrum |
| $\mathbf{g}(\mathbf{w}(n))$ | gradient of mean output power for given $\mathbf{w}(n)$ for uncorrelated sources |
| $\mathbf{g}_{st}(\mathbf{w}(n))$ | gradient of mean output power for given $\mathbf{w}(n)$ using structured method |
| $J$ | length of TDL structure |
| $L$ | number of elements used by processor, size of subarray |
| $L_0$ | number of subarrays |
| $M$ | number of sources |
| $m_k(t)$ | modulating function of kth source |
| $m_S(t)$ | unit variance, complex low-pass process of signal source |
| $m_I(t)$ | unit variance, complex low-pass process of interference source |
| $\mathbf{n}(t)$ | noise vector |
| $\mathbf{n}_k(t)$ | noise vector for kth subarray |
| $P$ | projection matrix |
| $\hat{P}$ | mean output power of optimal ESP |
| $P(\mathbf{w})$ | mean output power of ESP for given $\mathbf{w}$ |

| | |
|---|---|
| $P(w)$ | mean output power of PIC for given w |
| $P(\hat{w})$ | mean output power of optimal PIC |
| $P_S(\hat{w})$ | mean output signal power of optimal PIC |
| $P_I(\hat{w})$ | mean output interference power of optimal PIC |
| $P_n(\hat{w})$ | mean output uncorrelated noise power of optimal PIC |
| $P_N(\hat{w})$ | total mean noise output power of optimal PIC |
| $p_S$ | power of signal source |
| $p_I$ | power of interference source |
| $Q$ | array correlation matrix due to cross-correlation between signal source and interference |
| $\tilde{Q}$ | structured matrix of Q |
| $\tilde{Q}_m$ | mth diagonal of $\tilde{Q}$ |
| $R$ | array correlation matrix |
| $R_0$ | array correlation matrix when sources are not correlated |
| $R_0(m,n)$ | (m,n)th block of $R_0$ |
| $\tilde{R}_0(m,n)$ | (m,n)th block of $\tilde{R}$ |
| $R_k$ | array correlation matrix of kth subarray |
| $R_S$ | array correlation matrix due to signal |
| $R_I$ | array correlation matrix due to interference |
| $R_n$ | array correlation matrix due to white noise |
| $\bar{R}$ | spatially smoothed array correlation matrix |
| $\hat{\bar{R}}_N$ | estimate of spatially smoothed correlation matrix |
| $\tilde{R}$ | structured array correlation matrix |
| $S$ | source correlation matrix |
| $\bar{S}$ | smoothed sources covariance matrix |
| $S_{i,j}$ | correlation between ith and jth sources |
| $\bar{S}_{i,j}$ | smoothed correlation between ith and jth sources |
| $\mathbf{S}_0$ | steering vector in signal direction |
| $\mathbf{S}_I$ | steering vector in interference direction |
| $\mathbf{S}(\theta_k)$ | steering vector in direction $\theta_k$ |
| $\mathbf{S}_k(\theta)$ | steering vector for kth subarray in direction $\theta$ |
| $S(f)$ | power spectral density of desired signal |
| $\mathbf{s}(t)$ | vector of M modulating functions |
| $T_\ell(\theta_0)$ | steering delay on $\ell$th element |
| $T_c$ | correlation delay time measured at origin |
| $\mathbf{U}$ | fixed weights of interference beam of PIC |
| $\mathbf{V}$ | fixed weights of signal beam of PIC |
| $\mathbf{w}$ | weights of ESP |
| $\hat{\mathbf{w}}$ | weights of optimal ESP |
| $\mathbf{w}(n)$ | weights estimated by standard algorithm in absence of correlation |
| $\tilde{\mathbf{w}}(n)$ | weights estimated by structured method |

| | |
|---|---|
| $\hat{w}$ | weight of optimal PIC |
| $\mathbf{X}(t)$ | array signals across TDL structure |
| $\mathbf{x}(t)$ | array signal vector |
| $\mathbf{x}_k(t)$ | array signal vector of kth subarray |
| $\Phi$ | $M \times M$ diagonal matrix defined by (5.6.19) |
| $\Omega$ | complex scalar defined by (5.4.27) |
| $\Psi_p$ | phase of correlation coefficient measured at center of array |
| $\psi_c$ | correlation delay time measured at center of array |
| $\alpha$ | positive real scalar denoting magnitude of correlation |
| $\alpha_0$ | scalar defined by (5.7.37) |
| $\alpha_k$ | scalar defined by (5.7.42) |
| $\beta_0$ | scalar defined by (5.7.38) |
| $\beta$ | complex scalar defined by (5.2.8) |
| $\hat{\beta}$ | positive scalar to make $\hat{R}$ is positive definite in (5.8.21) |
| $\psi_{ij}$ | scalar defined by (5.6.27) |
| $\psi$ | scalar defined by (5.8.30) |
| $\gamma$ | real scalar defined by (5.4.22) |
| $\eta$ | scalar defined by (5.8.31) |
| $\sigma_n^2$ | uncorrelated noise power on each element |
| $\lambda$ | Lagrange multiplier |
| $\delta_{xy}(f)$ | correlation between two broadband signals x(t) and y(t) |
| $\delta$ | correlation between signal and interference |
| $\delta_p$ | phase of correlation coefficient $\delta$ |
| $\theta_k$ | direction of kth source |
| $\theta_0$ | direction of signal |
| $\theta_I$ | direction of interference |
| $\phi$ | scalar defined by (5.7.8) |
| $\tau_\ell(\theta_k)$ | delay on $\ell$th element for a source in direction $\theta_k$ |
| $\tau_\ell^k(\theta)$ | delay on $\ell$th element in kth subarray for source in direction $\theta$ |
| $\rho$ | complex scalar defined by (5.2.14) |
| $\rho_{xy}(\tau)$ | cross correlation function between x and y |
| $\rho_I(\tau)$ | autocorrelation functions of $m_I(t)$ |
| $\rho_S(\tau)$ | autocorrelation functions of $m_S(t)$ |

## References

Ali92 Ali, M.E. and Schreib, F., Adaptive single snapshot beamforming: a new concept for the rejection of nonstationary and coherent interferers, *IEEE Trans. Signal Process.*, 40, 3055–3058, 1992.

Car87 Carter, G.C., Coherence and time delay estimation, *IEEE Proc.*, 75, 236–255, 1987.

Cho87 Cho, K. and Ahmed, N., On a constrained LMS dither algorithm, *IEEE Proc.*, 75, 1338–1340, 1987.

Eva81 Evans, J.E., Johnson, J.R. and Sun, D.F., High resolution angular spectrum estimation techniques for terrain scattering analysis and angle of arrival estimation, in Proceedings of First ASSP Workshop on Spectral Estimation, Hamilton, Ontario, Canada, 134–139, 1981.

Faw64 Fawer, U., A coherent spread spectrum diversity receiver with AFC for multipath fading channels, *IEEE Trans. Commn.*, 42, 1300–1311, 1964.

Gab80 Gabriel, W.F., Spectral analysis and adaptive array super resolution techniques, *IEEE Proc.*, 68, 654–666, 1980.

God81 Godara, L.C. and Cantoni, A., Uniqueness and linear independence of steering vectors in array space, *J. Acoust. Soc. Am.*, 70, 467–475, 1981.

God90 Godara, L.C., Beamforming in the presence of correlated arrivals using structured correlation matrix, *IEEE Trans. Acoust. Speech Signal Process.*, 38, 1–15, 1990.

God91 Godara, L.C., Adaptive beamforming in the presence of correlated arrivals, *J. Acoust. Soc. Am.*, 89, 1730–1736, 1991.

God92 Godara, L.C., Beamforming in the presence of broadband correlated arrivals, *J. Acoust. Soc. Am.*, 92, 2702–2708, 1992.

God97 Godara, L.C., Application to antenna arrays to mobile communications. Part II: Beamforming and direction of arrival considerations, *IEEE Proc.*, 85, 1195–1247, 1997.

Gra77 Gray, R.M., Toeplitz and Circulant Matrices: II, Technical Report 6504–1, Information Systems Laboratory, Stanford University, Palo Alto, CA, 1977.

Han86 Hanna, M.T. and Simaan, M., Array filters for attenuating coherent interference in the presence of random noise, *IEEE Trans. Acoust. Speech Signal Process.*, 34, 661–668, 1986.

Han88 Hanna, M.T., Array filters for attenuating multiple coherent interference, *IEEE Trans. Acoust. Speech Signal Process.*, 36, 844–953, 1988.

Han92 Hanna, M.T., Kia, A. and Robinson, J.P., Digital filters for attenuating interference arriving from a wide range of angles, *IEEE Trans. Signal Process.*, 40, 1499–1507, 1992.

Lim 90 Lim, B.L., Hui, S.K. and Lim, Y.C., Bearing estimation of coherent sources by circular spatial modulation averaging (CSMA) technique, *Electron. Lett.*, 26, 343–345, 1990.

Lut86 Luthra, A.K., A solution to the adaptive nulling problem with a look direction constraint in the presence of coherent jammers, *IEEE Trans. Antennas Propagat.*, 34, 702–710, 1986.

Pri58 Price, R. and Green, P.E., A communication technique for multipath channels, *IRE Proc.*, 46, 555–570, 1958.

Qia95 Qian, F. and Van Veen, B.D., Quadratically constrained adaptive beamforming for coherent signals and interference, *IEEE Trans. Signal Process.*, 43, 1890–1900, 1995.

Red87 Reddy, V.U., Paulraj, A. and Kailath, T., Performance analysis of the optimum beamformer in the presence of correlated smoothing, *IEEE Trans. Acoust. Speech Signal Process.*, 35, 927–936, 1987.

Sha85 Shan, T.J. and Kailath, T., Adaptive beamforming for coherent signals and interference, *IEEE Trans. Acoust. Speech Signal Process.*, 33, 527–536, 1985.

Tak86 Takao, K., Kikuma, N. and Yano, Y., Toeplitization of correlation matrix in multipath environment, *Proc. ICASSP*, 1873–1876, 1986.

Tak87 Takao, K. and Kikuma, N., An adaptive array utilizing an adaptive spatial averaging technique for multipath environments, *IEEE Trans. Antennas Propagat.*, 35, 1389–1396, 1987.

Tur80 Turin, G.L., Introduction to spread-spectrum techniques and their application to urban digital radio, *IEEE Proc.*, 68, 328–353, 1980.

Vau88 Vaughan, R.G., On optimum combining at the mobile, *IEEE Trans. Vehicular Technol.*, 37, 181–188, 1988.

Wid82 Widrow, B., et al., Signal cancellation phenomena in adaptive antennas: causes and cures, *IEEE Trans. Antennas Propagat.*, 30, 469–478, 1982.

Wil88 Williamson, D., Teo, K.L. and Musumeci, P.C., Optimum FIR array filters, *IEEE Trans. Acoust. Speech Signal Process.*, 36, 1211–1222, 1988.

Zol88 Zoltowski, M.D., On the performance analysis of the MVDR beamformer in the presence of correlated interference, *IEEE Trans. Acoust. Speech Signal Process.*, 36, 945–947, 1988.

# 6

# *Direction-of-Arrival Estimation Methods*



The problem of localization of sources radiating energy by observing their signal received at spatially separated sensors is of considerable importance, occurring in many fields, including radar, sonar, mobile communications, radio astronomy, and seismology. In this chapter, an estimation of the direction of arrival (DOA) of narrowband sources of the same central frequency, located in the far field of an array of sensors is considered, and various DOA estimation methods are described, compared, and sensitivity to various perturbations is analyzed. The chapter also contains discussion of various preprocessing and source estimation methods [God96, God97]. Source direction is parameterized by the variable $\theta$. The DOA estimation methods considered include spectral estimation, minimum-variance distortionless response estimator, linear prediction, maximum entropy, and maximum likelihood. Various eigenstructure methods are also described, including many versions of MUSIC algorithms, minimum norm methods, CLOSEST method, ESPRIT method, and the weighted subspace fitting method.

## 6.1 Spectral Estimation Methods

These methods estimate DOA by computing the spatial spectrum $P(\theta)$, that is, the mean power received by an array as a function of $\theta$, and then determining the local maximas of this computed spatial spectrum [Cap69, Lac71, Nut74, Joh82a, Wag84, Zha95, Bar56]. Most of these techniques have their roots in time series analysis. A brief overview and comparison of some of these methods are found in [Lac71, Joh82a].

### 6.1.1 Bartlett Method

One of the earliest methods of spectral analysis is the Bartlett method [Lac71, Bar56], in which a rectangular window of uniform weighting is applied to the time series data to be analyzed. For bearing estimation problems using an array, this is equivalent to applying equal weighting on each element. Thus, by steering the array in $\theta$ direction this method estimates the mean power $P_B(\theta)$, an expression for which is given by

$$P_B(\theta) = \frac{\mathbf{S}_\theta^H \, R \, \mathbf{S}_\theta}{L^2} \tag{6.1.1}$$

where $\mathbf{S}_\theta$ denotes the steering vector associated with the direction $\theta$, L denotes the number of elements in the array, and R is the array correlation matrix.

A set of steering vectors $\{\mathbf{S}_\theta\}$ associated with various direction $\theta$ is often referred to as the array manifold in DOA estimation literature. In practice, it may be measured at the time of array calibration. From the array manifold and an estimate of the array correlation matrix, $P_B(\theta)$ is computed using (6.1.1). Peaks in $P_B(\theta)$ are then taken as the directions of the radiating sources.

The process is similar to that of mechanically steering the array in this direction and measuring the output power. Due to the resulting side-lobes, output power is not only contributed from the direction in which the array is steered but from the directions where the side-lobes are pointing. The processor is also known as the conventional beamformer and the resolving power of the processor depends on the aperture of the array or the beamwidth of the main lobe.

## 6.2 Minimum Variance Distortionless Response Estimator

The minimum variance distortionless response estimator (MVDR) is the maximum likelihood method (MLM) of spectrum estimation [Cap69], which finds the maximum likelihood (ML) estimate of the power arriving from a point source in direction $\theta$ assuming that all other sources are interference. In the beamforming literature, it is known as the MVDR beamformer as well as the optimal beamformer, since in the absence of errors, it maximizes the output SNR and passes the look direction signal undistorted as discussed in Chapter 2. For DOA estimation problems, MLM is used to find the ML estimate of the direction rather than the power [Mil90]. Following this convention, the current estimator is referred to as the MVDR estimator.

This method uses the array weights obtained by minimizing the mean output power subject to a unity constraint in the look direction. The expression for the power spectrum $P_{MV}(\theta)$ is

$$P_{MV}(\theta) = \frac{1}{\mathbf{S}_\theta^H R^{-1} \mathbf{S}_\theta} \tag{6.2.1}$$

This method has better resolution properties than the Bartlett method [Cox73], but does not have the best resolution properties of all methods [Joh82a].

## 6.3 Linear Prediction Method

The linear prediction (LP) method estimates the output of one sensor using linear combinations of the remaining sensor outputs and minimizes the mean square prediction error, that is, the error between the estimate and the actual output [Joh82a, Mak75]. Thus, it obtains the array weights by minimizing the mean output power of the array subject to the constraint that the weight on the selected sensor is unity. Expressions for the array weights $\hat{\mathbf{w}}$ and the power spectrum $P_{LP}(\theta)$, respectively, are

$$\hat{\mathbf{w}} = \frac{R^{-1}\mathbf{u}_1}{\mathbf{u}_1^H R^{-1} \mathbf{u}_1} \tag{6.3.1}$$

and

$$P_{LP}(\theta) = \frac{\mathbf{u}_1^H R^{-1} \mathbf{u}_1}{\left|\mathbf{u}_1^H R^{-1} \mathbf{S}_\theta\right|^2} \tag{6.3.2}$$

where $\mathbf{u}_1$ is a column vector such that one of its elements is unity and the remaining elements are zero [Joh82a].

The position of 1 in the column vector corresponds to the position of the selected element in the array for predicting its output. There is no criterion for proper choice of this element; however, choice of this element affects the resolution capability and bias in the estimate. These effects are dependent on the SNR and separation of directional sources [Joh82a]. LP methods perform well in moderately low SNR environments and are good compromises in situations where sources are of approximately equal strength and are nearly coherent [Kes85].

## 6.4 Maximum Entropy Method

The maximum entropy (ME) method finds a power spectrum such that its Fourier transform equals the measured correlation subjected to the constraint that its entropy is maximized [Bur67]. The entropy of a Gaussian band-limited time series with power spectrum S(f) is defined as

$$H(S) = \int_{-f_N}^{f_N} \ln S(f)df \tag{6.4.1}$$

where $f_N$ is the Nyquist frequency.

For estimating DOA from the measurements using an array of sensors, the ME method finds a continuous function $P_{ME}(\theta) > 0$ such that it maximizes the entropy function

$$H(P) = \int_0^{2\pi} \ln P_{ME}(\theta)d\theta \tag{6.4.2}$$

subject to the constraint that the measured correlation between the ith and the jth elements $r_{ij}$ satisfies

$$r_{ij} = \int_0^{2\pi} P_{ME}(\theta)\cos\left(2\pi\tau_{ij}(\theta)\right)d\theta \tag{6.4.3}$$

where $\tau_{ij}(\theta)$ denotes the differential delay between elements i and j due to a source in $\theta$ direction.

The solution to this problem requires an infinite dimensional search. The problem may be transformed to a finite dimensional search using the duality principle [McC83] leading to

$$P_{ME}(\theta) = \frac{1}{\hat{\mathbf{w}}^T\mathbf{q}(\theta)} \tag{6.4.4}$$

In (6.4.4), $\hat{\mathbf{w}}$ is obtained by minimizing

$$H(\mathbf{w}) = \int_0^{2\pi} \ln\left(\mathbf{w}^T\mathbf{q}(\theta)\right)d\theta \tag{6.4.5}$$

subject to

$$\mathbf{w}^T\mathbf{r} = 2\pi \tag{6.4.6}$$

and

$$\mathbf{w}^T\mathbf{q}(\theta) > 0 \quad \forall\theta \tag{6.4.7}$$

where $\mathbf{q}(\theta)$ and $\mathbf{r}$, respectively, are defined as

$$\mathbf{q}(\theta) = \left[1, \sqrt{2}\cos\left(2\pi f\,\tau_{12}(\theta)\right), \dots\right]^T \tag{6.4.8}$$

and

$$\mathbf{r} = \left[r_{11}, \sqrt{2}\,r_{12}, \dots\right]^T \tag{6.4.9}$$

It should be noted that the dimension of these vectors depends on the array geometry and is equal to the number of known correlations $r_{ij}$ for every possible i and j.

The minimization problem defined above may be solved iteratively using the standard gradient LMS algorithm. For more information on various issues of the ME method, see [Nag94, Ski79, Tho80, McC82, Lan83, Far85]. Suitability of the ME method for mobile communications in fast-fading signal conditions has been studied by [Nag94].

## 6.5 Maximum Likelihood Method

The MLM estimates the DOAs from a given set of array samples by maximizing the log-likelihood function [Mil90, Lig73, Sch68, Zis88, Sto90, Oh92, Lee94, Wu94a, She96]. The likelihood function is the joint probability density function of the sampled data given the DOAs and viewed as a function of the desired variables, which are the DOAs in this case. The method searches for those directions that maximize the log of this function. The ML criterion signifies that plane waves from these directions are most likely to cause the given samples to occur [Hay85].

Maximization of the log-likelihood function is a nonlinear optimization problem, and in the absence of a closed-form solution requires iterative schemes. There are many such schemes available in the literature. The well-known gradient descent algorithm using the estimated gradient of the function at each iteration as well as the standard Newton–Raphson method are well suited for the job [Wax83]. Other schemes, such as the alternating projection method [Zis88, Oh92] and the expectation maximization algorithm [Mil90, Dem77, Hin81], have been proposed for solving this problem in general as well as for specialized cases such as unknown polarization [Lee94a], unknown noise environments [Wu94], and contaminated Gaussian noise [Lig73]. A fast algorithm [Aba85] based on Newton's method developed for estimating frequencies of sinusoids may be modified to suit DOA estimation based on ML criteria.

The MLM provides superior performance compared to other methods particularly when SNR is small, the number of samples is small, or the sources are correlated [Zis88], and thus is of practical interest. For a single source, the estimates obtained by this method are asymptotically unbiased [Lee94a], that is, the expected values of the estimates approach their true values in the limit as the number of samples used in the estimate increase. In that sense, it may be used as a standard to compare the performance of other methods. The method normally assumes that the number of sources, M, is known [Zis88].

When a large number of samples is available, other computationally more efficient schemes may be used with performance almost equal to this method [Sto90]. Analysis of the method to estimate the direction of sources when the array and the source are in relative motion to each other indicates its potential for mobile communications [Wig95, Zei95].

## 6.6 Eigenstructure Methods

These methods rely on the following properties of the array correlation matrix: (1) The space spanned by its eigenvectors may be partitioned in two subspaces, namely the signal subspace and the noise subspace; and (2) The steering vectors corresponding to the directional sources are orthogonal to the noise subspace. As the noise subspace is orthogonal to the signal subspace, these steering vectors are contained in the signal subspace. It should be noted that the noise subspace is spanned by the eigenvectors associated with the smaller eigenvalues of the correlation matrix, and the signal subspace is spanned by the eigenvectors associated with its larger eigenvalues.

In principle, the eigenstructure-based methods search for directions such that the steering vectors associated with these directions are orthogonal to the noise subspace and are contained in the signal subspace. In practice, the search may be divided in two parts. First, find a weight vector **w** that is contained in the noise subspace or is orthogonal to the signal subspace, and then search for directions such that the steering vectors associated with these directions are orthogonal to this vector. The source directions correspond to the local minima of the function $|\mathbf{w}^H \mathbf{S}_\theta|$, where $\mathbf{S}_\theta$ denotes a steering vector.

When these steering vectors are not guaranteed to be in the signal subspace there may be more minima than the number of sources. The distinction between the actual source direction and a spurious minima in $|\mathbf{w}^H \mathbf{S}_\theta|$ is made by measuring the power in these directions.

Many methods have been proposed that utilize the eigenstructure of the array correlation matrix. These methods differ in the way that available array signals have been utilized, required array geometry, applicable signal model, and so on. Some of these methods do not require explicit computation of the eigenvalues and eigenvectors of the array correlation matrix, whereas in others it is essential. Effective computation of these quantities may be done by methods similar to those described in [Tuf86]. When the array correlation matrix is not available, a suitable estimate of the matrix is made from available samples.

One of the earliest DOA estimation methods based on the eigenstructure of covariance matrix was presented by Pisarenko [Pis73], and has better resolution than the minimum variance, ME, and LP methods [Wax84]. A critical comparison of this method with two other schemes [Red79, Can80] applicable for a correlated noise field has been presented in [Bor81] to show that the Pisarenko's method is an economized version of these schemes, restricted to equispaced linear arrays. The scheme presented in [Red79] is useful for off-line implementations similar to those presented in [Joh82, Bro83], whereas the method described in [Can80] is useful for real-time implementations and uses normalized gradient algorithm to estimate a vector in the noise subspace from available array signals. Other schemes suitable for real-time implementation are discussed in [Red82, Yan88, Lar83]. A scheme known as the matrix pencil method, shown by [Oui89] to be similar to Pisarenko's method, has been described in [Oui88].

Eigenstructure methods may also be used for finding DOAs when the background noise is not white but has a known covariance [Pau86] unknown covariance [Wax92], or when the sources are in the near field and/or the sensors have unknown gain patterns [Wei88]. For the latter case, the signals induced on all elements of the array are not of the equal intensity, as is the case when the array is in the far field of the directional sources. The effect of spatial coherence on resolution capability of the these methods is discussed in [Bie80], whereas the issue of the optimality of these methods is considered in [Bie83]. In the following, some popular schemes are described in detail.

## 6.7 MUSIC Algorithm

The multiple signal classification (MUSIC) method [Sch86] is a relatively simple and efficient eigenstructure variant of DOA estimation methods. It is perhaps the most studied method in its class and has many variations. Some of these are discussed in this section.

### 6.7.1 Spectral MUSIC

In its standard form, also known as spectral MUSIC, the method estimates the noise subspace from available samples. This can be done either by eigenvalue decomposition of the estimated array correlation matrix or singular value decomposition of the data matrix with its N columns being the N array signal vector samples, also known as snapshots. The latter is preferred for numerical reasons [DeG93].

Once the noise subspace has been estimated, a search for M directions is made by looking for steering vectors that are as orthogonal to the noise subspace as possible. This is normally accomplished by searching for peaks in the MUSIC spectrum given by

$$P_{MU}(\theta) = \frac{1}{\left|\mathbf{S}_{\theta}^{H} U_{N}\right|^{2}} \tag{6.7.1}$$

where $U_N$ denotes an L by L − M dimensional matrix, with L − M columns being the eigenvectors corresponding to the L − M smallest eigenvalues of the array correlation matrix and $\mathbf{S}_\theta$ denoting the steering vector that corresponds to direction θ.

It should be noted that instead of using the noise subspace and searching for directions with steering vectors orthogonal to this subspace, one could also use the signal subspace and search for directions with steering vectors contained in this space [Bar83]. This amounts to searching for peaks in

$$P_{MU}(\theta) = \left|U_{S}^{H} \mathbf{S}_{\theta}\right|^{2} \tag{6.7.2}$$

where $U_S$ denotes an L × M dimensional matrix with its M columns being the eigenvectors corresponding to the M largest eigenvalues of the array correlation matrix.

It is advantageous to use the one with smaller dimensions. For the case of a single source, the DOA estimate made by the MUSIC method asymptotically approaches the Cramer–Rao lower bound, that is, where the number of snapshots increases infinitely, the best possible estimate is made. For multiple sources, the same holds for large SNR cases, that is, when the SNR approaches infinity [Fri90, Por88]. The Cramer–Rao lower bound (CRLB) gives the theoretical lowest value of the covariance for an unbiased estimator.

In [Klu93], an application of the MUSIC algorithm to cellular mobile communications was investigated to locate land mobiles, and it is shown that when multipath arrivals are grouped in clusters the algorithm is able to locate the mean of each cluster arriving at a mobile. This information then may be used to locate line of sight. Its use for mobile satellite communications has been suggested in [Geb95].

### 6.7.2 Root-MUSIC

For a uniformly spaced linear array (ULA), the MUSIC spectra can be expressed such that the search for DOA can be made by finding the roots of a polynomial. In this case, the method is known as root-MUSIC [Bar83]. Thus, root-MUSIC is applicable when a ULA is used and solves the polynomial rooting problem in contrast to spectral MUSIC's identification and localization of spectral peaks. Root-MUSIC has better performance than spectral MUSIC [Rao89a].

### 6.7.3 Constrained MUSIC

This method incorporates the known source to improve estimates of the unknown source direction [DeG93]. The situation arises when some of the source directions are already known. The method removes signal components induced by these known sources from the data matrix and then uses the modified data matrix for DOA estimation. Estimation is achieved by projecting the data matrix onto a space orthogonal complement to a space spanned by the steering vectors associated with known source directions. A matrix operation, the process reduces the signal subspace dimension by a number equal to the known sources and improves estimate quality, particularly when known sources are strong or correlated with unknown sources.

### 6.7.4 Beam Space MUSIC

The MUSIC algorithms discussed so far process the snapshots received from sensor elements without any preprocessing, such as forming beams, and thus may be thought of as element space algorithms, which contrasts with the beamspace MUSIC algorithm in which the array data are passed through a beamforming processor before applying MUSIC or any other DOA estimation algorithms. The beamforming processor output may be thought of as a set of beams; thus, the processing using these data is normally referred to as beamspace processing. A number of DOA estimation schemes are discussed in [May87, Kar90], where data are obtained by forming multiple beams using an array.

The DOA estimation in beam space has a number of advantages such as reduced computation, improved resolution, reduced sensitivity to system errors, reduced resolution threshold, reduced bias in the estimate, and so on [Fri90, Lee90, Xu93, Zol93, Zol93a]. These advantages arise from the fact that a beamformer is used to form a number of beams that are less than the number of elements in the array; consequently, less data to process a DOA estimation are necessary.

This process may be understood in terms of array degrees of freedom. Element space methods have degrees of freedom equal to the number of elements in the array, whereas the degrees of freedom of beamspace methods are equal to the number of beams formed by the beamforming filter. Thus, the process reduces the array's degrees of freedom. Normally, only M + 1 degrees of freedom to resolve M sources are needed.

The root-MUSIC algorithm discussed for the element space case may also be applied to this case, giving rise to beamspace root-MUSIC [Zol93, Zol93a]. Computational savings for this method are the same as for beamspace methods compared to element space methods in general.

## 6.8 Minimum Norm Method

Minimum norm method [Red79, Kum83] is applicable for ULA, and finds the DOA estimate by searching for peak locations in the spectrum [Erm94], as in the following expression:

$$P_{MN}(\theta) = \frac{1}{\left|\mathbf{w}^H \mathbf{S}_\theta\right|^2} \tag{6.8.1}$$

where **w** denotes an array weight such that it is of the minimum norm, has first element equal to unity, and is contained in the noise subspace. The solution to the above problem leads to the following expression for the spectrum [Erm94, Nic88, Cle89]:

$$P_{MN}(\theta) = \frac{1}{\left|\mathbf{S}_\theta^H U_N U_N^H \mathbf{e}_1\right|^2} \tag{6.8.2}$$

where the vector $\mathbf{e}_1$ contains all zeros except the first element, which is equal to unity.

Given that the method is applicable for ULA, the optimization problem to solve for the array weight may be transformed to a polynomial rooting problem, leading to a root-minimum-norm method similar to root-MUSIC. A performance comparison [Kri92] indicated that the variance in the estimate obtained by root-MUSIC is smaller than or equal to that of the root-minimum-norm method. Schemes to speed up the DOA estimation algorithm of the minimum norm and to reduce computations are discussed in [Erm94, Ng90].

## 6.9 CLOSEST Method

The CLOSEST method is useful for locating sources in a selected sector. Contrary to beamspace methods, which work by first forming beams in selected directions, CLOSEST operates in the element space and in that sense it is an alternative to beamspace MUSIC. In a way, it is a generalization of the minimum-norm method. It searches for array weights in the noise subspace that are close to the steering vectors corresponding to DOAs in the sector under consideration, and thus its name. Depending on the definition of closeness, it leads to various schemes. A method referred to as FINE (First Principal Vector) selects an array weight vector by minimizing the angle between the selected vector and the subspace spanned by the steering vectors corresponding to DOAs in the selected sector. In short, the method replaces the vector $\mathbf{e}_1$ used in the minimum-norm method by a suitable vector depending on the definition of closeness used. For details about the selection of these vectors and the relative merits of the CLOSEST method, see [Buc90].

## 6.10 ESPRIT Method

Estimation of signal parameters via rotational invariance techniques (ESPRIT) [Roy89] is a computationally efficient and robust method of DOA estimation. It uses two identical arrays in the sense that array elements need to form matched pairs with an identical displacement vector, that is, the second element of each pair ought to be displaced by the same distance and in the same direction relative to the first element.

However, this does not mean that one has to have two separate arrays. The array geometry should be such that the elements could be selected to have this property. For example, a ULA of four identical elements with inter-element spacing d may be thought of as two arrays of three matched pairs, one with first three elements and the second with last three elements such that the first and the second elements form a pair, the second and the third elements form another pair, and so on. The two arrays are displaced by the distance d. The way ESPRIT exploits this subarray structure for DOA estimation is now briefly described.

Let the signals induced on the $\ell$th pair due to a narrowband source in direction $\theta$ be denoted by $x_\ell(t)$ and $y_\ell(t)$. The phase difference between these two signals depends on the time taken by the plane wave arriving from the source under consideration to travel from one element to the other. Assume that the two elements are separated by the displacement $\Delta_0$. Thus, it follows that

$$y_\ell(t) = x_\ell(t)e^{j2\pi\Delta_0\cos\theta} \tag{6.10.1}$$

where $\Delta_0$ is measured in wavelengths.

Note that $\Delta_0$ is the magnitude of the displacement vector. This vector sets the reference direction and all angles are measured with reference to this vector. Let the array signals received by the two K-element arrays be denoted by $\mathbf{x}(t)$ and $\mathbf{y}(t)$. These are given by

$$\mathbf{x}(t) = A\mathbf{s}(t) + \mathbf{n}_x(t) \tag{6.10.2}$$

and

$$\mathbf{y}(t) = A\Phi\mathbf{s}(t) + \mathbf{n}_y(t) \tag{6.10.3}$$

where A is a $K \times M$ matrix with its columns denoting the M steering vectors corresponding to M directional sources associated with the first subarray, $\Phi$ is an $M \times M$ diagonal matrix with its mth diagonal element given by

$$\Phi_{m,m} = e^{j2\pi\Delta_0\cos\theta_m} \tag{6.10.4}$$

$\mathbf{s}(t)$ denotes M source signals induced on a reference element, and $\mathbf{n}_x(t)$ and $\mathbf{n}_y(t)$, respectively, denote the noise induced on the elements of the two subarrays. Comparing the equations for $\mathbf{x}(t)$ and $\mathbf{y}(t)$, it follows that the steering vectors corresponding to M directional sources associated with the second subarray are given by $A\Phi$.

Let $U_x$ and $U_y$ denote two $K \times M$ matrices with their columns denoting the M eigenvectors corresponding to the largest eigenvalues of the two array correlation matrices $R_{xx}$ and $R_{yy}$, respectively. As these two sets of eigenvectors span the same M-dimensional signal space, it follows that these two matrices $U_x$ and $U_y$ are related by a unique nonsingular transformation matrix $\psi$, that is,

$$U_x\psi = U_y \tag{6.10.5}$$

Similarly, these matrices are related to steering vector matrices A and $A\Phi$ by another unique nonsingular transformation matrix T as the same signal subspace is spanned by these steering vectors. Thus,

$$U_x = AT \tag{6.10.6}$$

and

$$U_y = A\Phi T \tag{6.10.7}$$

Substituting for $U_x$ and $U_y$ and the fact that A is of full rank,

$$T\psi T^{-1} = \Phi \tag{6.10.8}$$

According to this statement, the eigenvalues of $\psi$ are equal to the diagonal elements of $\Phi$, and columns of T are eigenvectors of $\psi$.

This is the main relationship in the development of ESPRIT [Roy89]. It requires an estimate of $\psi$ from the measurement **x**(t) and y(t). An eigendecompositon of $\psi$ provides its eigenvalues, and by equating them to $\Phi$ leads to the DOA estimates,

$$\theta_m = \cos^{-1}\left\{\frac{\mathrm{Arg}(\lambda_m)}{2\pi\Delta_0}\right\}, \quad m = 1, \ldots, M \tag{6.10.9}$$

The ways in which estimates of $\psi$ were efficiently obtained from the array signal measurements led to many versions of ESPRIT [Roy89, Xu94, Ham94, Roy86, Pau86a, Wei91]. The one summarized below is referred to as total least squares (TLS) ESPRIT [Roy89, Xu94].

1. Make measurements from two identical subarrays that are displaced by $\Delta_0$. Estimate the two array correlation matrices from the measurement and find their eigenvalues and eigenvectors.
2. Find the number of directional sources M using available methods; some are described in Section 6.14.
3. Form the two matrices with their columns being the M eigenvectors associated with the largest eigenvalues of each correlation matrix. Let these be denoted by $U_x$ and $U_y$. For a ULA, this could be done by first forming an $L \times M$ matrix U, by selecting its columns as the M eigenvectors associated with the largest eigenvalues of the estimated array correlation matrix of the full array of L elements. Then select the first $K < L$ rows of U to form $U_x$ and the last of its K rows to form $U_y$.
4. Form a $2M \times 2M$ matrix

$$\begin{bmatrix} U_x^H \\ U_y^H \end{bmatrix}\begin{bmatrix} U_x & U_y \end{bmatrix} \tag{6.10.10}$$

and find its eigenvalues $\lambda_1 \geq \ldots \geq \lambda_{2M}$. Let $\Lambda$ be a diagonal matrix:

$$\Lambda = \begin{bmatrix} \lambda_1 & & 0 \\ & \ddots & \\ 0 & & \lambda_{2M} \end{bmatrix} \tag{6.10.11}$$

Let the eigenvectors associated with $\lambda_1 \geq \ldots \geq \lambda_{2M}$ be the columns of a matrix V such that

$$\begin{bmatrix} U_x^H \\ U_y^H \end{bmatrix}\begin{bmatrix} U_x & U_y \end{bmatrix} = V\Lambda V^H \tag{6.10.12}$$

5. Partition V into four matrices of dimension $M \times M$ as

$$V = \begin{pmatrix} V_{11} & V_{12} \\ V_{21} & V_{22} \end{pmatrix} \tag{6.10.13}$$

6. Calculate the eigenvalues $\lambda_m$, $m = 1, \ldots, M$ of the matrix $-V_{11}V_{22}^{-1}$.
7. Estimate the angle of arrival $\theta_m$, using

$$\theta_m = \cos^{-1}\left\{\frac{\text{Arg}(\lambda_m)}{2\pi\Delta_0}\right\}, \quad m = 1, \ldots, M \tag{6.10.14}$$

Other ESPRIT variations include beamspace ESPRIT [Xu94], beamspace ESPRIT for uniform rectangular array [Gan96], resolution-enhanced ESPRIT [Ham94], virtual interpolated array ESPRIT [Pau86a], multiple invariance ESPRIT [Swi92a], higher-order ESPRIT [Yue96], and procrustes rotation–based ESPRIT [Zol89]. Use of ESPRIT for DOA estimation employing an array at a base station in the reverse link of a mobile communication system has been studied in [Wan95].

## 6.11 Weighted Subspace Fitting Method

The weighted subspace fitting (WSF) method [Vib91, Vib91a] is a unified approach to schemes such as MLM, MUSIC, and ESPRIT. It requires that the number of directional sources be known. The method finds the DOA such that the weighted version of a matrix whose columns are the steering vectors associated with these directions is close to a data-dependent matrix. The data-dependent matrix could be a Hermitian square root of the array correlation matrix or a matrix whose columns are the eigenvectors associated with the largest eigenvalues of the array correlation matrix. The framework proposed in the method can be used for deriving common numerical algorithms for various eigenstructure methods as well as for their performance studies. WSF application for mobile communications employing an array at the base station has been investigated in [And91, Klo96].

## 6.12 Review of Other Methods

In this section, a brief review of methods not covered in detail is provided. A number of eigenstructure methods reported in the literature exploit specialized array structures or noise scenarios. Two methods using uniform circular arrays presented in [Mat94] extend beamspace MUSIC and ESPRIT algorithms for two-dimensional angle estimation, including an analysis of MUSIC to resolve two sources in the presence of gain, phase, and location errors. Properties of an array have also been exploited in [Swi93] to find the azimuth and elevation of a directional source. Two DOA estimation schemes in an unknown noise field using two separate arrays proposed in [Wu94a] appear to offer superior performance compared to their conventional counterparts.

Use of a minimum redundancy linear array offers several advantages as discussed in [Zol93]. By using such arrays, one may be able to resolve more than L sources using L elements, $L(L-1)/2$ being the upper limit. A minimum redundancy linear array has nonuniform spacing such that the number of sensor pairs measuring the same special correlation lag is minimized for a given number of elements. In designing such an array, having only one pair with spacing d, one pair with spacing 2d, and so on is perferred, such as a three-element array with element positions $x_1 = 0$, $x_2 = d$, and $x_3 = 3d$. The minimum redundancy linear arrays are also referred to as augmented arrays [God88].

The direction-finding methods applicable to unknown noise field are described in [Wax92, LeC89, Won92, Rei92, Ami92]. The MAP (maximum a posteriori) method presented in [Won92, Rei92] is based on Bayesian analysis, and estimated results are not asymptotically consistent, that is, the results may be biased [Wu94a]. The method in [Ami92], referred to as concurrent nulling and location (CANAL), may be implemented using analog hardware, thus eliminating the need for sampling, data storage, and so on. A DOA estimation method in the presence of correlated arrivals using an array of unrestricted geometry is discussed in [Cad88]. Several methods that do not require eigenvalue decomposition are discussed in [Rei87, Di85, Xu92, Wei93a, Fuc94, Che94, Yan94a, Sou95].

The method proposed in [Rei87] is applicable for a linear array of L elements. It forms a $K \times K$ correlation matrix from one snapshot with $K \geq M$, and is based on the QR orthonormal decomposition [Gol83] on this correlation matrix, with Q being a $K \times K$ unitary matrix and R being upper triangle. The last $K - M$ columns of Q define a set of orthonormal basis for the noise space. Denoting these columns by $U_N$, the source directions are obtained from power spectrum peaks:

$$P(\theta) = \frac{1}{\left|\mathbf{S}_\theta^H U_N\right|^2} \tag{6.12.1}$$

The method is computationally efficient and the performance is comparable to MUSIC [Rei87]. A multiple-source location method based on the matrix decomposition approach is presented in [Di85]. The method requires the knowledge of the noise power estimate, and is applicable for coherent as well as noncoherent arrivals. It does not require knowledge of number of sources.

The method discussed in [Xu92] exploits the cyclostationarity [Gar91] of data that may exist in certain situations. The method has significant implementation advantages and its performance is comparable with the other methods. A method is discussed in [Wei93a] that is based on polynomial rooting estimates DOA with high resolution and has low computation requirements; it exploits the diversity polarization of an array. Such arrays have the capability of separating signals based on polarization characteristics, and thus have an advantage over uniformly polarized arrays [Fer83, Zis90].

An adaptive scheme based on Kalman filtering to estimate noise subspace is presented in [Che94], which is then combined with root-MUSIC to estimate DOA. The method has good convergence characteristics. The method presented in [Fuc94] uses a deconvolution approach to the output of a conventional processor to localize sources, whereas those discussed in [Yan94a, Sou95] use a neural network approach to direction finding.

The discussion on DOA estimation so far has been concentrated on estimating the directions of stationary narrowband sources. Although extension of a narrowband direction-finding scheme to the broadband case is not trivial, some of the methods discussed here have been extended to estimate broadband source directions. For discussion of these and other schemes, see [Wax84, Su83, Wan85, Swi89, Kro89, Ott90, Cad90, Dor93, Gre94, Swi94, Buc88, Hun90]. The methods described in [Su83, Wan85, Swi89, Cad90, Buc88] are based on a signal subspace approach, whereas those discussed in [Ott90, Hun90] and [Dor93, Sch93] are related to the ESPRIT method and the ML method, respectively. Applications of high-resolution direction-finding methods to estimate the directions of moving sources and to track these sources are described in [Rao94, Yan94, Liu94, Eri94, Sas91]. The problem of estimating the mean DOA of spatially distributed sources such as those in base mobile communication systems has been examined in [Men96, Tru96].

## 6.13 Preprocessing Techniques

Several techniques are used to process data before using direction-finding methods for DOA estimation, particularly in situations where directional sources are correlated or coherent. Correlation of directional sources may exist due to multipath propagation, and tends to reduce the rank of the array correlation matrix as discussed in Chapter 5. The correlation matrix may be tested for source coherency by applying the rank profile test described in [Sha87]. Most preprocessing techniques either try to restore this rank deficiency in the correlation matrix or modify it to be useful for the DOA estimation methods. In this section, some of these techniques are reviewed.

One scheme referred to as the spatial smoothing method has been widely studied in the literature [Sha85, Wil88, Yeh89, Pil89, Lee90a, Mog91, Du91, Mog92, Yan92, Rao93, Lio89, Wei93, Eva81], and is applicable for a linear array. Details on spatial smoothing for beamforming are provided in Chapter 5. In its basic form, it decorrelates the correlated arrival by subdividing the array into a number of smaller overlapping subarrays and then averaging the array correlation matrix obtained from each subarray. The number of subarrays obtained from an array depends on the number of elements used in each subarray. For example, using K elements in each subarray, $L - (K - 1)$ subarrays can be formed from an array of L elements by forming the first subarray using elements 1 to K, the second subarray using elements 2 to $K + 1$ and so on. The number and size of subarrays are determined from the number of directional sources under consideration. For M sources, a subarray size of $M + 1$ and a subarray number greater than or equal to M are necessary [Sha85].

Thus, to estimate the directions of M sources, array size $L = 2M$ is required, which could be reduced to $3/2M$ by using the forward-backward spatial smoothing method [Wil88, Pil89]. This process uses the average of the correlation matrix obtained from the forward subarray scheme and the correlation matrix obtained from the backward subarray scheme.

The forward subarray scheme subdivides the array starting from one side of the array as discussed above, whereas the backward subarray scheme subdivides the array starting from the other side of the array. Thus, in the forward subarray scheme, the first subarray is formed using elements 1 to K, whereas in the backward subarray scheme the first subarray is formed using elements L to $L - (K - 1)$ and so on. The mth subarray matrix $\bar{R}_m$ of the backward method is related to the forward-method matrix $R_m$ by

$$\bar{R}_m = J_0 R_m^* J_0 \tag{6.13.1}$$

where $J_0$ is a reflection matrix with all its elements along the secondary diagonal being equal to unity and zero elsewhere, that is,

$$J_0 = \begin{bmatrix} 0 & & 1 \\ & \cdot^{\cdot^{\cdot}} & \\ 1 & & 0 \end{bmatrix} \quad (6.13.2)$$

The process is similar to that used by forward–backward prediction for bearing estimation [Lee90a].

An improved spatial smoothing method [Du91] uses correlation between all array elements, rather than correlation between subarray elements as is done in forward–backward spatial smoothing method. It estimates a cross-correlation matrix $R^{mj}$ from subarrays m and j, that is,

$$R^{mj} = E\left[\mathbf{x}_m(t)\mathbf{x}_j^H(t)\right] \quad (6.13.3)$$

with $\mathbf{x}_m(t)$ and $\mathbf{x}_j(t)$, respectively, denoting the array signal vector from the mth and jth subarrays.

The forward subarray matrix $R_m$ is then obtained using

$$R_m = \frac{1}{L_0}\sum_{j=1}^{L_0} R^{mj}R^{jm} \quad (6.13.4)$$

and $\bar{R}_m$ is obtained by substituting $R_m$ from (6.13.4) in (6.13.1). $L_0$ in (6.13.4) denotes the number of subarrays used.

A method described in [Mog91, Mog92] removes the effects of sensor noise to make spatial smoothing more effective in low SNR situations. This spatial filtering method is further refined in [Del96] to offer DOA estimates of coherent sources with reduced RMS errors.

A decorrelation analysis of spatial smoothing [Yan92] shows that there exists an upper bound on the number of subarrays and the maximum distance between the subarrays depends on the fractional bandwidth of the signals. A comprehensive analysis of the use of spatial smoothing as a preprocessing technique to weighted ESPRIT and MUSIC methods of DOA estimation presented in [Rao93] shows how their performance could be improved by proper choice of the number of subarrays and weighting matrices. An ESPRIT application to estimate the source directions and polarization shows improvement in its performance in the presence of coherent arrivals when it is combined with the spatial smoothing method [Li93].

Spatial smoothing methods using subarray arrangements reduce the effective aperture of the array as well as degrees of freedom, and thus more elements are needed to process correlated arrivals than would otherwise be required. The schemes that do not reduce effective array size include those that restore the structure of the array correlation matrix for the linear array to an uncorrelated one. These are referred to as structured methods [God90, Tak87].

Structured methods rely on the fact that for a linear equispaced array, the correlation matrix in the absence of correlated arrivals has a Toeplitz structure, that is, the elements of the matrix along its diagonals are equal. Correlation between sources destroys this structure. In [God90], the structure is restored by averaging the matrix obtained in the presence of correlated arrivals by simple averaging along the diagonals as detailed in Chapter 5, while in [Tak87] a weighted average is used. A DOA estimation method using the array correlation matrix structured by averaging along its diagonals discussed in [Fuc96] appears to offer computational advantages over similar methods.

Other preprocessing schemes to decorrelate sources include random permutation [Lio89], mechanical movement using a circular disk [Lim90], construction of a preprocessing matrix using approximate knowledge of DOA estimate [Wei94a], signal subspace transformation in the spatial domain [Par93], unitary transformation method [Hua91], and methods based on aperture interpolations [Wei93, Swi89a, Wei95].

## 6.14 Estimating Source Number

Many high-resolution direction-finding methods require that the number of directional sources, and their performance is dependent on perfect knowledge of these numbers. Selected methods for estimating the number of these sources are discussed in this section.

The most commonly referred method for detecting the number of sources was first introduced in [Wax85] based on Akaike's information criterion (AIC) [Aka74] and Rissanen's minimum description length (MDL) [Ris78] principle. The method was further analyzed in [Zha89, Wan86] and modified in [Yin87, Won90]. A variation of the method that is applicable to coherent sources is discussed in [Wax92, Wax89a, Wax91]. Briefly, the method works as follows [Wax85, Wan86]:

1. Estimate the array correlation matrix from N independent and identically distributed samples.
2. Find the L eigenvalues $\lambda_i$, i = 1, 2, ..., L of the correlation matrix such that $\lambda_1 > \lambda_2 > \ldots > \lambda_L$.
3. Estimate the number of sources M by solving

$$\underset{M}{\text{minimize}}\ N(L-M)\log\left(\frac{f_1(M)}{f_2(M)}\right)+f_3(M,N) \tag{6.14.1}$$

where

$$f_1(M)=\frac{1}{L-M}\sum_{i=M+1}^{L}\lambda_i \tag{6.14.2}$$

$$f_2(M)=\left(\prod_{i=M+1}^{L}\lambda_i\right)^{\frac{1}{L-M}} \tag{6.14.3}$$

and the penalty function

$$f_3(M,N)=\begin{cases}M(2L-M) & \text{for AIC}\\ \frac{1}{2}M(2L-M)\log N & \text{for MDL}\end{cases} \tag{6.14.4}$$

with L denoting the number of elements in the array.

A modification of the method based on the MDL principle applicable to coherent sources is discussed in [Wax89a], which is further refined in [Wax92, Wax91] to improve performance. A parametric method that does not require knowledge of eigenvalues of the array correlation matrix is discussed in [Wu91]. It has better performance than some other methods discussed and is computationally more complex.

All methods that partition the eigenvalues of the array correlation matrix rely on the fact that the M eigenvalues corresponding to M directional sources are larger than the rest of the $L - M$ eigenvalues corresponding to the background noise; they also select the threshold differently. One of the earliest methods uses a hypothesis-testing procedure based on the confidence interval of noise eigenvalues [And63]. Threshold assignment was subjective.

The eigenthreshold method uses a one-step prediction of the threshold for differentiating the smallest eigenvalues from the others. This method performs better than AIC and MDL. It has a threshold at a lower SNR value than MDL and a lower error rate than AIC at high SNRs [Che91].

An alternate scheme for estimating the number of sources discussed in [Lee94] uses the eigenvectors of the array correlation matrix; in contrast, other methods use the eigenvalues of the array correlation matrix. This method, referred to as the eigenvector detection technique, is applicable to a cluster of sources whose approximate directions are known, and is able to estimate the number of sources at a lower SNR than those by AIC and MDL.

In practice, the number of sources an array may be able to resolve not only depends on the number of elements in the array but also on array geometry, available number of snapshots, and spatial distribution of sources. For discussion of these and other issues related to array capabilities to uniquely resolve the number of sources, see [Fri91, Wax89, Bre86] and references therein.

## 6.15 Performance Comparison

Performance analysis of various direction finding-schemes has been carried out by many researchers [Joh82, Rao89, Lee90, Xu93, Pil89a, Sto89, Sto90a, Xu92a, Xu94a, Lee91, Zho91, Zho93, Zha95a, Kau95, Kav86, Sto90, Sto91, Ott91, Mat94, Ott92, Vib95, Wei93, Cap70]. The performance measures considered for analysis include bias, variance, resolution, CRLB, and probability of resolution. In this section, the performance of selected DOA estimation schemes is discussed.

The MUSIC algorithm has been studied in [Lee90, Xu93, Pil89a, Sto89, Sto90a, Xu92a, Xu94a, Lee91, Zho91, Zho93, Zha95a, Kau95, Kav86]. Most of these studies concentrate on its performance and performance comparisons with other methods when a finite number of samples is used for direction finding rather than their ensemble average.

A rigorous bias analysis of MUSIC shows [Xu92a] that the MUSIC estimates are biased. For a linear array in the presence of a single source, the bias increases as the source moves away from broadside. Interestingly, the bias also increases as the number of elements increases without changing the aperture. An asymptotic analysis of MUSIC with forward–backward spatial smoothing in the presence of correlated arrivals shows that to estimate two angles of arrival of equal power under identical conditions, more snapshots are required for correlated sources than for uncorrelated sources [Pil89a, Kav86].

Bias and the standard deviation (STD) are complicated functions of the array geometry, SNR, and number and directions of sources, and vary inversely proportional to the number of snapshots. A poorer estimate generally results using a smaller number of snapshots and sources with lower SNR. As shown in [Xu93, Xu92a], the performance of conventional MUSIC is poor in the presence of correlated arrivals, and it fails to resolve coherent sources.

Although the bias and STD both play important roles in direction estimation, the effect of bias near the threshold region is critical. A comparison of MUSIC performance with those of the minimum-norm and FINE for finite-sample cases shows [Xu94a] that in the low SNR range, the minimum-norm estimates have the largest STD and MUSIC estimates have the largest bias. These results are dependent on source SNR, and the performance of all three schemes approaches to the same limit as the SNR is increased. The overall performance of FINE is better than the other two in the absence of correlated arrivals.

The estimates obtained by MUSIC and ML methods are compared with the CRLB in [Sto89, Sto90a] for large-sample cases. The CRLB gives the theoretically lowest value of the covariance of an unbiased estimator; it decreases with the number of samples, number of sensors in the array, and source SNR [Sto89]. The study [Sto89] concluded that the MUSIC estimates are the large-sample realization of ML estimates in the presence of uncorrelated arrivals. Furthermore, it shows that the variance of the MUSIC estimate is greater than that of the ML estimate, and variance of the two methods approchaes each other, as the number of elements and snapshots increases. Thus, using an array with a large number of elements and samples, excellent estimates are possible of directions of uncorrelated sources with large SNRs using the MUSIC method [Sto89]. It should be noted that MLM estimates are unbiased [Vib95]. An unbiased estimate is also referred to as a consistent estimate.

An improvement in MUSIC DOA estimation is possible by beamspace MUSIC [Lee90, Xu93]. By properly selecting a beamforming matrix and then using the MUSIC scheme to estimate DOA, one is able to reduce the threshold level of the required SNR to resolve the closely spaced sources [Lee90]. Although the variance of this estimate is not much different from the element space case, it has less bias [Xu93]. The resolution threshold of beamspace MUSIC is lower than the conventional minimum-norm method. However, for two closely spaced sources, the beamspace MUSIC and beamspace minimum-norm provide identical performances when suitable beamforming matrices are selected [Lee90].

As shown in [Kau95], when beamforming weights have conjugate symmetry (useful only for arrays with particular symmetry), the beamspace MUSIC has decorrelation properties similar to backward–forward smoothing and thus is useful for estimation of correlated arrival source direction and offers performance advantages in terms of lower variance for the estimated angle.

The resolution property of MUSIC analyzed in [Lee91, Zho91, Zho93, Zha95a, Kav86] shows how it depends on SNR, number of snapshots, array geometry, and separation angle of the two sources. The two closely spaced sources are said to be resolved when two peaks in the spectrum appear in the vicinity of the two sources' directions. Analytical expressions of resolution probability and its variation as a function of various parameters are presented in [Zha95a], and could be used to predict the behavior of a MUSIC estimate for a given scenario.

A performance comparison of MUSIC and another eigenvector method, which uses noise eigenvectors divided by corresponding eigenvalues for DOA estimation, indicates [Joh82] that the former is more sensitive to the choice of assumed number of sources compared to actual number of sources.

Performance analysis of many versions of ESPRIT are considered in [Rao89, Sto91a, Ott91, Mat94a] and compared with other methods. Estimates obtained by subspace rotation methods that include the Toeplitz approximation method (TAM) and ESPRIT have greater variance than those obtained by MUSIC using large numbers of samples [Sto91a];

estimates by ESPRIT using a uniform circular array are asymptotically unbiased [Mat94a]; LS-ESPRIT and TAM estimates are statistically equivalent; LS-ESPRIT and TLS-ESPRIT have the same MSE [Rao89] and their performance depends on how subarrays are selected [Ott91]; the minimum-norm method is equivalent to TLS-ESPRIT [Dow91]; and root-MUSIC outperforms the ESPRIT [Rao89a]. TAM is based on the state space model, and finds DOA estimates from signal subspace. In spirit, its approach is similar to ESPRIT [Rao89]. The WSF and ML methods are efficient for Gaussian signals, as both attain CRLB asymptotically [Sto90b, Ott92]. A method is said to be efficient when it achieves CRLB.

A correlation between sources affects the capabilities of various DOA estimation algorithms differently [DeG85]. A study of the effect of the correlation between two sources on the accuracy of DOA-finding schemes presented in [Wei93b] shows that the correlation phase is more significant than correlation magnitude. Most performance analysis discussed assumes that the background noise is white. When this is not the case, the DOA schemes perform differently. In the presence of colored background noise, MUSIC performance is better than that of ESPRIT and the minimum-norm method over a wide range of SNRs. The performance of the minimum-norm method is worse than MUSIC and ESPRIT [Li92].

## 6.16 Sensitivity Analysis

Sensitivity analysis of MUSIC to various perturbations is presented in [Swi92, Rad94, Fri94, Wei94, Ng95, Soo92]. A compact expression for the error covariance of the MUSIC estimates given in [Swi92] may be used to evaluate the effect of various perturbation parameters including gain and phase errors, effect of mutual coupling, channel errors, and random perturbations in sensor locations. It should be noted that MUSIC estimates of DOA require knowledge of the number of sources, similar to certain other methods and underestimation of the source number may lead to inaccurate estimates of DOAs [Rad94]. A variance expression for the DOA estimate for this case has been provided in [Rad94].

Analysis of the effect of model errors on the MUSIC resolution threshold [Fri90, Wei94] and on the wave forms estimated using MUSIC [Fri94] indicate that the probability of resolution decreases [Wei94] with the error variance, and that the sensitivity to phase errors depends more on array aperture than the number of elements [Fri94] in a linear array. The effect of gain and phase error on the mean square error (MSE) of the MUSIC estimate of a general array is analyzed in [Sri91]. The problem of estimating gain and phase errors of sensors with known locations is considered in [Ng95].

An analysis [Soo92] of ESPRIT under random sensor uncertainties suggests that the MUSIC estimates generally give lower MSEs than ESPRIT estimates. The former is more sensitive to both sensor gain and phase errors, whereas the latter depends only on phase errors. The study further suggests that for a linear array with a large number of elements, the MSE of the ESPRIT estimate with maximum overlapping subarrays is lower than nonoverlapping subarrays.

The effect of gain and phase errors on weighted eigenspace methods including MUSIC, minimum-norm, FINE, and CLOSEST is studied in [Ham95] by deriving bias and variance expressions. This study indicates that the effect is gradual up to a point and then the increase in error magnitude causes the abrupt deterioration in bias and variance. The weighted eigenspace methods differ from the standard ones such that a weighting matrix is used in the estimate, and that matrix could be optimized to improve the quality of the estimate under particular perturbation conditions.

The effect of nonlinearity in the system on spectral estimation methods, including hard clipping common in digital beamformers, has been analyzed in [Tut81]. It shows that by using additional preprocessing such distortions could be eliminated.

Effects of various perturbations on DOA estimation methods emphasize the importance of precise knowledge of various array parameters. Selected techniques to calibrate arrays are discussed in [Wei89, Wyl94]. Schemes to estimate the steering vector, and in turn DOA from uncalibrated arrays, are discussed in [Tse95]. [Che91a] focus on a scheme to estimate DOA. Discussions on robustness issues of direction-finding algorithms are found in [Fli94, Wei90]. A summary of performance and sensitivity comparisons of various DOA estimation schemes is provided in Tables 6.1 to Table 6.12 [God97].

**TABLE 6.1**

Performance Summary of Bartlett Method

| Property | Comments and Comparison |
|---|---|
| Bias | Biased |
| | Bartlett > LP > MLM |
| Resolution | Depends on array aperture |
| Sensitivity | Robust to element position errors |
| Array | General array |

**TABLE 6.2**

Performance Summary of MVDR Method

| Property | Comments and Comparison |
|---|---|
| Bias | Unbiased |
| Variance | Minimum |
| Resolution | MVDR > Bartlett |
| | Does not have best resolution of any method |
| Array | General array |

**TABLE 6.3**

Performance Summary of Maximum Entropy Method

| Property | Comments and Comparison |
|---|---|
| Bias | Biased |
| Resolution | ME > MVDR > Bartlett |
| | Can resolve at lower SNR than Bartlett |

**TABLE 6.4**

Performance Summary of Linear Prediction Method

| Property | Comments and Comparison |
|---|---|
| Bias | Biased |
| Resolution | LP > MVDR |
| | >Bartlett |
| | >ME |
| Performance | Good in low SNR conditions |
| | Applicable for correlated arrivals |

**TABLE 6.5**

Performance Summary of ML Method

| Property | Comments and Comparison |
|---|---|
| Bias | Unbiased |
| | Less than LP, Bartlett, MUSIC |
| Variance | Less than MUSIC for small samples |
| | Asymptotically efficient for random signals |
| | Not efficient for finite samples |
| | Less efficient for deterministic signals than random signals |
| | Asymptotically efficient for deterministic signals using very large array |
| Computation | Intensive with large samples |
| Performance | Same for deterministic and random signals for large arrays |
| | Applicable for correlated arrivals |
| | Works with one sample |

**TABLE 6.6**

Performance Summary of Element Space MUSIC Method

| Property | Comments and Comparison |
|---|---|
| Bias | Biased |
| Variance | Less than ESPRIT and TAM for large samples, minimum norm |
| | Close to MLM, CLOSEST, FINE |
| | Variance of weighted MUSIC is more than unweighted MUSIC |
| | Asymptotically efficient for large array |
| Resolution | Limited by bias |
| Array | Applicable for general array |
| | Increasing aperture makes it robust |
| Performance | Fails to resolve correlated sources |
| Computation | Intensive |
| Sensitivity | Array calibration is critical, sensitivity to phase error depends more on array aperture than number of elements, preprocessing can improve resolution |
| | Correct estimate of source number is important |
| | MSE depends on both gain and phase errors and is lower than for ESPRIT |
| | Increase in gain and phase errors beyond certain value causes an abrupt deterioration in bias and variance |

**TABLE 6.7**

Performance Summary of Beam Space MUSIC Method

| Property | Comments and Comparison |
|---|---|
| Bias | Less than element space MUSIC |
| Variance | Larger than element space MUSIC |
| RMS Error | Less than ESPRIT, minimum norm |
| Resolution | Similar to beamspace minimum norm, CLOSEST |
| | Better than element space MUSIC, element space minimum norm |
| | Threshold SNR decreases as the separation between the sources increases |
| Computation | Less than element space MUSIC |
| Sensitivity | More robust than element space MUSIC |

**TABLE 6.8**

Performance Summary of Root-MUSIC Method

| Property | Comments and Comparison |
|---|---|
| Variance | Less than root minimum norm, ESPRIT |
| Resolution | Beamspace root-MUSIC has better probability of resolution than beamspace MUSIC |
| RMS error | Less than LS ESPRIT |
| Array | Equispaced linear array |
| Performance | Better than spectral MUSIC |
| | Similar to TLS ESPRIT at SNR lower than MUSIC threshold |
| | Beamspace root-MUSIC is similar to element space root MUSIC |

**TABLE 6.9**

Performance Summary of Minimum Norm Method

| Property | Comments and Comparison |
|---|---|
| Bias | Less than MUSIC |
| Resolution | Better than CLOSEST, element space MUSIC |
| Method | Equivalent to TLS |

**TABLE 6.10**

Performance Summary of CLOSEST Method

| Property | Comments and Comparison |
|---|---|
| Variance | Similar to element space MUSIC |
| Resolution | Similar to beamspace MUSIC |
| | Better than minimum norm |
| Performance | Good in clustered situation |
| Sensitivity | An increase in sensor gain and phase errors beyond certain value causes an abrupt deterioration in bias and variance |

**TABLE 6.11**

Performance Summary of ESPRIT Method

| Property | Comments and Comparison |
|---|---|
| Bias | TLS ESPRIT unbiased |
| | LS ESPRIT biased |
| RMS Error | Less than minimum norm |
| | TLS similar to LS |
| Variance | Less than MUSIC for large samples and difference increases with number of elements in array |
| Computation | Less than MUSIC |
| | Beam space ESPRIT needs less computation than beamspace root-MUSIC and ES ESPRIT |
| Method | LS ESPRIT is similar to TAM |
| Array | Needs doublets, no calibration needed |
| Performance | Optimum-weighted ESPRIT is better than uniform-weighted ESPRIT |
| | TLS ESPRIT is better than LS ESPRIT |
| Sensitivity | More robust than MUSIC and cannot handle correlated sources |
| | MSE robust for sensor gain errors |
| | MSE is lowest for maximum overlapping subarrays under sensor perturbation |

**TABLE 6.12**

Performance Summary of FINE Method

| Property | Comments and Comparison |
|---|---|
| Bias | Less than MUSIC |
| Resolution | Better than MUSIC and minimum norm |
| Variance | Less than minimum norm |
| Performance | Good at low SNR |

## Acknowledgments

An edited version of Chapter 6 is reprinted from Godara, L.C., Application of antenna arrays to mobile communications, I. Beamforming and DOA considerations, *Proc. IEEE.*, 85(8), 1195-1247, 1997.

## Notation and Abbreviations

| | |
|---|---|
| AIC | Akaike's information criterion |
| CANAL | concurrent nulling and location |
| CRLB | Cramer–Rao lower bound |
| DOA | direction of arrival |
| ESPRIT | estimation of signal parameters via rotational invariance technique |
| FINE | first principal vector |
| LMS | least mean square |
| LP | linear prediction |
| LS | least square |
| MAP | maximum a posteriori |
| MDL | minimum description length |
| ME | maximum entropy |
| ML | maximum likelihood |
| MLM | maximum likelihood method |
| MSE | mean square error |
| MVDR | minimum variance distortionless response |
| MUSIC | multiple signal classification |
| SNR | signal-to-noise ratio |
| STD | standard deviation |
| TAM | Toeplitz approximation method |
| TLS | total least square |
| ULA | uniformly spaced linear array |
| WSF | weighted subspace fitting |

| | |
|---|---|
| A | $L \times M$ matrix with columns being steering vectors |
| d | interelement spacing of linear equispaced array |
| $E[\cdot]$ | expectation operator |
| $\mathbf{e}_1$ | vector of all zeros except first element, which is equal to unity |
| $f_N$ | Nyquist frequency |
| H(s) | entropy function |
| $J_0$ | reflection matrix with all elements along secondary diagonal being equal to unity and zero elsewhere |
| K | number of elements in subarray |
| L | number of elements in array |
| $L_0$ | number of subarrays |
| M | number of directional sources |
| N | number of samples |
| $P_B(\theta)$ | power estimated by Bartlette method as function of θ |
| $P_{LP}(\theta)$ | power estimated by linear prediction method as function of θ |
| $P_{ME}(\theta)$ | power estimated by maximum entropy method as function of θ |
| $P_{MN}(\theta)$ | power estimated by minimum norm method as function of θ |
| $P_{MU}(\theta)$ | power estimated by MUSIC method as function of θ |
| $P_{MV}(\theta)$ | power estimated by MVDR method as function of θ |
| R | array correlation matrix |
| $R_m$ | mth subarray matrix of forward method |
| $\bar{R}_m$ | mth subarray matrix of backward method |
| $R^{mj}$ | cross correlation matrix of mth and jth subarrays |
| $r_{ij}$ | correlation between the ith and the jth elements |
| $\mathbf{S}_\theta$ | steering vector associated with the direction θ |
| $S(f)$ | power spectral density of signal s(t) |
| $\mathbf{s}(t)$ | vector of M source signals induced on reference element |
| T | transformation matrix |
| $U_N$ | matrix with its L – M columns being the eigenvectors corresponding to the L – M smallest eigenvalues of R |
| $U_S$ | matrix with M columns being eigenvectors corresponding to M largest eigenvalues |
| $\mathbf{u}_1$ | column vector of all zeros except one element that is equal to unity |
| $\mathbf{w}$ | array weight vector |
| $\hat{\mathbf{w}}$ | optimized array weights |
| Φ | diagonal matrix defined by (6.10.4) |
| $\Delta_0$ | magnitude of displacement vector |
| Λ | diagonal matrix defined by (6.10.11) |
| θ | direction of source |
| $\tau_{ij}(\theta)$ | differential delay between elements i and j due to source in direction θ |
| ψ | transformation matrix |

## References


Aba85 Abatzoglou, T.J., A fast maximum likelihood algorithm for frequency estimation of a sinusoid based on Newton's method, *IEEE Trans. Acoust. Speech Signal Process.*, 33, 77–89, 1985.

Aka74 Akaike, H., A new look at the statistical model identification, *IEEE Trans. Automat. Control*, 19, 716–723, 1974.

Ami92 Amin, M.G., Concurrent nulling and locations of multiple interferences in adaptive antenna arrays, *IEEE Trans. Signal Process.*, 40, 2658–2668, 1992.

And63 Anderson, T.W., Asymptotic theory for principal component analysis, *Ann J. Math. Stat.*, 34, 122–148, 1963.

And91 Anderson, S., et al., An adaptive array for mobile communication systems, *IEEE Trans. Vehicular Technol.*, 40, 230–236, 1991.

Bar56 Bartlett, M.S., *An Introduction to Stochastic Process*, Cambridge University Press, New York, 1956.

Bar83 Barabell, A., Improving the resolution of eigenstructured based direction finding algorithms, in *Proc. ICASSP*, 336–339, 1983.

Bie80 Bienvenu, G. and Kopp, L., Adaptive high resolution spatial discrimination of passive sources, in Underwater Acoustics and Signal Processing: Proceedings of the NATO Advanced Study Institute, Bjorno, L., Ed., D. Reidel, Dordrecht, 1980, pp. 509–515.

Bie83 Bienvenu, G. and Kopp, L., Optimality of high resolution array processing using eigensystem approach, *IEEE Trans. Acoust. Speech Signal Process.*, 31, 1235–1248, 1983.

Bor81 Bordelon, D.J., Complementarity of the Reddi method of source direction estimation with those of Pisarenko and Cantoni and Godara, I., *J. Acoust. Soc. Am.*, 69, 1355–1359, 1981.

Bre86 Bresler, Y. and Macovski, A., On the number of signals resolvable by a uniform linear array, *IEEE Trans. Acoust. Speech Signal Process.*, 34, 1361–1375, 1986.

Bro83 Bronez, T.P. and Cadzow, J.A., An algebraic approach to super resolution array processing, *IEEE Trans. Aerosp. Electron. Syst.*, 19, 123–133, 1983.

Buc88 Buckley, K.M. and Griffiths, L.J., Broad-band signal-subspace spatial-spectrum (BASS-ALE) estimation, *IEEE Trans. Acoust. Speech Signal Process.*, 36, 953–964, 1988.

Buc90 Buckley, K.M. and Xu, X.,L., Spatial spectrum estimation in a location sector, *IEEE Trans. Acoust. Speech Signal Process.*, 38, 1842–1852, 1990.

Bur67 Burg, J.P., Maximum Entropy Spectral Analysis, paper presented at 37th annual meeting, Society of Exploration in Geophysics, Oklahoma City, 1967.

Cad88 Cadzow, J.A., A high resolution direction-of-arrival algorithm for narrowband coherent and incoherent sources, *IEEE Trans. Acoust. Speech Signal Process.*, 36, 965–979, 1988.

Cad90 Cadzow, J.A., Multiple source location: the signal subspace approach, *IEEE Trans. Acoust. Speech Signal Process.*, 38, 1110–1125, 1990.

Can80 Cantoni, A. and Godara, L.C., Resolving the directions of sources in a correlated field incident on an array, *J. Acoust. Soc. Am.*, 67, 1247–1255, 1980.

Cap69 Capon, J., High-resolution frequency-wave number spectrum analysis, *IEEE Proc.*, 57, 1408–1418, 1969.

Cap70 Capon, J., Probability distribution for estimators of the frequency-wave number spectrum, *IEEE Proc.*, 58, 1785–1786, 1970.

Che91 Chen, W., Wong, K.M. and Reilly, J.P., Detection of the number of signals: a predicted eigenthreshold approach, *IEEE Trans. Signal Process.*, 39, 1088–1098, 1991.

Che91a Chen, Y.M., et al., Bearing estimation without calibration for randomly perturbed arrays, *IEEE Trans. Signal Process.*, 39, 194–197, 1991.

Che94 Chen, Y.H. and Chiang, C.T., Kalman-based estimators for DOA estimation, *IEEE Trans. Signal Process.*, 42, 3543–3547, 1994.

Cle89 Clergeot, H., Tresseus, S. and Ouamri, A., Performance of high resolution frequencies estimation methods compared to the Cramer–Rao bounds, *IEEE Trans. Acoust. Speech Signal Process.*, 37, 1703–1720, 1989.

Cox73 Cox, H., Resolving power and sensitivity to mismatch of optimum array processors, *J. Acoust. Soc. Am.*, 54, 771–785, 1973.

DeG85 DeGraaf, S.R. and Johnson, D.H., Capability of array processing algorithms to estimate source bearings, *IEEE Trans. Acoust. Speech Signal Process.*, 33, 1368–1379, 1985.

DeG93 DeGroat, R.D., Dowling, E.M. and Linebarger, D.A., The constrained MUSIC problem, *IEEE Trans. Signal Process.*, 41, 1445–1449, 1993.

Del96 Delis, A. and Papadopoulos, G., Enhanced forward/backward spatial filtering method for DOA estimation of narrowband coherent sources, *IEE Proc. Radar Sonar Navig.*, 143, 10–16, 1996.

Dem77 Dempster, A.D., Laird, N.M. and Rubin, D.B., Maximum likelihood from incomplete data via the EM algorithm, *J. R. Stat. Soc.*, 13–19, 1–37, 1977.

Di85 Di, A., Multiple source location: a matrix decomposition approach, *IEEE Trans. Acoust. Speech Signal Process.*, 33, 1086–1091, 1985.

Dor93 Doron, M.A., Weiss, A.J. and Messer, H., Maximum-likelihood direction finding of wide-band sources, *IEEE Trans. Signal Process.*, 41, 411–414, 1993.

Dow91 Dowling, E.M. and DeGroat, R.D., The equivalence of the total least squares and minimum norm methods, *IEEE Trans. Signal Process.*, 39, 1891–1892, 1991.

Du91 Du, W. and Kirlin, R.L., Improved spatial smoothing techniques for DOA estimation of coherent signals, *IEEE Trans. Signal Process.*, 39, 1208–1210, 1991.

Eri94 Eriksson, A., Stoica, P. and Soderstrom, T., On-line subspace algorithms for tracking moving sources, *IEEE Trans. Signal Process.*, 42, 2319–2330, 1994.

Erm94 Ermolaev, V.T. and Gershman, A.B., Fast algorithm for minimum-norm direction-of-arrival estimation, *IEEE Trans. Signal Process.*, 42, 2389–2394, 1994.

Eva81 Evans, J.E., Johnson, J.R. and Sun, D.F., High Resolution Angular Spectrum Estimation Techniques for Terrain Scattering Analysis and Angle of Arrival Estimation, in Proceedings 1st ASSP Workshop Spectral Estimation, Hamilton, Ontario, Canada, 1981, pp. 134–139.

Far85 Farrier, D.R., Maximum entropy processing of band-limited spectra. Part 1: Noise free case, *IEE Proc.*, 132, Pt. F, 491–504, 1985.

Fer83 Ferrara, E.R. and Parks, T.M., Direction finding with an array of antennas having diverse polarization, *IEEE Trans. Antennas Propagat.*, 31, 231–236, 1983.

Fli94 Flieller, A., Larzabal, P. and Clergeot, H., Robust self calibration of the maximum likelihood method in array processing, *Signal Processing VII: Theories and Applications*, Holt, M., et al., Eds., European Association for Signal Processing, Edinburgh, 1994, pp. 1293–1296.

Fri90 Friedlander, B., A sensitivity analysis of MUSIC algorithm, *IEEE Trans. Acoust. Speech Signal Process.*, 38, 1740–1751, 1990.

Fri91 Friedlander, B. and Weiss, A.J., On the number of signals whose directions can be estimated by an array, *IEEE Trans. Signal Process.*, 39, 1686–1689, 1991.

Fri94 Friedlander, B. and Weiss, A.J., Effects of model errors on waveform estimation using the MUSIC algorithm, *IEEE Trans. Signal Process.*, 42, 147–155, 1994.

Fuc94 Fuchs, J.J. and Chuberre, H., A deconvolution approach to source localization, *IEEE Trans. Signal Process.*, 42, 1462–1470, 1994.

Fuc96 Fuch,, J.J., Rectangular Pisarenko method applied to source localization, *IEEE Trans. Signal Process.*, 44, 2377–2383, 1996.

Gan96 Gansman, J.A., Zoltowski, M.D. and Krogmeier, J.V., Multidimensional multirate DOA estimation in beamspace, *IEEE Trans. Signal Process.*, 44, 2780–2792, 1996.

Gar91 Gardner, W.A., Exploitation of spectral redundancy in cyclostationary signals, *IEEE Signal Process. Mag.*, 8, 14–37, 1991.

Geb95 Gebauer, T. and Gockler, H.G., Channel-individual adaptive beamforming for mobile satellite communications, *IEEE J. Selected Areas Commn.*, 13, 439–48, 1995.

God90 Godara, L.C., Beamforming in the presence of correlated arrivals using structured correlation matrix, *IEEE Trans. Acoust. Speech Signal Process.*, 38, 1–15, 1990.

God88 Godara, L.C. and Gray, D.A., An algorithm for adaptive augmented array beamforming, *J. Acoust. Soc. Am.*, 83, 2261–2265, 1988.

God96 Godara, L.C., Limitations and Capabilities of Directions-of-Arrival Estimation Techniques Using an Array of Antennas: A Mobile Communications Perspective, paper presented at IEEE International Symposium on Phased Array Systems and Technology, Boston, 1996.

God97 Godara, L.C., Application to antenna arrays to mobile communications. Part II: Beamforming and direction of arrival considerations, *IEEE Proc.*, 85, 1195–1247, 1997.

Gol83 Golub, G.H. and Van Loan, C.F., *Matrix Computation*, John Hopkins University Press, Baltimore, 1983.

Gre94 Grenier, Y., Wideband source location through frequency-dependent modeling, *IEEE Trans. Signal Process.*, 42, 1087–1096, 1994.

Ham94 Hamza, R. and Buckley, K., Resolution enhanced ESPRIT, *IEEE Trans. Signal Process.*, 42, 688–691, 1994.

Ham95 Hamza, R. and Buckley, K., An analysis of weighted eigenspace methods in the presence of sensor errors, *IEEE Trans. Signal Process.*, 43, 1140–1150, 1995.

Hay85 Haykin, S., Radar array processing for angle of arrival estimation, in *Array Signal Processing*, Haykin, S., Ed., Prentice Hall, New York, 1985.

Hin81 Hinich, M.J., Frequency-wave number array processing, *J. Acoust. Soc. Am.*, 69, 732–737, 1981.

Hua91 Huarng, K.C. and Yeh, C.C., A unitary transformation method for angle-of-arrival estimation, *IEEE Trans. Signal Process.*, 39, 975–977, 1991.

Hun90 Hung, H. and Kaveh, M., Coherent wide-band ESPRIT method for direction-of-arrival estimation of multiple wide-band sources, *IEEE Trans. Acoust. Speech Signal Process.*, 38, 354–356, 1990.

Joh82 Johnson, D.H. and DeGraff, S.R., Improving the resolution of bearing in passive sonar arrays by eigenvalue analysis, *IEEE Trans. Acoust. Speech Signal Process.*, 29, 401–413, 1982.

Joh82a Johnson, D.H., The application of spectral estimation methods to bearing estimation problems, *IEEE Proc.*, 70, 1018–1028, 1982.

Kar90 Karasalo, I., A high-high-resolution postbeamforming method based on semidefinite linear optimization, *IEEE Trans. Acoust. Speech Signal Process.*, 38, 16–22, 1990.

Kau95 Kautz, G.M. and Zoltowski, M.D., Performance analysis of MUSIC employing conjugate symmetric beamformers, *IEEE Trans. Signal Process.*, 43, 737–748, 1995.

Kav86 Kaveh, M. and Barabell, A.J., The statistical performance of MUSIC and mini-norm algorithms in resolving plane wave in noise, *IEEE Trans. Acoust. Speech Signal Process.*, 34, 331–341, 1986.

Kes85 Kesler, S.B., Boodaghians, S., and Kesler, J., Resolving uncorrelated and correlated sources by linear prediction, *IEEE Trans. Antennas Propagat.*, 33, 1221–1227, 1985.

Klo96 Klouche-Djedid, A. and Fujita, M., Adaptive array sensor processing applications for mobile telephone communications, *IEEE Trans. Vehicular Technol.*, 45, 405–416, 1996.

Klu93 Klukas, R.W. and Fattouche, M., Radio Signal Direction Finding in the Urban Radio Environment, in Proceedings National Technical Meeting of the Institute of Navigation, 1993, pp. 151–160.

Kri92 Krim, H., Forster, P. and Proakis, J.G., Operator approach to performance analysis of root-MUSIC and root-min-norm, *IEEE Trans. Signal Process.*, 40, 1687–1696, 1992.

Kro89 Krolik, J. and Swingler, D., Multiple broad-band source location using steered covariance matrices, *IEEE Trans. Acoust. Speech Signal Process.*, 37, 1481–1494, 1989.

Kum83 Kumaresan, R. and Tufts, D.W., Estimating the angles of arrival of multiple plane waves, *IEEE Trans. Aerosp. Electron. Syst.*, 19, 134–139, 1983.

Lac71 Lacoss, R.T., Data adaptive spectral analysis method, *Geophysics*, 36, 661–675, 1971.

Lan83 Lang, S.W. and McClellan, J.H., Spectral estimation for sensor arrays, *IEEE Trans. Acoust. Speech Signal Process.*, 31, 349–358, 1983.

Lar83 Larimore, M.G., Adaptive convergence of spectral estimation based on Pisarenko harmonic retrieval, *IEEE Trans. Acoust. Speech Signal Process.*, 31, 955–962, 1983.

LeC89 LeCadre, J., Parametric methods for spatial signal processing in the presence of unknown colored noise fields, *IEEE Trans. Acoust. Speech Signal Process.*, 37, 965–983, 1989.

Lee90 Lee, H.B. and Wengrovitz, M.S., Resolution threshold of beamspace MUSIC for two closely spaced emitters, *IEEE Trans. Acoust. Speech Signal Process.*, 38, 1545–1559, 1990.

Lee90a Lee, W.C., et al., Adaptive spatial domain forward–backward predictors for bearing estimation, *IEEE Trans. Acoust. Speech Signal Process.*, 38, 1105–1109, 1990.

Lee91 Lee, H.B. and Wengrovitz, M.S., Statistical characterization of MUSIC null spectrum, *IEEE Trans. Signal Process.*, 39, 1333–1347, 1991.

Lee94 Lee, H. and Li, F., An eigenvector technique for detecting the number of emitters in a cluster, *IEEE Trans. Signal Process.*, 42, 2380–2388, 1994.

Lee94a Lee, H. and Stovall, R., Maximum likelihood methods for determining the direction of arrival for a single electromagnetic source with unknown polarization, *IEEE Trans. Signal Process.*, 42, 474–479, 1994.

Li92 Li, F. and Vaccaro, R.J., Performance degradation of DOA estimators due to unknown noise fields, *IEEE Trans. Signal Process.*, 40, 686–690, 1992.

Li93 Li, J. and Compton, Jr., R.T., Angle and polarization estimation in a coherent signal environment, *IEEE Trans. Aerosp. Electron. Syst.*, 29, 706–716, 1993.

Lig73 Ligget, W.S., Passive Sonar: Fitting Models to Multiple Time-Series, in *NATO ASI on Signal Processing*, Griffiths, J.W.R., et al., Eds., Academic Press, New York, 1973, pp. 327–345.

Lim90 Lim, B.L., Hui, S.K. and Lim, Y.C., Bearing estimation of coherent sources by circular spatial modulation averaging (CSMA) technique, *Electron. Lett.*, 26, 343–345, 1990.

Lio89 Liou, C.Y. and Liou, R.M., Spatial pseudorandom array processing, *IEEE Trans. Acoust. Speech Signal Process.*, 37, 1445–1449, 1989.

Liu94 Liu, K.J.R., et al., URV ESPRIT for tracking time-varying signals, *IEEE Trans. Signal Process.*, 42, 3441–3448, 1994.

Mak75 Makhoul, J., Linear prediction: a tutorial review, *IEEE Proc.*, 63, 561–580, 1975.

Mat94 Mathews, C.P. and Zoltowski, M.D., Eigenstructure techniques for 2-D angle estimation with uniform circular arrays, *IEEE Trans. Signal Process.*, 42, 2395–2407, 1994.

Mat94a Mathews, C.P. and Zoltowski, M.D., Performance analysis of the UCA-ESPRIT algorithm for circular ring arrays, *IEEE Trans. Signal Process.*, 42, 2535–2539, 1994.

May87 Mayhan, J.T. and Niro, L., Spatial spectral estimation using multiple beam antennas, *IEEE Trans. Antennas Propagat.*, 35, 897–906, 1987.

McC82 McClellan, J.H., Multidimensional spectral estimation, *IEEE Proc.*, 70, 1029–1039, 1982.

McC83 McClellan, J.H. and Lang, S.W., Duality for multidimensional MEM spectral analysis, *IEE Proc.*, 130, Pt. F, 230–235, 1983.

Men96 Meng, Y., Stocia, P. and Wong, K.M., Estimation of the directions of arrival of spatially dispersed signals in array processing, *IEE Proc. Radar Sonar Navig.*, 143, 1–9, 1996.

Mil90 Miller, M.I. and Fuhrmann, D.R., Maximum likelihood narrowband direction finding and the EM algorithm, *IEEE Trans. Acoust. Speech Signal Process.*, 38, 1560–1577, 1990.

Mog91 Moghaddamjoo, A. and Chang, T.C., Signal enhancement of the spatial smoothing algorithm, *IEEE Trans. Signal Process.*, 39, 1907–1911, 1991.

Mog92 Moghaddamjoo, A. and Chang, T.C., Analysis of spatial filtering approach to the decorrelation of coherent sources, *IEEE Trans. Signal Process.*, 40, 692–694, 1992.

Nag94 Nagatsuka, M., et al., Adaptive array antenna based on spatial spectral estimation using maximum entropy method, *IEICE Trans. Commn.*, E77-B, 624–633, 1994.

Ng90 Ng, B.P., Constraints for linear predictive and minimum-norm methods in bearing estimation, *IEE Proc.*, 137, Pt. F, 187–191, 1990.

Ng95 Ng, A.P.C., Direction-of-arrival estimates in the presence of wavelength, gain, and phase errors, *IEEE Trans. Signal Process.*, 43, 225–232, 1995.

Nic88 Nickel, U., Algebraic formulation of Kumaresan-Tufts superresolution method, showing relation to ME and MUSIC methods, *IEE Proc.*, Pt. F., 135, 7–10, 1988.

Nut74 Nuttall, A.H, Carter, G.C. and Montaron, E.M., Estimation of two-dimensional spectrum of the space-time noise field for a sparse line array, *J. Acoust. Soc. Am.*, 55, 1034–1041, 1974.

Oh92 Oh, S.K. and Un, C.K., Simple computational methods of the AP algorithm for maximum likelihood localization of multiple radiating sources, *IEEE Trans. Signal Process.*, 40, 2848–2854, 1992.

Ott90 Ottersten, B. and Kailath, T., Direction-of-arrival estimation for wide-band signals using the ESPRIT algorithm, *IEEE Trans. Acoust. Speech Signal Process.*, 38, 317–327, 1990.

Ott91 Ottersten, B., Viberg, M. and Kailath, T., Performance analysis of the total least squares ESPRIT algorithm, *IEEE Trans. Signal Process.*, 39, 1122–1135, 1991.

Ott92 Ottersten, B., Viberg, M. and Kailath, T., Analysis of subspace fitting and ML techniques for parameter estimation from sensor array data, *IEEE Trans. Signal Process.*, 40, 590–600, 1992.

Oui88 Ouibrahim, H., Weiner, D.D. and Sarkar, T.K., A generalized approach to direction finding, *IEEE Acoust. Speech Signal Process.*, 36, 610–613, 1988.

Oui89 Ouibrahim, H., Prony, Pisarenko and the matrix pencil: a unified presentation, *IEEE Trans. Acoust. Speech Signal Process.*, 37, 133–134, 1989.

Par93 Park, H.R. and Kim, Y.S., A solution to the narrow-band coherency problem in multiple source location, *IEEE Trans. Signal Process.*, 41, 473–476, 1993.

Pau86 Paulraj, A. and Kailath, T., Eigenstructure methods for direction of arrival estimation in the presence of unknown noise field, *IEEE Trans. Acoust. Speech Signal Process.*, 34, 13–20, 1986.

Pau86a Paulraj, A., Roy, R. and Kailath, T., A subspace rotation approach to signal parameter estimation, *IEEE Proc.*, 74, 1044–1045, 1986.

Pil89 Pillai, S.U. and Kwon, B.H., Forward/backward spatial smoothing techniques for coherent signal identification, *IEEE Trans. Acoust. Speech Signal Process.*, 37, 8–15, 1989.

Pil89a Pillai, S.U. and Kwon, B.H., Performance analysis of MUSIC-type high resolution estimators for direction finding in correlated and coherent scenes, *IEEE Trans. Acoust. Speech Signal Process.*, 37, 1176–1189, 1989.

Pis73 Pisarenko, V.F., The retrieval of harmonics from a covariance function, *Geophys. J. R. Astron. Soc.*, 33, 347–366, 1973.

Por88 Porat, B. and Friedlander, B., Analysis of the asymptotic relative efficiency of MUSIC algorithm, *IEEE Trans. Acoust. Speech Signal Process.*, 36, 532–544,1988.

Rad94 Radich, B.M. and Buckley, K.M., The effect of source number under estimation on MUSIC location estimates, *IEEE Trans. Signal Process.*, 42, 233–236, 1994.

Rao89 Rao, B.D. and Hari, K.V.S., Performance analysis of ESPRIT and TAM in determining the direction of arrival of plane waves in noise, *IEEE Trans. Acoust. Speech Signal Process.*, 37, 1990–1995, 1989.

Rao89a Rao, B.D. and Hari, K.V.S., Performance analysis of root-MUSIC, *IEEE Trans. Acoust. Speech Signal Process.*, 37, 1939–1949, 1989.

Rao93 Rao, B.D. and Hari, K.V.S., Weighted subspace methods and spatial smoothing: analysis and comparison, *IEEE Trans. Signal Process.*, 41, 788–803, 1993.

Rao94 Rao, C.R., Sastry, C.R. and Zhou, B., Tracking the direction of arrival of multiple moving targets, *IEEE Trans. Signal Process.*, 42, 1133–1144, 1994.

Red79 Reddi, S.S., Multiple source location: a digital approach, *IEEE Trans. Aerosp. Electron. Syst.*, 15, 95–105, 1979.

Red82 Reddy, V.U., Egardt, B. and Kailath, T., Least-squares type algorithm for adaptive implementation of Pisarenko's harmonic retrieval method, *IEEE Trans. Acoust. Speech Signal Process.*, 30, 399–405, 1982.

Rei87 Reilly, J.P., A real-time high-resolution technique for angle of arrival estimation, *IEEE Proc.*, 75, 1692–1694, 1987.

Rei92 Reilly, J.P. and Won, K.M., Estimation of the direction of arrival of signals in unknown correlated noise. Part II: Asymptotic behavior and performance of the MAP, *IEEE Trans. Signal Process.*, 40, 2018–2028, 1992.

Ris78 Rissanen, J., Modeling by the shortest data description, *Automatica*, 14, 465–471, 1978.

Roy86 Roy, R., Paulraj, A. and Kailath, T., ESPRIT — A subspace rotation approach to estimation of parameters of cisoids in noise, *IEEE Trans. Acoust. Speech Signal Process.*, 34, 1340–1342, 1986.

Roy89 Roy, R. and Kailath, T., ESPRIT: estimation of signal parameters via rotational invariance techniques, *IEEE Trans. Acoust. Speech Signal Process.*, 37, 984–995, 1989.

Sas91 Sastry, C.R., Kamen, E.W. and Simaan, M., An efficient algorithm for tracking the angles of arrival of moving targets, *IEEE Trans. Signal Process.*, 39, 242–246, 1991.

Sch68 Schweppe, F.C., Sensor array data processing for multiple signal sources, *IEEE Trans. Inform. Theory*, IT–14, 294–305, 1968.

Sch86 Schmidt, R.O., Multiple emitter location and signal parameter estimation, *IEEE Trans. Antennas Propagat.*, 34, 276–280, 1986.

Sch93 Schultheiss P.M. and Messer, H., Optimal and suboptimal broad-band source location estimation, *IEEE Trans. Signal Process.*, 41, 2752–2763, 1993.

Sha85 Shan, T.J., Wax, M. and Kailath, T., On spatial smoothing for directional of arrival estimation of coherent signals, *IEEE Trans. Acoust. Speech Signal Process.*, 33, 806–811, 1985.

Sha87 Shan, T.J., Paulraj, A. and Kailath, T., On smoothed rank profile tests in eigenstructure methods for directions-of-arrival estimation, *IEEE Trans. Acoust. Speech Signal Process.*, 35, 1377–1385, 1987.

She96 Sheinvald, J., Wax, M. and Weiss, A.J., On maximum-likelihood localization of coherent signals, *IEEE Trans. Signal Process.*, 44, 2475–2482, 1996.

Ski79 Skinner, D.P., Hedlicka, S.M. and Mathews, A.D., Maximum entropy array processing, *J. Acoust. Soc. Am.*, 66, 488–493, 1979.

Soo92 Soon, V.C. and Huans, Y.F., An analysis of ESPRIT under random sensor uncertainties, *IEEE Trans. Signal Process.*, 40, 2353–2358, 1992.

Sou95 Southall, H.L., Simmers, J.A. and O'Donnell, T.H., Direction finding in phased arrays with neural network beamformer, *IEEE Trans. Antennas Propagat.*, 43, 1369–1374, 1995.

Sri91 Srinath, H. and Reddy, V.U., Analysis of MUSIC algorithm with sensor gain and phase perturbations, *Signal Process.*, 23, 245–256, 1991.

Sto89 Stoica, P. and Nehorai, A., MUSIC, maximum likelihood, and Cramer–Rao bound, *IEEE Trans. Acoust. Speech Signal Process.*, 37, 720–741, 1989.

Sto90 Stoica, P. and Sharman, K.C., Maximum likelihood methods for direction of arrival estimation, *IEEE Trans. Acoust. Speech Signal Process.*, 38, 1132–1143, 1990.

Sto90a Stoica, P. and Nehorai, A., MUSIC, maximum likelihood, and Cramer–Rao bound: further results and comparisons, *IEEE Trans. Acoust. Speech Signal Process.*, 38, 2140–2150, 1990.

Sto90b Stoica, P. and Nehorai, A., Performance study of conditional and unconditional direction-of-arrival estimate, *IEEE Trans. Signal Process.*, 38, 1783–1795, 1990.

Sto91 Stoica, P. and Nehorai, A., Comparative performance study of element-space and beam-space MUSIC estimators, *Circuits Syst. Signal Process.*, 10, 285–292, 1991.

Sto91a Stoica, P. and Nehorai, A., Performance comparison of subspace rotation and MUSIC methods of direction estimation, *IEEE Trans. Signal Process.*, 39, 446–453, 1991.

Su83 Su, G. and Morf, M., The signal subspace approach for multiple wide-band emitter location, *IEEE Trans. Acoust. Speech Signal Process.*, 31, 1502–1522, 1983.

Swi89 Swingler, D.N. and Krolik, J., Source location bias in the coherently focussed high-resolution broad-band beamformer, *IEEE Trans. Acoust. Speech Signal Process.*, 37, 143–145, 1989.

Swi89a Swingler, D.N. and Walker, R.S., Line-array beamforming using linear prediction for aperture interpolation and extrapolation, *IEEE Trans. Acoust. Speech Signal Process.*, 37, 16–30, 1989.

Swi92 Swindlehurst, A.L. and Kailath, T., A performance analysis of subspace-based methods in the presence of model errors. Part I: The MUSIC algorithm, *IEEE Trans. Signal Process.*, 40, 1758–1774, 1992.

Swi92a Swindlehurst, A.L., et al., Multiple invariance ESPRIT, *IEEE Trans. Signal Process.*, 40, 867–881, 1992.

Swi93 Swindlehurst, A.L. and Kailath, T., Azimuth/elevation direction finding using regular array geometries, IEEE *Trans. Aerosp. Electron. Syst.*, 29, 145–156, 1993.

Swi94 Swingler, D.N., An approximate expression for the Cramer–Rao bound on DOA estimates of closely spaced sources in broadband line-array beamforming, *IEEE Trans. Signal Process.*, 42, 1540–1543, 1994.

Tak87 Takao, K. and Kikuma, N., An adaptive array utilizing an adaptive spatial averaging technique for multipath environments, *IEEE Trans. Antennas Propagat.*, 35, 1389–1396, 1987.

Tho80 Thorvaldsen, T., Maximum entropy spectral analysis in antenna spatial filtering, *IEEE Trans. Antennas Propagat.*, 28, 99.556–562,1980.

Tru96 Trump, T. and Ottersten, B., Estimation of nominal direction of arrival and angular spread using an array of sensors, *Signal Process.*, 50, 57–69, 1996.

Tse95 Tseng, C.Y., Feldman, D.D. and Griffiths, L.J., Steering vector estimation in uncalibrated arrays, *IEEE Trans. Signal Process.*, 43, 1397–1412, 1995.

Tuf86 Tufts, D.W. and Melissinos, C.D., Simple, effective computation of principal eigenvectors and their eigenvalues and application to high-resolution estimation of frequencies, *IEEE Trans. Acoust. Speech Signal Process.*, 34, 1046–1053, 1986.

Tut81 Tuteur, F.B. and Presley Jr., J.A., Spectral estimation of space-time signals with a DIMUS array, *J. Acoust. Soc. Am.*, 70, 80–89, 1981.

Vib91 Viberg, M. and Ottersten, B., Sensor array processing based on subspace fitting, *IEEE Trans. Signal Process.*, 39, 1110–1121, 1991.

Vib91a Viberg, M., Ottersten, B. and Kailath, T., Detection and estimation in sensor arrays using weighted subspace fitting, *IEEE Trans. Signal Process.*, 39, 2436–2449, 1991.

Vib95 Viberg, M., Ottersten, B. and Nehorai, A., Performance analysis of direction finding with large arrays and finite data, *IEEE Trans. Signal Process.*, 43, 469–477, 1995.

Wag84 Wagstaff, R.A. and Berrou, J.L., A fast and simple nonlinear technique for high-resolution beamforming and spectral analysis, *J. Acoust. Soc. Am.*, 75, 1133–1141, 1984.

Wan85 Wang, H. and Kaveh, M., Coherent signal-subspace processing for the detection and estimation of angle of arrival of multiple wide-band sources, *IEEE Trans. Acoust. Speech Signal Process.*, 33, 823–831, 1985.

Wan86 Wang, H. and Kaveh, M., On the performance of signal-subspace processing: Part I: Narrowband systems, *IEEE Trans. Acoust. Speech Signal Process.*, 34, 1201–1209, 1986.

Wan95 Wang, Y. and Cruz, J.R., Adaptive antenna arrays for cellular CDMA cellular communication systems, *IEEE ICASSP*, 1725–1728, 1995.

Wax83 Wax, M. and Kailath, T., Optimum localization of multiple sources by passive arrays, *IEEE Trans. Acoust. Speech Signal Process.*, 31, 1210–11221, 1983.

Wax84 Wax, M., Shan, T.J. and Kailath, T., Spatio-temporal spectral analysis by eigenstructure methods, *IEEE Trans. Acoust. Speech Signal Process.*, 32, 817–827, 1984.

Wax85 Wax, M. and Kailath, T., Detection of signals by information theoretic criteria, *IEEE Trans. Acoust. Speech Signal Process.*, 33, 387–392, 1985.

Wax89 Wax, M. and Ziskind, I., On unique localization of multiple sources by passive sensor arrays, *IEEE Trans. Acoust. Speech Signal Process.*, 37, 996–1000, 1989.

Wax89a Wax, M. and Ziskind, I., Detection of the number of coherent signals by the MDL principle, *IEEE Trans. Acoust. Speech Signal Process.*, 37, 1190–1196, 1989.

Wax91 Wax, M., Detection and localization of multiple sources via the stochastic signal model, *IEEE Trans. Signal Process.*, 39, 2450–2456, 1991.

Wax92 Wax, M., Detection and localization of multiple sources in noise with unknown covariance, *IEEE Trans. Signal Process.*, 40, 245–249, 1992.

Wei88 Weiss, A.J., Willsky, A.S. and Levy, B.C., Eigenstructure approach for array processing with unknown intensity coefficients, *IEEE Trans. Acoust. Speech Signal Process.*, 36, 1613–1617, 1988.

Wei89 Weiss, A.J. and Friedlander, B., Array shape calibration using sources in unknown locations: a maximum likelihood approach, *IEEE Trans. Acoust. Speech Signal Process.*, 37, 1958–1966, 1989.

Wei90 Weiss, A.J. and Friedlander, B., Eigenstructure methods for direction finding with sensor gain and phase uncertainties, *Circuits Syst. Signal Process.*, 9, 271–300, 1990.

Wei91 Weiss, A.J. and Gavish, M., Direction finding using ESPRIT with interpolated arrays, *IEEE Trans. Signal Process.*, 39, 1473–1478, 1991.

Wei93 Weiss, A.J. and Friedlander, B., Performance analysis of spatial smoothing with interpolated data, *IEEE Trans. Signal Process.*, 41, 1881–1892, 1993.

Wei93a Weiss, A. and Friedlander, B., Direction finding for diversely polarized signals using polynomial rooting, *IEEE Trans. Signal Process.*, 41, 1893–1905, 1993.

Wei93b Weiss, A.J. and Friedlander, B., On the Cramer–Rao bound for direction finding of correlated signals, *IEEE Trans. Signal Process.*, 41, 495–499, 1993.

Wei94 Weiss, A.J. and Friedlander, B., Effect of modeling errors on the resolution threshold of the MUSIC algorithm, *IEEE Trans. Signal Process.*, 42, 1519–1526, 1994.

Wei94a Weiss, A.J. and Friedlander, B., Preprocessing for direction finding with minimal variance degradation, *IEEE Trans. Signal Process.*, 42, 1478–1485, 1994.

Wei95 Weiss, A.J., Friedlander, B. and Stoica, P., Direction-of-arrival estimation using MODE with interpolated arrays, *IEEE Trans. Signal Process.*, 43, 296–300, 1995.

Wig95 Wigren, T. and Eriksson, A., Accuracy aspects of DOA and angular velocity estimation in sensor array processing, *IEEE Signal Process. Lett.*, 2, 60–62, 1995.

Wil88 Williams, R.T., et al., An improved spatial smoothing technique for bearing estimation in multipath environment, *IEEE Trans. Acoust. Speech Signal Process.*, 36, 425–432, 1988.

Won90 Wong, K.M., et al., On information theoretic criteria for determining the number of signals in high resolution array processing, *IEEE Trans. Acoust. Speech Signal Process.*, 38, 1959–1971, 1990.

Won92 Wong, K.M., et al., Estimation of the direction of arrival of signals in unknown correlated noise. Part I: The MAP approach and its implementation, *IEEE Trans. Signal Process.*, 40, 2007–2017, 1992.

Wu91 Wu, Q. and Fuhrmann, D.R., A parametric method for determining the number of signals in narrowband direction finding, *IEEE Trans. Signal Process.*, 39, 1848–1857, 1991.

Wu94 Wu, Q., Wong, K.M. and Reilly, J.P., Maximum likelihood direction finding in unknown noise environments, *IEEE Trans. Signal Process.*, 42, 980–983, 1994.

Wu94a Wu, Q. and Wong, K.M., UN-MUSIC and UN-CLE: An application of generalized correlation to the estimation of the direction of arrival of signals in unknown correlated noise, *IEEE Trans. Signal Process.*, 42, 2331–2343, 1994.

Wyl94 Wylie, M.P., Roy, S. and Messer, H., Joint DOA estimation and phase calibration of linear equispaced (LES) arrays, *IEEE Trans. Signal Process.*, 42, 3449–3459, 1994.

Xu92 Xu, G. and Kailath, T., Direction-of-arrival estimation via exploitation of cyclostationarity: a combination of temporal and spatial processing, *IEEE Trans. Signal Process.*, 40, 1775–1786, 1992.

Xu92a Xu, X.L. and Buckley, K.M., Bias analysis of the MUSIC location estimator, *IEEE Trans. Signal Process.*, 40, 2559–2569, 1992.

Xu93 Xu, X.L. and Buckley, K., An analysis of beam-space source localization, *IEEE Trans. Signal Process.*, 41, 501–504, 1993.

Xu94 Xu, G., et al., Beamspace ESPRIT, *IEEE Trans. Signal Process.*, 42, 349–356, 1994.

Xu94a Xu, X.L. and Buckley, K.M., Bias and variance of direction-of-arrival estimate from MUSIC, MIN-NORM, and FINE, *IEEE Trans. Signal Process.*, 42, 1812–1816, 1994.

Yan88 Yang, J.R. and Kaveh, M., Adaptive eigensubspace algorithms for direction or frequency estimation and tracking, *IEEE Trans. Acoust. Speech Signal Process.*, 36, 241–251, 1988.

Yan92 Yang, J.F. and Tsai, C.J., A further analysis of decorrelation performance of spatial smoothing techniques for real multipath sources, *IEEE Trans. Signal Process.*, 40, 2109–2112, 1992.

Yan94 Yang, J.F. and Lin, H.J., Adaptive high resolution algorithms for tracking nonstationary sources without the estimation of source number, *IEEE Trans. Signal Process.*, 42, 563–571, 1994.

Yan94a Yang, W.H., Chan, K.K. and Chang, P.R., Complexed-valued neutral network for direction of arrival estimation, *Electron. Letters*, 30, 574–575, 1994.

Yeh89 Yeh, C.C., Lee, J.H. and Chen, Y.M., Estimating two-dimensional angles of arrival in coherent source environment, *IEEE Trans. Acoust. Speech Signal Process.*, 37, 153–155, 1989.

Yin87 Yin, Y. and Krishnaiah, P., On some nonparametric methods for detection of number of signals, *IEEE Trans. Acoust. Speech Signal Process.*, 35, 1533–1538, 1987.

Yue96 Yuen, N. and Friedlander, B., Asymptotic performance analysis of ESPRIT, higher order ESPRIT, and virtual ESPRIT algorithms, *IEEE Trans. Signal Process.*, 44, 2537–2550, 1996.

Zei95 Zeira, A. and Friedlander, B., On the performance of direction finding with time varying arrays, *Signal Process.*, 43, 133–147, 1995.

Zha89 Zhang, Q.T., et al., Statistical analysis of the performance of information theoretical criteria in the detection of the number of signals in an array, *IEEE Trans. Acoust. Speech Signal Process.*, 37, 1557–1567, 1989.

Zha95 Zhang, Q.T., A statistical resolution theory of the beamformer-based spatial spectrum for determining the directions of signals in white noise, *IEEE Trans. Signal Process.*, 43, 1867–1873, 1995.

Zha95a Zhang, Q.T., Probability of resolution of MUSIC algorithm, *IEEE Trans. Signal Process.*, 43, 978–987, 1995.

Zho91 Zhou, C., Haber, F. and Jaggard, D.L., A resolution measure for the MUSIC algorithm and its application to plane wave arrivals contaminated by coherent interference, *IEEE Trans. Signal Process.*, 39, 454–463, 1991.

Zho93 Zhou, C.G., Haber, F. and Jaggard, D.L., The resolution threshold of MUSIC with unknown spatially colored noise, *IEEE Trans. Signal Process.*, 41, 511–516, 1993.

Zis88 Ziskind, I. and Wax, I., Maximum likelihood localization of multiple sources by alternating projection, *IEEE Trans. Acoust. Speech Signal Process.*, 36, 1553–1560, 1988.

Zis90 Ziskind, I. and Wax, M., Maximum likelihood localization of diversely polarized source by simulated annealing, *IEEE Trans. Antennas Propagat.*, 38, 1111–1114, 1990.

Zol89 Zoltowski, M.D. and Stavrinides, D., Sensor array signal processing via a procrustes rotations based eigenanalysis of the ESPRIT data pencil, *IEEE Trans. Acoust. Speech Signal Process.*, 37, 832–861, 1989.

Zol93 Zoltowski, M.D., Silverstein, S.D. and Mathews, C.P., Beamspace root-MUSIC for minimum redundancy linear arrays, *IEEE Trans. Signal Process.*, 41, 2502–2507, 1993.

Zol93a Zoltowski, M.D., Kautz, G.M. and Silverstein, S.D., Beamspace root-MUSIC, *IEEE Trans. Signal Process.*, 41, 344–364, 1993.

# 7

# *Single-Antenna System in Fading Channels*



In previous chapters, it is assumed that the directional signals arrive from point sources as plane wave fronts. In mobile communication channels, the received signal is a combination of many components arriving from various directions as a result of multipath propagation. Depending on terrain conditions and local buildings and structures, the power of the received signal fluctuates randomly as a function of distance. Fluctuations on the order of 20 dB are common within the distance of one wavelength. This phenomenon is called fading.

In this chapter, a brief review of fading channels is presented with a view to introduce notation and to develop mathematical equations to be used for analyzing the behavior of communication systems. The chapter also contains analyses of a single antenna system under various fading conditions. The methodology presented in this chapter would be helpful in analyzing the performance of various diversity-combining schemes discussed in Chapter 8, and results would serve as a reference for comparison.

A detailed treatment of fading channels is presented in [Skl02]. For an introduction to mobile communications, see [God02]. For details on digital communications and the required probability theory, see [Pro95].

## 7.1 Fading Channels

Let a transmitted signal s(t) be expressed in complex notation as

$$s(t) = \mathrm{Re}\left[g(t)e^{j2\pi f_c t}\right] \tag{7.1.1}$$

where Re[.] denotes the real part of a complex quantity, $f_c$ is the carrier frequency, and g(t) is the complex envelope of s(t) that can be expressed in the magnitude and phase form as

$$g(t) = |g(t)|e^{j\phi(t)} \tag{7.1.2}$$

where $|g(t)|$ is the magnitude and $\phi(t)$ is the phase of the complex baseband wave form.

For the frequency and phase-modulated signals, $|g(t)|$ is constant. Without any loss of generality, it is assumed in the present discussion that it is equal to unity.

In the mobile communications environment, the transmitted signal undergoes fading. There are two kinds of fading, namely large-scale and small-scale.

Large-scale fading, also known as shadowing, is caused by hills and large buildings. It determines the local mean signal power at distance R from the transmitter. Let $\bar{S}$ denote this power. It is a random quantity and the random variable $\bar{S}$ has a log-normal distribution.

Let $S_d$ denote the mean signal power in decibels. Thus, $S_d$ and $\bar{S}$ are related by

$$S_d = 10 \log \bar{S} \tag{7.1.3}$$

where log denotes $\log_{10}(.)$. The random variable (RV) $S_d$ has a normal distribution.

Small-scale fading, on the other hand, is a local phenomenon caused by multipath propagation. It causes in a rapid fluctuation of the signal around the slowly varying local mean.

Let $x_0(t)$ denote the signal component induced on an antenna. It is given by

$$x_0(t) = \mathrm{Re}\left[\tilde{x}_0(t)\right] \tag{7.1.4}$$

where $\tilde{x}_0(t)$ is the signal component in the complex form. Following (7.1.1) and (7.1.2) $\tilde{x}_0(t)$ can be expressed as

$$\tilde{x}_0(t) = r(t)e^{j\theta(t)}\, e^{j2\pi f_c t}\, e^{j\phi(t)} \tag{7.1.5}$$

In the above equation, the complex random quantity $r(t)e^{j\theta(t)}$ accounts for channel fading with r(t) denoting the signal amplitude and $\theta(t)$ representing the random phase process uniformly distributed in $[0,2\pi)$.

It is convenient to think r(t) as a product of two variables, that is,

$$r(t) = m(t)r_0(t) \tag{7.1.6}$$

where m(t) is a slowly varying quantity and denotes the local mean value of the signal. It accounts for large-scale fading and the effect of shadowing, and determines the local mean power $\bar{S}$ given by

$$\bar{S} = m^2(t) \tag{7.1.7}$$

The complex quantity $r_0(t)e^{j\theta(t)}$ is the result of small-scale fading and causes rapid fluctuations about the local mean signal m(t).

### 7.1.1 Large-Scale Fading

In free-space propagation, the received signal power $P_R$ at distance R from the transmitter and the transmitted power $P_T$ of an isotropic source are linked by the well-known relation

$$P_R = \frac{P_T}{4\pi R^2} \tag{7.1.8}$$

Let $P_{R_0}$ denote the received signal power at a reference distance $R_0$ from the transmitter, that is,

$$P_{R_0} = \frac{P_T}{4\pi R_0^2} \tag{7.1.9}$$

It follows from (7.1.8) and (7.1.9) that the received powers $P_R$ and $P_{R_0}$ are related by

$$P_R = P_{R_0}\left(\frac{R_0}{R}\right)^2 \tag{7.1.10}$$

In mobile radio channels, the mean path loss between a transmitter and a receiver is proportional to the nth power of distance R relative to a reference distance $R_0$ rather than 2, as is the case for free-space propagation. In urban areas, a typical value of the path loss exponent n is four. Denoting the received signal power at distance $R_0$ from the transmitter by $S(R_0)$, and the received signal power at a distance R from the transmitter by $\overline{S}(R)$, it follows from (7.1.10) that for mobile radio channels,

$$\overline{S}(R) = \overline{S}(R_0)\left(\frac{R_0}{R}\right)^n \tag{7.1.11}$$

Let $\overline{S}_d$ denote $\overline{S}(R)$ in decibels, that is,

$$\overline{S}_d = 10\log\overline{S}(R) \tag{7.1.12}$$

Substituting for $\overline{S}(R)$ from (7.1.11) in (7.1.12) leads to

$$\overline{S}_d = 10\log\overline{S}(R_0) + 10n\log\left(\frac{R_0}{R}\right) \tag{7.1.13}$$

Note that the signal power $\overline{S}(R)$ received at R meters away from the transmitter is an average value and is referred to as the area mean. Thus, $\overline{S}_d$ is the area mean signal power in decibels. It is different from the mean signal power that we previously also referred to as the local mean signal power and denoted by $\overline{S}$ (and $S_d$ in decibels). The relationship between the two is now described.

The mean signal power $S_d$ is site dependent and for a given transmitter–receiver separation, it differs from location to location due to the shadowing effect. It is a random quantity with a normal distribution, and this randomness is reflected by adding a random quantity to the area mean power $\overline{S}_d$ to yield an expression for the received mean power in decibels, yielding

$$S_d = \bar{S}_d + X_{\sigma_s} \tag{7.1.14}$$

where $X_{\sigma_s}$ is a zero-mean, Gaussian random variable (in decibels) with standard deviation $\sigma_s$ (also in decibels). The parameter $\sigma_s$ is called the decibel spread and is a site-dependent quantity. It may take on values between 6 to 12 dB depending on the severity of shadowing [Fre79].

It follows from (7.1.4) that $S_d$ is a random variable with a mean value equal to the area mean $\bar{S}_d$, that is,

$$\bar{S}_d = E[S_d] \tag{7.1.15}$$

Thus, $S_d$ is a random variable having normal distribution with the mean value equal to $\bar{S}_d$ and the standard deviation equal to $\sigma_s$. An expression for its probability density function (pdf) $f_{s_d}(S_d)$ is given by [Yeh84]:

$$f_{s_d}(S_d) = \frac{1}{\sqrt{2\pi}\,\sigma_s} \exp\left[-\frac{(S_d - \bar{S}_d)^2}{2\sigma_s^2}\right] \tag{7.1.16}$$

Due to the fact that the log value of $\bar{S}$ has a normal distribution, $\bar{S}$ is said to have a log-normal distribution.

Note that the cumulative distribution function (cdf) $F_x$ of an RV x is related to its pdf by

$$F_x(y) = \int_0^y f_x(x)dx, \quad x \ge 0 \tag{7.1.17}$$

or alternately,

$$f_x(y) = \frac{dF_x(y)}{dy} \tag{7.1.18}$$

It follows from (7.1.16) and (7.1.17) that

$$F_{S_d}(y) = \int_0^y \frac{1}{\sqrt{2\pi}\,\sigma_s} \exp\left[-\frac{(S_d - \bar{S}_d)^2}{2\sigma_s^2}\right] dS_d \tag{7.1.19}$$

Since

$$\begin{aligned} S_d &= 10 \log \bar{S} \\ &= 10 \frac{\ln \bar{S}}{\ln 10} \end{aligned} \tag{7.1.20}$$

a differentiation on both sides with respect to $\bar{S}$ results in

$$dS_d = \frac{10}{\ln 10} \frac{d\bar{S}}{\bar{S}} \tag{7.1.21}$$

By substituting for $S_d$ and $dS_d$, it follows from (7.1.19) that the cdf of $\bar{S}$ is given by

$$F_{\bar{S}}(z) = \int_0^z \frac{10}{\sqrt{2\pi}\,\sigma_s \bar{S} \ln 10} \exp\left[-\frac{\left(10\log\bar{S} - \bar{S}_d\right)^2}{2\sigma_s^2}\right] d\bar{S} \tag{7.1.22}$$

This along with (7.1.18) implies that the pdf of $\bar{S}$ can be expressed as

$$f_{\bar{S}}(\bar{S}) = \frac{10}{\sqrt{2\pi}\,\sigma_s \bar{S} \ln 10} \exp\left[-\frac{\left(10\log\bar{S} - \bar{S}_d\right)^2}{2\sigma_s^2}\right] \tag{7.1.23}$$

### 7.1.2 Small-Scale Fading

For the discussion on small-scale fading, assume that the large-scale fading component m(t) and thus the local mean signal power $\bar{S}$ remain constant. This would be the case when the receiving antenna remains within a limited trajectory such that shadowing effects may be ignored.

Under the assumption of m(t) being constant, it follows from (7.1.6) that the quantity $r(t)e^{j\theta(t)}$ may be thought of as representing the small-scale fading effect similar to $r_0(t)e^{j\theta(t)}$. This quantity is the resultant sum of many scattered multipath components of varying amplitude and phase arriving at the receiving antenna. Denote this in terms of its orthogonal components a(t) and b(t), that is,

$$r(t)e^{j\theta(t)} \equiv a(t) + jb(t) \tag{7.1.24}$$

The variables a(t) and b(t) result from the addition of many multipath components. When the number of such components is large, these variables at a given time are statistically independent, Gaussian random variables with a zero mean and equal variance $\sigma^2$. Dropping the reference to time for ease of notation, one thus writes expressions for pdfs of a and b as

$$f_a(z) \equiv f_b(z) = \frac{1}{\sqrt{2\pi}\,\sigma} \exp\left[-\frac{z^2}{2\sigma^2}\right] \tag{7.1.25}$$

Writing an expression for the signal envelope from (7.1.24) as

$$r(t) = \sqrt{a^2(t) + b^2(t)} \tag{7.1.26}$$

it follows that $r(t) \geq 0$ for all t. Furthermore, at a given time, it is a RV with

$$\begin{aligned} E[r^2] &= E[a^2 + b^2] \\ &= E[a^2] + E[b^2] \\ &= 2\sigma^2 \end{aligned} \tag{7.1.27}$$

Next, an expression for the pdf of r is derived. Consider

$$Y = \sum_{i=1}^{N} x_i^2 \tag{7.1.28}$$

where $x_i$, i = 1, 2, …, N are statistically independent, Gaussian random variables with a zero mean and equal variance $\sigma^2$. The pdf of Y is given by [Pro95]

$$f_Y(y) = \frac{1}{\sigma^N 2^{N/2} \Gamma(N/2)} y^{N/2-1} \exp\left[-\frac{y}{2\sigma^2}\right], \quad y \geq 0 \tag{7.1.29}$$

where $\Gamma(p)$ is the gamma function, defined as

$$\Gamma(p) = \int_0^{\infty} t^{p-1} e^{-t} \, dt, \quad p > 0 \tag{7.1.30}$$

It has the following properties [Abr72]:

$$\begin{aligned} &\Gamma(p) = (p-1)!, \ \text{p an integer}, \quad p > 0 \\ &\Gamma(p)\Gamma(1-p) = -p\Gamma(p)\Gamma(-p) = \frac{\pi}{\sin \pi p}, \quad 0 < p < 1 \end{aligned} \tag{7.1.31}$$

and

$$\Gamma\left(\frac{1}{2}\right) = \sqrt{\pi}, \quad \Gamma\left(\frac{3}{2}\right) = \frac{1}{2}\sqrt{\pi} \tag{7.1.32}$$

The pdf given by (7.1.29) is called a chi-square or gamma pdf with N degrees of freedom. In the present case,

$$Y \equiv r^2 = a^2 + b^2 \tag{7.1.33}$$

Thus, N = 2 and the pdf of $Y \equiv r^2$ becomes

$$f_Y(y) = \frac{1}{2\sigma^2} \exp\left[-\frac{y}{2\sigma^2}\right], \quad y \geq 0 \tag{7.1.34}$$

Since the cdf of a RV x is related to its pdf by (7.1.17), it follows from (7.1.34) and (7.1.17) that the cdf of Y, $F_Y$, is given by

$$F_Y(r) = \int_0^{r} \frac{1}{2\sigma^2} \exp\left[-\frac{y}{2\sigma^2}\right] dy \tag{7.1.35}$$

Define

$$x = y^{1/2} \tag{7.1.36}$$

Thus,

$$y = x^2 \tag{7.1.37}$$

and

$$dy = 2xdx \tag{7.1.38}$$

Substituting from (7.1.37) and (7.1.38) in (7.1.35) and using $r = Y^{1/2}$ leads to the cdf of r given by

$$F_r(r) = \int_0^r \frac{x}{\sigma^2} \exp\left[-\frac{x^2}{2\sigma^2}\right] dx \tag{7.1.39}$$

This along with the relation (7.1.18) yields

$$f_r(r) = \frac{r}{\sigma^2} \exp\left[-\frac{r^2}{2\sigma^2}\right] \quad r \geq 0 \tag{7.1.40}$$

This is the pdf of a Rayleigh-distributed RV. Thus, r is an RV with Rayleigh distribution.

When the received signal has a significant nonfading component (line-of-sight component) other than the reflective multipath component, the received signal envelope has a Rice distribution with pdf given by [Rap96]

$$f_r(r) = \begin{cases} \dfrac{r}{\sigma^2} \exp\left[-\dfrac{r^2 + A^2}{2\sigma^2}\right] I_0\left(\dfrac{rA}{\sigma^2}\right) & r \geq 0,\ A \geq 0 \\ 0 & \text{otherwise} \end{cases} \tag{7.1.41}$$

where A denotes the peak amplitude of the line-of-sight (LOS) component and $I_0(\ )$ is the modified Bessel function of the first kind and zero order. The Rice distribution is often characterized by a parameter $K_0$ defined as

$$K_0 = \frac{A^2}{2\sigma^2} \tag{7.1.42}$$

As the magnitude of the LOS component *A* goes to zero, the Rice pdf approaches the Rayleigh pdf given by (7.1.40).

Another distribution that is frequently used to describe the statistics of signals transmitted through multipath fading channels is the Nakagami m-distribution with pdf given by [Pro95]:

$$f_r(r) = \frac{2}{\Gamma(m)} \left(\frac{m}{\Omega}\right)^m r^{2m-1} e^{-\frac{mr^2}{\Omega}} \tag{7.1.43}$$

where

$$\Omega = E\left[r^2\right] \tag{7.1.44}$$

The fading parameter m is defined as

$$m = \frac{\Omega^2}{E\left[\left(r^2 - \Omega\right)^2\right]}, \quad m \geq \frac{1}{2} \tag{7.1.45}$$

and $\Gamma(m)$ is the gamma function.

Note that when m = 1, (7.1.43) reduces to the Rayleigh pdf given by (7.1.40). For values of m in the range of $1/2 \leq m \leq 1$, (7.1.43) results in pdfs having larger tails than a Rayleigh pdf, whereas for m > 1 the tails of this pdf decay faster than the Rayleigh pdf. The term $m = \infty$ denotes no fading.

The discussion presented thus far relates to distribution of the received signal amplitude. Now the distribution of signal power is considered.

### 7.1.3 Distribution of Signal Power

The instantaneous power S of the received signal $x_0(t)$ is given by

$$S = x_0^2(t) \tag{7.1.46}$$

It follows from (7.1.4), (7.1.5) and (7.1.46) that the expression for the instantaneous power averaged over one radio frequency cycle reduces to

$$S = \frac{r^2}{2} \tag{7.1.47}$$

Equation (7.1.47) along with (7.1.27) implies that the local mean signal power $\bar{S}$ is given by

$$\begin{aligned} \bar{S} &= E[S] \\ &= \sigma^2 \end{aligned} \tag{7.1.48}$$

Now, an expression for $f_S$, the pdf of S is derived for the case when the signal amplitude has the Rayleigh distribution given by (7.1.40). For this case, the cdf of r, $F_r$, is given by (7.1.39).

To transform $f_r$ into $f_S$ define a new variable:

$$y = \frac{x^2}{2} \tag{7.1.49}$$

Thus,

$$dy = x\,dx \tag{7.1.50}$$

Using (7.1.47) to (7.1.50) in (7.1.39), the cdf of S is given by

$$F_S(S) = \int_0^S \frac{1}{\bar{S}} e^{-\frac{y}{\bar{S}}} dy \tag{7.1.51}$$

This along with the relation

$$f_S(S) = \frac{dF_S(S)}{dS} \tag{7.1.52}$$

yields

$$f_S(S) = \frac{1}{\bar{S}} e^{-\frac{S}{\bar{S}}} \tag{7.1.53}$$

Similarly, the pdf of S in a Nakagami and Rice fading environment may be obtained.

For the case of Nakagami distributed environment, S is a gamma-distributed RV with pdf $f_S(S)$ given by [Abu91, Woj86]:

$$f_S(S) = \left(\frac{m}{\bar{S}}\right)^m \frac{S^{m-1}}{\Gamma(m)} \exp\left[-\frac{mS}{\bar{S}}\right], \quad S \geq 0, \ m \geq 0.5 \tag{7.1.54}$$

## 7.2 Channel Gain

In this section, the concept of channel gain, sometimes also referred to as channel attenuation [Pro95] (to be used later in the book), is introduced and selected signal and power variables are expressed using the channel gain. Define a real variable

$$\alpha(t) = \frac{r(t)}{\sqrt{\bar{S}}} \tag{7.2.1}$$

to denote the signal envelope normalized with respect to the square root of the mean signal power $\bar{S}$ and a complex variable C(t)

$$C(t) = \alpha(t)e^{j\theta(t)} \tag{7.2.2}$$

to denote the channel gain.

Thus, the received signal in the complex form given by (7.1.5) can be expressed as

$$\tilde{x}_0(t) = C(t)\sqrt{\bar{S}}g(t)e^{j2\pi f_c t} \tag{7.2.3}$$

where g(t) is the transmitted signal normalized to have a unit energy, that is,

$$\int_0^T [g(t)]^2 dt = 1 \tag{7.2.4}$$

Assume that the channel varies slowly such that α(t) and θ(t) may be regarded as constant over a time duration T of interest, such as a bit or symbol duration. Thus, over this time the channel gain can be regarded as constant and the reference to time t may be dropped, yielding

$$\tilde{x}_0(t) = C\sqrt{\bar{S}}\, g(t)e^{j2\pi f_c t}, \quad 0 \le t \le T \tag{7.2.5}$$

The instantaneous signal power S averaged over time T is then given by

$$S = \frac{\alpha^2 \bar{S}}{2} \tag{7.2.6}$$

In view of (7.2.1) and the fact that the mean signal power is constant for a given large-scale fading, the channel gain is a complex RV and has the same statistics as $r(t)e^{j\theta(t)}$.

## 7.3 Single-Antenna System

In this section, a single-channel system is considered, and its performance is examined by studying the outage probability and the average bit error rate (BER) in the presence of frequency nonselective slow-fading channels. Both noise-limited and interference-limited systems are considered. The methodology to evaluate outage probability and average BER presented here should be helpful in evaluating these parameters for various diversity schemes discussed in Chapter 8.

### 7.3.1 Noise-Limited System

In a noise-limited system, system performance is limited by noise and the effect of co-channel interference is negligible. Consider a noise-limited system with a single source in the presence of an additive white Gaussian noise (AWGN) of zero mean and variance N. Let $\gamma$ denote the instantaneous signal power to the mean noise power ratio, that is,

$$\gamma = \frac{S}{N} \tag{7.3.1}$$

As discussed in Section 7.1, S is an RV and thus $\gamma$ is an RV. Let $\Gamma$ denote its mean value, that is,

$$\Gamma = E(\gamma) \tag{7.3.2}$$

Substituting from (7.3.1) in (7.3.2) and noting that $\bar{S}$ denotes the mean value of S,

$$\Gamma = \frac{\bar{S}}{N} \tag{7.3.3}$$

For a receiver to function properly in a noise-limited system the SNR at its input must be above a certain threshold $\gamma_0$. When SNR drops below this threshold, the communication link does not remain operational. The probability of this happening is referred to as the outage probability, denoted by $P^o$. Denoting P[x] as the probability of an event x, $P^o$ may be expressed as

$$\begin{aligned} P^o &= P[\gamma \le \gamma_0] \\ &= \int_0^{\gamma_0} f_\gamma(\gamma)d\gamma \\ &= F_\gamma(\gamma_0) \end{aligned} \tag{7.3.4}$$

where $f_\gamma$ and $F_\gamma$, respectively, denote the pdf and cdf of $\gamma$.

It follows from (7.3.4) that evaluation of outage probability requires knowledge of the pdf or the cdf of $\gamma$, which depends on the fading environment. The pdf of $\gamma$ is also useful in determining the average BER in the fading environment. When the conditional probability of bit error for a given value of the SNR, $P_e(\gamma)$, is known, the average BER, $P_e$, may be obtained by averaging over $\gamma$, that is,

$$P_e = \int_0^\infty P_e(\gamma) f_\gamma(\gamma) d\gamma \tag{7.3.5}$$

#### 7.3.1.1 Rayleigh Fading Environment

In this case, the signal amplitude has a Rayleigh distribution. First, we derive the pdf and cdf of $\gamma$. The pdf of S is given by (7.1.53), that is,

$$f_S(S) = \frac{1}{\overline{S}} e^{-\frac{S}{\overline{S}}} \tag{7.3.6}$$

It follows from (7.1.17) and (7.3.6) that the cdf of S is given as

$$F_S(S) = \int_0^S \frac{1}{\overline{S}} e^{-\frac{x}{\overline{S}}} dx \tag{7.3.7}$$

Define a new variable:

$$y = \frac{x}{N} \tag{7.3.8}$$

Thus,

$$x = Ny \tag{7.3.9}$$

and

$$dx = Ndy \tag{7.3.10}$$

Substituting from (7.3.9) and (7.3.10) in (7.3.7) and using (7.3.1) and (7.3.3),

$$\begin{aligned} F_\gamma(\gamma) &= \int_0^\gamma \frac{1}{\Gamma} e^{-\frac{y}{\Gamma}} dy \\ &= 1 - e^{-\frac{\gamma}{\Gamma}} \end{aligned} \tag{7.3.11}$$

Equations (7.1.17) and (7.3.11) imply that

$$f_\gamma(\gamma) = \frac{1}{\Gamma} e^{-\frac{y}{\Gamma}} \tag{7.3.12}$$

Using (7.3.4), the outage probability then becomes

$$\begin{aligned} P^o &= F_\gamma(\gamma_0) \\ &= 1 - e^{-\frac{\gamma_0}{\Gamma}} \end{aligned} \tag{7.3.13}$$

#### 7.3.1.2 Nakagami Fading Environment

For this case, the pdf of S is given by (7.1.54). Following the procedure of the previous section, the following expression for the pdf of $\gamma$ is obtained:

$$f_\gamma(\gamma) = \left(\frac{m}{\Gamma}\right)^m \frac{\gamma^{m-1}}{\Gamma(m)} e^{-\frac{m\gamma}{\Gamma}} \tag{7.3.14}$$

The cumulative distribution function of $\gamma$ then is given by

$$\begin{aligned} F_\gamma(\gamma) &= \int_0^\gamma f_\gamma(x)dx \\ &= \int_0^\gamma \left(\frac{m}{\Gamma}\right)^m \frac{x^{m-1}}{\Gamma(m)} e^{-\frac{mx}{\Gamma}} dx \end{aligned} \tag{7.3.15}$$

Defining

$$y = mx \tag{7.3.16}$$

and noting that for integer m, $\Gamma(m) = (m-1)!$, (7.3.15) becomes

$$\begin{aligned} F_\gamma(\gamma) &= \frac{1}{\Gamma^m (m-1)!} \int_0^{m\gamma} y^{m-1} e^{-\frac{y}{\Gamma}} dy \\ &= 1 - e^{-m\frac{\gamma}{\Gamma}} \sum_{k=1}^{m} \frac{(m\gamma/\Gamma)^{k-1}}{(k-1)!} \end{aligned} \tag{7.3.17}$$

yielding

$$P^o = F_\gamma(\gamma_0) \tag{7.3.18}$$

### 7.3.2 Interference-Limited System

In an interference-limited system, system performance is limited by total interference power and not noise power, as is the case for noise-limited systems. In this case, the effect of noise is negligible and thus is ignored.

In this section, the effect of co-channel interference is examined by deriving the expression for the probability of signal-to-interference ratio $\mu$ being less than the desired threshold value $\mu_0$ in a Nakagami fading environment [Abu91]. Rayleigh fading is treated as a special case.

Assume that the received signal amplitude is an RV with a Nakagami distribution. Then the signal power S is a gamma-distributed RV with pdf given by (7.1.54).

Assume that there are K co-channel interferences present. Let $q_i$, i = 1, 2, ..., K be i.i.d. RVs with Nakagami distribution, and denote the amplitude of these interferences. Let $I_i$, i = 1, 2, ..., K denote their instantaneous powers, that is,

$$I_i = \frac{q_i^2}{2}, \quad i = 1, 2, \ldots, K \tag{7.3.19}$$

Let $\psi_i$, i = 1, 2, ..., K be i.i.d. random-phase processes associated with K interferences. It is assumed that r, $\theta$, $q_i$ and $\psi_i$, i = 1, 2, ..., K are mutually independent. Note that r and $\theta$ denote the amplitude and phase of the signal, respectively.

The interference power $I_i$ is a gamma-distributed RV with pdf given by (following 7.1.54)

$$f_{I_i}(I_i) = \left(\frac{m_i}{\bar{I}_i}\right)^{m_i} \frac{I_i^{m_i-1}}{\Gamma(m_i)} \exp\left[\frac{m_i I_i}{\bar{I}_i}\right] \tag{7.3.20}$$

where $m_i$ is the fading parameter and $\bar{I}_i$ is the mean power of the ith interference.

### *7.3.2.1 Identical Interferences*

Consider that all interferences have identical statistics with the fading parameter denoted by m (same as the signal) and equal mean power denoted by $\bar{I}$. As interferences are independent, the total interference power I is given by

$$I = \sum_{i=1}^{K} I_i \tag{7.3.21}$$

I is a gamma-distributed RV with pdf given by [Abu91, Fel66]

$$f_I(I) = \left(\frac{m}{\bar{I}}\right)^{mK} \frac{I^{mK-1}}{\Gamma(mK)} \exp\left[-\frac{mI}{\bar{I}}\right] \tag{7.3.22}$$

Note that I denotes the total interference power and $\bar{I}$ denotes the mean power of each interference.

Assume that the channel becomes inoperable when the ratio of signal power S to total interference power I becomes less than some desired value $\mu_0$. Thus, the outage probability with K interference $P_K^o$ can be written as

$$\begin{aligned} P_K^o &= P\left[\frac{S}{I} \le \mu_0\right] \\ &= P\left[S \le \mu_0 I\right] \end{aligned} \tag{7.3.23}$$

This could be solved by letting S = x, finding the probability that $\mu_0 I \geq x$, and integrating over x. Thus, (7.3.23) becomes [Fre79]

$$P_K^o = \int_0^\infty P[S = x]\, P\left[I \geq \frac{x}{\mu_0}\right] dx \tag{7.3.24}$$

Since

$$P(S = x) = f_S(x) \tag{7.3.25}$$

and

$$P\left[I \geq \frac{x}{\mu_0}\right] = \int_{\frac{x}{\mu_0}}^\infty f_I(y)\,dy \tag{7.3.26}$$

it follows from (7.3.24) that

$$P_K^o = \int_0^\infty f_S(x) \int_{\frac{x}{\mu_0}}^\infty f_I(y)\,dy\,dx \tag{7.3.27}$$

Substituting for $f_S$ and $f_I$ from (7.1.54) and (7.3.22), respectively, and evaluating the integral [Abu91],

$$P_K^o = I_x(m, mK) \quad m \geq 0.5 \tag{7.3.28}$$

where

$$x = \frac{1}{1 + \frac{\mu}{\mu_0}} \tag{7.3.29}$$

$\tilde{\mu}$ denotes the average signal power to the average power of a single interference ratio, that is,

$$\tilde{\mu} = \frac{\overline{S}}{\overline{I}} \tag{7.3.30}$$

and $I_x(m,mK)$ is the incomplete beta function, given by

$$I_x(m, mK) = \frac{\Gamma(m + mK)}{\Gamma(m)\Gamma(mK)} \int_0^x u^{m-1}(1-u)^{mK-1}\,du \tag{7.3.31}$$

For a special case where m is an integer, (7.3.28) reduces to

$$P_K^o = \left(\frac{\mu_0}{\tilde{\mu} + \mu_0}\right)^m \sum_{k=0}^{mK-1} \binom{m + k - 1}{k} \left(\frac{\tilde{\mu}}{\tilde{\mu} + \mu_0}\right)^k \tag{7.3.32}$$

where

$$\binom{N}{M} = \frac{N!}{M!(N-M)!} \tag{7.3.33}$$

For the Rayleigh fading environment, m = 1 and $P_K^o$ becomes

$$P_K^o = \left(\frac{\mu_0}{\tilde{\mu}+\mu_0}\right)\sum_{k=0}^{K-1}\left(\frac{\tilde{\mu}}{\tilde{\mu}+\mu_0}\right)^k \tag{7.3.34}$$

which for a single interference case (K = 1) reduces to

$$P_1^o = \frac{\mu_0}{\tilde{\mu}+\mu_0} \tag{7.3.35}$$

Thus, (7.3.35) gives the outage probability in the presence of one interference with the same statistics as the signal in the Rayleigh fading environment.

#### *7.3.2.2 Signal and Interference with Different Statistics*

Consider the case of an interference with fading statistics different from those of the signal [Abu91]. Assume that only one interference exists. Let $q_1$ denote the amplitude of the interference with Nakagami distribution of parameter $m_1$, whereas the signal amplitude is assumed to be Nakagami distributed with parameter m. The interference power

$$I = \frac{q_1^2}{2} \tag{7.3.36}$$

is a gamma-distributed RV with pdf given by

$$f_I(I) = \left(\frac{m_1}{\bar{I}}\right)^{m_1-1}\frac{I^{m_1-1}}{\Gamma(m_1)}\exp\left(-\frac{m_1 I}{\bar{I}}\right) \tag{7.3.37}$$

where $\bar{I}$ denotes the mean interference power. Substituting for $f_S$ and $f_I$ from (7.1.54) and (7.3.37), respectively, in (7.3.27), and evaluating the integrals for integer values of m and $m_1$,

$$P_1^o = \sum_{k=0}^{m_1-1}\binom{m+k-1}{k}\frac{\left(\frac{m}{m_1}\frac{\mu_0}{\tilde{\mu}}\right)^m}{\left(1+\frac{m}{m_1}\frac{\mu_0}{\tilde{\mu}}\right)^{m+k}} \tag{7.3.38}$$

For details on (7.3.38) and results when there is more than one interference with different statistics, see [Abu91].

### 7.3.3 Interference with Nakagami Fading and Shadowing

In this section, the analysis is extended to a scenario where both desired signal and interference experience Nakagami fading in the presence of log-normal shadowing [Abu91, Fre79]. The analysis in the previous section was without shadowing, and thus the mean values of signals and interferences were assumed to be constant. In the presence of shadowing, the mean signal value $\bar{S}$ and the mean interference value $\bar{I}$ are log-normal distributed random variables with respective pdfs given by

$$f_{S_d}(S_d) = \frac{1}{\sqrt{2\pi}\,\sigma_s} \exp\left[ -\frac{(S_d - \bar{S}_d)^2}{2\sigma_s^2} \right] \tag{7.3.39}$$

and

$$f_{I_d}(I_d) = \frac{1}{\sqrt{2\pi}\,\sigma_I} \exp\left[ -\frac{(I_d - \bar{I}_d)^2}{2\sigma_I^2} \right] \tag{7.3.40}$$

where

$$S_d = 10\log \bar{S} \tag{7.3.41}$$

$$I_d = 10\log \bar{I} \tag{7.3.42}$$

$$\bar{S}_d = E[S_d] \tag{7.3.43}$$

and

$$\bar{I}_d = E[I_d] \tag{7.3.44}$$

and $\sigma_S$ and $\sigma_I$ are decibel spread parameters for signal and interference, respectively.

As discussed in previous sections, calculation of the outage probability requires unconditional pdfs of the signal power and interference power, $f_S$ and $f_I$, respectively. These may be obtained by combining the pdfs of the mean signal power and mean interference power given by (7.3.39) and (7.3.40), respectively, with the corresponding conditional pdfs of the signal power and interference power given by (7.1.54) and (7.3.37), respectively. Note that the pdfs given by (7.1.54) and (7.3.37) are conditional that the mean signal power and mean interference power are constant. Denoting the conditional pdf of the signal power given by (7.1.54) as $f_{S/S_d}$, it follows that

$$f_S(S) = \int f_{S/S_d}(S) f_{S_d}(S_d)\, dS_d \tag{7.3.45}$$

Similarly, denoting the conditional pdf of the interference power given by (7.3.37) as $f_{I/I_d}$, it follows that

$$f_I(I) = \int f_{I/I_d}(I) f_{I_d}(I_d)\, dI_d \tag{7.3.46}$$

Let the outage probability $P_I^o$ denote the probability that $I \geq S/\mu_0$ in the presence of a single interference. The probability may be evaluated using (7.3.27), that is,

$$P_I^o = \int_0^{\infty} f_S(S) \int_{\frac{S}{\mu_0}}^{\infty} f_I(I)\,dI\,dS \tag{7.3.47}$$

Substituting for $f_S$ and $f_I$ from (7.3.45) and (7.3.46) in (7.3.47), and evaluating the resulting double integral leads to an expression for the outage probability, $P_I^o$. It can be achieved by following a procedure similar to that used by [Fre79] for converting a double integral into a single integral to evaluate the outage probability in the presence of Rayleigh fading and shadowing.

In [Abu91], a slightly different approach was used to obtain $P_I^o$. It uses the expression for $P_I^o$ in the absence of shadowing given by (7.3.38), and averages it using the joint pdf of the signal power and interference power to include the effect of shadowing. Denoting the joint pdf of $S_d$ and $I_d$ by $f_{S_dI_d}$, it follows from (7.3.39) and (7.3.40) that

$$f_{S_dI_d}(S_d, I_d) = \frac{1}{2\pi\sigma_I\sigma_s} \exp\left[-\frac{\left(I_d - \bar{I}_d\right)^2}{2\sigma_I^2} - \frac{\left(S_d - \bar{S}_d\right)^2}{2\sigma_s^2}\right] \tag{7.3.48}$$

Noting from (7.3.30) and (7.1.3) that

$$\tilde{\mu} = \frac{\bar{S}}{\bar{I}} = 10^{\frac{S_d - I_d}{10}} \tag{7.3.49}$$

substituting for $\tilde{\mu}$ from (7.3.49) in (7.3.38) and averaging the result using (7.3.48) yields

$$P_I^o = \left(\frac{\mu_0 m}{m_1 \tilde{\mu}_a}\right)^m \sum_{k=0}^{m_1-1} \binom{m+k-1}{k} \frac{1}{\sqrt{\pi}} \int_{-\infty}^{\infty} \frac{10^{\frac{\sigma m u}{5}} \exp\left(-u^2\right)}{\left(1 + \frac{m\mu_0}{m_1\tilde{\mu}_a} 10^{\frac{\sigma u}{5}}\right)^{m+k}} du \tag{7.3.50}$$

where

$$\sigma = \sigma_s = \sigma_I \tag{7.3.51}$$

and m and $m_1$ are the signal and interference fading parameters, respectively, and $\tilde{\mu}_a$ is the ratio of the area mean signal power to the area mean interference power. Using $\bar{S}_d$ and $\bar{I}_d$, $\tilde{\mu}_a$ can be expressed as

$$\tilde{\mu}_a = 10^{\left(\frac{\bar{S}_d - \bar{I}_d}{10}\right)} \tag{7.3.52}$$

For the Rayleigh fading case,

$$m = m_1 = 1 \tag{7.3.53}$$

and thus (7.3.50) becomes

$$P_I^o = \frac{1}{\sqrt{\pi}} \int_{-\infty}^{\infty} \frac{\exp(-u^2)}{1+\frac{1}{\mu_0} 10^{\frac{\bar{S}_d - \bar{S}_I - 2\sigma u}{10}}} du \tag{7.3.54}$$

The integrals in (7.3.50) and (7.3.54) can be evaluated using numerical methods for a suitable choice of $\mu_0$.

### 7.3.4 Error Rate Performance

Let $P_e(\gamma)$ denote the conditional probability of error for a given SNR $\gamma$ for a particular modulation technique, while $P_e$ denotes the average BER. In fading conditions when $\gamma$ is a random variable with pdf $f_\gamma$, $P_e$ is obtained from $P_e(\gamma)$ by averaging over all $\gamma$ using (7.3.5). For various modulation schemes, $P_e(\gamma)$ in additive white Gaussian noise is given below [Skl01].

For coherent binary phase shift keying (BPSK),

$$\begin{aligned} P_e(\gamma) &= Q\left(\sqrt{2\gamma}\right) \\ &= \frac{1}{2}\operatorname{erfc}\sqrt{\gamma} \end{aligned} \tag{7.3.55}$$

differentially coherent binary phase shift keying (DPSK),

$$P_e(\gamma) = \frac{1}{2}\exp(-\gamma) \tag{7.3.56}$$

coherent orthogonal frequency shift keying (CFSK),

$$\begin{aligned} P_e(\gamma) &= Q\left(\sqrt{\gamma}\right) \\ &= \frac{1}{2}\operatorname{erfc}\sqrt{\frac{\gamma}{2}} \end{aligned} \tag{7.3.57}$$

and noncoherent orthogonal frequency shift keying (NCFSK),

$$P_e(\gamma) = \frac{1}{2}\exp\left(-\frac{\gamma}{2}\right) \tag{7.3.58}$$

where

$$Q(x) = \frac{1}{\sqrt{2\pi}} \int_x^{\infty} e^{\frac{-z^2}{2}} dz \tag{7.3.59}$$

and related to erfc(x) as

$$Q(x) = \frac{1}{2}\operatorname{erfc}\left(\frac{x}{\sqrt{2}}\right) \tag{7.3.60}$$

For example, consider a NCFSK system in additive white Gaussian noise and Rayleigh fading conditions. For this case, $P_e(\gamma)$ is given by (7.3.58). In the Rayleigh fading environment with pdf of $\gamma$, $f_\gamma$ is given by (7.3.12). Substituting for $f_\gamma$ and $P_e(\gamma)$ in (7.3.5) and denoting the BER for this case by $P_{e,NCFSK}$, it follows that

$$P_{e,NCFSK} = \int_0^\infty \frac{1}{2\Gamma} e^{-\frac{\gamma}{2}} e^{-\frac{\gamma}{\Gamma}} d\gamma \tag{7.3.61}$$

where $\Gamma$ is the mean signal-power to noise-power ratio.

Carrying out the integral, (7.3.61) yields

$$P_{e,NCFSK} = \frac{1}{\Gamma + 2} \tag{7.3.62}$$

Similarly, (7.3.5) may be used to evaluate the BER for other modulation schemes in fading conditions when conditional BER for a given modulation scheme and the pdf of $\gamma$ in fading conditions are known.

In Rayleigh fading channels, expressions for BER for coherent BPSK, CFSK, and DPSK are given by [Pro95]

$$P_{e,PBSK} = \frac{1}{2}\left[1 - \sqrt{\frac{\Gamma}{1+\Gamma}}\right] \tag{7.3.63}$$

$$P_{e,CFSK} = \frac{1}{2}\left[1 - \sqrt{\frac{\Gamma}{2+\Gamma}}\right] \tag{7.3.64}$$

and

$$P_{e,DPSK} = \frac{1}{2(1+\Gamma)} \tag{7.3.65}$$

It follows from (7.3.62) and (7.3.65) that the average BER for NCFSK may be obtained from the average BER for DPSK by replacing $\Gamma$ with $\Gamma/2$. Similarly, it follows from (7.3.63) and (7.3.64) that the average BER for CFSK may be obtained from the average BER for BPSK by replacing $\Gamma$ with $\Gamma/2$.

## Notation and Abbreviations

| | |
|---|---|
| BER | bit error rate |
| BPSK | binary phase shift keying |
| DPSK | differential phase shift keying |
| FSK | frequency shift keying |
| NCFSK | noncoherent FSK |
| A | amplitude of line-of-sight component |

| | |
|---|---|
| $a(t)$ | real part of $r(t)e^{j\theta(t)}$ |
| $b(t)$ | imaginary part of $r(t)e^{j\theta(t)}$ |
| $C, C(t)$ | channel gain |
| cdf | cumulative distribution function |
| $F_x(\ )$ | cumulative distribution function of RV x |
| $f_c$ | carrier frequency |
| $f_x$ | probability density function of random variable x |
| $f_{x/y}$ | conditional pdf of x for given y |
| $g(t)$ | complex baseband signal |
| $I$ | total power from all interference |
| $\bar{I}$ | mean power of single interference |
| $I_0(\ )$ | modified Bessel function of the first kind and zero order |
| $I_d$ | interference power in dB |
| $\bar{I}_d$ | mean value of $I_d$ |
| $I_i$ | instantaneous power of ith interference |
| $I_x(m_1,m_2)$ | incomplete beta function |
| $\bar{I}_i$ | mean power of ith interference |
| $K$ | number of interferences |
| $K_0$ | Rice distribution parameter |
| $m(t)$ | accounts for large-scale fading |
| $m$ | Nakagami fading parameter |
| $m_i$ | Nakagami fading parameter for ith interference |
| $N$ | mean noise power |
| $n$ | path loss exponent in mobile communications |
| pdf | probability density function |
| $P[\ ]$ | probability of an event [ ] |
| $P_e$ | average probability of error |
| $P_e(\gamma)$ | probability of error for given $\gamma$ |
| $P_{e,x}$ | BER for modulation method x |
| $P_R$ | received power in free space |
| $P_T$ | transmitted power of an isotropic source |
| $P^o$ | outage probability |
| $P_K^o$ | outage probability when K interferences are present |
| $p_S$ | signal power |
| $q_i$ | amplitude of the ith interference |
| $R$ | distance between transmitter and receiver |
| RV | random variable |
| $R_0$ | reference distance |
| $r_0(t)$ | accounts for small-scale fading |
| $r(t)$ | received signal amplitude |
| $\bar{S}$ | local mean received signal power |

| | |
|---|---|
| $\bar{S}(R)$ | area mean, mean signal power received at R distance from transmitter |
| $S_d$ | local mean power in dB |
| $\bar{S}_d$ | area mean power in dB and mean value of $S_d$ |
| $x_0(t)$ | received signal |
| $\tilde{x}_0(t)$ | received signal in complex form |
| $x_{\sigma_s}$ | zero mean, Gaussian random variable with standard deviation $\sigma_s$ (dB) |
| $\Gamma$ | mean signal power to noise power ratio |
| $\Gamma(p)$ | gamma function |
| $\alpha$, $\alpha(t)$ | normalized signal envelope |
| $\gamma$ | instantaneous signal power to mean noise power ratio |
| $\gamma_0$ | threshold value of $\gamma$ for outage |
| $\phi(t)$ | phase of complex baseband signal |
| $\theta(t)$ | phase delay introduced by channel |
| $\sigma$ | standard deviation of a(t) and b(t) |
| $\sigma_I$ | dB spread parameter of interference |
| $\sigma_S$ | dB spread parameter of signal |
| $\mu$ | signal to interference power ratio |
| $\mu_0$ | threshold value of $\mu$ for outage |
| $\tilde{\mu}$ | ratio of mean powers of signal and interference |
| $\tilde{\mu}_a$ | ratio of area mean powers of signal and interference |
| $\psi_i$ | phase of the ith interference |
| $\Omega$ | $E[r^2]$ |

## References


Abr72 Abramowitz, M. and Segun, I.A., Eds., *Handbook of Mathematical Functions with Formulas, Graphs, and Mathematical Tables*, Dover, New York, 1972.

Abu91 Abu-Dayya, A.A. and Beaulieu, N.C., Outage probabilities of cellular mobile radio systems with multiple Nakagami interferers, *IEEE Trans. Vehicular Technol.*, 40, 757–768, 1991.

Fel66 Feller,W., *An Introduction to Probability Theory and Its Applications*, Vol. 2, Wiley, New York, 1966.

Fre79 French, R.C., The effect of fading and shadowing on channel reuse in mobile radio, *IEEE Trans. Vehicular Technol.*, 28, 171–181, 1979.

God02 Godara, L.C., Cellular systems, in *Handbook of Antennas in Wireless Communications*, Godara, L.C., Ed., CRC Press, Boca Raton, FL, 2002.

Pro95 Proakis, J.G., *Digital Communications*, 3rd Ed., McGraw-Hill, New York, 1995.

Rap96 Rappaport, T.S., *Wireless Communications: Principles and Practice*, Prentice Hall, New York, 1996, chap. 4.

Skl01 Sklar, B., *Digital Communications: Fundamentals and Applications*, Prentice Hall, New York, 2001.

Skl02 Sklar, B., Fading channels, in *Handbook of Antennas in Wireless Communications*, Godara, L.C., Ed., CRC Press, Boca Raton, FL, 2002, chap. 4.

Woj86 Wojnar, A., Unknown bounds on performance in Nakagami channels, *IEEE Trans. Commn.*, 34, 22–24, 1986.

Yeh84 Yeh, Y.S., Outage probability in mobile telephony due to multiple log-normal interferers, *IEEE Trans. Commn.*, 32, 380–388, 1984.

# 8

# *Diversity Combining*

In the presence of multipath fading channels, the received signal experiences great attenuation while the channel is in deep fade, resulting in the loss of transmitted information. This loss can be reduced by combining signals received over several independent fading channels. The reason for the reduction in information loss is that the likelihood of all signals experiencing deep fade simultaneously is considerably less than that experienced by individual signals.

The process of combining several signals with independent fading statistics to reduce large attenuation of the desired signal in the presence of multipath channels is referred to as diversity combining [Jak74, Bre59]. There are many ways by which several independent fading copies of a signal may be provided to a receiver for diversity combining. Some of these are described below.

*Frequency diversity*: A signal may be transmitted using several carriers such that the separation between successive carrier frequencies is longer than the coherence bandwidth of the channel to ensure that the fading associated with different frequencies is uncorrelated.

*Time diversity*: In this method, several copies of the signal are transmitted using different time slots such that the separation between successive time slots is more than the coherence time of the channel.

*Space diversity*: This method uses multiple antennas and is the subject of this chapter. The method requires that the separation between multiple antennas should be sufficient enough for the various signals to be uncorrelated. The multiple antennas may be used at a transmitter, at a receiver, or at both places depending on the application. This chapter considers various diversity combining schemes using single transmitting antennas and multiple receiving antennas.

*Predetection and postdetection schemes*: A diversity-combining method may be classified as a predetection or postdetection method. Predetection diversity-combining methods combine the received signals prior to detection and use single detectors to receive the information. Postdetection diversity methods, on the other hand, employ separate detectors on each branch and then combine the signals from different branches.

In this chapter, various diversity schemes are described, and their performance is analyzed and compared with that of a system using single receiving antennas.

There are basically two performance parameters, namely the outage probability and the average bit error rate (BER), to denote the performance of a diversity combiner. The outage probability is the probability that the SNR $\gamma$ is below some threshold value $\gamma_0$. It is given by (7.3.4). The average BER is determined by averaging the conditional BER $P_e(\gamma)$ for a given SNR over all values of $\gamma$. An expression for the average BER is given by (7.3.5). The conditional BER $P_e(\gamma)$ is a modulation dependent quantity and is available in most textbooks on digital communications such as [Cou95, Pro95, Skl01]. Expressions for conditional BER for coherent binary phase shift keying (BPSK), differentially coherent binary phase shift keying (DPSK), coherent orthogonal frequency shift keying (CFSK), and noncoherent orthogonal frequency shift keying (NCFSK) are given by (7.3.55), (7.3.56), (7.3.57), and (7.3.58), respectively.

The outage probability $P^o$ is a predetection parameter, and thus requires the pdf of $\gamma$ at the input to the receiver. The average BER $P_e$ for predetection diversity schemes may be determined using the pdf of $\gamma$ at the input to the receiver. For postdetection diversity schemes it may be determined using the pdf of $\gamma$ at the output of the receiver.

In view of the above discussion, it is clear that we need to determine the pdf of $\gamma$ at the input to the receiver to determine $P^o$ and $P_e$ for the predetection diversity schemes and pdf of $\gamma$ at the output of the receiver to determine $P_e$ for the postdetection diversity schemes.

Consider a diversity-combining system consisting of L antennas as shown in Figure 8.1. It is assumed that the signal is transmitted using a single antenna. Thus, the system consists of L diversity channels carrying the same information. It is assumed that these channels are slow fading and frequency nonselective. Furthermore, the fading processes among these channels are mutually statistically independent.

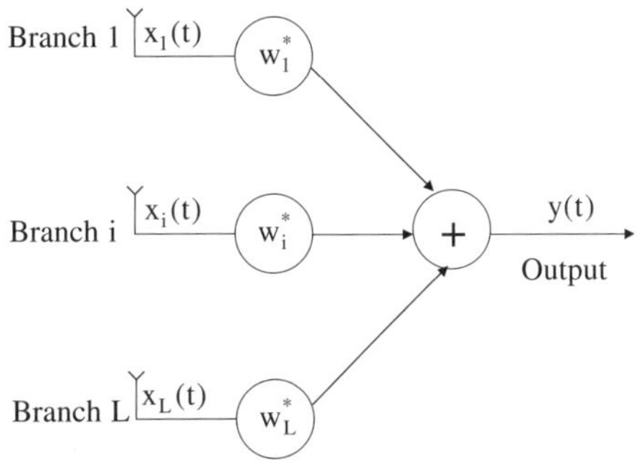


**FIGURE 8.1**
Block diagram of a diversity combining system.

The signal received on each antenna is weighted and summed to produce the output y(t) of a diversity combiner. Let $x_i(t)$ denote the signal induced on the ith antenna due to a desired signal source, and K, cochannel interference and uncorrelated noise. In the complex form and omitting the carrier terms for ease of notation,

$$x_i(t) = r_i(t)g(t)e^{j\theta_i(t)} + \sum_{j=1}^{K} q_{ij}(t)g_j(t)e^{j\psi_{ij}(t)} + n_i(t) \tag{8.1}$$

where $r_i(t)$ and $\theta_i(t)$, respectively, denote the amplitude and phase of the desired signal received on the ith branch; $q_{ij}(t)$ and $\psi_{ij}(t)$, respectively, denote the amplitude and phase of the jth interference received on the ith branch, g(t) denotes the designed message, $g_j(t)$ denotes the jth interference message, and $n_i(t)$ denotes the zero mean, Gaussian noise of variance (noise power) N present on the ith channel.

It is assumed that for any time t $r_i$, i = 1, 2, …, L are i.i.d random variables (RVs) with a specified distribution; $q_{ij}$, i = 1, 2, …, L and j = 1, 2, …, K are i.i.d RVs with a specified distribution; $\theta_i$, i = 1, 2, …, L are i.i.d RVs uniformly distributed in [0,2π); $\psi_{ij}$, i = 1, 2, …, L and j = 1, 2, …, K are i.i.d. RVs uniformly distributed in [0,2π); and $r_i$, $n_i$, $q_{ij}$, $\theta_i$, and $\psi_{ij}$, i = 1, 2, …, L and j = 1, 2, …, K are mutually independent.

Fading on various channels is assumed to be independent and represents the effect of small-scale fading unless stated otherwise. In other words, the results presented are for a given large-scale fading. It is assumed that all channels have the same mean signal power $\bar{S}$ and the mean interference power $\bar{I}^j$ due to jth interference. For identical interference, it is denoted by $\bar{I}$, that is,

$$\bar{I} = \bar{I}^j, \quad j = 1, 2, \ldots, K \tag{8.2}$$

The instantaneous signal power and interference power on the ith branch is denoted by $S_i$ and $I_i$, respectively. These are related to $r_i$ and $q_{ij}$ as follows:

$$S_i = \frac{r_i^2}{2} \tag{8.3}$$

and

$$I_i = \frac{1}{2}\sum_{j=1}^{K} q_{ij}^2 \tag{8.4}$$

The received signal from each channel is multiplied by a complex weight before combining. Let $w_i^*$ denote the weight of the ith channel. It then follows from Figure 8.1 that the combiner output y(t) is given by

$$y(t) = \sum_{i=1}^{L} w_i^* x_i(t) \tag{8.5}$$

where * denotes the conjugate of the complex quantity.

Define an L-dimensional complex vector $\mathbf{C}_S$, referred to as the signal channel gain vector, to denote the instantaneous channel gains for the desired signals received on L branches, that is,

$$\mathbf{C}_{S_i} = \alpha_i e^{j\theta_i}, \quad i = 1, 2, \ldots, L \tag{8.6}$$

where $\alpha_i$ and $\theta_i$ are the magnitude and the phase of the channel gain of the ith branch, as discussed in Section 7.2.

Similarly, define interference channel gain vectors $\mathbf{C}_{Ij}$, j = 1, 2, …, K to denote KL channel gains. It should be noted that due to independent assumptions for the signal and various interference envelopes, $\mathbf{C}_S$ and $\mathbf{C}_{Ij}$, j = 1, 2, …, K are mutually independent.

Let an L-dimensional vector be defined as

$$\mathbf{x}(t) = \left[x_1(t), x_2(t), \ldots, x_L(t)\right]^T \tag{8.7}$$

It follows from (8.1) and the discussion in Section 7.2 that $\mathbf{x}(t)$ can be written as

$$\mathbf{x}(t) = \sqrt{p_S}\, g(t)\mathbf{C}_S + \sum_{j=1}^{K} \sqrt{p_{ij}}\, g_j(t)\mathbf{C}_{Ij} + \mathbf{n}(t) \tag{8.8}$$

where the L-dimensional vector $\mathbf{n}(t)$ denotes the noise on L channels, $p_S$ denotes the mean signal power, and $p_{ij}$ denotes the mean power of the jth interference.

Let R denote the array correlation matrix. It follows from (8.8) that for given $\mathbf{C}_S$ and $\mathbf{C}_{Ij}$, j = 1, 2, …, K, it is given by

$$R = p_S \mathbf{C}_S \mathbf{C}_S^H + \sum_{j=1}^{K} p_{Ij} \mathbf{C}_{Ij} \mathbf{C}_{Ij}^H + NI \tag{8.9}$$

Defining

$$\mathbf{w} = \left[w_1, w_2, \ldots, w_L\right]^T \tag{8.10}$$

it follows from (8.5) that the output of an L-branch combiner can be written in vector notation as

$$y(t) = \mathbf{w}^H \mathbf{x}(t) \tag{8.11}$$

The way that the weights on various branches are selected determines the type of diversity combiner being employed. Several of these diversity schemes are now considered.

## 8.1 Selection Combiner

In this case, one of the L-diversity signals is selected for further processing. Thus,

$$w_\ell = \begin{cases} 1 & \text{if } \ell = \ell_0 \\ 0 & \text{otherwise} \end{cases} \tag{8.1.1}$$

where $\ell_0$ denotes the selected branch. Theoretically, one would like to select the branch with the highest signal to noise ratio or in the interference limited system, with the highest signal to co-channel interference ratio. In practice, however, it is easy to implement a scheme that selects a branch with the largest power.

Now we analyze the performance of a system using L branch selection combining scheme. Both noise limited and interference limited systems are considered [Abu92, Abu94, Abu94b, Cha79, Sim99].

### 8.1.1 Noise-Limited Systems

The analysis of noise-limited systems consists of deriving expressions for the outage probability, the mean signal-to-noise ratio (SNR) and the average BER.

#### *8.1.1.1 Rayleigh Fading Environment*

First, consider that the system operates in the Rayleigh fading environment.

##### *8.1.1.1.1 Outage Probability*

Denoting the instantaneous SNR at the $\ell$th branch by $\gamma_\ell$, it follows from (7.3.13) that

$$P(\gamma_\ell \le \gamma_0) = 1 - e^{-\frac{\gamma_0}{\Gamma}} \tag{8.1.2}$$

where $\Gamma$ denotes the mean SNR at each branch.

Let $P^o_{sc}$ denote the outage probability of the selection combiner (SC). Then $P^o_{sc}$ is the probability that the instantaneous SNR in all L branches is simultaneously less than or equal to $\gamma_0$. Assuming that the fading on each branch is independent, it follows from (8.1.2) that

$$\begin{aligned} P^o_{sc} &= P(\gamma_1, \gamma_2, \ldots, \gamma_L \le \gamma_0) \\ &= P(\gamma_1 \le \gamma_0)P(\gamma_2 \le \gamma_0)\ldots P(\gamma_L \le \gamma_0) \\ &= \left(1 - e^{-\frac{\gamma_0}{\Gamma}}\right)^L \end{aligned} \tag{8.1.3}$$

##### *8.1.1.1.2 Mean SNR*

Let $\Gamma_{SC}$ denote the mean SNR of an L-branch selection combiner. An expression for the mean SNR $\Gamma_{SC}$ may be obtained as follows.

The mean SNR is given by

$$\Gamma_{sc} = \int_0^\infty \gamma f_{\gamma SC}(\gamma) d\gamma \tag{8.1.4}$$

where $f_{\gamma SC}$ denotes the pdf of the instantaneous SNR of the received signal using the L-branch SC. It is related to the cdf of $\gamma$ by

$$f_{\gamma SC}(\gamma) = \frac{dF_{\gamma SC}(\gamma)}{d\gamma} \tag{8.1.5}$$

Noting from (7.3.4) that $P^o = F_\gamma(\gamma_0)$, it follows from (8.1.3) that

$$F_{\gamma SC}(\gamma) = \left(1 - e^{-\frac{\gamma}{\Gamma}}\right)^L \tag{8.1.6}$$

From (8.1.5) and (8.1.6),

$$\begin{aligned} f_{\gamma SC}(\gamma) &= \frac{L}{\Gamma}\left(1 - e^{-\frac{\gamma}{\Gamma}}\right)^{L-1} e^{-\frac{\gamma}{\Gamma}} \\ &= \frac{L}{\Gamma}\sum_{k=0}^{L-1}(-1)^k \binom{L-1}{k} e^{-\frac{\gamma}{\Gamma}(k+1)} \end{aligned} \tag{8.1.7}$$

The second step follows using the binomial expansion, that is,

$$(1-x)^n = \sum_{k=0}^{n}(-1)^k \binom{n}{k} x^k \tag{8.1.8}$$

Substituting from (8.1.7) in (8.1.4) and evaluating the integral [Jak74],

$$\Gamma_{sc} = \Gamma \sum_{\ell=1}^{L} \frac{1}{\ell} \tag{8.1.9}$$

It follows from (8.1.9) that the mean SNR of the processor becomes improved by using an L-branch SC. The improvement factor is given by $\sum_{\ell=1}^{L} \frac{1}{\ell}$.

#### 8.1.1.1.3 *Average BER*

The average BER at the output of SC can be obtained by averaging the conditional BER for a given $\gamma$ over all values of $\gamma$. Thus,

$$P_e^{sc} = \int_0^\infty P_e(\gamma) f_{\gamma SC}(\gamma)\, d\gamma \tag{8.1.10}$$

where $P_e(\gamma)$ denotes the conditional BER at the output of SC for a given value of $\gamma$ for a particular modulation scheme, and $f_{\gamma SC}(\gamma)$ denotes the pdf of $\gamma$.

Consider an example of a coherent BPSK system. For this case, $P_e(\gamma)$ is given by (7.3.55) and in Rayleigh fading environment with independent fading, $f_{\gamma SC}(\gamma)$ is given by (8.1.7). Substituting these values and evaluating the integral using the identities

$$\begin{aligned} \int_0^\infty \operatorname{erfc}\left(\sqrt{x}\right) e^{-\alpha x} dx &= \frac{1}{\alpha}\left[1 - \frac{1}{\sqrt{1+\alpha}}\right] \\ \int_0^\infty \operatorname{erfc}\left(\sqrt{x}\right) x e^{-\alpha x} dx &= \frac{1}{\alpha^2}\left[1 - \frac{1}{\sqrt{1+\alpha}} - \frac{\alpha}{2(1+\alpha)\sqrt{1+\alpha}}\right] \end{aligned} \tag{8.1.11}$$

one obtains [Eng96]

$$P^{sc}_{e,BPSK} = \frac{L}{2}\sum_{k=0}^{L-1}\binom{L-1}{k}(-1)^k \frac{1}{1+k}\left[1-\frac{1}{\sqrt{1+\frac{1+k}{\Gamma}}}\right] \tag{8.1.12}$$

Similarly, the result for DPSK may be obtained and is given by [Eng96]

$$P^{sc}_{e,DPSK} = \frac{L}{2}\sum_{k=0}^{L-1}\binom{L-1}{k}(-1)^k \frac{1}{\Gamma+1+k} \tag{8.1.13}$$

Note that the results for CFSK and NCFSK may be obtained by replacing $\Gamma$ by $\Gamma/2$ in (8.1.12) and (8.1.13), respectively.

#### *8.1.1.2 Nakagami Fading Environment*

First, consider the pdf of the SNR at the output of the selection combiner.

##### *8.1.1.2.1 Output SNR pdf*

An expression for the pdf of $\gamma$ at a single branch is given by (7.3.14). Thus, it follows that (7.3.14) denotes the pdf of $\gamma$ at a branch of the SC. Rewrite (7.3.14):

$$f_\gamma(\gamma) = \left(\frac{m}{\Gamma}\right)^m \frac{\gamma^{m-1}}{\Gamma(m)} e^{-\frac{m\gamma}{\Gamma}} \tag{8.1.14}$$

It follows from (7.3.17) that the corresponding expression for $F_\gamma(\gamma)$ is given by

$$F_\gamma(\gamma) = 1 - e^{-m\frac{\gamma}{\Gamma}}\sum_{k=0}^{m-1}\frac{(m\gamma/\Gamma)^k}{k!} \tag{8.1.15}$$

Assuming that the fading on each branch is independent, the cdf at the output of the SC is the product of individual cdfs, as denoted in (8.1.6) for the Rayleigh fading case. Thus,

$$F_{\gamma SC}(\gamma) = (F_\gamma(\gamma))^L \tag{8.1.16}$$

The pdf of $\gamma_{SC}$ then is

$$f_{\gamma SC}(\gamma) = \frac{dF_{\gamma SC}(\gamma)}{df} \tag{8.1.17}$$

Substituting from (8.1.16), it follows that

$$f_{\gamma SC}(\gamma) = \frac{d(F_\gamma(\gamma))^L}{df} \tag{8.1.18}$$

$$= L\left(F_\gamma(\gamma)\right)^{L-1}\frac{dF_\gamma(\gamma)}{df}$$

$$= L\left(F_\gamma(\gamma)\right)^{L-1} f_\gamma(\gamma)$$

which, along with (8.1.14) and (8.1.15), implies that

$$f_{\gamma SC}(\gamma) = L\left(\frac{m}{\Gamma}\right)^m \frac{\gamma^{m-1}}{\Gamma(m)} e^{-\frac{m\gamma}{\Gamma}}\left(1 - e^{-m\frac{\gamma}{\Gamma}}\sum_{k=0}^{m-1}\frac{(m\gamma/\Gamma)^k}{k!}\right)^{L-1} \tag{8.1.19}$$

Using binomial expansion (8.1.8), (8.1.19) can be expressed as [Abu94b]

$$\begin{aligned} f_{\gamma CD}(\gamma) &= L\left(\frac{m}{\Gamma}\right)^m \frac{\gamma^{m-1}}{\Gamma(m)} e^{-m\frac{\gamma}{\Gamma}}\sum_{i=0}^{L-1}\binom{L-1}{i}(-1)^i\left(e^{-m\frac{\gamma}{\Gamma}}\sum_{k=0}^{m-1}\frac{\left(m\frac{\gamma}{\Gamma}\right)^k}{k!}\right)^i \\ &= L\left(\frac{m}{\Gamma}\right)^m \frac{1}{\Gamma(m)}\sum_{i=0}^{L-1}\binom{L-1}{i}(-1)^i\sum_{j\epsilon B} e^{-m(i+1)\frac{\gamma}{\Gamma}} d_{ji}\left(\frac{m}{\Gamma}\right)^{C_{ji}}\frac{\gamma^{C_{ji}+m-1}}{A_{ji}} \end{aligned} \tag{8.1.20}$$

where B is a set of all possible nonnegative integer combinations such that

$$\sum_{k=0}^{m-1} n_k = i \tag{8.1.21}$$

$$C_{ji} = n_1 + 2n_2 + \ldots + (m-1)n_{m-1} \tag{8.1.22}$$

$$A_{ji} = (2!)^{n_2}(3!)^{n_3}\ldots((m-1)!)^{n_{m-1}} \tag{8.1.23}$$

and

$$d_{ji} = \frac{i!}{n_0!n_1!n_2!\ldots n_{m-1}!} \tag{8.1.24}$$

Now, consider a dual diversity system (L = 2) operating in a Nakagami fading environment with fading parameter m [Sim99]. For this case, an expression for the pdf of the SNR at the output of the SC is given by

$$\begin{aligned} f_{\gamma SC}(\gamma) &= \frac{m^m}{\Gamma(m)\Gamma_1}\left(\frac{\gamma}{\Gamma_1}\right)^{m-1}\exp\left(-\frac{m\gamma}{\Gamma_1}\right)\left[1 - \frac{\Gamma\left(m, \frac{m\gamma}{\Gamma_2}\right)}{\Gamma(m)}\right] \\ &+ \frac{m^m}{\Gamma(m)\Gamma_2}\left(\frac{\gamma}{\Gamma_2}\right)^{m-1}\exp\left[-\frac{m\gamma}{\Gamma_2}\right]\left[1 - \frac{\Gamma\left(m, \frac{m\gamma}{\Gamma_1}\right)}{\Gamma(m)}\right] \end{aligned} \tag{8.1.25}$$

where $\Gamma_1$ and $\Gamma_2$ denote the mean SNR at Branch 1 and Branch 2, respectively, and $\Gamma(m,x)$ is the incomplete gamma function.

For the integer value of the fading parameter m, $\Gamma(m,x)$ has a closed-form solution, and thus (8.1.25) becomes

$$\begin{aligned} f_{\gamma SC}(\gamma) = &\frac{m^m}{(m-1)!\Gamma_1}\left(\frac{\gamma}{\Gamma_1}\right)^{m-1}\exp\left(-\frac{m\gamma}{\Gamma_1}\right)\left[1-H(\gamma,\Gamma_2,m)\right] \\ &+\frac{m^m}{(m-1)!\Gamma_2}\left(\frac{\gamma}{\Gamma_2}\right)^{m-1}\exp\left(-\frac{m\gamma}{\Gamma_2}\right)\left[1-H(\gamma,\Gamma_1,m)\right] \end{aligned} \tag{8.1.26}$$

where

$$H(\gamma,\Gamma_i,m) = \exp\left(-\frac{m\gamma}{\Gamma_i}\right)\sum_{k=0}^{m-1}\frac{\left(\frac{m\gamma}{\Gamma_i}\right)^k}{k!}, \quad i = 1, 2 \tag{8.1.27}$$

### 8.1.1.2.2 *Outage Probability*

It follows from (8.1.15) and (8.1.16) that the outage probability in the presence of Nakagami fading is given by

$$\begin{aligned} P^o_{\gamma SC} &= F_{\gamma SC}(\gamma_0) \\ &= \left(F_\gamma(\gamma_0)\right)^L \\ &= \left(1 - e^{-m\frac{\gamma_0}{\Gamma}}\sum_{k=0}^{m-1}\frac{(m\gamma_0/\Gamma)^k}{k!}\right)^L \end{aligned} \tag{8.1.28}$$

For a dual-diversity system, an expression can be derived for $F_{\gamma SC}(\gamma)$, the cdf of the SNR at the output of the SC in the Nakagami fading environment for integer values of m, by integrating (8.1.26). It is given by [Sim99]

$$\begin{aligned} F_{\gamma SC}(\gamma) = &1 - H(\gamma,\Gamma_1,m) + 1 - H(\gamma,\Gamma_2,m) \\ &-\sum_{n=0}^{m-1}\frac{(n+m-1)!}{n!(m-1)!}\left[\frac{(\Gamma_1)^n(\Gamma_2)^m + (\Gamma_1)^m(\Gamma_2)^n}{(\Gamma_1+\Gamma_2)^{n+m}}\right]\left[1-H_n\left(\gamma,\frac{\Gamma_1\Gamma_2}{\Gamma_1+\Gamma_2},m\right)\right] \end{aligned} \tag{8.1.29}$$

where

$$H_n(\gamma,\Gamma,m) = \exp\left(-\frac{m\gamma}{\Gamma}\right)\sum_{k=0}^{m+n-1}\frac{\left(\frac{m\gamma}{\Gamma}\right)^k}{k!} \tag{8.1.30}$$

and $H_0(\gamma,\Gamma,m)$ is equal to $H(\gamma,\Gamma,m)$ given by (8.1.27). The outage probability $P^o_{sc}$ is then given by

$$P^o_{sc} = F_{\gamma SC}(\gamma_0) \tag{8.1.31}$$

*8.1.1.2.3 Average BER*

Using the conditional BER given by (7.3.58) for an NCFSK system and the pdf given by (8.1.20) in Nakagami fading environment (integer values of m), the average BER can be obtained using (7.3.5). An expression is given by [Abu94b]

$$P_e^{SC} = \frac{L}{2}\frac{1}{\Gamma(m)}\sum_{i=0}^{L-1}(-1)^i\binom{L-1}{i}\sum_{j\in B} d_{ji}\frac{\Gamma(C_{ji}+m)}{A_{ji}}\left(\frac{m}{m(i+1)+.5\Gamma}\right)^{C_{ji+m}} \tag{8.1.32}$$

The result for a DPSK system is given by replacing $\Gamma$ with $2\Gamma$ in (8.1.32). For a dual-diversity system, the average BER can also be obtained using the pdf given by (8.1.26). An expression for the average BER for a DPSK system is given by [Sim99]

$$P_e^{sc} = \frac{1}{2}\sum_{i=1}^{2}\left(\frac{m}{m+\Gamma_i}\right)^m - \frac{1}{2}\sum_{n=0}^{m-1}\frac{(n+m-1)!}{n!(m-1)!}\left[\frac{(\Gamma_1)^n(\Gamma_2)^m + (\Gamma_1)^m(\Gamma_2)^n}{(\Gamma_1+\Gamma_2)^{n+m}}\right]\left(\frac{m}{m+\frac{\Gamma_1\Gamma_2}{\Gamma_1+\Gamma_2}}\right)^{m+n} \tag{8.1.33}$$

The result for NCFSK can be obtained from (8.1.33) by replacing $\Gamma_i$ with $\Gamma_i/2$, i = 1, 2.

### 8.1.2 Interference-Limited Systems

Assume that a desired signal and K co-channel interferences are present in a Nakagami fading environment. Assume that all interferences have the same statistics with the fading parameter denoted by m, the same as the desired signal, and have equal mean power denoted by $\bar{I}$.

Now, selection combiner performance using three possible selection algorithms [Abu92] — desired signal power algorithm, total power algorithm and signal-to-interference (SIR) power algorithm — is presented.

#### *8.1.2.1 Desired Signal Power Algorithm*

In this algorithm, the selection combiner selects the branch with the largest desired signal power. Let $S_i$ denote the signal power of the ith branch, that is,

$$S_i = \frac{r_i^2}{2} \tag{8.1.34}$$

When all branches experience independent fading, it follows from (7.1.54) that the pdf of $S_i$ is given by

$$F_{S_i}(S) = \left(\frac{m}{\bar{S}}\right)^m \frac{S^{m-1}}{\Gamma(m)}\exp\left(-\frac{mS}{\bar{S}}\right) \tag{8.1.35}$$

where $\bar{S}$ denotes the mean signal power of each branch.

Let $I_i$ denote the total interference power received on the ith branch, that is,

$$I_i = \frac{1}{2}\sum_{j=1}^{K} q_{ij}^2 \tag{8.1.36}$$

The pdf of $I_i$, as seen in (7.3.22), is given by

$$f_{I_i}(I) = \left(\frac{m}{\bar{I}}\right)^{mK} \frac{I^{mK-1}}{\Gamma(mK)} \exp\left(-\frac{mI}{\bar{I}}\right) \tag{8.1.37}$$

Let $\mu_{sc}$ denote the SIR power ratio at the output of the selection combiner. As the SC selects only one branch for processing, $\mu_{sc}$ also denotes the SIR at the selected branch. The probability that $\mu_{sc}$ is less than or equal to the threshold $\mu_0$ is given by

$$\begin{aligned} P_{sc}^{o} &= P[\mu_{sc} \le \mu_0] \\ &= \int_0^{\mu_0} f_{\mu_{sc}}(\mu) d\mu \end{aligned} \tag{8.1.38}$$

where $f_{\mu_{sc}}(\mu)$ denotes the pdf of the SIR at the output of the SC, and is given by [Abu92]

$$f_{\mu_{sc}}(\mu) = \frac{LA}{\mu^{mK+1}} \sum_{i=0}^{L-1} \sum_{j \in B} (-1)^i \binom{L-1}{i} \frac{d_{ji}}{A_{ji}} \left(\frac{m}{\bar{S}}\right)^{C_{ji}} \frac{\Gamma(m + mK + C_{ji})}{\left[\frac{m}{\bar{I}\mu_0} + \frac{m(i+1)}{\bar{S}}\right]^{m+mK+C_{ji}}} \tag{8.1.39}$$

where

$$A = \frac{1}{\Gamma(m)\Gamma(mK)} \left(\frac{m}{\bar{S}}\right)^m \left(\frac{m}{\bar{I}}\right)^{mK} \tag{8.1.40}$$

B, $C_{ji}$, $A_{ji}$, and $d_{ji}$ are given by (8.1.21) to (8.1.24).

Substituting for $f_{\mu_{sc}}(\mu)$ in (8.1.38) and carrying out the integral [Abu92],

$$\begin{aligned} P_{sc}^{o} = \frac{L}{\Gamma(m)\Gamma(mK)} \sum_{i=0}^{L-1} \sum_{j \in B} \sum_{t=0}^{m+C_{ij}-1} (-1)^{-i} \binom{L-1}{i} \frac{d_{ji}}{A_{ji}} \Gamma(m + mK + C_{ij}) \\ \binom{m + C_{ij} - 1}{t} \frac{(-1)^t}{(i+1)^{m+mK+C_{ij}}} \left(\frac{\tilde{\mu}}{\mu_0}\right)^{mk} \frac{(\eta)^t}{t+mK} \left\{\frac{1}{(\eta)^{mK+t}} - \frac{1}{(1+\eta)^{mK+t}}\right\} \end{aligned} \tag{8.1.41}$$

where

$$\eta = \frac{\tilde{\mu}}{\mu_0(i+1)} \tag{8.1.42}$$

with $\tilde{\mu}$ denoting the average signal power $\bar{S}$ to average interference power $\bar{I}$ ratio at one branch.

For the Rayleigh fading environment (m = 1), (8.1.41) becomes

$$P_{sc}^{o} = L\sum_{i=0}^{L-1}(-1)^{i}\binom{L-1}{i}\frac{1}{(i+1)^{K+1}}\left(\frac{\tilde{\mu}}{\mu_0}\right)^{K}\left\{\frac{1}{(\eta)^{K}}-\frac{1}{(1+\eta)^{K}}\right\} \tag{8.1.43}$$

which, for a single interference (K = 1), reduces to

$$P_{sc}^{o} = L\sum_{i=0}^{L-1}(-1)^{i}\binom{L-1}{i}\frac{1}{(i+1)+\dfrac{\tilde{\mu}}{\mu_0}} \tag{8.1.44}$$

Note that for L = 1, (8.1.44) reduces to (7.3.35), the result for the single-antenna system.

### 8.1.2.2 Total Power Algorithm

In this algorithm, the branch with the largest total power is selected. Thus, Branch 1 is selected if

$$S_1 + I_1 \geq S_j + I_j \qquad j = 2, \ldots, L \tag{8.1.45}$$

For this case, the outage probability is given by

$$\begin{aligned} P_{sc}^{o} &= P[\mu_{sc} \leq \mu_0] \\ &= LP\left[\frac{S_1}{I_1} \leq \mu_0,\ S_1 + I_1 \geq S_j + I_j,\ \ j = 2, \ldots, L\right] \end{aligned} \tag{8.1.46}$$

This expression is evaluated in [Abu 92], resulting in

$$P_{sc}^{o} = L\int_0^{\mu_0} du \int_0^{\infty} f_{uv}(u,v)\left(P[S_i + I_i \leq v]\right)^{L-1} dv \tag{8.1.47}$$

where $f_{uv}$ is the joint pdf of u and v, given by

$$f_{uv}(u,v) = A\frac{u^{m-1}v^{m+mK-1}}{(1+u)^{m+mK}}\exp\left[\frac{-mv}{1+u}\left(\frac{u}{\bar{S}}+\frac{1}{\bar{I}}\right)\right] \tag{8.1.48}$$

with A given by (8.1.40) and

$$u = \frac{S_1}{I_1} \tag{8.1.49}$$

$$v = S_1 + I_1 \tag{8.1.50}$$

and

$$P[S_i + I_i \le v] = 1 - \exp\left(-\frac{mv}{\bar{I}}\right)\sum_{t=0}^{mK-1}\left\{\frac{(mv/\bar{I})^t}{t!}\right.$$

$$-\exp\left(-\frac{mv}{\bar{S}}\right)\sum_{j=0}^{m-1}\sum_{n=0}^{j}\left\{\frac{(-1)^{j-n}}{j!}\binom{j}{n}\left(\frac{mv}{\bar{S}}\right)^n\right.$$

$$\left.\cdot\frac{\Gamma(mK+j-n)}{\Gamma(mK)}\frac{(\tilde{\mu})^{mK}}{(\tilde{\mu}-1)^{mK+j-n}}\right\}$$

$$+\exp\left(-\frac{mv}{\bar{I}}\right)\sum_{j=0}^{m-1}\sum_{n=0}^{j}\sum_{z=0}^{mK+j-n-1}\binom{j}{n}$$

$$\cdot\left\{\frac{(-1)^{j-n}}{j!}\left(\frac{mv}{\bar{S}}\right)^{n+z}\frac{\Gamma(mK+j-n)}{\Gamma(mK)z!}\frac{(\tilde{\mu})^{mK}}{(\tilde{\mu}-1)^{mK+j-n-z}}\right\} \tag{8.1.51}$$

For dual-branch diversity systems, (8.1.47) reduces to

$$P_{sc}^{o} = 2(R_1 - R_2 - R_3 + R_4) \tag{8.1.52}$$

where

$$R_1 = \frac{\Gamma(m+mK)}{\Gamma(m)\Gamma(mK)}\sum_{i=0}^{m-1}\binom{m-1}{i}\frac{(-1)^{m-1-i}}{i-m-mK+1}\left[\left(\frac{\mu_0}{\tilde{\mu}}+1\right)^{i+1-m-mK}-1\right] \tag{8.1.53}$$

$$R_2 = \frac{1}{\Gamma(m)\Gamma(mK)}\sum_{t=0}^{mK-1}\frac{\Gamma(m+mK+t)}{t!}(\tilde{\mu})^{-m}\int_0^{\mu_0}\frac{u^{m-1}(u+1)^t}{\left[u\left(1+\frac{1}{\tilde{\mu}}\right)+2\right]^{m+mK+t}}du \tag{8.1.54}$$

$$R_3 = \frac{1}{\Gamma(m)\Gamma(mK)}\sum_{j=0}^{m-1}\sum_{n=0}^{j}\frac{\Gamma(mK+j-n)}{j!\Gamma(mK)}\Gamma(n+mK+m)(\tilde{\mu})^{2mK}(-1)^{j-n}$$

$$\cdot\binom{j}{n}(\tilde{\mu}-1)^{n-j-mK}\int_0^{\mu_0}\frac{u^{m-1}(u+1)^n}{[2u+\tilde{\mu}+1]^{n+mK+m}}du \tag{8.1.55}$$

and

$$R_4 = \frac{1}{\Gamma(m)\Gamma(mK)}\sum_{j=0}^{m-1}\sum_{n=0}^{j}\sum_{z=0}^{mK+j-n-1}\frac{\Gamma(j+mK-n)}{j!z!\Gamma(mK)}\Gamma(n+z+mK+m)$$

$$\cdot(-1)^{j-n}\binom{j}{n}(\tilde{\mu})^{mK-m-z-n}(\tilde{\mu}-1)^{z-mK-j+n}\int_0^{\mu_0}\frac{u^{m-1}(u+1)^{n+z}}{\left[u\left(1+\frac{1}{\tilde{\mu}}\right)+2\right]^{m+mK+n+z}}du \tag{8.1.56}$$

#### *8.1.2.3 SIR Power Algorithm*

In this algorithm, a diversity branch with the highest signal power to interference power ratio is selected and an expression for outage probability is given by [Abu 92]

$$\begin{aligned} P_{sc}^{o} &= P\left[\mu_{sc} \le \mu_0\right] \\ &= \left\{P\left[\mu_{sc} \le \mu_0 \text{ and } L = 1\right]\right\}^{L} \\ &= \left[\left(\frac{\tilde{\mu}}{\mu_0}+1\right)^{-m} \sum_{i=0}^{mK-1} \binom{m+i-1}{i} \left(1+\frac{\mu_0}{\tilde{\mu}}\right)^{-i}\right]^{L} \end{aligned} \tag{8.1.57}$$

For the Rayleigh fading environment (m = 1) and one interference (K = 1), this reduces to

$$P_{sc}^{o} = \left(\frac{\mu_0}{\mu_0+\tilde{\mu}}\right)^{L} \tag{8.1.58}$$

## 8.2 Switched Diversity Combiner

The switched diversity scheme, also known as scanning diversity, is similar to the selection diversity discussed in the previous section except that in this scheme signals received on L branches are continuously scanned in a fixed sequence until one is found above a given threshold, rather than using the best one as is done in selection diversity. For example, when the total received power is considered, the received power of the selected branch is continuously compared with a given threshold value, $\xi_0$. Until the received power remains above $\xi_0$, no switching is done. When it drops below $\xi_0$, the next branch is examined and switched to the receiver if the received power on this branch is found to be above $\xi_0$; otherwise, the search continues.

In this section, expressions for the outage probability and average BER in the Nakagami fading environment are derived, and the effect of correlation on average BER is examined [Abu92, Abu94a, Sch72].

### 8.2.1. Outage Probability

The outage probability for this scheme in the presence of Nakagami distributed interferences with statistics similar to those described in Section 8.1 can be written as [Abu92, Sch72]

$$\begin{aligned} P_{sw}^{o} &= P\left[\mu_{SW} \le \mu_0\right] \\ &= P\left[\mu_{SW} \le \mu_0 \middle| S_i + I_i < \xi_0, \text{ for all } i\right] P\left[S_i + I_i < \xi_0, \text{ for all } i\right] \\ &\quad + P\left[\mu_{SW} \le \mu_0 \middle| S_i + I_i \ge \xi_0, \text{ for at least one } i\right] P\left[S_i + I_i \ge \xi_0, \text{ for at least one } i\right] \end{aligned} \tag{8.2.1}$$

where $\mu_{SW}$ denotes the SIR at the output of a switched diversity combiner (SDC), which is the same as the SIR at the selected branch.

Now, consider the various terms on the right side of (8.2.1). When $S_i + I_i \le \xi_0$ for all i,

$$P[\mu_{SW} \le \mu_0] = 1 \tag{8.2.2}$$

as none of the branches is selected, effectively representing an outage.

From the independence of the desired signal and interferences, it follows that

$$P[S_i + I_i < \xi_0, \text{ for all i}] = \{P[S_1 + I_1 < \xi_0]\}^L \tag{8.2.3}$$

and

$$\begin{aligned} P[S_i + I_i \ge \xi_0 \text{ for at least one i}] &= 1 - P[S_i + I_i < \xi_0, \text{ for all i}] \\ &= 1 - \{P[S_1 + I_1 < \xi_0]\}^L \end{aligned} \tag{8.2.4}$$

In writing (8.2.4), it is assumed without any loss of generality that Branch 1 is selected.

It follows from (8.2.1) to (8.2.4) that

$$\begin{aligned} P_{sw}^{o} &= \{P[S_1 + I_1 < \xi_0]\}^L \\ &+ \left[\frac{S_1}{I_1} \le \mu_0 \middle| S_1 + I_1 \ge \xi_0\right]\left(1 - \{P[S_1 + I_1 < \xi_0]\}^L\right) \end{aligned} \tag{8.2.5}$$

A further manipulation of (8.2.5) leads to [Abu92]

$$P_{sw}^{o} = \{P[S_1 + I_1 < \xi_0]\}^L + P_1 \frac{\left\{1 - \left(P[S_1 + I_1 < \xi_0]\right)^L\right\}}{1 - P[S_1 + I_1 < \xi_0]} \tag{8.2.6}$$

where

$$\begin{aligned} P_1 = &\frac{\Gamma(m + mK)}{\Gamma(m)\Gamma(mK)} (\tilde{\mu})^{mK} \sum_{j=0}^{m+mK-1} \left(\frac{m\xi_0}{2\bar{S}}\right)^j \frac{1}{j!} \\ &\cdot \int_0^{\mu_0} \frac{(u + \tilde{\mu})^{j-m-mK} u^{m-1}}{(1+u)^K} \exp\left(\frac{-m\xi_0}{(1+u)2\bar{S}}(u + \tilde{\mu})\right) du \end{aligned} \tag{8.2.7}$$

and $P[S_1 + I_1 < \xi_0]$ is given by (8.1.51).

### 8.2.2 Average Bit Error Rate

In this section, average BER is examined by considering an example of noncoherent detection of binary FSK (NCFSK) signals in additive white Gaussian noise (AWGN) for a two-branch diversity system in a slow Nakagami fading environment [Abu94a].

Assume that the switching is done at discrete intervals of time t = nT, where n denotes an integer and T is the time interval between samples. Let $a_n$ and $b_n$ denote the samples of the signal envelopes at two antennas at time t = nT, and $X_n = 1/2\ a_n^2$ and $Y_n = 1/2\ b_n^2$

denote their respective local signal powers. In the Nakagami fading environment, $a_n$ and $b_n$ are Nakagami distributed RVs, and $X_n$ and $Y_n$ are gamma distributed.

Assume the following switching scheme is employed. Let the antenna selected at t = (n – 1)T be Number 1. Switching to antenna Number 2 is done iff $X_n < \xi_0$. Next, let $S_n$ denote the local signal power at the output of the switched diversity system. It follows from the above switching strategy that

$$S_n = X_n \text{ iff} \begin{cases} S_{n-1} = X_{n-1} \text{ and } X_n \geq \xi_0 \\ \text{or} \\ S_{n-1} = Y_{n-1} \text{ and } Y_{n-1} < \xi_0 \end{cases} \tag{8.2.8}$$

$S_n = Y_n$ as above with X and Y interchanged.

When the fading at two antennas is independent, the pdf of $S_n$ is given by [Abu94a]

$$f_{S_n}(u) = \begin{cases} B\dfrac{m}{\bar{S}}\dfrac{\left(\dfrac{m}{\bar{S}}u\right)^{m-1}}{(m-1)!}e^{-\frac{m}{\bar{S}}u} & u \leq \xi_0 \\ (1+B)\dfrac{m}{\bar{S}}\dfrac{\left(\dfrac{m}{\bar{S}}u\right)^{m-1}}{(m-1)!}e^{-\frac{m}{\bar{S}}u} & u > \xi_0 \end{cases} \tag{8.2.9}$$

where

$$B = 1 - e^{-\frac{m}{\bar{S}}\xi_0}\sum_{i=0}^{m-1}\frac{\left(\dfrac{m\xi_0}{\bar{S}}\right)^i}{i!} \tag{8.2.10}$$

Let $\gamma_n$ denote the instantaneous SNR at the output of the system at time t = nT, defined as

$$\gamma_n = \frac{S_n}{N} \tag{8.2.11}$$

with N denoting the variance (noise power) of zero mean AWGN.

Following a procedure similar to that used in Section 7.3.1, it can easily be shown that when the fading at two antennas is independent, the pdf of $\gamma_n$ obtained from the pdf of $S_n$ is given by

$$f_{\gamma_n}(\gamma) = \begin{cases} B\dfrac{m}{\Gamma}\dfrac{\left(\dfrac{m\gamma}{\Gamma}\right)^{m-1}}{(m-1)!}e^{-\frac{m\gamma}{\Gamma}} & \gamma \leq \dfrac{\xi_0}{N} \\ (1+B)\dfrac{m}{\Gamma}\dfrac{\left(\dfrac{m\gamma}{\Gamma}\right)^{m-1}}{(m-1)!}e^{-\frac{m\gamma}{\Gamma}} & \gamma > \dfrac{\xi_0}{N} \end{cases} \tag{8.2.12}$$

Substituting for the pdf of the instantaneous SNR given by (8.2.12) and the conditional probability of error for the NCFSK system given by (7.3.58) in (7.3.5), and evaluating the integral for the average BER of the NCFSK in the Nakagami fading environment becomes [Abu94a]

$$P_e^{sw} = \frac{1}{2\left(\frac{\Gamma}{2m}+1\right)^m}\left[1 - e^{-\frac{m\xi_0}{\Gamma}}\sum_{i=0}^{m-1}\frac{\left(\frac{m\xi_0}{\Gamma}\right)^i}{i!} + e^{-\xi_0\left(\frac{1}{2}+\frac{m}{\Gamma}\right)}\sum_{j=0}^{m-1}\frac{\xi_0^j\left(\frac{1}{2}+\frac{m}{\Gamma}\right)^j}{j!}\right] \tag{8.2.13}$$

It should be noted that $P_e^{sw}$ depends on the threshold value of signal power $\xi_0$ used for switching. The optimum value of the threshold $\ddot{\xi}_0$ may be obtained by solving

$$\left.\frac{dP_e^{sw}}{d\xi_0}\right|_{\xi_0=\ddot{\xi}_0} = 0 \tag{8.2.14}$$

Substituting from (8.2.13) in (8.2.14) and solving for $\hat{\xi}_0$ yields

$$\hat{\xi}_0 = 2m\ln\left(\frac{\Gamma}{2m}+1\right) \tag{8.2.15}$$

For the Rayleigh fading environment, m = 1. Thus, substituting m = 1 in (8.2.13) and (8.2.15), expressions for $P_e^{sw}$ and $\hat{\xi}_0$ in Rayleigh fading channels become

$$P_e^{sw} = \frac{1}{2+\Gamma}\left(1 - e^{-\frac{\xi_0}{\Gamma}} + e^{-\xi_0\left(\frac{1}{2}+\frac{1}{\Gamma}\right)}\right) \tag{8.2.16}$$

and

$$\hat{\xi}_0 = 2\ln\left(\frac{\Gamma}{2}+1\right) \tag{8.2.17}$$

### 8.2.3 Correlated Fading

The expression for average BER given by (8.2.13) is derived when the fading at two branches is independent and there is no correlation between the signal envelopes. Now assume that the two are correlated with the power correlation coefficient $k^2$ defined as

$$k^2 = \frac{E\left[\left(a_n^2 - E\left[a_n^2\right]\right)\left(b_n^2 - E\left[b_n^2\right]\right)\right]}{\sqrt{E\left[a_n^2 - E\left[a_n^2\right]\right]E\left[b_n^2 - E\left[b_n^2\right]\right]}} \tag{8.2.18}$$

The average BER for this case is a function of $k^2$ and is given by [Abu94a]

$$P_e^{sw} = C\frac{\Gamma(m)}{D^m}\left[1 - e^{\xi_0 D}\sum_{i=0}^{m-1}\frac{(\xi_0 D)^i}{i!} + \frac{e^{-\xi_0\left(\frac{1}{2}+\frac{m}{\Gamma}\right)}}{2\left(\frac{\Gamma}{2m}+1\right)^m}\sum_{j=0}^{m-1}\frac{\xi_0^j\left(\frac{1}{2}+\frac{m}{\Gamma}\right)^j}{j!}\right] \tag{8.2.19}$$

where constants C and D are defined as

$$C = \frac{\left(\frac{m}{\Gamma}\right)^{2m}}{2\Gamma(m)\left(\frac{m}{\Gamma} + \frac{1}{2}\left(1-k^2\right)\right)^m} \tag{8.2.20}$$

and

$$D = \frac{m}{\left(1-k^2\right)\Gamma}\left(1 - \frac{mk^2}{m + \frac{1}{2}\left(1-k^2\right)\Gamma}\right) \tag{8.2.21}$$

The optimum threshold $\hat{\xi}_0$, which minimizes the $P_e^{sw}$ for this case, becomes

$$\hat{\xi}_0 = \frac{2}{1 + \frac{2m}{\Gamma} - 2D} \ln\left(\frac{\left(\frac{m}{\Gamma}\right)^m}{2C\Gamma(m)}\right) \tag{8.2.22}$$

For the Rayleigh fading case, substituting m = 1 in (8.2.19) and (8.2.22), expressions for $P_e^{sw}$ and $\hat{\xi}_0$ become

$$P_e^{sw} = \frac{G}{F}\left(1 - e^{-F\xi_0}\right) + \frac{1}{2+\Gamma} e^{-\xi_0\left(\frac{1}{2}+\frac{1}{\Gamma}\right)} \tag{8.2.23}$$

and

$$\hat{\xi}_0 = \frac{2}{1 + \frac{2}{\Gamma} - 2F} \ln\left(\frac{1}{2\Gamma G}\right) \tag{8.2.24}$$

where constants G and F are defined as

$$G = \frac{1}{2\Gamma + \Gamma^2\left(1-k^2\right)} \tag{8.2.25}$$

and

$$F = \frac{1}{\Gamma\left(1-k^2\right)}\left(1 - \frac{k^2}{1 + \frac{1}{2}\left(1-k^2\right)\Gamma}\right) \tag{8.2.26}$$

These equations can be used to evaluate the average BER and optimal threshold values for various fading parameters. Using these equations to plot the optimal threshold as a function of SNR, it is reported in [Abu94a] that $\hat{\xi}_0$ is an increasing function of SNR and the fading parameter m, and a decreasing function of the correlation coefficient.

## 8.3 Equal Gain Combiner

In equal gain combining, the desired signals on all branches are co-phased and equally weighted before summing to produce the output. Without loss of generality, assume that each channel has a unity gain. Thus, the weights of an equal gain combiner (EGC) are given by

$$w_i = e^{j\theta_i} \quad i = 1, 2, \ldots, L \tag{8.3.1}$$

and the signal envelope at the output of the EGC is sum of L signal envelopes, that is,

$$r = \sum_{i=1}^{L} r_i \tag{8.3.2}$$

In this section, the performance of an EGC in both noise-limited and interference-limited environments in the presence of Nakagami fading is analyzed.

### 8.3.1 Noise-Limited Systems

First, consider a noise-limited system [Abu92, Abu94, Bea91, Zha97, Zha99]. In this section, expressions for the mean SNR, outage probability, and average BER for the EGC are derived.

#### *8.3.1.1 Mean SNR*

Let $S_{EG}$ denote the instantaneous signal power at the output of the EGC. It follows from (8.3.2) that it is given by

$$S_{EG} = \frac{r^2}{2} \tag{8.3.3}$$

When each branch has the same noise power N, the total noise power at the output of the EGC is equal to NL, as each channel has a unity gain and the output SNR $\gamma$ is given by

$$\begin{aligned} \gamma &= \frac{S_{EG}}{LN} \\ &= \frac{r^2}{2LN} \end{aligned} \tag{8.3.4}$$

Let $\Gamma_{EG}$ denote the mean SNR at the output of the EGC. Thus, (8.3.4) implies that

$$\begin{aligned} \Gamma_{EG} &= E[\gamma] \\ &= \frac{E[r^2]}{2LN} \end{aligned} \tag{8.3.5}$$

It follows from (8.3.2), assuming independent fading, that

$$
\begin{aligned}
E\left[r^2\right] &= E\left[\left(\sum_{i=1}^{L} r_i\right)^2\right] \\
&= \sum_{i,\,j=1}^{L} E\left[r_i\, r_j\right] \\
&= \sum_{i=1}^{L} E\left[r_i^2\right] + \sum_{i \neq j=1}^{L} E\left[r_i\right] E\left[r_j\right]
\end{aligned}
\tag{8.3.6}
$$

where the last step follows from the independent fading assumption.

Denoting the mean signal power at each branch by $\bar{S}$, it follows that

$$
E\left[r_i^2\right] = 2\bar{S}, \quad i = 1, 2, \ldots, L \tag{8.3.7}
$$

The second term on the RHS of (4.3.6) can be evaluated by noting that

$$
E\left[r_i\right] = \int_0^{\infty} r_i\, f_{r_i}(r_i)\, dr_i\,, \quad i = 1, 2, \ldots, L \tag{8.3.8}
$$

where $f_{r_i}$ denotes the pdf of the signal envelope at the ith branch.

For Nakagami distributed signals, the mean SNR is given by (7.1.43). Substituting in (8.3.8) and evaluating the integral [Pro95],

$$
E\left[r_i\right] = \frac{\Gamma\left(m + \frac{1}{2}\right)}{\Gamma(m)} \left(\frac{2\bar{S}}{m}\right)^{\frac{1}{2}}, \quad i = 1, 2, \ldots, L \tag{8.3.9}
$$

Substituting from (8.3.7) and (8.3.9) in (8.3.6) and using (8.3.5), one obtains the following expression for the mean SNR of the EGC for independent Nakagami fading:

$$
\begin{aligned}
\Gamma_{EG} &= \frac{1}{2LN}\left\{2L\bar{S} + (L^2 - L)\left[\frac{\Gamma\left(m + \frac{1}{2}\right)}{\Gamma(m)}\left(\frac{2\bar{S}}{m}\right)^{\frac{1}{2}}\right]^2\right\} \\
&= \Gamma\left(1 + \left[\frac{\Gamma\left(m + \frac{1}{2}\right)}{\Gamma(m)}\right]^2 \frac{L-1}{m}\right)
\end{aligned}
\tag{8.3.10}
$$

where $\Gamma = \dfrac{\bar{S}}{N}$ denotes the mean SNR of a single branch.

For the Rayleigh fading case, m = 1. Substituting for m = 1 and noting that $\Gamma\left(\frac{3}{2}\right)=\frac{\sqrt{\pi}}{2}$, (8.3.10) becomes

$$\Gamma_{EG}=\Gamma\left(1+(L-1)\frac{\pi}{4}\right) \tag{8.3.11}$$

yielding an expression for the mean SNR of the EGC for independent Rayleigh fading.

#### *8.3.1.2 Outage Probability*

Estimation of outage probability requires knowledge of $F_\gamma$, the cdf of $\gamma$. It can be obtained from $F_r$, the cdf of r as follows. Let $F_X$ and $F_Y$ denote the cdfs of two RVs X and Y. If X and Y are related via

$$Y=X^2, X\geq 0 \tag{8.3.12}$$

then

$$\begin{aligned} F_Y(q) &= P[Y\leq q] \\ &= P[X^2\leq q] \\ &= P[X\leq\sqrt{q}] \\ &= \int_0^{\sqrt{q}} f_x(x)dx \\ &= F_x(\sqrt{q}) \end{aligned} \tag{8.3.13}$$

Thus, it follows from (8.3.4) that $F_\gamma(\gamma)$ is given by

$$F_\gamma(\gamma)=F_r\left(\sqrt{2LN\gamma}\right) \tag{8.3.14}$$

where $F_r(x)$ denotes the cdf of r, the sum of L independent RVs $r_i$, i = 1, 2, ..., L.

When $r_i$, i = 1, 2, ..., L are i.i.d. RVs, and are Nakagami distributed with parameter m, $F_r(x)$ can be computed within a determined accuracy using the following infinite series [Bea90, Bea91]:

$$F_r(x)\cong\frac{1}{2}+\frac{2}{\pi}\sum_{\substack{n=1\\ n\,\text{odd}}}^{\infty}\frac{(A_n)\sin(\theta_n)}{n} \tag{8.3.15}$$

where

$$A_n=\left(\Phi_R^2+\Phi_I^2\right)^{\frac{L}{2}} \tag{8.3.16}$$

$$\theta_n=L\tan^{-1}\left\{\frac{\Phi_I\cos(n\omega\varepsilon)-\Phi_R\sin(n\omega\varepsilon)}{\Phi_R\cos(n\omega\varepsilon)+\Phi_I\sin(n\omega\varepsilon)}\right\} \tag{8.3.17}$$

$$\Phi_R=E[\cos(n\omega x)] \tag{8.3.18}$$

$$\Phi_I = E[\sin n\omega x] \tag{8.3.19}$$

$$\varepsilon = \frac{x}{L} \tag{8.3.20}$$

and

$$\omega = \frac{2\pi}{T} \tag{8.3.21}$$

with x denoting the signal envelope at one of the branches and T denoting the period of the square wave used in deriving the series. T determines the accuracy of the results. A value of T between 40 and 80 has been suggested in [Bea91].

For RVs $r_i$, i = 1, 2, …, L, the Nakagami distributed with parameter m expectations in (8.3.18) and (8.3.19) become

$$\Phi_R = {}_1F_1\left(m; \frac{1}{2}; \frac{-n^2\omega^2\bar{S}}{2m}\right) \tag{8.3.22}$$

and

$$\Phi_I = \sqrt{\frac{2\bar{S}}{m}} \frac{\Gamma(m+0.5)}{\Gamma(m)} n\omega \; {}_1F_1\left(m+\frac{1}{2}; \frac{3}{2}; \frac{-n^2\omega^2\bar{S}}{2m}\right) \tag{8.3.23}$$

where ${}_1F_1(.;.;.)$ denotes the confluent hypergeometric function [Abr72] which is defined as

$${}_1F_1(a;b;x) = \sum_{n=0}^{\infty} \frac{\Gamma(a+n)\Gamma(b)x^n}{\Gamma(a)\Gamma(b+n)n!} \tag{8.3.24}$$

This function can be calculated as follows [Bea91]:

$${}_1F_1\left(m+\frac{1}{2}; \frac{3}{2}; -a\right) = e^{-a} \sum_{k=0}^{m-1} \frac{(-1)^k \binom{m-1}{k} 2^k a^k}{(2k+1)!!} \tag{8.3.25}$$

where

$$(2k+1)!! = (2k+1)(2k-1)\cdots(3)(1)$$

${}_1F_1\left(m; \frac{1}{2}; -a\right)$ can be calculated recursively as follows:

$$\begin{aligned} {}_1F_1\left(m; \frac{1}{2}; -a\right) &= \frac{1}{m-1}\left\{\left(-a + \frac{4m-5}{2}\right) {}_1F_1\left(m-1; \frac{1}{2}; -a\right)\right. \\ &\quad \left. + \frac{3-2m}{2} \, {}_1F_1\left(m-2; \frac{1}{2}; -a\right)\right\}, \qquad m \geq 2 \end{aligned} \tag{8.3.26}$$

$$ {}_1F_1\left(1;\frac{1}{2};-a\right)=e^{-a}\sum_{i=0}^{\infty}\frac{a^i}{(2i-1)i!} \tag{8.3.27} $$

when the parameter a is a small quantity,

$$ {}_1F_1\left(1;\frac{1}{2};-a\right)=\frac{-1}{2a}\left[1+\frac{1.3}{a}+\frac{1.3.5}{(2a)^2}+\frac{1.3.5.7}{(2a)^3}+\dots\right] \tag{8.3.28} $$

when the parameter a is a medium to large quantity, and

$$ {}_1F_1\left(0;\frac{1}{2};-a\right)=1 \tag{8.3.29} $$

For a discussion of the diversity gain obtainable using L = 2, 4, and 8, and m = 1, 2, 3, and 4, see [Bea91]. The discussion concludes that a transmitter requires 11 dB less power using a dual-diversity system in a Rayleigh fading condition (m = 1). The required power decreases as diversity branches (L) increase, whereas an increase in power is required in more severe fading as m increases.

#### 8.3.1.3 Average BER

Calculation of the average BER using (7.3.5) requires knowledge of the SNR pdf for the EGC, which is not available. However, it can be obtained using

$$ P_e=\int_0^{\infty}P_e(r)f_r(r)dr \tag{8.3.30} $$

where r denotes the amplitude of the signal envelope, $f_r$ denotes the pdf of r, and $P_e(r)$ denotes the conditional BER for a given value of r.

When r is the sum of i.i.d. RV, as is assumed to be the case for EGC, $f_r$ is given by [Abu92]

$$ f_r(x)=\frac{2}{T}\sum_{\substack{n=1\\ n\ \text{odd}}}^{\infty}A_n\,e^{j\tau_n}\,e^{-jn\omega x}+A_n\,e^{-j\tau_n}\,e^{jn\omega x} \tag{8.3.31} $$

where

$$ \tau_n=L\tan^{-1}\frac{\Phi_I}{\Phi_R} \tag{8.3.32} $$

$A_n$, $\Phi_R$, $\Phi_I$, and $\omega$ were defined previously.

Substituting for $\gamma$ from (8.3.4), in (7.3.55) to (7.3.58) expressions for the conditional BER as a function of the signal component become

$$ P_e(r)=Q\left(\sqrt{\frac{g}{LN}}\,r\right),\quad \begin{cases} g=1 & \text{BPSK}\\ g=.5 & \text{CFSK}\end{cases} \tag{8.3.33} $$

and

$$P_e(r) = \frac{1}{2}\exp\left(-\frac{gr^2}{2LN}\right), \quad \begin{cases} g = 1 & \text{DPSK} \\ g = .5 & \text{NCFSK} \end{cases} \tag{8.3.34}$$

Substituting for $P_e(r)$ from (8.3.33) and $f_r(r)$ from (8.3.31) in (8.3.30) and carrying out the integral, the average BER becomes [Bea91]

$$P_e = \frac{2}{T}\sum_{\substack{n=0 \\ n\ \text{odd}}}^{\infty} A_n\, B_n \cos(\tau_n - \alpha_n) \tag{8.3.35}$$

where

$$B_n = \left\{\frac{1}{(n\omega)^2}\left[1 - \exp\left(\frac{-n^2\omega^2}{4b^2}\right)\right]^2 + \frac{1}{\pi b^2}\left[{}_1F_1\left(1; \frac{3}{2}; \frac{-n^2\omega^2}{4b^2}\right)\right]^2\right\}^{\frac{1}{2}} \tag{8.3.36}$$

$$\alpha_n = \tan^{-1}\left\{\frac{\sqrt{\pi}\, b\left[1 - \exp\left(\frac{-n^2\omega^2}{4b^2}\right)\right]}{n\omega\ {}_1F_1\left(1; \frac{3}{2}; \frac{-n^2\omega^2}{4b^2}\right)}\right\} \tag{8.3.37}$$

and

$$b = \sqrt{\frac{g}{2NL}} \tag{8.3.38}$$

where $g = 1$ for coherent BPSK and $g = 0.5$ for CFSK.

Similarly, substituting for $P_e(r)$ from (8.3.34) and $f_r(r)$ from (8.3.31) in (8.3.30), and carrying out the integral the average BER for DPSK and NCFSK systems becomes [Bea91]

$$P_e = \frac{2}{T}\sum_{\substack{n=1 \\ n\ \text{odd}}}^{\infty} A_n\, D_n \cos(\tau_n - \beta_n) \tag{8.3.39}$$

where

$$D_n = \frac{1}{2}\left\{\frac{\pi}{b^2}\exp\left(\frac{-n^2\omega^2}{2b^2}\right) + \frac{n^2\omega^2}{b^4}\ {}_1F_1^2\left(1; \frac{3}{2}; \frac{-n^2\omega^2}{4b^2}\right)\right\}^{\frac{1}{2}} \tag{8.3.40}$$

$$\beta_n = \tan^{-1}\left\{\frac{n\omega\ {}_1F_1\left(1; \frac{3}{2}; \frac{-n^2\omega^2}{4b^2}\right)}{\sqrt{\pi}\, b \exp\left(\frac{-n^2\omega^2}{4b^2}\right)}\right\} \tag{8.3.41}$$

b is given by (8.3.38), $g = 1$ for DPSK, and $g = 0.5$ for NCFSK.

The results presented here are for the case in which there is no gain unbalance on different branches of EGC. For the effect of gain unbalance on the performance of EGC, see [Bea91].

The error rate performance of an EGC in Rician fading channels is discussed in [Abu94], and is compared with that of the MRC and SC for BPSK and NCFSK signals.

#### 8.3.1.4 Use of Characteristic Function

Calculation of the average BER by the above method involved two steps. First, determine the pdf of the required variable and then use it to obtain the average BER. The average BER can also be determined using a one-step procedure using the characteristic function (CF) of the decision variable, as in [Zha97, Zha99]

$$P_e = \frac{1}{2} - \frac{1}{2\pi}\int_{-\infty}^{\infty} \frac{1}{t}\,\mathrm{Im}\left[\psi_r(t)\right]dt \tag{8.3.42}$$

where Im[x] denotes the imaginary part of x, and $\psi_r(t)$ denotes the characteristic function of r (the decision variable at the output of EGC), defined as

$$\psi_r(t) = E\left[e^{jrt}\right] \tag{8.3.43}$$

It provides a general formula for evaluating the average BER for an EGC with coherent detection, and applies to arbitrary fading channels as long as their CF exists. The solution relies on numerical methods to estimate the integral. An algorithm to evaluate the integral using the Hermit method is discussed in [Zha99] for coherent detection of BPSK signals in Rayleigh fading channels. More discussion on the use of CFs may be found in Section 8.6.

### 8.3.2 Interference-Limited Systems

Consider a desired signal and K co-channel interferences in a Nakagami fading environment with fading parameter m. The interference power $I_{EG}$ at the output of EGC is given by

$$I_{EG} = \frac{Y}{2} \tag{8.3.44}$$

with

$$Y = \sum_{i=1}^{L}\sum_{j=1}^{K} q_{ij}^2 \tag{8.3.45}$$

Then the signal power to interference power ratio $\mu_{EG}$ at the output of EGC is given by

$$\begin{aligned}\mu_{EG} &= \frac{S}{I}\\ &= \frac{r^2}{Y}\end{aligned} \tag{8.3.46}$$

#### 8.3.2.1 Outage Probability

Let $P_{EG}^{o}$ denote the outage probability for EGC. It is given by

$$P_{EG}^{o} = P\left[\mu_{EG} \le \mu_0\right] \tag{8.3.47}$$

Substituting from (8.3.46) and using the argument in deriving (7.3.27), it follows that

$$P_{EG}^{o} = P\left[\frac{r^2}{Y} \le \mu_0\right] = \int_0^{\infty} f_r(x) \int_{\frac{x^2}{\mu_0}}^{\infty} f_Y(y)\, dy\, dx \tag{8.3.48}$$

where $f_Y(y)$ denotes the pdf of Y given by

$$f_Y(y) = \left(\frac{m}{\Omega_y}\right)^{mKL} \frac{y^{mKL-1}}{\Gamma(mKL)} \exp\left(-\frac{my}{\Omega_y}\right) \tag{8.3.49}$$

with

$$\Omega_y = E\left[q_{ij}^2\right] \tag{8.3.50}$$

and $f_r$ is given by (8.3.31). Substituting for $f_r$ and $f_y(y)$ in (8.3.48), it becomes [Abu92]

$$P_{EG}^{o} = \frac{4}{T} \sum_{\substack{n=1 \\ n\ \text{odd}}}^{\infty} \left\{ A_n \sum_{i=0}^{MKL-1} \left(\frac{m}{\Omega_y \mu_o}\right)^{i} \frac{B_{n_i}}{i!} \cos\left(\tau_n - \alpha_{ni}\right) \right\} \tag{8.3.51}$$

where

$$B_{ni} = \sqrt{a_{ni}^2 + b_{ni}^2} \tag{8.3.52}$$

$$\alpha_{ni} = \tan^{-1} \frac{b_{ni}}{a_{ni}} \tag{8.3.53}$$

$$a_{ni} = \frac{\Gamma\left(i + \frac{1}{2}\right)}{2\left(m/\left(\Omega_y \mu_o\right)\right)^{i+\frac{1}{2}}}\ {}_1F_1\left(i + \frac{1}{2}; \frac{1}{2}; -\frac{n^2\omega^2\Omega_y\mu_o}{4m}\right) \tag{8.3.54}$$

and

$$b_{n_i} = \frac{\Gamma(i+1)n\omega}{2\left(m/\Omega_y\mu_o\right)^{i+1}}\ {}_1F_1\left(i+1; \frac{3}{2}; -\frac{n\omega^2\Omega_y\mu_o}{4m}\right) \tag{8.3.55}$$

The function ${}_1F_1(\ )$ in (8.3.54) and (8.3.55) can be computed as follows:

$${}_1F_1\left(i + \frac{1}{2}; \frac{1}{2}; -a\right) = \begin{cases} e^{-a} - 2ae^{-a} \displaystyle\sum_{j=1}^{i} \sum_{t=0}^{j-1} \frac{(-2)^t}{(2t+1)!!} \binom{j-1}{t} a^t, & i \ge 1 \\ e^{-a}, & i = 0 \end{cases} \tag{8.3.56}$$

and

$$ {}_1F_1\left(i+1;\frac{3}{2};-a\right)=e^{-a}\left\{1-\frac{(1-2i)a}{3}+\frac{(1-2i)(3-2i)}{3\cdot 5}\frac{a^2}{2!}+\frac{(1-2i)(3-2i)(5-2i)}{3\cdot 5\cdot 7}\frac{a^3}{3!}+\cdots\right\} \tag{8.3.57}$$

#### 8.3.2.2 Mean Signal Power to Mean Interference Power Ratio

In this section, an expression for the mean signal power to the mean interference power ratio $\tilde{\mu}_{EG}$ at the output of the EGC is derived in Rayleigh fading channels. Let $\bar{S}_{EG}$ and $\bar{I}_{EG}$ denote the mean signal power and mean interference power at the output of the EGC. Thus,

$$\tilde{\mu}_{EG}=\frac{\bar{S}_{EG}}{\bar{I}_{EG}} \tag{8.3.58}$$

Noting that $\Gamma=\bar{S}/N$ and $\Gamma_{EG}=\bar{S}_{EG}/LN$, it follows from (8.3.11) that for Rayleigh fading channels (m = 1), the mean output signal power is given by

$$\bar{S}_{EG}=\bar{S}L\left[1+(L-1)\frac{\pi}{4}\right] \tag{8.3.59}$$

Using the fact that

$$E\left[\frac{q_{ij}^2}{2}\right]=\bar{I}\qquad i=1,\ldots,L,\quad j=1,2,\ldots,K \tag{8.3.60}$$

(8.3.44) and (8.3.45) imply

$$\bar{I}_{EG}=\bar{I}LK \tag{8.3.61}$$

Thus, it follows from (8.3.58), (8.3.59), and (8.3.61) that the ratio of the mean signal power to the mean interference power at the output of EGC becomes

$$\begin{aligned}\tilde{\mu}_{EG}&=\frac{\bar{S}}{\bar{I}}\frac{1+(L-1)\frac{\pi}{4}}{K}\\&=\tilde{\mu}\frac{1+(L-1)\frac{\pi}{4}}{K}\end{aligned} \tag{8.3.62}$$

The ratio increases with the number of branches in the combiner.

## 8.4 Maximum Ratio Combiner

In maximal ratio combining, the signals on all branches are co-phased and the gain on each branch is set equal to the signal amplitude to the mean noise power ratio [Jak74]. Thus, the branch weights are given by

$$w_i=a_i e^{j\theta_i}\qquad i=1,2,\ldots,L \tag{8.4.1}$$

with

$$a_i = \frac{r_i}{N_i} \tag{8.4.2}$$

where $N_i$ denotes the mean noise power on the ith branch.

When the mean noise power on all branches is identical, that is, $N_i = N$, $i = 1, 2, \ldots, L$, the gain on each branch becomes proportional to the signal amplitude. The difference between a maximal ratio combiner (MRC) and an EGC is that in EGC, $a_i = 1$, $i = 1, 2, \ldots, L$.

In this section, the performance of an MRC is evaluated. Both noise-limited and interference-limited systems are considered [Sha00a, Tom99, Zha99a].

### 8.4.1 Noise-Limited Systems

In this section, expressions for the mean signal to noise ratio, outage probability, and average BER are derived.

#### *8.4.1.1 Mean SNR*

First, consider the mean SNR at the output of the MRC. It follows from (8.5) and (8.4.1) that the signal envelope at the output of the MRC is given by

$$r = \sum_{i=1}^{L} a_i r_i \tag{8.4.3}$$

Thus, the output signal envelope is the sum of individual signal envelopes weighted with respective branch gains. Similarly, the total noise power $N_T$ at the output is given by

$$N_T = \sum_{i=1}^{L} N_i a_i^2 \tag{8.4.4}$$

where the mean noise power of each channel has been weighted by the branch power gain, namely the square of the branch gain, before summing.

The instantaneous SNR $\gamma$ at the output of the combiner is given by

$$\begin{aligned} \gamma &= \frac{r^2}{2N_T} \\ &= \frac{\left(\sum_{i=1}^{L} \frac{r_i^2}{N_i}\right)^2}{2\sum_{i=1}^{L} \frac{r_i^2}{N_i}} \\ &= \sum_{i=1}^{L} \frac{r_i^2}{2N_i} \\ &= \sum_{i=1}^{L} \gamma_i \end{aligned} \tag{8.4.5}$$

Thus, the SNR at the output the combiner is the sum of the branch SNRs.

Let $\Gamma_{MR}$ denote the mean SNR of the MRC. It follows from (8.4.5) that $\Gamma_{MR}$ is given by

$$\begin{aligned}\Gamma_{MR} &= E[\gamma] \\ &= \sum_{i=1}^{L} E[\gamma_i] \\ &= L\Gamma\end{aligned} \tag{8.4.6}$$

where the last step follows from the assumption that the noise power on each branch is N, and $\Gamma$ denotes the mean SNR at each branch. Thus, the mean SNR at the output of an MRC varies linearly with number of branches in the combiner.

#### *8.4.1.2 Rayleigh Fading Environment*

For the Rayleigh fading environment, first consider the pdf of the SNR at the output of the MRC, and then the outage probability and average BER.

##### *8.4.1.2.1 PDF of Output SNR*

When $r_i$, i = 1, 2, …, L are Rayleigh distributed, the pdf of $\gamma$ may be estimated as follows. Consider an RV y given by

$$y = \sum_{i=1}^{n} x_i^2 \tag{8.4.7}$$

where $x_i$ denotes a Gaussian RV of zero mean and variance $\sigma^2$. The RV y has a chi-squared distribution with n degrees of freedom with pdf $f_y$ given by (7.1.29). Rewriting,

$$f_y(y) = \frac{1}{\left(\sqrt{2}\sigma\right)^n \Gamma(n/2)} y^{\frac{n}{2}-1} e^{-\frac{y}{2\sigma^2}} \tag{8.4.8}$$

Now, using (7.1.33) and (7.1.47), (8.4.5) can be expressed as

$$\gamma = \sum_{i=1}^{L} \frac{1}{2N}\left(a_i^2 + b_i^2\right) \tag{8.4.9}$$

where $a_i$ and $b_i$ denote two Gaussian RVs with zero mean and variance equal to $\bar{S}$.

As $a_i/\sqrt{2N}$, and $b_i/\sqrt{2N}$, i = 1, 2, …, L are 2L Gaussian RVs with zero mean and variance equal to $\bar{S}/(2N) \equiv \Gamma/2$, it follows from a comparison of (8.4.9) and (8.4.7) that $\gamma$ has a chi-squared distribution with 2L degrees of freedom. Substituting for n = 2L and $\sigma^2 = \Gamma/2$ in (8.4.8) and noting that $\Gamma(L) = (L-1)!$,

$$f_\gamma(\gamma) = \frac{1}{\Gamma^L (L-1)!} \gamma^{L-1} e^{-\frac{\gamma}{\Gamma}} \tag{8.4.10}$$

Alternately, an expression for $f_\gamma(\gamma)$ may be derived using CFs as discussed in Section 8.7.

*8.4.1.2.2 Outage Probability*

The outage probability is given by

$$\begin{aligned} P^{o}_{MR} &= P[\gamma \le \gamma_0] \\ &= F_\gamma(\gamma_0) \end{aligned} \tag{8.4.11}$$

with

$$F_\gamma(\gamma) = \int_0^\gamma f_\gamma(x)dx \tag{8.4.12}$$

Substituting for $f_\gamma(\gamma)$ from (8.4.10) in (8.4.12) and carrying out the integral, one obtains the following expression for the distribution of $\gamma$ [Jak74]:

$$\begin{aligned} F_\gamma(\gamma) &= \int_0^\gamma \frac{1}{\Gamma^L (L-1)!} y^{L-1} e^{-\frac{y}{\Gamma}} dy \\ &= 1 - e^{-\frac{\gamma}{\Gamma}} \sum_{k=1}^{L} \frac{\left(\frac{\gamma}{\Gamma}\right)^{k-1}}{(k-1)!} \end{aligned} \tag{8.4.13}$$

The outage probability at the output of MRC in Rayleigh distributed channels is then given by $F_\gamma(\gamma_0)$.

*8.4.1.2.3 Average BER*

Let $P_e^{MR}$ denote the average BER at the output of the MRC. It can be obtained by averaging the conditional BER for a fixed SNR $\gamma$ over the pdf of $\gamma$, that is,

$$P_e^{MR} = \int_0^\infty P_e(\gamma) f_\gamma(\gamma) d\gamma \tag{8.4.14}$$

where $P_e(\gamma)$ is the BER for a fixed $\gamma$ at the output of the MRC for an arbitrary modulation scheme. For coherent BPSK, coherently detected orthogonal FSK, and DPSK, is given by [Pro95]

$$P_e(\gamma) = Q\left(\sqrt{2\gamma}\right) \quad \text{BPSK} \tag{8.4.15}$$

$$P_e(\gamma) = Q\left(\sqrt{\gamma}\right) \quad \text{CFSK} \tag{8.4.16}$$

and

$$P_e(\gamma) = \left(\frac{1}{2}\right)^{2L-1} e^{-\gamma} \sum_{k=0}^{L-1} b_k \gamma^2 \quad \text{DPSK} \tag{8.4.17}$$

where

$$b_k = \frac{1}{k!}\sum_{n=0}^{L-1-k}\binom{2L-1}{n} \tag{8.4.18}$$

The pdf of $\gamma$ at the output of the MRC in the Rayleigh fading environment is given by (8.4.10). Substituting in (8.4.14) and evaluating the integral, the expression for the average BER is obtained [Pro95]. For BPSK, it becomes

$$P_e^{MR} = \left[\frac{1}{2}\left(1-\Gamma_0\right)\right]^L \sum_{k=0}^{L-1}\binom{L-1+k}{k}\left[\frac{1}{2}\left(1+\Gamma_0\right)\right]^k \tag{8.4.19}$$

where

$$\Gamma_0 = \sqrt{\frac{\Gamma}{1+\Gamma}} \tag{8.4.20}$$

For CFSK, it is also given by (8.4.19), with $\Gamma_0$ defined as

$$\Gamma_0 = \sqrt{\frac{\Gamma}{2+\Gamma}} \tag{8.4.21}$$

For DPSK, it becomes

$$P_e^{MR} = \frac{1}{2^{2L-1}(L-1)!(1+\Gamma)^L}\sum_{k=0}^{L-1} b_k (L-1+k)!\left(\frac{\Gamma}{1+\Gamma}\right)^k \tag{8.4.22}$$

where $b_k$ is given by (8.4.18). The average BER for NCFSK can be obtained by replacing $\Gamma$ with $\Gamma/2$ in (8.4.22).

It should be noted that these results are for a slow fading environment, such that the $r_i e^{j\theta_i}$ $i = 1, 2, \ldots, L$ are constant over the bit duration. For DPSK modulation, these are assumed to be constant over the duration of two bits.

Evaluating $P_e^{MR}$ as a function of L and $\Gamma$ using the above expressions, one can determine the effect of diversity on the average BER.

#### *8.4.1.3 Nakagami Fading Environment*

The average BERs for coherent BPSK and CFSK in the Nakagami fading environment with integer m are given by [Zha99a]

$$P_e^{MR} = \left(\frac{1-\Gamma_0}{2}\right)^{mL} \sum_{k=0}^{mL-1}\binom{mL-1+k}{k}\left(\frac{1+\Gamma_0}{2}\right)^k \tag{8.4.23}$$

where $\Gamma_0$ is defined for BPSK as

$$\Gamma_0 = \sqrt{\frac{\Gamma}{m+\Gamma}} \tag{8.4.24}$$

and for CFSK as

$$\Gamma_0 = \sqrt{\frac{\Gamma}{2m+\Gamma}} \tag{8.4.25}$$

Results for noninteger m and correlated fading may be found in [Zha99a].

### *8.4.1.4 Effect of Weight Errors*

The effect of weight errors introduced by incorrect estimates of the channel gain is examined in [Tom99]. Let $\rho$ denote the normalized correlation between the actual complex channel gain $C_i = \alpha_i e^{j\theta_i}$, i = 1, 2, ..., L and its estimate $\hat{C}_i$ at some time t, with squared correlation given by

$$\rho^2 = \left| \frac{E\left[\hat{C}_i^* C_i\right]}{\sqrt{E\left[\hat{C}_i^* \hat{C}_i\right] E\left[C_i^* C_i\right]}} \right|^2 \tag{8.4.26}$$

Note that $\rho^2 = 1$ corresponds to the estimate with no error.

#### *8.4.1.4.1 Output SNR pdf*

When estimated channel gain $\hat{\alpha}_i$ differs from the actual channel gain $\alpha_i$, the weights used in the MRC differ from those given by (8.4.1) by an error component. Assuming that these weight errors are complex Gaussian distributed RVs, it can be shown that the pdf of the output SNR in the presence of weight errors is given by [Tom99, Gan71]

$$\hat{f}_\gamma(\gamma) = \sum_{k=1}^{L} A(k) \frac{\gamma^{k-1}}{(k-1)!\Gamma^k} e^{-\frac{\gamma}{\Gamma}} \tag{8.4.27}$$

where

$$A(k) = \binom{L-1}{k-1} \left(1-\rho^2\right)^{L-k} \rho^{2(k-1)} \tag{8.4.28}$$

This is an interesting result and shows that the pdf of $\gamma$ in the presence of error is the weighted error of the pdf of $\gamma$ in the absence of error with the weighting co-efficient A(k) given by (8.4.28), that is,

$$\hat{f}_\gamma(\gamma) = \sum_{k=1}^{L} A(k) f_\gamma(\gamma) \tag{8.4.29}$$

where $f_\gamma(\gamma)$ is given by (8.4.10). The pdf of $\gamma$ in the presence of errors may be used to estimate the effect of errors on the outage probability and average BER.

*8.4.1.4.2 Outage Probability*

The outage probability $P_{MR}^{o}$ in the absence of errors is given by (8.4.11) with $F_\gamma(\gamma)$ denoting the distribution function of $\gamma$ in the absence of errors. Thus, it follows that the outage probability in the presence of errors $\hat{P}_{MR}^{o}$ is given by

$$\hat{P}_{MR}^{o} = \hat{F}_\gamma(\gamma_0) \tag{8.4.30}$$

with the distribution function in the presence of errors $\hat{F}(\gamma)$ given by

$$\hat{F}(\gamma) = \int_0^\gamma \hat{f}_\gamma(x)dx \tag{8.4.31}$$

Substituting for $\hat{f}_\gamma(\gamma)$ from (8.4.27) it becomes [Tom99, Gan71]

$$\hat{F}_\gamma(\gamma) = 1 - \left( \sum_{k=1}^{L} A(k) e^{-\frac{\gamma}{\Gamma}} \sum_{n=1}^{k} \frac{\left(\frac{\gamma}{\Gamma}\right)^{n-1}}{(n-1)!} \right) \tag{8.4.32}$$

Note that in the absence of errors $\rho^2 = 1$, (8.4.32) reduces to (8.4.10) as only A(L) is nonzero. For $\rho^2 = 0$, when the channel estimate is completely uncorrelated with the actual channel parameters, only A(1) is nonzero, and the distribution function reduces to that of the single branch case. Hence, no diversity advantage is available [Tom99, Gan71].

*8.4.1.4.3 Average BER*

Similarly, the effect of errors on the average BER may be obtained by replacing the pdf of $\gamma$ in the absence of errors $f_\gamma(\gamma)$ with $\hat{f}_\gamma(\gamma)$ in (8.4.14). For this case, the average BER becomes

$$\hat{P}_e^{MR} = \int_0^\infty P_e(\gamma)\hat{f}_\gamma(\gamma)d\gamma \tag{8.4.33}$$

Substituting for $\hat{f}_\gamma(\gamma)$ from (8.4.29) and using (8.4.14), it follows that

$$\begin{aligned} \hat{P}_e^{MR} &= \int_0^\infty P_e(\gamma) \sum_{k=1}^{L} A(k) f_\gamma(\gamma) d\gamma \\ &= \sum_{k=1}^{L} A(k) \int_0^\infty P_e(\gamma) f_\gamma(\gamma) d\gamma \\ &= \sum_{k=1}^{L} A(k) P_e^{MR} \end{aligned} \tag{8.4.34}$$

Thus, the average BER in the presence of errors is the weighted sum of the average BER in the absence of errors.

### 8.4.2 Interference-Limited Systems

Expressions for the mean SIR, outage probability, and average BER are examined in this section.

#### 8.4.2.1 *Mean Signal Power to Interference Power Ratio*

Assume that $a_i = \alpha_i$ in (8.4.2), with $\alpha_i$ denoting the amplitude of the signal channel gain on the ith channel. When the mean noise power is identical on all channels, the weight vector for MRC can be expressed as

$$\mathbf{w} = \mathbf{C}_S \tag{8.4.35}$$

For an interference-limited system, the array signal $\mathbf{x}(t)$ due to a desired signal and K identical interferences is given by

$$\mathbf{x}(t) = \sqrt{p_S}\mathbf{C}_S g(t) + \sqrt{p_I}\sum_{k=1}^{K}\mathbf{C}_{Ik} g_k(t) \tag{8.4.36}$$

and the output of the MRC becomes

$$\begin{aligned} y(t) &= \mathbf{w}^H\mathbf{x}(t) \\ &= \sqrt{p_S}\mathbf{C}_S^H\mathbf{C}_S g(t) + \sqrt{p_I}\sum_{k=1}^{K}\mathbf{C}_S^H\mathbf{C}_{Ik} g_k(t) \end{aligned} \tag{8.4.37}$$

It then follows that the signal power $S_{MR}$ and the interference power $I_{MR}$ for given $\mathbf{C}_S$ and $\mathbf{C}_{Ik}$ are, respectively, given by

$$S_{MR} = p_S\left(\mathbf{C}_S^H\mathbf{C}_S\right)^2 \tag{8.4.38}$$

and

$$I_{MR} = p_I\sum_{k=1}^{K}\left|\mathbf{C}_S^H\mathbf{C}_{Ik}\right|^2 \tag{8.4.39}$$

and the SIR at the output of MRC, $\mu$, becomes

$$\mu = \frac{\tilde{\mu}\left(\mathbf{C}_S^H\mathbf{C}_S\right)^2}{\sum_{k=1}^{K}\left|\mathbf{C}_S^H\mathbf{C}_{Ik}\right|^2} \tag{8.4.40}$$

with

$$\tilde{\mu} = \frac{p_S}{p_I} \tag{8.4.41}$$

[Sha00a] shows that when the desired signal is Rice distributed and the interferences are Rayleigh distributed, the pdf of μ is given by

$$f_\mu(\mu) = e^{-LK_0}\,{}_1F_1\left(K+L;\,L;\,LK_0\frac{\mu}{\tilde{\mu}+\mu}\right)\tilde{\mu}^K\frac{\Gamma(K+L)}{\Gamma(K)\Gamma(L)}\frac{\mu^{L-1}}{(\tilde{\mu}+\mu)^{L+K}} \tag{8.4.42}$$

where $K_0$ denotes the Rice distribution parameter defined by (7.1.42). For $K_0 = 0$, the Rice distribution becomes the Rayleigh distribution, and the pdf of μ for the Rayleigh distributed signal case becomes

$$f_\mu(\mu) = \frac{\Gamma(L+K)}{\Gamma(L)\Gamma(K)}\tilde{\mu}^L\frac{\mu^{L-1}}{(\tilde{\mu}+\mu)^{L+K}} \tag{8.4.43}$$

yielding

$$\begin{aligned}\bar{\mu}_{MR} &= \int_0^\infty \mu f_\mu(\mu)d\mu \\ &= \frac{L\tilde{\mu}}{K-1}, \quad K>1\end{aligned} \tag{8.4.44}$$

where $\bar{\mu}_{MR}$ denotes the mean value of the signal power to the interference power at the output of the MRC.

For $K_0 = \infty$, the desired signal becomes nonfading and the pdf for the nonfading signal case becomes

$$f_\mu(\mu) = \frac{(L\tilde{\mu})^K}{\Gamma(K)}\mu^{-K-1}e^{-L\frac{\tilde{\mu}}{\mu}} \tag{8.4.45}$$

yielding

$$\begin{aligned}\bar{\mu}_{MR} &= \int_0^\infty \mu f_\mu(\mu)d\mu \\ &= \frac{L\tilde{\mu}}{K-1}, \quad K>1\end{aligned} \tag{8.4.46}$$

Thus, the mean signal power to interference power is the same for both cases.

#### *8.4.2.2 Outage Probability*

The outage probability is defined as

$$\begin{aligned}P^o &= P[\mu \le \mu_0] \\ &= \int_0^{\mu_0} f_\mu(\mu)d\mu\end{aligned} \tag{8.4.47}$$

For MRC, when the signal envelope has a Rayleigh distribution, (8.4.47) and (8.4.43) yield [Sha00a]

$$P_{MR}^{o} = \frac{\Gamma(L+K)}{\Gamma(L+1)\Gamma(K)}\left(\frac{\mu_0}{\tilde{\mu}}\right)^{L} {}_2F_1\left(L+K,\ L;\ L+1;\ -\frac{\mu_0}{\tilde{\mu}}\right) \tag{8.4.48}$$

where ${}_2F_1(a;\ b;\ c;\ x)$ denotes the Gauss hypergeometric function defined as [Abr72]

$$ {}_2F_1(a,\ b;\ c;\ x) = \sum_{n=0}^{\infty} \frac{(a)_n(b)_n}{(c)_n}\frac{x^n}{n!} \tag{8.4.49}$$

with

$$(x)_n = \frac{\Gamma(x+n)}{\Gamma(x)} \tag{8.4.50}$$

When the signal envelope is nonfading, (8.4.45) and (8.4.47) yield

$$P_{MR}^{o} = \frac{\Gamma(K, L\tilde{\mu}/\mu_o)}{\Gamma(K)} \tag{8.4.51}$$

where $\Gamma(K,\ x)$ is the incomplete gamma function.

#### *8.4.2.3 Average BER*

Assuming that the interference term in (8.4.37) is Gaussian distributed, the conditional $P_e(\mu)$ for coherent BPSK is given by [Sha00a]

$$\begin{aligned} P_e(\mu) &= Q\left(\sqrt{2\mu}\right) \\ &= \frac{1}{2}\operatorname{erfc}\sqrt{\mu} \end{aligned} \tag{8.4.52}$$

The average BER may be obtained by averaging all values of $\mu$ (the SIR) as

$$P_e^{MR} = \int_0^{\infty} P_e(\mu) f_{\mu}(\mu)\, d\mu \tag{8.4.53}$$

Substituting for $P_e(\mu)$ from (8.4.52) and $f_{\mu}(\mu)$ from (8.4.43) in (8.4.53), $P_e^{MR}$ for the Rayleigh fading environment becomes

$$P_e^{MR} = \frac{\Gamma(L+K)}{2\Gamma(L)\Gamma(K)}\tilde{\mu}^{L}\int_0^{\infty} \operatorname{erfc}\left(\sqrt{\mu}\right)\frac{\mu^{L-1}}{(\tilde{\mu}+\mu)^{L+K}}\, d\mu \tag{8.4.54}$$

Evaluation of the integral leads to [Sha00a]

$$P_e^{MR} = \frac{1}{2\sqrt{\pi}\Gamma(L)\Gamma(K)}\left[\frac{\tilde{\mu}^L\Gamma\left(\frac{1}{2}-K\right)\Gamma(L+K)}{(-K)}\,{}_2F_2\left(L+K, K; K+\frac{1}{2}, K+1; \tilde{\mu}\right)\right.$$
$$+\frac{\tilde{\mu}^{\frac{1}{2}}\Gamma\left(K-\frac{1}{2}\right)}{\Gamma\left(\frac{1}{2}\right)}\Gamma\left(-\frac{1}{2}\right)\Gamma\left(L+\frac{1}{2}\right){}_2F_2\left(L+\frac{1}{2}, \frac{1}{2}; \frac{3}{2}-K; \tilde{\mu}\right) \tag{8.4.55}$$
$$\left.+\Gamma(K)\Gamma(L)\Gamma\left(\frac{1}{2}\right)\right]$$

where ${}_2F_2(.)$ is a hypergeometric function. An expression for the generalized hypergeometric function is given by [And85]

$$ {}_pF_q\left(a_1, \ldots a_p; b_1, \ldots b_q; x\right) = \sum_{n=0}^{\infty}\frac{(a_1)_n \cdots (a_p)_n}{(b_1)_n \cdots (b_q)_n}\frac{x^n}{n!} \tag{8.4.56}$$

with

$$(x)_n = \frac{\Gamma(x+n)}{\Gamma(x)} \tag{8.4.57}$$

## 8.5 Optimal Combiner

An optimal combiner (OC) or beamformer as discussed in Chapter 2 maximizes the output SNR at the output of the combiner, and is useful in canceling unwanted interferences in nonfading and uncorrelated environments when the system has more degrees of freedom than the number of interferences present. In mobile communications, the situation is different than that assumed in Chapter 2. There are generally more co-channel interferences than the number of elements in the array; these interferences may not be as strong as the desired signal and fading conditions prevail. Under these conditions, the OC is not able to fully cancel all interferences. However, it is able to achieve performance improvement by combating the effect of fading and causing some reduction in the power of co-channel interferences entering the receiver [Win84, Win87, Win87a].

In this section, OC performance is examined when there are more co-channel interferences than the number of elements in the array in fading conditions. Expressions for the average BER and the probability of errors are derived using the procedure presented in [Sha98, Sha00]. For analytical simplicity, it is assumed that all interferences are of equal power and that the system is interference limited. Thus, the effect of noise is ignored.

Let $\mathbf{w}_{oc}$ denote the weights of the OC given by (2.4.1). For an interference-limited system when the effect of noise is ignored, the noise-only array correlation matrix $R_N$ in (2.4.1) is identical to the correlation matrix due to interference $R_I$. Thus, $\mathbf{w}_{oc}$ can be estimated using

$$\mathbf{w}_{oc} = \alpha_0 R_I^{-1}\mathbf{C}_s \tag{8.5.1}$$

where $\alpha_0$ is an arbitrary constant and the steering vector in the look direction has been replaced by the signal channel gain vector. Let $R_I$ be estimated using

$$R_I = p_I \sum_{j=1}^{K} \mathbf{C}_{Ij} \mathbf{C}_{Ij}^H \tag{8.5.2}$$

with $p_I$ denoting the power of each interference, and $\mathbf{C}_{Ij}$ denoting the channel gain vector for the jth interference.

It can be shown [Gir77] that for $K \geq L$, $R_I^{-1}$ exists with probability one if

$$\Sigma = E\left[\mathbf{C}_{Ij} \mathbf{C}_{Ij}^H\right]$$

is positive definite. Thus, it is assumed here that $R_I^{-1}$ exists.

### 8.5.1 Mean Signal Power to Interference Power Ratio

Let $S_{OC}$ and $I_{OC}$ denote the signal power and the interference power at the output of OC, respectively. These are given by

$$S_{OC} = \mathbf{w}_{oc}^H R_S \mathbf{w}_{oc} \tag{8.5.3}$$

and

$$I_{OC} = \mathbf{w}_{oc}^H R_I \mathbf{w}_{oc} \tag{8.5.4}$$

where

$$R_S = p_S \mathbf{C}_S \mathbf{C}_S^H \tag{8.5.5}$$

denotes an estimate of the signal array correlation matrix.

Substituting for $R_S$ and $\mathbf{w}_{oc}$, it follows that

$$S_{OC} = \alpha_0^2 p_S \left(\mathbf{C}_S^H R_I^{-1} \mathbf{C}_S\right)^2 \tag{8.5.6}$$

and

$$I_{OC} = \alpha_0^2 \left(\mathbf{C}_S^H R_I^{-1} \mathbf{C}_S\right) \tag{8.5.7}$$

Let $\mu$ denote the signal power to interference power at the output of the OC. Thus, it follows from (8.5.6) and (8.5.7) that

$$\begin{aligned} \mu &= \frac{S_{oc}}{I_{oc}} \\ &= p_S\left(\mathbf{C}_S^H R_I^{-1} \mathbf{C}_S\right) \end{aligned} \tag{8.5.8}$$

In Rayleigh fading channels, the pdf of $\mu$, $f_\mu(\mu)$ is given by [Sha98]

$$f_\mu(\mu)=\frac{\Gamma(K+1)(\tilde{\mu})^{K+1-L}}{\Gamma(L)\Gamma(K+1-L)}\frac{\mu^{L-1}}{(\tilde{\mu}+\mu)^{K+1}},\quad \mu\geq 0,\ 1\leq L\leq K \tag{8.5.9}$$

and the mean value of SIR at the output of the OC $\bar{\mu}_{oc}$ becomes

$$\begin{aligned}\bar{\mu}_{oc}&=E[\mu]=\int_0^\infty \mu f_\mu(\mu)d\mu\\&=\frac{L}{K-L}\tilde{\mu},\quad 1\leq L<K\end{aligned} \tag{8.5.10}$$

where

$$\tilde{\mu}=\frac{p_s}{p_I} \tag{8.5.11}$$

Note that for K = L, E($\mu$) does not exist. It follows from (8.5.10) that for $K \gg L$, the mean SIR is proportional to number of branches in the combiner.

### 8.5.2. Outage Probability

The outage probability $P_{oc}^{o}$ is given by

$$\begin{aligned}P_{oc}^{o}&=F_\mu(\mu_0)\\&=\int_0^{\mu_0} f_\mu(\mu)d\mu\end{aligned} \tag{8.5.12}$$

Substituting for $f_\mu(\mu)$ from (8.5.9), and evaluating the integral (8.5.12) becomes [Sha98]

$$P_{oc}^{o}=\frac{\Gamma(K+1)}{\Gamma(L+1)\Gamma(K+1-L)}\left(\frac{\mu_0}{\tilde{\mu}}\right)^L {}_2F_1\left(K+1,\ L;\ L+1;\ -\frac{\mu_0}{\tilde{\mu}}\right) \tag{8.5.13}$$

where ${}_2F_1(a, b; c; x)$ is a hypergeometric function given by (8.4.49).

### 8.5.3 Average Bit Error Rate

In this section, an expression for average BER for BPSK signals is presented in a slow Rayleigh fading environment employing a coherent receiver. The computation of average BER requires the distribution of interferences at the array output. When interferences are not identically distributed, the central limit theorem could not strictly be invoked to claim the Gaussian property for the sum of the binary RVs, but for large numbers of interferences the Gaussian property is often assumed for such analyses and the results presented here are under this assumption.

The conditional BER (for a given μ) for a coherent BPSK is given by (8.4.52). The average BER is then obtained by averaging all values of μ. Thus,

$$P_e^{OC} = \frac{1}{2}\int_0^{\infty} \text{erfc}\sqrt{\mu}\, f_{\mu}(\mu)\, d\mu \tag{8.5.14}$$

Substituting for $f_\mu$ in (8.5.14) from (8.5.9) and evaluating the integral [Sha98],

$$\begin{aligned} P_e^{OC} = \frac{1}{2\sqrt{\pi}\Gamma(L)\Gamma(K+1-L)} \Bigg[ & \tilde{\mu}^{K+1-L} \frac{\Gamma\left(L-K-\frac{1}{2}\right)\Gamma(K+1)}{(L-K-1)} \\ & \cdot {}_2F_2\left(K+1, K+1-L; K-L+\frac{3}{2}, K-L+2; \tilde{\mu}\right) \\ & + \frac{1}{\sqrt{\pi}}\tilde{\mu}^{\frac{1}{2}}\Gamma\left(K-L+\frac{1}{2}\right)\Gamma\left(-\frac{1}{2}\right)\Gamma\left(L+\frac{1}{2}\right){}_2F_2\left(L+\frac{1}{2}, \frac{1}{2}; L-K+\frac{1}{2}, \frac{3}{2}; \tilde{\mu}\right) \\ & + \Gamma(K+1-L)\sqrt{\pi}\Gamma(L)\Bigg] \end{aligned} \tag{8.5.15}$$

where ${}_2F_2(.)$ is a hypergeometric function given by (8.4.56).

## 8.6 Generalized Selection Combiner

A conventional selection combiner discussed in Section 8.1 selects the signal from a branch with the strongest signal, normally with the largest instantaneous SNR. A generalized selection combiner (GSC), on the other hand, selects strongest signals from more than one branch and combines these selected signals coherently using an MRC or an EGC. In maximum ratio combining, the selected signals are combined coherently with a gain proportional to the amplitude of the signal received on respective branches as discussed in Section 8.4, whereas in equal gain combining the gain on each branch is the same as discussed in Section 8.3.

Thus, a GSC is a two-stage processor as shown in Figure 8.2, where the first stage selects the $L_C$ strongest signals from L branches and the second stage combines these using an MRC (or EGC). It becomes an MRC (or EGC) when all branches are selected at the first stage, and becomes an SC when only one branch is selected.

Since the MRC and EGC select all branches, including those with poor SNR that provide only a marginal contribution to the information as well as a possible source of errors. Thus, a GSC is expected to be more robust than the MRC and EGC in the presence of channel gain errors. On the other hand, an SC only processes one branch, and thus may be losing too much information in the process. Thus, an GSC is expected to offer advantages over an SC. For more details on the GSC and its performance under various noise environments, see [Alo00, Eng96, Kon98, Roy96].

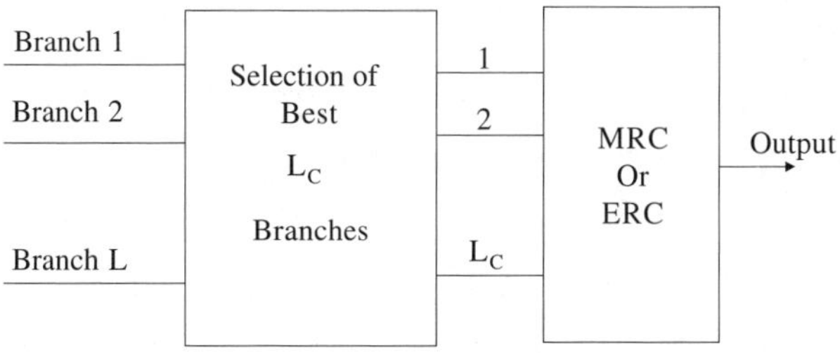


**FIGURE 8.2**
Block diagram of a generalized selection combiner.

In this section, the performance of a GSC is studied and the use of moment-generating functions and the CFs to evaluate various performance measures is explained.

### 8.6.1 Moment-Generating Functions

The moment-generating function (MGF) of an RV x, $M_x(S)$, is defined as [Alo00]

$$\begin{aligned} M_x(S) &= E\left[e^{Sx}\right] \\ &= \int_{-\infty}^{\infty} e^{Sx} f_x(x)dx \end{aligned} \tag{8.6.1}$$

It is related to the Laplace transform of the pdf of x, $f_x$, by

$$L(f_x) = M_x(-S) \tag{8.6.2}$$

and thus one is able to obtain the pdf of x from $M_x$ by taking the inverse Laplace transform on both sides of (8.6.2).

The MGF is related to the CF of x, $\psi_x(j\omega)$, by [Pro95]

$$\psi_x(j\omega) = M_x(S)\Big|_{S=j\omega} \tag{8.6.3}$$

and the cumulant function of x by [Abr72]

$$\phi_x(S) = \ln(M_x(S)) \tag{8.6.4}$$

The mean value of x may be obtained from $\phi_x$ using

$$\bar{x} = \frac{d\phi_x(S)}{dS}\bigg|_{S=0} \tag{8.6.5}$$

or alternately,

$$\bar{x} = -j\frac{d\phi_x(j\omega)}{d\omega}\bigg|_{\omega=0} \tag{8.6.6}$$

The use of CF to evaluate average BER is discussed in Section 8.3.

Now the performance of an L-branch GSC is examined by deriving expressions for the mean output SNR, outage probability, and average BER when it selects $L_c$ signals from branches with the largest instantaneous SNR in a noise-limited Rayleigh fading environment and combines them using an MRC.

### 8.6.2 Mean Output Signal-to-Noise Ratio

Section 8.4 shows that the SNR at the output of the MRC is the sum of the SNR at each branch. Let $\gamma(\ell)$ denote the ordered SNR at the $\ell$th branch, such that $\gamma(1) \geq \gamma(2) \geq \ldots \geq \gamma(L)$. Thus, it follows that the SNR at the output of the GSC is the sum of the SNR at selected branches. Hence,

$$\gamma_{GS} = \sum_{\ell=1}^{L_c} \gamma(\ell) \tag{8.6.7}$$

It should be noted that even when $\gamma_\ell$, $\ell = 1, 2, \ldots, L$ are i.i.d. RVs, RVs $\gamma(\ell)$, $\ell = 1, 2, \ldots, L_c$ are not i.i.d. RVs [Alo00].

The mean SNR at the output of the GSC, $\Gamma_{GS}$, is then given by

$$\begin{aligned}\Gamma_{GS} &= E[\gamma_{GS}] \\ &= \sum_{\ell=1}^{L_C} \bar{\gamma}(\ell)\end{aligned} \tag{8.6.8}$$

where $\bar{\gamma}(\ell)$ is the mean value of the ordered SNR on a selected branch, and is given by

$$\bar{\gamma}(\ell) = \int_0^\infty \gamma f_{\gamma(\ell)}(\gamma) d\gamma \tag{8.6.9}$$

with $f_{\gamma(\ell)}$ denoting the pdf of the ordered SNR $\gamma(\ell)$.

When $\gamma_\ell$, $\ell = 1, 2, \ldots, L$ are i.i.d. RVs, $f_{\gamma(\ell)}$ can be expressed in terms of $f_\gamma$, the pdf of the unordered SNR. It is given by [Alo00]

$$f_{\gamma(\ell)}(\gamma) = \frac{L!}{(L-1)!(\ell-1)!}\left(F_\gamma(\gamma)\right)^{L-1}\left(1 - F_\gamma(\gamma)\right)^{\ell-1} f_\gamma(\gamma) \tag{8.6.10}$$

where $F_\gamma$ is the cdf of $\gamma$.

For Rayleigh fading channels, $f_\gamma$ and $F_\gamma$ are, respectively, given by (7.3.12) and (7.3.11). Substituting in (8.6.10),

$$f_{\gamma(\ell)}(\gamma) = \frac{1}{\Gamma}\frac{L!}{(L-1)!(\ell-1)!}\left(1 - e^{-\gamma/\Gamma}\right)^{L-1} e^{-\ell\gamma/\Gamma} \tag{8.6.11}$$

Substituting this in (8.6.9) with u denoting $e^{-\gamma/\Gamma}$, the expression for $\bar{\gamma}(\ell)$ becomes

$$\bar{\gamma}(\ell) = \frac{-L!\Gamma}{(L-1)!(\ell-1)!}\int_0^1 (1-u)^{L-1}u^{\ell-1}\ln u\,du$$
$$= \Gamma\sum_{k=0}^{L-1}\frac{1}{\ell+k} \tag{8.6.12}$$

which, along with (8.6.8), yields

$$\Gamma_{GS} = \Gamma\sum_{\ell=1}^{L_c}\sum_{k=0}^{L-1}\frac{1}{\ell+k}$$
$$= \left(1+\sum_{\ell=L_c+1}^{L}\frac{1}{\ell}\right)L_c\Gamma \tag{8.6.13}$$

Details on the evaluation of (4.6.12) and (4.6.13) are provided in [Alo00]. A derivation of (8.6.13) is also provided in [Roy96]. It can easily be seen that for $L_c = 1$, (8.6.13) reduces to (8.1.9), an expression for $\Gamma_{SC}$. For $L_c = L$, it becomes (8.4.6), an expression for $\Gamma_{MR}$. Thus, (8.6.13) is a generalization of the two results. It can be shown that $\Gamma_{GS}$ is a monotonically increasing function of the selected branches, resulting in $\Gamma_{SC} \le \Gamma_{GS} \le \Gamma_{MR}$ [Kon98]. Thus, the mean SNR of the GSC is bounded below by $\Gamma_{SC}$ and above by $\Gamma_{MR}$.

An alternative derivation of (8.6.13) using moment-generating functions is presented below. An expression for $M_{\gamma_{GS}}(S)$, the MGF of $\gamma_{GS}$ in the Rayleigh fading environment, is given by [Alo00]:

$$M_{\gamma GS}(S) = (1-S\Gamma)^{-L_c+1}\sum_{\ell=L_c}^{L}\left(1-\frac{S\Gamma L_c}{\ell}\right)^{-1} \tag{8.6.14}$$

Alternately, in summation form the expression becomes

$$M_{\gamma_{GS}}(S) = (1-S\Gamma)^{-L_c+1}\sum_{\ell=0}^{L-L_c}(-1)^{\ell}\frac{\binom{L}{L_c}\binom{L-L_c}{\ell}}{1+\dfrac{\ell}{L_c}-S\Gamma} \tag{8.6.15}$$

Substituting for $M_{\gamma_{GS}}$ from (8.6.14) in (8.6.4), one obtains an expression for the cumulant function:

$$\phi_{\gamma_{GS}}(S) = (-L_c+1)\ln(1-S\Gamma) - \sum_{\ell=L_c}^{L}\ln\left(1-\frac{S\Gamma L_c}{\ell}\right)$$
$$= -L_c\ln(1-S\Gamma) - \sum_{\ell=L_c+1}^{L}\ln\left(1-\frac{S\Gamma L_c}{\ell}\right) \tag{8.6.16}$$

which, along with (8.6.5), yields (8.6.13).

To derive the pdf of $\gamma_{GS}$, rewrite (8.6.14) by replacing $\ell$ with $\ell + L_c$ to shift the range of product terms from $\ell = L_c, \ldots, L$ to $\ell = 0, \ldots, L - L_c$:

$$\begin{aligned} M_{\gamma_{GS}}(S) &= (1-S\Gamma)^{-L_c+1} \prod_{\ell=0}^{L-L_c} \left(1 - \frac{S\Gamma L_c}{\ell + L_c}\right)^{-1} \\ &= (1-S\Gamma)^{-L_c+1} \prod_{\ell=0}^{L-L_c} \left(\frac{\ell + L_c}{L_c}\right)\left(1 - \frac{\ell}{L_c} - S\Gamma\right)^{-1} \\ &= \frac{L!}{L_c! L_c^{L-L_c}} (1-S\Gamma)^{-L_c+1} \prod_{\ell=0}^{L-L_c} \left(1 - \frac{\ell}{L_c} - S\Gamma\right)^{-1} \end{aligned} \tag{8.6.17}$$

The required $f_{\gamma_{GS}}$ may be obtained by substituting $M_{\gamma_{GS}}$ from (8.6.17) in (8.6.2), and then taking the inverse Laplace transforms. An expression for $f_{\gamma_{GS}}$ becomes [Roy96]

$$f_{\gamma_{GS}}(\gamma) = \sum_{\ell=0}^{L_c-1} \frac{a_\ell \gamma^\ell e^{-\frac{\gamma}{\Gamma}}}{\Gamma^{\ell+1} \ell!} + \sum_{k=1}^{L-L_c} b_k e^{-\left(1+\frac{k}{L_c}\right)\frac{\gamma}{\Gamma}} \tag{8.6.18}$$

where $a_\ell$ are the coefficients of the partial fraction expansion of (8.6.17) and are given by

$$a_{L_c-\ell-1} = \binom{L}{L_c} \sum_{k=1}^{L-L_c} \binom{L-L_c}{k} \frac{(-1)^{k+\ell+1} L_c^\ell}{k^\ell} \tag{8.6.19}$$

and

$$b_k = \binom{L}{L_c} \binom{L-L_c}{k} \frac{(-1)^{L_c+k} L_c^{L_c-1}}{k^{L_c-1}} \tag{8.6.20}$$

The expression also has a compact form given by [Alo00]:

$$\begin{aligned} f_{\gamma_{GS}}(\gamma) = \binom{L}{L_c} \Bigg[ & \frac{\gamma^{L_c-1} e^{-\gamma/\Gamma}}{\Gamma^{L_c}(L_c-1)!} \\ & + \frac{1}{\Gamma} \sum_{\ell=1}^{L-L_c} (-1)^{L_c+\ell-1} \binom{L-L_c}{\ell} \left(\frac{L_c}{\ell}\right)^{L_c-1} e^{-\gamma/\Gamma} \left( e^{\frac{-\ell\gamma}{\Gamma L_c}} - \sum_{m=0}^{L_c-2} \frac{1}{m!} \left(\frac{-\ell\gamma}{L_c \Gamma}\right)^m \right) \Bigg] \end{aligned} \tag{8.6.21}$$

### 8.6.3 Outage Probability

Let $P_{GS}^o$ denote the outage probability for GSC. It is defined as

$$P_{GS}^o = F_{\gamma_{GS}}(\gamma_0) \tag{8.6.22}$$

An expression of $F_{\gamma_{GS}}(\gamma)$ may be obtained by integrating (8.6.18) [Roy96]:

$$F_{\gamma_{GS}}(\gamma) = 1 - \sum_{\ell=0}^{L_c-1} \frac{A_\ell \gamma^\ell e^{\frac{-\gamma}{\Gamma}}}{\Gamma^\ell \ell!} + \sum_{k=1}^{L-L_c} B_K e^{-\left(1+\frac{k}{L_c}\right)\frac{\gamma}{\Gamma}} \tag{8.6.23}$$

where

$$A_\ell = \sum_{i=0}^{L_c-\ell-1} a_{\ell+i} \tag{8.6.24}$$

and

$$B_k = \frac{L_c b_k}{k + L_c} \tag{8.6.25}$$

An expression for $P^o_{GS}$ follows from (8.6.22) and (8.6.23). A closed-form expression for $P^o_{GS}$ is obtained by integrating (8.6.21) and using (8.6.22) [Alo00]:

$$P^o_{GS} = \binom{L}{L_c}\left[1 - e^{-\frac{\gamma_0}{\Gamma}} \sum_{\ell=0}^{L_c-1} \frac{\left(\frac{\gamma_0}{\Gamma}\right)^\ell}{\ell!} + \sum_{\ell=1}^{L-L_c} (-1)^{L_c+\ell-1} \binom{L-L_c}{\ell}\binom{L_c}{\ell}^{L_c-1} \right.$$
$$\left. \cdot\left(\frac{1 - e^{-\left(1+\frac{\ell}{L_c}\right)\frac{\gamma_0}{\Gamma}}}{1+\frac{\ell}{L_c}} - \sum_{m=0}^{L_c-2}\left(\frac{-\ell}{L_c}\right)^m \left(1 - e^{-\frac{\gamma_0}{\Gamma}} \sum_{k=0}^{m} \frac{\left(\frac{\gamma_0}{\Gamma}\right)^k}{k!}\right)\right)\right] \tag{8.6.26}$$

### 8.6.4 Average Bit Error Rate

The average BER for the GSC can be obtained by averaging the conditional BER for a given $\gamma$ over all $\gamma$ using an expression for $f_{\gamma_{GS}}$. Using (8.6.18) and (7.3.55), an expression for the average BER for coherent BPSK becomes [Roy96]

$$P_e^{GS} = \frac{1}{2} - \sum_{\ell=0}^{L_c-1} \binom{2\ell}{\ell} \frac{A_\ell \Gamma^{\frac{1}{2}}}{4(\Gamma+1)^{\ell+\frac{1}{2}}} - \sum_{k=1}^{L-L_c} \frac{B_k \Gamma^{\frac{1}{2}}}{2\left(\Gamma + 1 + \frac{k}{L_c}\right)^{\frac{1}{2}}} \tag{8.6.27}$$

Replacing $\Gamma$ by $\Gamma/2$ in (8.6.27), an expression for the average BER for CFSK is obtained.

The MGF may also be used to directly derive expressions for the average BER. For various modulation schemes, the average BER using the MGF is given in [Alo00]. For coherent BPSK and CFSK,

$$P_e^{GS} = \frac{1}{\pi}\int_0^{\frac{\pi}{2}} M_{\gamma_{GS}}\left(-\frac{g}{\mathrm{Sin}^2\phi}\right) d\phi, \quad \begin{cases} g = 1 & \text{BPSK} \\ g = 0.5 & \text{CFSK} \end{cases} \tag{8.6.28}$$

Substituting for $M_{\gamma GS}$ from (8.6.15) and carrying out the integral,

$$P_e^{GS} = \binom{L}{L_c} \sum_{\ell=0}^{L-L_c} \frac{(-1)^{\ell} \binom{L-L_c}{\ell}}{1+\frac{\ell}{L_c}} I_{L_c-1}\left( \frac{\pi}{2}; g\Gamma, \frac{g\Gamma}{1+\frac{1}{L_c}} \right) \quad (8.6.29)$$

where

$$I_n(\theta; C_1, C_2) = \frac{1}{\pi} \int_0^{\theta} \left( \frac{\sin^2\phi}{\sin^2\phi + C_1} \right)^n \left( \frac{\sin^2\phi}{\sin^2\phi + C_2} \right) d\phi \quad (8.6.30)$$

and can be evaluated using the following equations[Alo00].

For $C_1 = C_2 = C$,

$$\begin{aligned} I_n(\theta; C) = \frac{\theta}{\pi} - \left( \frac{1+\operatorname{Sgn}(\theta-\pi)}{2} + \frac{T}{\pi} \right) \sqrt{\frac{C}{1+C}} \sum_{k=0}^{n-1} \binom{2k}{k} \frac{1}{[4(1+C)]^k} \\ - \frac{2}{\pi} \sqrt{\frac{C}{1+C}} \sum_{k=0}^{n-1} \sum_{j=0}^{k-1} \binom{2k}{j} \frac{(-1)^{j+k}}{[4(1+C)]^k} \frac{\sin[(2k-2j)T]}{2k-2j}, \quad 0 \le \theta \le 2\pi \end{aligned} \quad (8.6.31)$$

with

$$T = \frac{1}{2} \arctan\left( \frac{N}{D} \right) + \frac{\pi}{2} \left[ -\operatorname{Sgn}N \left( \frac{1+\operatorname{Sgn}D}{2} \right) \right] \quad (8.6.32)$$

$$N = 2\sqrt{C(1+C)} \sin(2\theta) \quad (8.6.33)$$

$$D = (1+2C)\cos(2\theta) - 1 \quad (8.6.34)$$

and sgn(x) denoting the sign of x.

For $C_1 \neq C_2$,

$$\begin{aligned} I_n(\theta; C_1, C_2) = I_n(\theta; C_1) \\ - \left( \frac{1+\operatorname{Sgn}(\theta-\pi)}{2} + \frac{T_2}{\pi} \right) \sqrt{\frac{C_2}{1+C_2}} \left( \frac{C_2}{C_2 - C_1} \right)^n \\ + \left( \frac{1+\operatorname{Sgn}(\theta-\pi)}{2} + \frac{T_1}{\pi} \right) \sqrt{\frac{C_1}{1+C_1}} \sum_{k=0}^{n-1} \left( \frac{C_2}{C_2 - C_1} \right)^{n-k} \binom{2k}{k} \frac{1}{[4(1+C_1)]^k} \\ + \frac{2}{\pi} \sqrt{\frac{C_1}{1+C_1}} \sum_{k=0}^{n-1} \sum_{j=0}^{k-1} \left( \frac{C_2}{C_2 - C_1} \right)^{n-k} \binom{2k}{j} \frac{(-1)^{j+k}}{[4(1+C_1)]^k} \frac{\sin[(2k-2j)T_1]}{2k-2j} \\ 0 \le \theta \le 2\pi \end{aligned} \quad (8.6.35)$$

with $T_1$ and $T_2$ corresponding to T of (8.6.32), with C replaced by $C_1$ and $C_2$, respectively.

The expression for the average BER for BPSK (g = 1) given by (8.6.28) yields the same numerical results as given by [Eng96] for $L_c = 2$ and $L_c = 3$ using the conventional approach of averaging the conditional BER over all values of $\gamma$. These expressions are given below.

For $L_c = 2$, the average BER $P_e^{SC2}$ for coherent BPSK is given by

$$P_e^{SC2} = \frac{L(L-1)}{2}\left\{\frac{1}{2}\left[1-\frac{1}{\sqrt{1+\alpha}}-\frac{\alpha}{2(1+\alpha)\sqrt{1+\alpha}}\right]+\sum_{k=1}^{L-2}\binom{L-2}{k}(-1)^k V(k)\right\} \quad (8.6.36)$$

with

$$V(k)=\frac{1}{2+k}-\frac{1}{k\sqrt{1+\alpha}}+\frac{2}{k(2+k)\sqrt{1+\frac{\alpha(2+k)}{2}}} \quad (8.6.37)$$

and

$$\alpha=\frac{1}{\Gamma} \quad (8.6.38)$$

For $L_c = 3$, the expression is given by

$$P_e^{SC3} = \frac{L(L-1)(L-2)}{4}\left\{\frac{1}{3}\left[1-\mu\sum_{k=0}^{2}\binom{2k}{k}\left(\frac{1-\mu^2}{4}\right)^k\right]+\sum_{k=1}^{L-3}\binom{L-3}{k}(-1)^k G(k)\right\} \quad (8.6.39)$$

with

$$G(k)=\frac{k-3}{k^2}(1-\mu)-\frac{\mu}{2k}\left(1-\mu^2\right)+\frac{9}{k^2(3+k)}\left(1-\frac{1}{\sqrt{1+\alpha\left(1+\frac{k}{3}\right)}}\right) \quad (8.6.40)$$

and

$$\mu=\sqrt{\frac{\Gamma}{1+\Gamma}} \quad (8.6.41)$$

## 8.7 Cascade Diversity Combiner

A cascade diversity combiner (CDC) is similar to the GSC discussed in the previous section in that it employs a two-stage diversity combining [Roy96]. However, there are some differences. The CDC divides the L branches in $L_c$ groups of M branch each, and then uses a selection combiner to select one best signal from each group at the first stage. At the second stage, it uses an MRC to combine the $L_c$ signals selected at the first stage. The

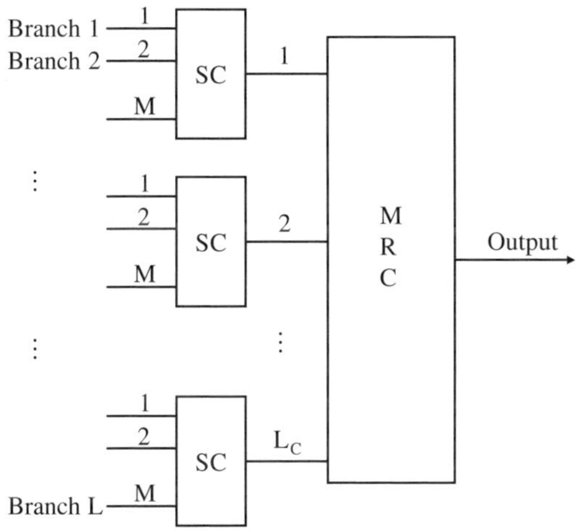


**FIGURE 8.3**
Block diagram of a cascade diversity combiner.

$L_c$ selected signals at the first stage are not necessarily the best $L_c$ signals, as is the case for the GSC. However, the combiner is perhaps easy to implement and analyze, as the SNR at different branches may be assumed to be i.i.d. RVs. In addition, having equal numbers of M inputs in different selection combiners helps in implementation. For $M = 1$, there is no selection; it is equivalent to an MRC. For $L_c = 1$, there is no combiner and it is equivalent to a conventional SC. Figure 8.3 shows a block diagram of a predetection CDC.

In this section, an analysis of a CDC is presented. Both Rayleigh fading and Nakgami fading environments are considered [Cho00, Roy96].

### 8.7.1 Rayleigh Fading Environment

First, consider the pdf of $\gamma_{CD}$, the SNR at the output of the CDC.

#### *8.7.1.1 Output SNR pdf*

It follows from (8.1.7) that when SNRs on different branches are i.i.d. RVs, the pdf of the SNR at the output of an M branch selection combiner in a Rayleigh fading environment is given by

$$f_\gamma(\gamma) = \frac{M}{\Gamma} e^{-\frac{\gamma}{\Gamma}} \left(1 - e^{-\frac{\gamma}{\Gamma}}\right)^{M-1} \tag{8.7.1}$$

The characteristic function of $\gamma$ is given by

$$\psi_\gamma(j\omega) = \int_0^\infty e^{j\omega\gamma} f_\gamma(\gamma) d\gamma, \quad \gamma \geq 0 \tag{8.7.2}$$

Substituting for $f_\gamma$ from (8.7.1) and carrying out the integral, it becomes

$$\psi_\gamma(j\omega) = \frac{M!}{\prod_{k=1}^{M}(k - j\omega\Gamma)} \tag{8.7.3}$$

The second stage uses an MRC to combine the $L_c$ signals. As the SNR at the output of the MRC is the sum of individual SNRs, it follows that the instantaneous SNR $\gamma_{CD}$ at the output of the CDC is given by

$$\gamma_{CD} = \sum_{\ell=1}^{L_c} \gamma_\ell \tag{8.7.4}$$

where $\gamma_\ell$ denotes the SNR at the output of the $\ell$th SC.

Since $\gamma_\ell$, $\ell = 1, \ldots, L_c$ are i.i.d. RVs, it implies that characteristic function of $\gamma_{CD}$ is the product of individual characteristic functions, that is,

$$\begin{aligned}\psi_{\gamma_{CD}}(j\omega) &= \prod_{\ell=1}^{L_c} \psi_{\gamma_\ell}(j\omega) \\ &= \left[\frac{M!}{\prod_{k=1}^{M}(k - j\omega\Gamma)}\right]^{L_c}\end{aligned} \tag{8.7.5}$$

One observes from (8.7.5) that $\psi_{\gamma_{CD}}(j\omega)$ has M poles of order $L_c$ each. Thus, it can be expressed in summation form suitable for inverse transformation to obtain pdf of $\gamma_{CD}$ as follows:

$$\psi_{\gamma_{CD}}(j\omega) = \sum_{k=1}^{M}\sum_{\ell=0}^{L_c-1} \frac{a_{k,\ell}}{(k - j\omega\Gamma)^{\ell+1}} \tag{8.7.6}$$

where $a_{k,\ell}$ are the coefficients of the partial fraction expression of $\psi_{\gamma_{CD}}(j\omega)$. For a technique to compute these coefficients, see [Gil81].

Taking the inverse transformation of (8.7.6), the pdf of $\gamma_{CD}$ then becomes [Roy96]

$$f_{\gamma_{CD}} = \sum_{k=1}^{M}\sum_{\ell=0}^{L_c-1} \frac{a_{k,\ell}\,\gamma^\ell}{\ell!\,\Gamma^{\ell+1}} e^{-k\frac{\gamma}{\Gamma}} \tag{8.7.7}$$

### *8.7.1.2 Outage Probability*

Let $P^o_{\gamma_{CD}}$ denote the outage probability for a CDC. Integrating (8.7.7) yields

$$F_{\gamma_{CD}}(\gamma) = 1 - \sum_{k=1}^{M}\sum_{\ell=0}^{L_c-1} \frac{A_{k,\ell}\,\gamma^\ell}{\ell!\,\Gamma^\ell} e^{-k\frac{\gamma}{\Gamma}} \tag{8.7.8}$$

where

$$A_{k,\ell} = \sum_{i=0}^{L_c-\ell-1} \frac{a_{k,\ell+i}}{k^{i+1}} \tag{8.7.9}$$

An expression for outage probability is obtained by substituting $\gamma_0$ for $\gamma$ in $F_{\gamma_{CD}}$:

$$P^o_{\gamma_{CD}} = F_{\gamma_{CD}}(\gamma_0) \tag{8.7.10}$$

#### 8.7.1.3 Mean SNR

The mean value of SNR at the output of the CDC can be obtained using (8.6.3) to (8.6.5) with S = $j\omega$, that is,

$$\Gamma_{CD} = -j \left. \frac{d\phi_{\gamma_{CD}}(j\omega)}{d\omega} \right|_{\omega=0} \tag{8.7.11}$$

where

$$\phi_{\gamma_{CD}}(j\omega) = \ln \psi_{\gamma_{CD}}(j\omega) \tag{8.7.12}$$

Substituting for $\psi_{\gamma_{CD}}(j\omega)$ from (8.7.5) in (8.7.12),

$$\begin{aligned} \phi_{\gamma_{CD}}(j\omega) &= L_c \ln \frac{M!}{\prod_{k=1}^{M}(k - j\omega\Gamma)} \\ &= L_c \ln M! - L_c \sum_{k=1}^{M} \ln(k - j\omega\Gamma) \end{aligned} \tag{8.7.13}$$

Differentiating on both sides of (8.7.13) with respect to $\omega$ yields

$$\frac{d\phi_{\gamma_{CD}}(j\omega)}{d\omega} = L_c \sum_{k=1}^{M} \frac{j\Gamma}{(k - j\omega\Gamma)} \tag{8.7.14}$$

Using this in (8.7.11) it follows that

$$\Gamma_{CD} = L_c \Gamma \sum_{k=1}^{M} \frac{1}{k} \tag{8.7.15}$$

Thus, the mean SNR increases by $L_c \sum_{k=1}^{M} \frac{1}{k}$ from the single branch mean SNR. In fact, the gain in mean SNR is the product of the gain by an M branch SC and an $L_c$ branch MRC.

#### *8.7.1.4 Average BER*

The average BER may be obtained by averaging the conditional BER over all values of $\gamma$, that is,

$$P_e^{CD} = \int_0^\infty P_e(\gamma) f_{\gamma_{CD}}(\gamma) d\gamma \tag{8.7.16}$$

Consider an example of a coherent BPSK system. Substituting for $P_e(\gamma)$ from (7.3.55) in (8.7.16),

$$\begin{aligned} P_e^{CD} &= \int_0^\infty f_{\gamma_{CD}}(\gamma) \frac{\operatorname{erfc}\sqrt{\gamma}}{2} d\gamma \\ &= \int_0^\infty f_{\gamma_{CD}}(\gamma) \int_{\sqrt{\gamma}}^\infty \frac{e^{-u^2}}{\sqrt{\pi}} du\, d\gamma \quad \left(\text{substituting for } \operatorname{erfc}\sqrt{\gamma}\right) \end{aligned} \tag{8.7.17}$$

Changing the order of integration, it becomes

$$\begin{aligned} P_e^{CD} &= \int_0^\infty \frac{e^{-u^2}}{\sqrt{\pi}} \int_0^{u^2} f_{\gamma_{CD}}(\gamma) d\gamma\, du \\ &= \int_0^\infty \frac{e^{-u^2}}{\sqrt{\pi}} F_{\gamma_{CD}}\left(u^2\right) du \end{aligned} \tag{8.7.18}$$

Using (8.7.8), the formulas

$$\int_0^\infty u^a e^{-bu^2} du = \frac{\Gamma\left(\frac{a+1}{2}\right)}{2b^{\frac{a+1}{2}}} \tag{8.7.19}$$

and

$$2^{2x-1}\Gamma(x)\Gamma\left(x+\frac{1}{2}\right) = \sqrt{\pi}\Gamma(2x) \tag{8.7.20}$$

and carrying out the integral [Roy96], the expression becomes

$$P_e^{CD} = \frac{1}{2} - \sum_{k=1}^{M} \sum_{\ell=0}^{L_c-1} \binom{2\ell}{\ell} \frac{A_{k,\ell}\Gamma^{\frac{1}{2}}}{\left[4(\Gamma+k)\right]^{\ell+\frac{1}{2}}} \tag{8.7.21}$$

Now consider a case of $M = 1$ and $L_c = L$. In this situation, the CDC becomes an MRC. It follows from comparing the terms in (8.7.5) and (8.7.6) that

$$a_\ell = \begin{cases} 1 & \ell = L-1 \\ 0 & \text{otherwise} \end{cases} \tag{8.7.22}$$

Thus, from (8.7.9) using $L_c = L$ and (8.7.22):

$$A_\ell = 1,\ \ell = 0, \ldots, L-1 \tag{8.7.23}$$

Substituting this in (8.7.21),

$$P_e^{CD} = \frac{1}{2}\left(1 - \sqrt{\frac{\Gamma}{\Gamma+1}}\sum_{\ell=0}^{L-1}\binom{2\ell}{\ell}\frac{1}{[4(\Gamma+1)]^\ell}\right) \tag{8.7.24}$$

It is left an exercise for the reader to show that this is the same as (8.4.19).

Note that using $M = 1$, $L_c = L$, (8.7.22) and (8.7.23) in (8.7.7) and (8.7.8), respectively, leads to (8.4.10) and (8.4.13).

Now, consider an example DPSK system. For DPSK, $P_\ell(\gamma)$ is given by (8.4.17). Substituting (8.4.17) and (8.7.7) in (8.7.16) and carrying out the integral [Roy96],

$$P_e^{CD} = \sum_{i=0}^{L_c-1}\sum_{j=0}^{L_c-1-i}\sum_{k=1}^{L_c-1}\sum_{\ell=0}^{L_c-1}\binom{2L_c-1}{j}\binom{i+\ell}{\ell}\frac{a_{k,\ell}\,\Gamma^i}{2^{2L_c-1}(\Gamma+k)^{i+\ell+1}} \tag{8.7.25}$$

where $a_{k,\ell}$ are the same as in (4.7.6). The above result also applies to noncoherent orthogonal FSK when $\Gamma$ is replaced by $\Gamma/2$.

### 8.7.2 Nakagami Fading Environment

In the Nakagami fading environment with the fading parameter m acquiring integer values, the pdf of the SNR at the output of an L branch selection combiner when signals on all channels are i.i.d. RVs is given by (8.1.20). The characteristic function of $\gamma_{SC}$ is given by

$$\psi_{\gamma_{SC}}(j\omega) = \int_0^\infty e^{j\omega\gamma} f_{\gamma_{SC}}(\gamma)d\gamma, \quad \gamma \ge 0 \tag{8.7.26}$$

Substituting for $f_{\gamma_{SC}}(\gamma)$ from (8.1.20) in (8.7.26) and evaluating the integral, the CF of $\gamma_{SC}$ for an M branch selection combiner becomes [Cho00]

$$\psi_{\gamma_{SC}}(j\omega) = \left(\frac{m}{\Gamma}\right)^m \frac{M}{\Gamma(m)}\sum_{i=0}^{m-1}\binom{M-1}{i}(-1)^i\sum_{j\in B}\left(\frac{m}{\Gamma}\right)^{C_{ji}}\frac{d_{ji}(C_{ji}+m-1)!}{A_{ji}\left(\dfrac{m(i+1)}{\Gamma}+j\omega\right)^{C_{ji}+m}} \tag{8.7.27}$$

The characteristic function of $\gamma_{CD}$ is the product of the individual characteristic functions; thus, it is given by

$$\psi_{\gamma_{CD}}(j\omega) = \left\{\psi_{\gamma_{SC}}(j\omega)\right\}^{L_c} \tag{8.7.28}$$

Following a procedure similar to the Rayleigh fading case described in Section 8.7, the pdf of the SNR of the cascade receiver is given by [Cho00]

$$f_{\gamma_{CD}}(\gamma) = \sum_{i=0}^{M-1}\sum_{\ell=0}^{L'-1} a_{i,\ell}\,\gamma^\ell e^{-m(i+1)\frac{\gamma}{\Gamma}} \tag{8.7.29}$$

where

$$L' = \left(\max\left\{C_{ji}\right\} + m\right)L_c \tag{8.7.30}$$

and $a_{i,\ell}$ are the partial fraction coefficients.

### *8.7.2.1 Average BER*

Consider an example of differential QPSK [Cho00]. The conditional BER for an $L_c$ branch MRC using the differential QPSK in AWGN channels is given by

$$\begin{aligned} P_e(\gamma) &= e^{-2\gamma}\sum_{n=1}^{\infty}\left[\sqrt{2}-1\right]^n I_n\left(\sqrt{2}\,\gamma\right) \\ &+ e^{-2\gamma} I_0\left(\sqrt{2}\,\gamma\right) R_{L_c} \\ &+ e^{-2\gamma}\sum_{n=1}^{L_c-1} I_m\left(\sqrt{2}\,\gamma\right) R_{n,L_c} \end{aligned} \tag{8.7.31}$$

where

$$R_L = \frac{1}{2^{2L-1}}\sum_{j=0}^{L-1}\binom{2L-1}{j} \tag{8.7.32}$$

$$R_{n,L} = \frac{1}{2^{2L-1}}\sum_{j=0}^{L-1-n}\binom{2L-1}{j}\left[\left(\sqrt{2}+1\right)^n - \left(\sqrt{2}-1\right)^n\right] \tag{8.7.33}$$

and $I_n(.)$ is the modified Bessel function of the first kind and order n.

The average BER then becomes

$$\begin{aligned} P_e &= \int_0^{\infty} P_e(\gamma) f_{\gamma_{CD}}(\gamma)\, d\gamma \\ &= \sum_{i=0}^{M-1}\sum_{\ell=0}^{L'-1} a_{i,\ell}\Gamma^{\ell+1} G(m,i,\Gamma)^{\frac{-(\ell+1)}{2}} \\ &\left\{\sum_{n=0}^{\infty}\left(\sqrt{2}-1\right)^n\left(\frac{x-1}{x+1}\right)^{\frac{n}{2}}\frac{(n+\ell)!}{n!}F\left(-\ell;\,\ell+1;\,n+1;\,\frac{1-x}{2}\right)\right. \\ &+\sum_{n=1}^{L_c-1} R_{n,L_c}\frac{(n+\ell)!}{n!}\left[\frac{x-1}{x+1}\right]^{\frac{n}{2}} F\left(-\ell;\,\ell+1;\,n+1;\,\frac{1-x}{2}\right) \\ &\left.+\,\ell! R_{L_c} F\left(-\ell;\,\ell+1;\,1;\,\frac{1-x}{2}\right)\right\} \end{aligned} \tag{8.7.34}$$

where

$$G(m,i,\Gamma) = m^2(i+1)^2 + 4\Gamma m(i+1) + 2\Gamma^2 \tag{8.7.35}$$

$$x = \frac{m(i+1) + 2\Gamma}{\sqrt{G(m,i,\Gamma)}} \tag{8.7.36}$$

and F(a, b; c; x) = ${}_2F_1$(a, b; c; x) is a hypergeometric function given by (8.4.49).

## 8.8 Macroscopic Diversity Combiner

The signal envelope undergoes fast fluctuations due to local phenomena, and superimposed on these fluctuations is a slow varying mean signal level due to shadowing as discussed in Chapter 7. The fast-varying signal components received on spatially separated antennas may be regarded as uncorrelated with antenna spacing of the order of half a carrier wavelength. However, this is not the case for the slow-varying mean levels. The various space-diversity techniques discussed in previous sections required independent fading components. These space-diversity techniques are normally referred to as microdiversity techniques, and are only useful in combating the effect of fast fading.

A space-diversity technique referred to as macrodiversity is employed to overcome the effect of shadowing. In macrodiversity, a cell is served by a group of geographically separated base stations, and a base station receiving a strongest mean signal is used to establish a link with a mobile [Tur91, Abu94b, Abu95, Jak74].

### 8.8.1 Effect of Shadowing

In this section, the effect of shadowing on the performance of a system using a microscopic selection combiner and microscopic maximal ratio combiner schemes in the Rayleigh fading environment is considered [Tur91].

#### *8.8.1.1 Selection Combiner*

Let $f_{\gamma_{SC}}$ denote the pdf of the SNR at the output of a system using L-branch SC for a given mean SNR level, and let $f_\Gamma$ denote the pdf of the mean SNR at the site employing the SC system. Let $P_e(\gamma)$ denote the BER for a particular modulation scheme for a SNR $\gamma$. Then the BER at the output of the SC system is the average over all values of the SNR given by

$$P_e(\Gamma) = \int_0^\infty P_e(\gamma) f_{\gamma_{SC}}(\gamma)\, d\gamma \tag{8.8.1}$$

This quantity is dependent of the mean SNR $\Gamma$. When $\Gamma$ is not constant, the average of all $\Gamma$ needs to be carried out to evaluate the average BER. It is given by

$$P_e = \int_0^\infty P_e(\Gamma) f_\Gamma(\Gamma) d\Gamma \qquad (8.8.2)$$

An expression for $f_{\gamma_{SC}}$ in the Rayleigh fading environment is given by (8.1.7). Rewrite in the following from

$$f_{\gamma_{SC}}(\gamma) = -\sum_{k=1}^{L} (-1)^k \binom{L}{k} \frac{k}{\Gamma} \exp\left(-\frac{k\gamma}{\Gamma}\right) \qquad (8.8.3)$$

Substituting this in (8.8.1) gives $P_e(\Gamma)$, and $P_e$ then may be obtained using the pdf of $\Gamma$ in (8.8.2). The pdf of $\Gamma$ has a log-normal distribution. It follows from (7.1.23) that it is given by

$$f_\Gamma(\Gamma) = \frac{10}{\sigma\Gamma\sqrt{2\pi}\ln(10)} \exp\left[-\frac{\{10 \log \Gamma - \bar{\Gamma}_d\}^2}{2\sigma^2}\right] \qquad (8.8.4)$$

where $\bar{\Gamma}_d$ is the mean value of $\Gamma$ in decibels and $\sigma^2$ is its variance in decibels. If you know the BER for a particular modulation scheme, the average BER can be calculated using above procedure.

Consider an example of the CFSK scheme. For CFSK, $P_e(\gamma)$ is given by (7.3.57), that is,

$$P_e(\gamma) = \frac{1}{2} \operatorname{erfc}\sqrt{\frac{\gamma}{2}} \qquad (8.8.5)$$

In [Tur91], the average BER for a minimum shift keying (MSK) receiver is derived using

$$P_e(\gamma) = \frac{1}{2M} \sum_{i=1}^{M} \operatorname{erfc}\sqrt{d_i^2 \frac{\gamma}{2}} \qquad (8.8.6)$$

For $M = 1$ and $d_i = 1$, (8.8.6) reduces to (8.8.5). Thus, the results derived for MSK reduce to that for CFSK when $M = 1$ and $d_i = 1$.

Substituting (8.8.3) and (8.8.5) in (8.8.1) and carrying out the integrals [Tur91],

$$P_e(\Gamma) = \sum_{k=0}^{L} (-1)^k \binom{L}{k} \frac{1}{2\sqrt{\left(1 + \frac{2k}{\Gamma}\right)}} \qquad (8.8.7)$$

which, along with (8.8.4) and (8.8.2), results in the average BER in Rayleigh and log-normal fading,

$$P_e = \sum_{k=0}^{L} (-1)^k \binom{L}{k} \frac{5}{\sigma\sqrt{2\pi}\ln(10)} \int_0^\infty \frac{1}{\Gamma\sqrt{\left(1 + \frac{2k}{\Gamma}\right)}} \exp\left[-\frac{\left(10\log\Gamma - \bar{\Gamma}_d\right)^2}{2\sigma^2}\right] d\Gamma \qquad (8.8.8)$$

### *8.8.1.2 Maximum Ratio Combiner*

Now, consider the MRC under a similar environment. The SNR pdf at the output of the MRC is given by (8.4.10), that is,

$$f_{\gamma_{MR}}(\gamma) = \frac{\gamma^{L-1} e^{-\frac{\gamma}{\Gamma}}}{\Gamma^L (L-1)!} \tag{8.8.9}$$

Using (8.8.9) and (8.8.5), the average BER for a given mean SNR becomes

$$\begin{aligned} P_e(\Gamma) &= \int_0^\infty P_e(\gamma) f_{\gamma_{MR}}(\gamma) d\gamma \\ &= \frac{1}{2} - \sum_{k=1}^{L} \frac{\Gamma\left(L-k+\frac{1}{2}\right)\sqrt{\Gamma/2\pi}}{2(L-k)!(1+0.5\Gamma)^{L-k+\frac{1}{2}}} \end{aligned} \tag{8.8.10}$$

and the average BER after taking shadowing into consideration using (8.8.4) becomes [Tur91]

$$\begin{aligned} P_e &= \int_0^\infty P_e(\Gamma) f_\Gamma(\Gamma) d\Gamma \\ &= \frac{1}{2} - \sum_{k=1}^{L} \frac{5\Gamma\left(L-k+\frac{1}{2}\right)}{2\ln(10)\pi\sigma(L-k)!} \int_0^\infty \frac{\sqrt{\Gamma}}{(1+\Gamma/2)^{L-k+\frac{1}{2}}} \exp\left[-\frac{\left(10\log\Gamma - \bar{\Gamma}_d\right)^2}{2\sigma^2}\right] d\Gamma \end{aligned} \tag{8.8.11}$$

## 8.8.2 Microscopic Plus Macroscopic Diversity

Figure 8.4 shows a block diagram of a composite microscopic-plus-macroscopic diversity system in which transmission from a mobile is received by N different base stations. Each station employs an L-branch microscopic diversity system, which may employ any of the diversity-combining techniques discussed previously, and produces one output per base station. Thus N base stations produce a total of N outputs. A macroscopic diversity scheme is then used to produce one output. In principle, the macroscopic diversity scheme may use any one of the previous diversity-combining schemes to produce one output from N branches.

In this section, a scheme in which a selection diversity is employed to select one of the N branches is analyzed [Tur91]. Assuming that the signals on N branches are log-normally distributed, the pdf of the N-branch selection-diversity scheme is given by

$$f_{\Gamma_{SD}}(\Gamma) = \frac{10N}{\sigma\Gamma\ln(10)\sqrt{2\pi}} \exp\left[-\frac{\left(10\log\Gamma - \bar{\Gamma}_d\right)^2}{2\sigma^2}\right] F\left(\frac{10\log\Gamma - \bar{\Gamma}_d}{\sigma}\right)^{N-1} \tag{8.8.12}$$

where F(.) is the cumulative normal distribution function.

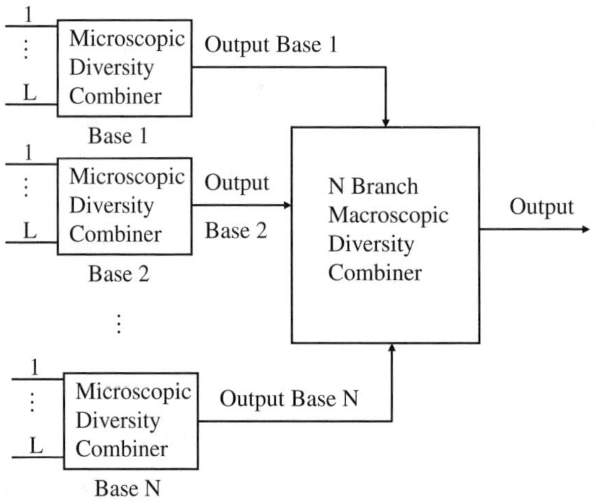


**FIGURE 8.4**
Block diagram of a macroscopic diversity combiner.

The average BER to include the shadowing effect may be calculated by averaging the conditional BER at the output of microscopic diversity combiner, that is,

$$P_e = \int_0^\infty P_e(\Gamma) f_{\Gamma_{SD}}(\Gamma)\, d\Gamma \tag{8.8.13}$$

where $P_e(\Gamma)$ denotes the average BER at the output of microscopic diversity combiner for a given mean SNR. For CFSK system operating in the Rayleigh fading environment, for SC and MRC, it is given by (8.8.7) and (8.8.10), respectively.

Let $P_e^{SCM}$ and $P_e^{MRM}$ denote the average BER when a composite system uses SC as macroscopic diversity with SC and MRC as microscopic diversity , respectively. Using (8.8.7) in (8.8.13) along with (8.8.12) yields [Tur91]

$$P_e^{SCM} = \sum_{k=0}^{L} (-1)^k \binom{L}{k} \frac{5N}{\sigma \ln 10 \sqrt{2\pi}} \int_0^\infty \frac{1}{\Gamma\sqrt{1+\frac{2k}{\Gamma}}} \exp\left[-\frac{\left(10\log\Gamma - \bar{\Gamma}_d\right)^2}{2\sigma^2}\right]\left[F\left(\frac{10\log\Gamma - \bar{\Gamma}_d}{\sigma}\right)\right]^{N-1} d\Gamma \tag{8.8.14}$$

and using (8.8.10) in (8.8.13) yields [Tur91]

$$P_e^{MRM} = \frac{1}{2} - \sum_{k=1}^{L} \frac{5N\Gamma\left(L-k+\frac{1}{2}\right)}{2\pi\sigma \ln 10 (L-k)!} \int_0^\infty \frac{\sqrt{\Gamma}}{(1+\Gamma/2)^{L-k+\frac{1}{2}}} \exp\left[-\frac{\left(10\log\Gamma - \bar{\Gamma}_d\right)^2}{2\sigma^2}\right]\left[F\left(\frac{10\log\Gamma - \bar{\Gamma}_d}{\sigma}\right)\right]^{N-1} d\Gamma \tag{8.8.15}$$

## Notation and Abbreviations

| | |
|---|---|
| AWGN | additive white Gaussian noise |
| BER | bit error rate |
| BPSK | binary phase shift keying |
| CDC | cascade diversity combiner |
| CFSK | coherent orthogonal frequency shift keying |
| DPSK | differentially binary phase shift keying |
| EGC | equal gain combiner |
| GSC | generalized selection combiner |
| MGF | moment generating function |
| MRC | maximum ratio combiner |
| NCFSK | noncoherent orthogonal frequency shift keying |
| OC | optimal combiner |
| cdf | cumulative distribution function |
| pdf | probability density function |
| RV | random variable |
| SC | selection combiner |
| SDC | switched diversity combiner |
| SIR | signal power to interference power ratio |
| $\mathbf{C}_S$ | channel gain vector for signal |
| $\mathbf{C}_{Ij}$ | channel gain vector for jth interference |
| $F_\gamma$ | cdf of $\gamma$ |
| $f_\gamma$ | pdf of $\gamma$ |
| $\hat{f}_\gamma$ | pdf of $\gamma$ in weight errors |
| $I_i$ | total interference power |
| $I_{OC}$ | total interference power at the output of OC |
| $I_{EG}$ | interference power at the output of EGC |
| $I_{MR}$ | interference power at the output of MRC |
| $\bar{I}^j$ | mean power due to jth interference, identical on all branches |
| $\bar{I}$ | mean interference power due to identical interferences on all branches |
| $I_i^j$ | instantaneous power on ith branch due to jth interference |
| K | number of interferences |
| L | number of branches |
| $L_C$ | number of selected branches |
| L(f) | Laplace transform of f |
| $M_x$ | MGF of x |
| m | Nakagami fading parameter |
| N | uncorrelated noise power |

| | |
|---|---|
| $n_i$ | noise on ith channel |
| $\mathbf{n}(t)$ | noise vector |
| P( ) | probability of ( ) |
| $P_e$ | average BER |
| $P_e(\gamma)$ | conditional BER |
| $P_e^{SC}$ | average BER at output of SC |
| $P_e^{GS}$ | average BER at output of GSC |
| $P_e^{SC2}$ | average BER at output of two-branch GSC |
| $P_e^{SC3}$ | average BER at output of three-branch GSC |
| $P_e^{CD}$ | average BER in CDC |
| $P_e^{MR}$ | average BER in MRC |
| $\hat{P}_e^{MR}$ | average BER in MRC with weight errors |
| $P_e^{SW}$ | average BER in SDC |
| $P_e^{OC}$ | average BER in OC |
| $P_{CD}^o$ | outage probability of CDC |
| $P_{EG}^o$ | outage probability of EGC |
| $P_{SC}^o$ | outage probability of SC |
| $P_{SW}^o$ | outage probability of SDC |
| $P_{GS}^o$ | outage probability of GSC |
| $P_{MR}^o$ | outage probability of MRC |
| $\hat{P}_{MR}^o$ | outage probability of MRC in weight errors |
| $P_{OC}^o$ | outage probability of OC |
| $p_{Ij}$ | power of jth interference source |
| $p_S$ | power of signal source |
| $q_{ij}$ | amplitude of the jth interference received on ith branch |
| $R$ | array correlation matrix |
| $R_S, R_I, R_N$ | array correlation matrix of signal, interference, and noise only, respectively |
| r | signal amplitude |
| $r_i$ | signal amplitude received on ith branch |
| SC2 | selection combiner with two branches selected |
| SC3 | selection combiner with three branches selected |
| $\bar{S}$ | mean signal power identical on all branches |
| $S_i$ | signal power on ith branch |
| $S_{OC}$ | signal power at output of OC |
| $S_{EG}$ | signal power at output of EGC |
| $S_{MR}$ | signal power at output of MRC |
| $w_i$ | weight on the ith branch |
| $\mathbf{w}$ | weight vector |
| $\mathbf{w}_{oc}$ | weight vector of optimal combiner |
| $x_i(t)$ | received signal on ith branch |
| $\mathbf{x}(t)$ | array signal vector |
| y(t) | combiner output. |

| | |
|---|---|
| $\Gamma$ | mean SNR at a branch |
| $\Gamma_{EG}$ | mean SNR of EGC |
| $\Gamma_{CD}$ | mean SNR at output of CDC |
| $\Gamma_{SC}$ | mean SNR at output of SC |
| $\Gamma_{GS}$ | mean SNR at output of GSC |
| $\Gamma_{MR}$ | mean SNR at output of MRC |
| $\Gamma_i$ | mean SNR at ith branch |
| $\alpha$ | inverse of $\Gamma$ |
| $\alpha_i, \theta_i$ | channel attenuation and phase on ith branch |
| $\alpha_0$ | an arbitrary constant |
| $\psi_{ij}$ | phase of jth interference received on ith branch |
| $\psi_r$ | characteristic function of an RV r |
| $\theta_i(t)$ | signal phase on ith branch |
| $\phi_x$ | cumulant generating function of x |
| $\gamma$ | SNR |
| $\gamma_0$ | threshold value of SNR |
| $\gamma_\ell$ | SNR of $\ell$th branch |
| $\gamma(\ell)$ | ordered SNR of $\ell$th branch |
| $\bar{\gamma}(\ell)$ | mean value of $\gamma(\ell)$ |
| $\gamma_{CD}$ | SNR of CDC |
| $\gamma_{GS}$ | SNR of GSC |
| $\gamma_{SC}$ | SNR of SC |
| $\xi_0$ | threshold value of power |
| $\hat{\xi}_0$ | optimum value of threshold power |
| $\rho$ | correlation coefficient |
| $\mu$ | signal power to interference power ratio |
| $\mu_0$ | threshold value of SIR |
| $\mu_{SC}$ | SIR of SC |
| $\mu_{EG}$ | SIR of EGC |
| $\mu_{SW}$ | SIR of SDC |
| $\tilde{\mu}$ | average signal power to average interference power ratio |
| $\tilde{\mu}_{EG}$ | average signal power to average interference power ratio of EGC |
| $\bar{\mu}_{MR}$ | mean SIR of MRC |
| $\bar{\mu}_{OC}$ | mean SIR of OC |

## References

Abr72 Abramowitz, M. and Segun, I.A., Eds., *Handbook of Mathematical Functions with Formulas, Graphs, and Mathematical Tables*, Dover, New York, 1972.

Abu92 Abu-Dayya, A.A. and Beaulieu, N.C., Outage probabilities of diversity cellular systems with cochannel interference in Nakagami fading, *IEEE Trans. Vehicular Technol.*, 41, 343–355, 1992.

Abu94 Abu-Dayya, A.A. and Beaulieu, N.C., Microdiversity on Rician fading channels, *IEEE Trans. Commn.*, 42, 2258–2267, 1994.

Abu94a Abu-Dayya, A.A. and Beaulieu, N.C., Switched diversity on microcellular Ricean channels, *IEEE Trans. Vehicular Technol.*, 43, 970–976, 1994.

Abu94b Abu-Dayya, A.A. and Beaulieu, N.C., Micro- and macrodiversity NCFSK (DPSK) on shadowed Nakagami-fading channels, *IEEE Trans. Commn.*, 42, 2693–2702, 1994.

Abu95 Abu-Dayya, A.A. and Beaulieu, N.C., Micro- and macrodiversity MDPSK on shadowed frequency-selective channels, *IEEE Trans. Commn.*, 43, 2334–2343, 1995.

Alo00 Alouini, M.S. and Simon, M.K., An MGF-based performance analysis of generalized selection combining over Rayleigh fading channesl, *IEEE Trans. Commn.*, 48, 401–415, 2000.

And85 Andrews, L.C., *Special Functions for Engineers and Applied Mathematicians*, MacMillan, New York, 1985.

Bea90 Beaulieu, N.C., An infinite series for the computation of the complimentary probability distribution function of a sum of independent random variables, *IEEE Trans. Commn.*, 38, 1463–1474, 1990.

Bea91 Beaulieu, N.C. and Abu-Dayya, A.A., Analysis of equal gain diversity on Nakagami fading channels, *IEEE Trans. Commn.*, 39, 225–234, 1991.

Bre59 Brennan, D.G., Linear diversity combining techniques, *Proc. IRE* ,47, 1075–1102, 1959.

Cou95 Couch, L.W., *Digital and Analog Communication Systems*, Prentice Hall, New York, 1995.

Cha79 Charash, U., Reception through Nakagami fading multipath channels with n random delays, *IEEE Trans. Commn.*, 27, 657–670, 1979.

Cho00 Choo, L.C. and Tjhung, T.T., BER performance of DQPSK in Nakagami fading with selection diversity and maximum ratio combining, *IEEE Trans. Commn.*, 48, 1618–1621, 2000.

Eng96 Eng, T., Kong, N. and Milstein, L.B., Comparison of diversity combining techniques for Rayleigh-fading channels, *IEEE Trans. Commn.*, 44, 1117–1129, 1996.

Gan71 Gans, M.J., The effect of Gaussian error in maximal ratio combiners, *IEEE Trans. Commn.*, 19, 492–500, 1971.

Gil81 Gille, J.G., et al., *Dynamique de la Commande Lineaire*, 6th ed., Bordas, Paris, 1981.

Gir77 Giri, N.C., *Multivariate Statistical Inference*, Academic, New York, 1977.

Jak74 Jakes Jr., W.C., Eds., *Microwave Mobile Communications*, Wiley, New York, 1974.

Kon98 Kong, N. and Milstein, L.B., Average SNR of a Generalized Diversity Selection Combining Scheme, in Proceedings of ICC, pp. 1556–1558, 1998.

Pro95 Proakis, J.G., *Digital Communications*, 3rd ed., McGraw-Hill, New York, 1995.

Roy96 Roy, Y., Chouinard, J.Y. and Mahmoud, S.A., Selection diversity combining with multiple antennas for MM-wave indoor wireless channels, *IEEE J. Selected Areas Commn.*, 14, 674–682, 1996.

Sch72 Schiff, L., Statistical suppression of interference with with diversity in a mobile-radio environment, *IEEE Trans. Vehicular Technol.*, 21, 121–128, 1972.

Sha98 Shah, A. and Haimovich, A.M., Performance analysis of optimum combining in wireless communications with Rayleigh fading and cochannel interference, *IEEE Trans. Commn.*, 46, 473–479, 1998.

Sha00 Shah, A., et al., Exact bit-error probability for optimum combining with a Rayleigh fading Gaussian cochannel interferer, *IEEE Trans. Commn.*, 48, 908–912, 2000.

Sha00a Shah, A. and Haimovich, A.M., Performance analysis of maximum ratio combining and comparison with optimum combining for mobile radio communications with cochannel interference, *IEEE Trans. Vehicular Technol.*, 49, 1454–1463, 2000.

Sim99 Simon, M. K. and Alouini, M.S., A unified performance analysis of digital communication with dual selective combining diversity over correlated Rayleigh and Nakagami-m fading channels, *IEEE Trans. Commn.*, 47, 33–43, 1999.

Skl01 Sklar, B., *Digital Communications: Fundamentals and Applications*, Prentice Hall, New York, 2001.

Tom99 Tomiuk, B.R., Beaulieu, N.C. and Abu-Dayya, A.A., General forms for maximal ratio diversity with weighting errors, *IEEE Trans. Commn.*, 47, 488–492, 1999.

Tur91 Turkmani, A.M.D., Performance evaluation of a composite microscopic plus macroscopic diversity system, *IEE Proc.* I, 138, 15–20, 1991.

Win84 Winters, J.H., Optimum combining in digital mobile radio with cochannel interference, *IEEE J. Selected Areas Commn.*, 2, 528–539, 1984.

Win87 Winters, J.H., On the capacity of radio communication systems with diversity in a Rayleigh fading environment, *IEEE Trans. Selected Areas Commn.*, 5, 5, 871–878, 1987.

Win87a Winters, J.H., Optimum combining for indoor radio systems with multiple users, *IEEE Trans. Commn.*, 35, 1222–1230, 1987.

Zha97 Zhang, Q.T., Probability of error for equal-gain combiners over Rayleigh channels: some closed-form solutions, *IEEE Trans. Commn.*, 45, 270–273, 1997.

Zha99 Zhang, Q.T., A simple approach to probability of error for equal gain combiners over Rayleigh channels, *IEEE Trans. Vehicular Technol.*, 48, 1151–1154, 1999.

Zha99a Zhang, Q.T., Maximal-ratio combining over Nakagami fading channels with an arbitrary branch covariance matrix, *IEEE Trans. Vehicular Technol.*, 48, 1141–1150, 1999.

# Index

## D



## E



## F



## G



## H



## I



## L



## M

## N



## O



## P

## Q



## R



## S



## T



## U



## W



## Z